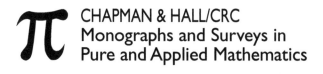

CHAPMAN & HALL/CRC
Monographs and Surveys in
Pure and Applied Mathematics   **140**

DIFFERENCE METHODS

FOR SINGULAR

PERTURBATION PROBLEMS

CHAPMAN & HALL/CRC
Monographs and Surveys in Pure and Applied Mathematics

CHAPMAN & HALL/CRC
Monographs and Surveys in
Pure and Applied Mathematics   140

DIFFERENCE METHODS

FOR SINGULAR

PERTURBATION PROBLEMS

Grigory I. Shishkin

Lidia P. Shishkina

CRC Press
Taylor & Francis Group
Boca Raton London New York

CRC Press is an imprint of the
Taylor & Francis Group, an **informa** business

A CHAPMAN & HALL BOOK

CRC Press
Taylor & Francis Group
6000 Broken Sound Parkway NW, Suite 300
Boca Raton, FL 33487-2742

First issued in paperback 2019

© 2009 by Taylor & Francis Group, LLC
CRC Press is an imprint of Taylor & Francis Group, an Informa business

No claim to original U.S. Government works

ISBN-13: 978-1-58488-459-0 (hbk)
ISBN-13: 978-0-367-38682-5 (pbk)

Library of Congress Cataloging-in-Publication Data

Shishkin, G. I.
    Difference methods for singular perturbation problems / Grigory I. Shishkin, Lidia P. Shishkina.
      p. cm. -- (Chapman & Hall/CRC monographs and surveys in pure and applied mathematics)
    Includes bibliographical references and index.
    ISBN 978-1-58488-459-0 (hardback : alk. paper)
    1. Singular perturbations (Mathematics) 2. Difference equations--Numerical solutions. 3. Algebra, Abstract. I. Shishkina, Lidia P. II. Title. III. Series.

QC20.7.P47S55 2008
515'.392--dc22
                                                2008025636

Visit the Taylor & Francis Web site at
http://www.taylorandfrancis.com

and the CRC Press Web site at
http://www.crcpress.com

Dedication

Dedicated to the memory of academicians
Alexandr Andreevich Samarskii and
Nikolai Sergeevich Bakhvalov

Contents

Preface

The present book is devoted to the *development of difference schemes* that *converge ε-uniformly* in the maximum norm for a representative class of *singularly perturbed problems*. It also deals with the *justification* of their *convergence*, and *surveys new directions and approaches developed recently*, which are of importance for further progress in numerical methods.

The book was intended to be an English translation of the Russian book [138] (Shishkin G.I. (1992). *Discrete Approximations of Singularly Perturbed Elliptic and Parabolic Equations*. Russian Academy of Sciences, Ural Section, Ekaterinburg (in Russian)) that was initiated by John J.H. Miller. The translation was made by Zora Uzelac, but we decided not to publish this version of the book. The very dense nature of this book, that allowed us to cover a large class of singularly perturbed boundary value problems in little space, was too difficult for most readers and also created problems in the implementation of the results. Since the appearance of the book [138], new results and ideas have appeared that are dealt with in the present book.

First, I would like to thank my teachers. My scientific interests in computational mathematics were formed and matured under the influence of the scientific schools of the Academicians of the Russian Academy of Science. A.M. Il'in, A.A. Samarskii, N.S. Bakhvalov, G.I. Marchuk and their influence led to the appearance of my second doctoral thesis. This thesis became the basis of [138], and is a continuing influence on my work.

It is with pleasure that I note the long-term and fruitful collaboration with the Irish and Dutch mathematicians in the groups of J. Miller and P. Hemker. This collaboration began in 1990, and yielded progress in the development of numerical methods for problems with boundary layers, and led to new results that were published in numerous joint papers and in two books [87] and [33].

The Russian scientists K.V. Emelianov, V.D. Liseikin, P.N. Vabishchevich, V.B. Andreev, V.F. Butuzov, A.V. Gulin, I.G. Belukhina, N.V. Kopteva, V.V. Shaidurov, B.M. Bagaev, E.D. Karepova, M.M. Lavrentiev, Jr, Yu.M. Laevsky, A.I. Zadorin, A.D. Ljashko and I.B. Badriev also influenced much of the detail of the approaches initiated in [138].

The idea to translate into English the book [138] began during my collaboration over the last dozen years with mathematicians and their students, namely, J.J.H. Miller, E. O'Riordan, A.F. Hegarty, M. Stynes, A. Ansari (Ireland), P.W. Hemker, J. Maubach, P. Wesseling (the Netherlands), P.A. Farrell (USA), F. Lisbona, C. Clavero, J.L. Gracia, J.C. Jorge (Spain), D. Creamer, Lin Pin (Singapore), and through discussions of papers (based on ideas from

[138]) on international conferences with, among others, I.P. Boglaev (New Zealand), R.E. O'Malley, R.B. Kellogg (USA), L. Tobiska, H.-G. Roos, G. Lube, T. Linß (Germany), Wang Song (Australia), L.G. Vulkov, I.A. Braianov (Bulgaria), R. Čiegis (Lithuania), and P.P. Matus (Belarus). Numerous ideas from [138] were extended and published in many papers.

My thanks especially to L.P. Shishkina, my better half, and main assistant-colleague and mathematician for participation as co-author in writing this book, for enormous scientific and technical support. She has prepared the present book including all stages: the clarification of results by numerous discussions, preparation in LaTeX, the translation, compiling the Index, and reviewing the page-proofs.

Significant assistance in the preparation of the English version of the present book, in the translation from Russian-English to idiomatic English, was made by M. Stynes (Part I, and fragments of Part II) and M. Mortell (the Preface and the Introduction) to whom I would like to express my deepest thanks.

My thanks to our assistant-colleague I.V. Tselishcheva for support in the process of preparing the book, participation in the translation of some chapters from Part II, of the Introduction, of the Survey, and many other tasks.

I am grateful for financial and material support (scientific books, computational technique) to the

Institute of Mathematics and Mechanics, Ural Branch of the Russian Academy of Sciences, Yekaterinburg, Russia;

Institute for Numerical Computation and Analysis, Dublin (INCA), Ireland;

Department of Mathematics at Trinity College Dublin, Ireland;

CWI (Research Institute of the Stichting Mathematisch Centrum), Amsterdam, the Netherlands;

School of Mathematical Sciences, Dublin City University, Ireland;

Department of Mathematics and Statistics at the University of Limerick, Ireland;

National Research Institute for Mathematics and Computer Science (NUS), Singapore;

School of Mathematical Sciences at the National University of Ireland in Cork (UCC), Ireland;

Boole Centre for Research in Informatics at the UCC, Ireland;

Mathematics Applications Consortium for Science and Industry in Ireland (MACSI) under the Science Foundation Ireland (SFI) Mathematics Initiative;

and to Chapman & Hall for friendly cooperation.

In particular, the research work of G.I. Shishkin and L.P. Shishkina was supported by the

Russian Foundation for Basic Research under grants
No. 01-01-01022, 04-01-00578, 07-01-00729;

grant RFBR-NWO (RFFI-NWO) 04-01-89007-NWO_a);

Dutch Research Organisation NWO under grants
No. 047.008.007, 047.016.008.

The research work of G.I. Shishkin was also supported by the

International Collaboration Programme of Forbairt, Dublin, Ireland No. IC/97/057;

Enterprise Ireland Basic Research grants
SC–98–612, SC–2000–070.

G.I. Shishkin

Part I

Grid approximations of singular perturbation partial differential equations

Chapter 1

Introduction

1.1 The development of numerical methods for singularly perturbed problems

The wide use of computing techniques, combined with the demands of scientific and technical practices, has stimulated the development of numerical methods to a great extent, and in particular, methods for solving differential equations. The efficiency of such methods is governed by their accuracy, simplicity in computing the discrete solution and also their relative insensitivity to parameters in the problem. At present, numerical methods for solving partial differential equations, in particular, finite difference schemes, are well developed for wide classes of boundary value problems (see, for example, [79, 108, 100, 214, 91, 216]).

Among boundary value problems, a considerable class includes problems for singularly perturbed equations, i.e., differential equations whose highest-order derivatives are multiplied by a (perturbation) parameter ε. The perturbation parameter ε may take arbitrary values in the open-closed interval $(0, 1]$ (see, e.g., [211, 210, 57, 94, 62]). Solutions of singularly perturbed problems, unlike regular problems, have boundary and/or interior layers, that is, narrow subdomains specified by the parameter ε on which the solutions vary by a finite value. The derivatives of the solution in these subdomains grow without bound as ε tends to zero.

In the case of singularly perturbed problems, the use of numerical methods developed for solving regular problems leads to errors in the solution that depend on the value of the parameter ε. Errors of the numerical solution depend on the distribution of mesh points and become small only when the effective mesh-size in the layer is much less than the value of the parameter ε (see, e.g., [138, 87, 106, 33]). Such numerical methods turn out to be inapplicable for singularly perturbed problems.

Due to this, there is an interest in the development of special numerical methods where solution errors are independent of the parameter ε and defined only by the number of nodes in the meshes used, i.e., *numerical methods* (in particular, *finite difference schemes*) that *converge ε-uniformly*. When the solutions by such methods are *ε-uniformly convergent*, we will call these methods and solutions *robust* (as in [33]). At present, only several books are devoted to the *development* of numerical methods for solving singularly

3

perturbed problems. Grid methods for boundary value problems for partial differential equations are considered in the books [138, 87, 33, 75]; see also [26, 13, 14, 76] for ordinary differential equations. In the book [106], the authors give a number of results and also a comprehensive bibliography on numerical methods for solving singularly perturbed problems for partial differential equations and for ordinary differential equations.

The present book was intended to be an English translation of the book [138]. A variety of ideas and approaches from [138] have since been further developed. New approaches and trends appear, which require further investigation. In the present book, we elaborate on approaches to the development of ε-uniformly convergent numerical methods for several boundary value problems from [138] and discuss some new trends in the development of other methods, which have appeared recently.

Quite often solutions of boundary value problems, their grid solutions, and also their convergence are considered using maximum norms. The use of either the energy norm or L_1, L_2-norms is inadequate to describe the solutions of singularly perturbed problems and their approximations. For example, in the case of problems with a parabolic boundary layer, the boundary-layer function (that is finite in the maximum norm) tends to zero in the norms mentioned above as $\varepsilon \to 0$ [87, 33]. In this book, *maximum norms* are consistently used. As a rule, we avoid reference to works where problems for singularly perturbed ordinary differential equations are considered since such results and techniques cannot, in general, be carried over to problems for partial differential equations.

The first ε-uniformly convergent difference schemes constructed for singularly perturbed problems used two main approaches: *fitted operator method* and *condensing mesh (grid) method/fitted mesh (grid) method*. Schemes based on the fitted operator method were constructed in [2] and, independently, constructed and justified in [56] (for ordinary differential equations); in [15], a scheme for the condensing mesh method was constructed and justified (for an elliptic equation). For schemes using condensing meshes, ε-uniform convergence of the solution of a difference scheme to the solution of the boundary value problem is guaranteed by a special choice of the distribution of mesh points (for the given number of nodes). Restrictions on the choice of difference equations approximating singularly perturbed problems (for ensuring the ε-uniform convergence of the scheme) are, in general, not imposed. In fitted operator methods, ε-uniform convergence of the solution of the difference scheme is achieved by a special choice of coefficients of the difference equations approximating the differential problem. Restrictions on the distribution of mesh points for ensuring the ε-uniform convergence of the scheme are not imposed.

We mention also an approach related to *additive splitting of a singularity* suggested for singularly perturbed problems in [12] (see also [11]). In this method, basic functions include special functions approximating the singular component of the solution of the problem. In the case of singularly perturbed problems for partial differential equations, this approach was not widely used because the singular components of solutions of boundary value problems have the form too complicated for the effective construction of a system of basis functions. The difference schemes based on the method of additive splitting of a singularity and constructed in [12, 11] converge ε-uniformly in the energy norm.

After the publications [15, 56], there was a large effort to develop fitted operator methods. The first book [26] is completely devoted to the development of such methods for ordinary differential equations. Later, fitted operator methods continued to be intensively developed (see, for example, a series of variants of the fitted operator schemes for elliptic equations in [103]). A comprehensive bibliography on numerical methods for singularly perturbed problems is given in [106]. After [15, 56], in the case of partial differential equations, the first finite difference schemes that converge ε-uniformly in the maximum norm are constructed in [29] (see also [30, 1] for the fitted operator scheme, and [74, 118] for the schemes on condensing meshes).

Note that fitted operator methods (see their description, e.g., in [26, 87, 33, 106]) have an advantage in simplicity because meshes used are uniform, and this contributed to their more rapid progress compared with condensing mesh methods. However, fitted operator methods have a restricted domain of applicability for constructing ε-uniformly convergent numerical methods.

It was first established in [124] that there are *no ε-uniformly convergent schemes based on the fitted operator method* in the case for singularly perturbed elliptic convection-diffusion equations in domains where parts of the boundary are characteristics of a reduced equation and *parabolic boundary layers* appear. In the same paper, a scheme was constructed that converges ε-uniformly, using both the fitted operator method for the approximation of derivatives along characteristics of the reduced equation and the condensing mesh method for the approximation of derivatives in the direction orthogonal to the characteristics. The resulting discrete solutions also make it possible to approximate the normalized derivatives ε-uniformly.

For parabolic equations with *parabolic layers* it is proved that there *exist no schemes* using the fitted operator method that converge ε-uniformly in any of the papers:

[130] in the case of a *parabolic boundary layer* and
[148] in the case of a *parabolic initial layer*.

When constructing schemes for *nonlinear problems*, the situation is much more complicated. In [137] (for reaction-diffusion equations) and [139] (for convection-diffusion equations), it was established that even for semilinear ordinary differential equations there *exist no schemes based on the fitted operator method* that converge ε-uniformly. Similar difficulties related to the use of fitted operator methods in numerical methods are discussed in later publications (see, e.g., [138, 87, 86] for partial differential equations and [84, 32] for semilinear ordinary differential equations). Numerical experiments that demonstrate the inconsistency of the fitted operator method for semilinear ordinary differential equation are considered in [33]. Thus, for a wide class of boundary value problems there are no schemes using fitted operator methods that converge ε-uniformly in the maximum norm; independently of these, schemes may be constructed using classical finite difference approximations or finite element or finite volume methods.

Having summarizing the approaches to the construction of ε-uniformly convergent numerical methods, we make one more remark. In works [134, 140] (see also discussions in [87, 83]), a class of problems is distinguished for elliptic and parabolic equations that degenerate on the boundary of the domain whose solutions contain initial and parabolic boundary layers. It is shown that for such problems *there are no schemes of the condensing mesh method* (for rectangular meshes) that *converge ε-uniformly*. But the application of *both approaches—fitted mesh and fitted operator methods—* makes it possible to construct schemes that converge ε-uniformly.

Up to now, practically all singularly perturbed partial differential equations, for which difference schemes that converge ε-uniformly in the maximum norm have been constructed, do not contain mixed derivatives. Boundary value problems have been considered only for the simplest subdomains of dimension not higher than two, and the elliptic operator in the differential equations is the Laplace operator. In the case of an elliptic equation with mixed derivatives considered only on a rectangle, the ε-uniformly convergent scheme obtained appears to be too complex and can not be extended to other dimensions in geometry [28]. Problems in domains with curvilinear boundaries are considered only in few publications.

It is one of the goals of the present book to overcome such an existing unsatisfactory state in the area of development of ε-uniformly convergent difference schemes. Another goal of the book is, on model problems, to consider some modern trends in the development of numerical methods for singularly perturbed problems that require further investigation.

1.2   Theoretical problems in the construction of difference schemes

We discuss some *basic principles* related to the foundations of the theory of finite difference schemes, which arise in the development of ε-uniformly convergent numerical methods.

The behaviour of derivatives of a solution to singularly perturbed boundary value problems motivates the type of grid and the difference schemes on that. The derivatives of the solution of the boundary value problem in a domain with a sufficiently smooth boundary are ε-uniformly bounded in most of the domain and grow without bound only in a narrow subdomain (the boundary layer) whose width tends to zero with ε. The essential *anisotropy of the directional derivatives* is exhibited. The maximal values of derivatives of the solution in a neighborhood of the boundary layer are found only for the derivative along the normal to the boundary of the domain; the k-th normal derivative in the neighborhood of the boundary is of order $\mathcal{O}\left(\varepsilon^{-k}\right)$, and moreover, the derivative decreases exponentially to finite values as the value $\varepsilon^{-k}\rho$ grows, where ρ is the distance to the boundary. The behaviour of the solution of the boundary value problem is particularly complicated in the case of domains with piecewise-smooth boundaries.

The observed anisotropy in the behaviour of the derivatives of the solution and the *unbounded growth of the derivatives* for $\varepsilon \to 0$ motivates the use of *essentially anisotropic meshes* condensing in a neighborhood of the boundary layer.

Note that the terms of a differential convection-diffusion equation, involving spatial derivatives across the boundary layer, become unbounded as $\varepsilon \to 0$. So, very close to the boundary, in order to approximate the solution of the boundary value problem by a solution of a finite difference scheme, it is necessary to use meshes whose step-size along the normal to the boundary is of order $o(\varepsilon)$ [138]. Under this requirement on the mesh-size, *errors* resulting in the approximation of the terms of the differential equation, when the derivatives in the equation are replaced by difference derivatives, also grow without bound [138]. The difference scheme no longer approximates the boundary value problem ε-uniformly. This unbounded behaviour of terms in the differential equation and their difference approximations gives rise to difficulties in the proof of ε-uniform convergence of the relevant schemes.

The violation of the *monotonicity property* (see [108]) of a boundary value problem when grid approximations are constructed even for the simplest problems can lead to both large errors and nonphysical results. For example, in the case of an ordinary differential convection-diffusion equation, when centered difference approximations on uniform meshes are used, the error in the discrete solution oscillates and grows without bound if the parameter ε is much less than the mesh-size [33].

In the presence of mixed derivatives in differential equations and also in the case of problems in domains with curvilinear boundaries, significant difficulties arise in the construction of monotone discrete approximations when *meshes that condense gradually in the boundary layer* are used (for example, Bakhvalov's meshes [15]). Also, for difference schemes that do not approximate the boundary value problems ε-uniformly, additional difficulties appear when justifying the ε-uniform convergence of solutions in the maximum norm.

Conventional approaches to overcoming the specific problems mentioned above that arise in the construction of ε-uniformly convergent numerical methods turn out to be ineffective.

So, in convection-diffusion problems the use of an artificial viscosity in a numerical method in order to suppress oscillations in the grid solution leads to essential errors in a neighborhood of the boundary layer [214, 55].

In the case of singularly perturbed problems the methods for the derivation

of *a priori estimates* developed *for regular problems* (see, e.g., [67, 68, 69, 37])
result in estimates for the derivatives that do not distinguish the different
behaviours of derivatives in different directions. The requirements imposed
on the data of a regular problem for obtaining *a priori* estimates by this
approach are close to being necessary. But estimates obtained in this way
for singularly perturbed problems turn out to be unable to justify ε-uniform
convergence.

Methods based on *asymptotic expansions in powers of the parameter ε* allow
one to construct approximations to the solutions with an accuracy up to a
sufficiently high order of the parameter ε (see, e.g., [211, 210, 208, 59, 62, 209,
94, 57] and references therein). But the detailed behaviour of the derivatives of
the solution in various directions is, in general, not considered. To construct
asymptotic expansions it is required that the data of the boundary value
problem possess sufficient smoothness. This then restricts the class of boundary value problems for which ε-uniformly convergent numerical methods can
be developed and justified. Thus, the above approaches turn out to be of
little use for the construction of *a priori* estimates required for developing
ε-uniformly convergent numerical methods.

1.3   The main principles in the construction of special schemes

We now outline the *main principles* underlying the development of the ε-
uniformly convergent difference schemes which are used in this book. Even
when numerical methods are constructed for regular problems that have a
complicated solution, it is necessary to use grid approximations that inherit
the monotonicity property of the boundary value problem [108] in order to
prevent the appearance of nonphysical effect in the solution. In the case of
singularly perturbed problems such a natural requirement leads to complicated difference schemes (in the case when mixed derivatives are present in
the equations, see, e.g., schemes in [112] for regular problems and in [138] for
singularly perturbed boundary value problems). For the classes of boundary
value problems studied here the problems that appear are to find appropriate
classes of grid approximations and also to obtain sufficient conditions, close
to necessary ones, for the ε-uniform convergence of the schemes constructed.

The study of sufficiently wide classes of boundary value problems requires
the development of new approaches to the construction of numerical methods
and, in general, leads to more complicated schemes. Primarily, we need *a
priori* estimates for the solutions that adequately reflect the behaviour of such
singularities as a boundary layer in a neighborhood of different parts of the
domain boundary. To construct special schemes, we use the simplest standard
ε-uniformly monotone approximations to the boundary value problem. The

schemes are constructed on the simplest piecewise uniform meshes that ensure the ε-uniform convergence of the grid solution.

The technique for constructing *a priori estimates based on a decomposition of the solution* first conceived in [118] and developed in [138] (see also [129, 131, 87] and the discussions, for example, in [105, 96, 95] for a two-dimensional problem) was called the Shishkin decomposition (see, e.g., [72, 73]). This technique uses *the decomposition of the solution into regular and singular components* and, in the case of problems in domains with piecewise smooth boundaries, the *singular component*, in turn, is *represented in terms of its components as the sum of regular and corner layers of different dimension.* The specific behaviour of the regular component, as well as of constituents of the singular components, is revealed using the *first terms of their asymptotic expansions* with respect to the parameter ε. Such a technique allows us to detect the distinctions in the behaviour of derivatives of the singular component in different directions, and to obtain estimates that are necessary to prove the ε-uniform convergence of the difference schemes being constructed.

In the method of condensing grids conceived in [128], which uses *piecewise uniform meshes*, the distribution of mesh points in the direction across the boundary layer, normal to the boundary, is defined by one transition parameter, that is, by the point at which the mesh size changes. In the literature, such meshes are called Shishkin's meshes (see, e.g., [83]). For difference schemes on these meshes, the difficulties in the development of ε-uniformly convergent grid methods are essentially reduced that allows the construction of ε-*uniformly monotone difference schemes* which converge ε-*uniformly* for representative classes of singularly perturbed boundary value problems. The first numerical results on piecewise uniform meshes were obtained in [88] for an ordinary differential equation, in [89] for a parabolic partial differential equation with a parabolic boundary layer, and in [40] for an elliptic convection-diffusion equation on a rectangle with a parabolic boundary layer that is generated on the characteristic parts of the boundary.

An ε-uniformly convergent scheme for a problem in a two-dimensional domain with a curvilinear boundary and an equation with mixed derivatives was constructed in 1989 for the first time in [130] in the case of a linear parabolic reaction-diffusion equation. For some boundary layer problems in n-dimensional domains, such similar schemes were constructed in [131], and in [136, 137] for quasilinear equations in an n-dimensional slab (see also [139] (1992)). The ε-uniformly convergent schemes in all these cases were constructed on piecewise uniform meshes.

The appearance of piecewise uniform meshes (see, e.g., [127, 128]) that simplified the construction and study of ε-uniformly convergent grid methods and also the progress in techniques for obtaining *a priori* estimates, as well as the identification of a large class of problems for which the approach based on a fitted operator method is restricted, contributed to the development of ε-uniformly convergent difference schemes for partial differential equations (see some results related to the development of ε-uniformly convergent schemes, for example, in the books [138, 87, 106, 33], in the survey [204], and in the references therein).

1.4 Modern trends in the development of special difference schemes

In the book we point out some problems related to numerical methods for singularly perturbed problems that require further investigation.

- As a rule, the order of ε-uniform convergence of difference schemes is low; therefore these schemes are too inefficient for practical use. Thus, the development of *schemes with a high convergence order* is an important problem.

- *A priori* estimates for solutions of boundary value problems largely "dictate the structure" of special robust schemes. Therefore, the development of a technique for obtaining *a priori* estimates for new classes of boundary value problems, in particular, *problems that have several perturbation parameters*, or *problems for systems of equations* is an important task.

- In a number of boundary value problems, for example, for parabolic equations with moving boundary and/or interior layers (see, e.g., [159]), the complicated solutions bring too complicated ε-uniformly convergent schemes under the construction. A problem of current interest is to find *necessary and sufficient conditions for ε-uniform convergence* of such special schemes.

- Even in regular problems, the conditioning number of classical difference schemes grows without bound when the mesh-size tends to zero. In the condensing grid method, the mesh-size in the layer for a fixed number of nodes can be arbitrarily small depending on the parameter ε. Thus, the study of conditioning, as well as *ε-uniform conditioning of special schemes*, is an important problem.

These problems relate to the foundations of the theory of numerical methods (difference schemes) for singularly perturbed problems with complicated singularities.

1.5   The contents of the present book

The contents of the book can be divided into two parts. Part I is *"translation"* of certain parts of the book [138] published in 1992 in Russian, and Part II is a *survey* of some recent results in the development of robust methods for boundary value problems with boundary layers and other singularities. The work involved in the *"translation"* resulted in the development in detail of principles for constructing monotone difference schemes for some classes of singularly perturbed problems.

In Part I (Chapters 2–7) boundary value problems are considered for elliptic and parabolic reaction-diffusion and convection-diffusion equations *in n-dimensional domains with smooth and piecewise smooth boundaries* in the case when the *differential equations contain mixed derivatives*. Here our goal is *to develop a technique for constructing and justifying ε-uniformly convergent difference schemes*, based on classical discrete approximations, for boundary value problems with less restrictions on the problem data. In this situation we are not too concerned about the rate of ε-uniform convergence of schemes constructed. Significant attention is given to the study of *conditions* that are *necessary and sufficient* for the *ε-uniform convergence* of the relevant schemes. When constructing *special difference schemes that converge ε-uniformly*, we use the techniques developed for *canonical problems on a slab and a parallelepiped.*

Part II contains material published mainly in the last four years. These are problems with *boundary layers*, and *additional singularities* generated by nonsmooth data, unboundedness of the domain, and also by the presence of the perturbation vector parameter. Another aspect of the considerations in this part is that we *find both the solution and its derivatives* with errors that are independent of the perturbation parameters.

In Chapters 8 and 13, boundary value problems for equations with perturbation vector parameters are studied. In Chapter 8, problems are considered for a scalar parabolic reaction-diffusion equation with three scalar parameters, and in Chapter 13, a problem is studied for a system of two elliptic reaction-diffusion equations with two scalar parameters. In these problems, layers of different types arise depending on the relation between the scalar parameters.

For grid approximations of boundary value problems *with a moving boundary layer*, necessary and sufficient conditions for the schemes to be robust are obtained in Chapter 9. Taking into account these conditions, difference schemes are constructed for a parabolic reaction-diffusion equation in a domain with moving boundaries.

Methods of *improving the accuracy of grid solutions* while preserving their robustness are examined in Chapter 10. *Conditioning* and *correctness* (or *well-posedness*) of a difference scheme that is constructed based on a classical grid approximation of a boundary value problem for a parabolic convection-diffusion equation on piecewise-uniform meshes are discussed in Chapter 12.

In Chapter 11, difference schemes on *a priori* adapted meshes are developed.

Promising approaches to the development of robust methods based on *a posteriori adapted techniques* are discussed in Chapter 14. In the case of singularly perturbed problems, the approaches known for regular problems improve the convergence only for finite values of the parameter ε, and *do not provide for the robustness of the method*. Here, difference schemes are discussed that converge "almost ε-uniformly".

In the survey Chapter 14 we give comments to some approaches in the construction of special difference schemes and research directions. In particular, the following problems are discussed:

- some "applied" problems whose solutions require robust numerical methods;

- approaches to the construction of robust numerical methods for parabolic equations with smooth and piecewise smooth boundary conditions;

- approximation of solutions and derivatives;

- specific boundary value problems for elliptic convection-diffusion equations such as a problem on an unbounded domain for an equation with a scalar parameter and a problem on a rectangle for an equation with a vector parameter;

- in the case of an unbounded domain, the concept of the domain of essential dependence of the solution that allows robust schemes to be constructed whose solutions converge on bounded domains by increasing the number of nodes in the meshes used;

- compatibility conditions for a problem on a rectangle that guarantee the smoothness of the solution and its regular and singular components that are required for the construction and justification of robust schemes.

1.6   The present book

We discuss *the existing state of the art* in the area of the development of difference schemes that converge ε-uniformly in the maximum norm.

The books [138, 87, 33], as well as the present book, are focussed on the development of a *technology for constructing finite difference schemes that converge ε-uniformly in the maximum norm* for a wide class of singularly perturbed boundary value problems. In [106], ε-*uniformly convergent numerical methods* are considered that use various techniques and methods developed for regular and singularly perturbed problems. The convergence of discrete methods is considered in *norms corresponding to the applied numerical method*. In [106] the basic results developed in [138] are given in a condensed form, and *problems for equations with mixed derivatives* and *multi-dimensional problems are not discussed*.

The books [138, 87, 33] are devoted to the development of finite difference schemes for singularly perturbed elliptic and parabolic equations. Difference schemes are constructed based on *standard monotone finite difference approximations of differential problems on piecewise uniform meshes*. The books [138, 87] emphasize the theory, and [33] is strictly numerical.

Appearing in 1988, piecewise-uniform meshes [127, 128] allow us to overcome a number of essential difficulties in the development and justification of ε-uniformly convergent finite difference schemes for representative classes of singularly perturbed equations with partial derivatives. The development of techniques to derive *a priori* estimates on the basis of the decomposition method applied to the solution of boundary value problems, and also techniques to construct monotone discrete approximations (of boundary value problems) constructed on piecewise uniform meshes, allow the development of special finite difference schemes that are convergent ε-uniformly. As a result of such a progress, the book [138] was written.

In [138] *rather wide classes of elliptic and parabolic equations* as reaction-diffusion, and convection-diffusion, have been considered in n-dimensional domains with smooth and piecewise smooth boundaries, and also in composite domains in the presence of concentrated sources. For these problems, *techniques to construct ε-uniformly convergent schemes have been developed, such schemes have been constructed and their convergence has been studied*.

The presentation of results derived in [138] has been given in a concentrated form as in a reference book, so that it was possible to cover large classes of boundary value problems. However, such a form of exposition makes it difficult to use them in the development of numerical methods for equations with partial derivatives. It could be that this circumstance is one of the reasons for the delay in using results from [138], as well as a rather long delay in the development of robust numerical methods for multidimensional problems.

The book [87] is a thorough introduction to robust finite difference schemes. In [87], *basic ideas and the techniques* that are used to construct and justify ε-uniformly convergent difference schemes were considered using model problems as examples. Here, difference *schemes* based on *the fitted operator method* and *the condensing mesh method* have been constructed that converge ε-uniformly. Difficulties that arise when constructing robust numerical methods for parabolic equations with parabolic boundary layers have been considered.

In the book [33], for a number of model problems with boundary layers, extensive *numerical studies* have been performed for robust difference schemes, and in particular, for those that have been considered in [138, 87]. For the quantitative analysis of numerical methods, an experimental technique has been developed which allows us to justify ε-uniform convergence, and also to find the parameters in the estimates of the convergence rate. A *technique for experimental study* of special difference schemes has been developed in [33]. This technique was tested by numerical experiments carried out for linear and nonlinear problems, in particular for elliptic convection-diffusion equations whose solutions have parabolic boundary layers that degenerate on the domain boundary. Details of the *solvers* that are used in the computations have also been given.

In the concluding chapters of the book [33], the efficiency of the techniques that have been developed in the first ten chapters is demonstrated when solving the *classical Prandtl problem* for flow of an incompressible fluid past a semi-infinite flat plate [113]. For the nonlinear Prandtl problem, a difference scheme was constructed using a nonlinear solver; in this problem, $\varepsilon = Re^{-1}$ is a perturbation parameter, and Re is the Reynolds number. The numerical method gives *the problem solution* (that is components of the flow velocity) and *the normalized first-order derivatives* with an accuracy close to order one. *Computational expanses for the solution* (up to a logarithmic multiplier, are proportional to the number of mesh points used) *are independent of the parameter* ε.

In the book [106], numerical methods have been considered for singularly perturbed boundary value problems to ordinary differential equations and also to one-dimensional parabolic and two-dimensional elliptic equations (with the Laplace operator as the main part of the differential operator). Various ideas and techniques that are used in the numerical analysis for singularly perturbed differential equations, and also approaches developed for regular problems, are discussed. *Finite difference methods* and *finite element methods* have also been considered for singularly perturbed boundary value problems. Approximations of solutions to boundary value problems in *various norms*, in particular in *the maximum* and *the energy norms*, have been discussed. Significant attention has been devoted to ε-uniformly convergent numerical methods in which *the fitted operator method* and *the fitted mesh method* are applied using *Bakhvalov meshes* and *piecewise-uniform meshes*. A separate chapter is devoted to *incompressible Navier-Stokes equations*. The book contains an extensive bibliography on numerical methods for singularly perturbed boundary value problems.

The present book is devoted to *a systematic detailed development of approaches* to the construction of ε-uniformly convergent finite difference schemes for some classes of singularly perturbed *boundary value problems considered* in [138]. Problems for *multi-dimensional elliptic and parabolic equations that are reaction-diffusion and convection-diffusion* are studied in the case when *mixed derivatives are involved in the differential equations*. The problems are considered in *domains with piecewise smooth and curvilinear boundaries*. Boundary value problems from [138] such as *problems for equations with dis-*

continuous coefficients in the presence of concentrated sources, and also *problems for convection-diffusion equations with characteristic parts on the boundary*, are not considered.

The *approach* presented here includes:

- *a technique to construct a priori estimates for solutions* of boundary value problems which are *necessary* to construct and justify finite difference schemes ε-uniformly convergent in the maximum norm,

- *a technique to construct finite difference schemes* based on standard discrete approximations of boundary value problems, preserving monotonicity of differential problems,

- *a technique to construct special piecewise uniform meshes (grids)* providing ε-uniform convergence of the schemes under construction.

Furthermore, significant attention is devoted to the consideration of *modern trends* in the development of ε-uniformly convergent numerical methods for singularly perturbed problems. The investigations of a series of model problems can be considered as representative models of more general classes, for which approaches from the first part of the present book could be applied to the development of appropriate numerical methods.

Approaches developed in the book, and results derived, should promote further development of efficient numerical methods for *singularly perturbed problems with different types of singularities*, and at the same time, preserve the same qualities already well-established in numerical methods for regular boundary value problems.

In the present book, as well in the books [138, 87, 33], ε-uniformly convergent schemes are constructed on *simplest* piecewise uniform meshes, i.e., meshes having only one *transition point* in a neighborhood of the boundary layer. Schemes on Bakhvalov meshes, as well as on meshes with few *transition points* in a neighborhood of the boundary layer (see, e.g., [169]), are not considered here.

The present book and the books [138, 87, 106, 33], considered as one package, cover a wide *spectrum of problems* that arise when constructing special numerical methods for problems with boundary layers. Nevertheless, there is a large *gap* between complicated systems of nonlinear singularly perturbed equations with partial derivatives which *need to be solved*, e.g., from mathematical modelling, and those singularly perturbed problems for which *theoretically justified numerical methods exist*. We hope that the results given here could be helpful *in the development and analysis of new numerical methods* for problems that have yet to be solved.

1.7   The audience for this book

The present book, together with the books [138, 87, 106, 33], could be useful for scientists-researchers (from students up to professionals) in the development of numerical methods for singularly perturbed problems and also for specialist-researchers who are interesting in mathematical modelling and where problems with boundary and interior layers arise naturally.

In the first part of the book devoted to *multi-dimensional problems*, techniques and ideas exploited to *derive a priori estimates for the solutions* and to *construct the simplest discrete constructions*, i.e., piecewise uniform meshes and standard monotone finite difference operators on them, are carefully discussed for each type of boundary value problem. Conditions imposed on discrete constructions are discussed, and it is shown that such *conditions are close to necessary conditions*. Techniques and principles given in the book allow the reader to construct independently robust finite difference schemes for new classes of boundary value problems, in particular, for problems from [138] that are not treated in the present book.

In the second part of the book, a consideration of modern trends are carried out on model problems, which allow to check on ideas and techniques used in the construction. Here, we give references to original sources where details are given in the exposed bibliography. The discussion of modern trends allows the reader to be more confident in this fast growing area of research. The first and second parts of the book can be studied independently. The book can be also helpful for researchers studying singularly perturbed ordinary differential equations.

If the reader wishes to develop new difference methods for particular singularly perturbed problems, it is only necessary to read Part I.

If the reader wishes to apply difference methods already developed for singularly perturbed problems, the reader should pass to Part II.

Chapter 2

Boundary value problems for elliptic reaction-diffusion
equations in domains with smooth boundaries

In this chapter boundary value problems are considered for elliptic reaction-
diffusion equations in domains with smooth boundaries. The construction
of finite difference schemes and the justification of their convergence is carried out using an analog of the well-known sufficient condition for convergence of schemes for regular boundary value problems that is a consequence
of approximation by a stable finite difference scheme (its description see, e.g.,
[79, 108, 100]). When applied to singularly perturbed boundary value problems, this *sufficient condition for ε-uniform convergence* can be stated in the
following way:

> *ε-uniform convergence of a finite difference scheme follows from*
> *ε-uniform approximation of the boundary value problem by an*
> *ε-uniformly stable finite difference scheme.*

Our finite difference schemes are constructed based on classical difference approximations of differential equations. $\mathcal{E}$-uniform approximation of boundary
value problems by finite difference schemes is obtained by a special distribution of mesh points. $\mathcal{E}$-uniform monotonicity (and, hence, stability) of the
finite difference schemes is achieved by means of spatial meshes whose distri-
butions of nodes in each coordinate direction are consistent with each other.

2.1 Problem formulation. The aim of the research

On an n-dimensional domain D with sufficiently smooth boundary $\Gamma = \Gamma(D)$ and

$$\overline{D} = D \bigcup \Gamma, \tag{2.1}$$

where D is, in general, an unbounded domain, we consider the Dirichlet problem for elliptic *reaction-diffusion equations*

$$L\,u(x) = f(x), \quad x \in D, \tag{2.2a}$$

$$u(x) = \varphi(x), \quad x \in \Gamma. \tag{2.2b}$$

Here [1]

$$L \equiv \varepsilon^2 L_2 + L_0, \quad L_0 \equiv -c^0(x),$$

$$L_2 = L_{2(2.2c)} \equiv \sum_{s,\,k=1}^{n} a_{sk}(x) \frac{\partial^2}{\partial x_s \partial x_k} + \sum_{s=1}^{n} b_s(x) \frac{\partial}{\partial x_s} - c(x). \tag{2.2c}$$

The coefficients in (2.2c) are bounded and satisfy the *ellipticity condition*

$$a_0 \sum_{s=1}^{n} \xi_s^2 \le \sum_{s,k=1}^{n} a_{sk}(x) \xi_s \xi_k \le a^0 \sum_{s=1}^{n} \xi_s^2, \quad x \in \overline{D}, \quad a_0 > 0. \tag{2.3a}$$

They also satisfy the condition

$$c^0(x) \ge c^0 > 0, \quad c(x) \ge 0, \quad x \in \overline{D}. \tag{2.3b}$$

The perturbation parameter ε takes arbitrary values in the open-closed interval $(0, 1]$. The coefficients, the right-hand side f, and the boundary function φ in (2.2) are assumed to be sufficiently smooth.

As well as the problem (2.2) on the arbitrary domain (2.1), we consider also the boundary value problem on the slab

$$\overline{D}^{(0)} = D^{(0)} \bigcup \Gamma^{(0)}, \tag{2.4}$$

$$D^{(0)} = \{x : d_{*1} < x_1 < d_1^*, \ |x_s| < \infty, \ s = 2, \ldots, n\}.$$

For a periodic boundary value problem on the slab $\overline{D}^{(0)}$ the coefficients, the right-hand side f of the differential equation, and the boundary function φ are assumed to be 2π-periodic functions in x_s for $s = 2, \ldots, n$.

In the problems (2.2), (2.1) and (2.2), (2.4), a boundary layer appears in a neighborhood of the boundary as the parameter ε tends to zero.

We give some **definitions**. A *finite difference scheme* [2] means a set of grids and systems of difference equations defined on these grids. The solution error of a difference scheme depends, in general, on the number of nodes in the meshes used and the value of the parameter ε. In the case when the solution error is independent of the value of the parameter ε, we say that *the solution of the difference scheme (or, briefly, the scheme itself) converges ε-uniformly.* Convergence of a finite difference scheme is considered *in the discrete maximum norm*. For singularly perturbed boundary value problems, by *classical finite difference schemes* (classical difference approximations of boundary value problems) we mean schemes (approximations) such that have no special

---

[1] The notation $L_{(j.k)}$ ($M_{(j.k)}$, $m_{(j.k)}$, $G_{h(j.k)}$, $f_{(j.k)}(x)$) means that these operators (constants, grids, functions) are introduced in formula (j.k).

[2] Further, briefly, we shall use terms "difference scheme" or "scheme" instead of "finite difference scheme".

imposed restrictions (other than standard ones) on the choice of grids and difference equations that ensure ε-uniform convergence of the scheme. We say that a scheme is *a fitted operator method* when its ε-uniform convergence is ensured by a special choice of approximating difference equations. When constructing these schemes we use grids with arbitrary distribution of nodes. *The condensing mesh method* (or *adapted mesh method*) comprises schemes whose ε-uniform convergence is ensured by the choice of an appropriate distribution of mesh points. Such schemes are classical finite difference schemes on condensing/adapted meshes. When constructing such schemes we use systems of difference equations whose coefficients are, in general, functionals of the coefficients in the differential equation and the problem data (see, e.g., [79, 108]).

Our **aim** for the boundary value problem (2.2), (2.1) is to construct a finite difference scheme that converges ε-uniformly on condensing meshes.

2.2 Estimates of solutions and derivatives

For the construction and investigation of difference schemes, we need estimates for the solution of the boundary value problem and its derivatives for $\varepsilon \in (0, 1]$. These *bounds* are derived by using *a priori estimates* for regular problems, i.e., Schauder-type interior estimates and estimates up to the boundary [69, 37]. Here an *a priori estimate* means an estimate of the solution under the assumption that this solution exists, that depends on the problem data, i.e., the coefficients in the equation, its right-hand side, and the boundary conditions.

We denote by $C^{l+\alpha}(\overline{D})$, where $l \geq 0$ and $\alpha \in (0, 1)$, the Banach space whose elements are continuous functions in $\overline{D}$ that have in $\overline{D}$ continuous derivatives of order up to l which are in the sense of Hölder continuous with exponent α. For these elements a finite value is taken by the norm

$$\overline{|u|}_{l+\alpha} = \overline{|u|}_l + \sum_{\substack{k_1+\ldots+k_n=k \\ k=l}} \overline{H}_\alpha \left(\frac{\partial^k}{\partial x_1^{k_1} \ldots \partial x_n^{k_n}} u \right).$$

Here

$$\overline{|u|}_l = \sum_{k=0}^{l} \sum_{k_1+\ldots+k_n=k} \left| \frac{\partial^k}{\partial x_1^{k_1} \ldots \partial x_n^{k_n}} u \right|_0, \quad \overline{|u|}_0 = \max_{\overline{D}} |u(x)|,$$

$$\overline{H}_\alpha(u) = \sup_{x,\, x' \in \overline{D}} \frac{|u(x) - u(x')|}{|x - x'|^\alpha},$$

and $|x - x'|$ is the distance between the points x and x', where x, $x' \in \overline{D}$, and $\overline{H}_\alpha(u)$ is Hölder coefficient of the function u in D.

Solutions of singularly perturbed problems change sharply and their derivatives grow without boundary as $\varepsilon \to 0$, however, in thin subdomains, i.e., in neighborhoods of boundary and interior layers. We are interested in estimates of the solutions and derivatives depending on the value of the parameter ε, and the behaviour of the derivatives in a neighborhood of the layer.

When studying boundary value problems, *a maximum principle* is often used in the following form of a *comparison lemma* (see, e.g., [69, 37, 99, 205]).

Lemma 2.2.1 *Let* v, $w \in C^2(D) \cap C(\overline{D})$, *and let the condition*

$$|L\,v(x)| \le -L\,w(x), \quad x \in D, \quad |v(x)| \le w(x), \quad x \in \Gamma,$$

hold, where $L = L_{(2.2a)}$. *Then* $|v(x)| \le w(x), \quad x \in \overline{D}$.

Here $w(x)$, $x \in D$ is a *majorant (barrier) function* for $v(x)$.

We consider the boundary value problem (2.2) on the slab (2.4). Using the maximum principle, we deduce the inequality [3]

$$|u(x)| \le M \max \left[\max_{\overline{D}} |f(x)|, \max_\Gamma |\varphi(x)| \right], \quad x \in \overline{D}. \tag{2.5}$$

For $a_{sk}, b_s, c, c^0, f \in C^{l+\alpha}(\overline{D})$, $\varphi \in C^{l+2+\alpha}(\Gamma)$, where l is an integer number, $l \ge k^0$, $k^0 \ge 0$, $\alpha \in (0, 1)$, we have $u \in C^{l+2+\alpha}(\overline{D})$ for each $\varepsilon \in (0, 1]$ [69].

Next, we pass to the boundary value problem in the variables $\xi_s = \xi_s(x) = \varepsilon^{-1}x_s$, where $s = 1, \ldots, n$. In these variables $(\xi = \xi(x))$, the differential equation on the set $\widetilde{D}$ becomes regular; here $\overline{\widetilde{D}} = \widetilde{D} \cup \widetilde{\Gamma}$, where $\widetilde{D} = \{\xi = \xi(x) : x \in D\}$. Taking into account the *a priori estimates* up to the boundary [69], we find

$$\left| \frac{\partial^k}{\partial \xi_1^{k_1} \ldots \partial \xi_n^{k_n}} \widetilde{u}(\xi) \right| \le M, \quad 0 \le k \le k^0 + 2, \quad \xi \in \overline{\widetilde{D}}, \quad k = k_1 + \ldots + k_n,$$

where $\widetilde{u}(\xi) = u(x(\xi))$. Returning to the variables x, we obtain the following estimate for the function $u(x)$:

$$\left| \frac{\partial^k}{\partial x_1^{k_1} \ldots \partial x_n^{k_n}} u(x) \right| \le M \varepsilon^{-k}, \quad x \in \overline{D}, \quad 0 \le k \le K, \tag{2.6}$$

where $K = k^0 + 2$.

---

[3] Here and below, M, M_i, M^i (or m, m_i, m^i) denote sufficiently large (small) positive constants that are independent of the parameter ε and of the discretization parameters.

Lemma 2.2.2 *Let* a_{sk}, b_s, c, c^0, $f \in C^{l+\alpha}(\overline{D})$, $s, k = 1, \ldots, n$, $\varphi \in$ $C^{l+2+\alpha}(\Gamma)$, $l \geq k^0$, $k^0 \geq 0$, $\alpha \in (0, 1)$. *Then the solution of the problem* (2.2), (2.4) *satisfies the estimate* (2.6).

Write the function $u(x)$, i.e., the solution of the problem (2.2), as the *decomposition*, i.e., the sum of the functions:

$$u(x) = U(x) + V(x), \tag{2.7}$$

where $U(x)$ and $V(x)$ are the regular and singular parts (components) of the solution of the problem. The function $U(x)$ is the restriction to $\overline{D}$ of the function $U^e(x)$, $x \in D^e$, where $U^e(x)$ is the solution of a problem which is extended beyond the set Γ:

$$L^e U^e(x) = f^e(x), \quad x \in D^e, \quad U^e(x) = \varphi^e(x), \quad x \in \Gamma^e. \tag{2.8}$$

Here the domain D^e involves the domain D together with its m-neighborhood; $\Gamma^e = \Gamma(D^e)$ is sufficiently smooth boundary of the domain D^e; the coefficients of the operator L^e and the right-hand side $f^e(x)$ are sufficiently smooth extensions on $\overline{D}^e$ of corresponding data of the problem (2.2) and $\varphi^e(x)$, $x \in \Gamma^e$, is a sufficiently smooth function. The functions $f^e(x)$, $\varphi^e(x)$, are assumed to be equal to zero outside an m_1-neighborhood of the set $\overline{D}$, where $m_1 < m$. The function $V(x)$ is the solution of the problem

$$L V(x) = 0, \quad x \in D, \quad V(x) = \varphi(x) - U(x), \quad x \in \Gamma. \tag{2.9}$$

Write the function $U^e(x)$, $x \in \overline{D}^e$, as the sum of the functions

$$U^e(x) = U_0^e(x) + v_1^e(x), \quad x \in \overline{D}^e. \tag{2.10a}$$

Here the function $U_0^e(x)$ is the solution of a degenerated problem and is defined by the relation

$$-c^{0e}(x) U_0^e(x) = f^e(x), \quad x \in \overline{D}^e; \tag{2.10b}$$

the function $v_1^e(x)$ is the remainder term in the representation of the function $U^e(x)$ and is the solution of the problem

$$L^e v_1^e(x) = -\varepsilon^2 L_2^e U_0^e(x), \quad x \in D^e, \quad v_1^e(x) = \varphi^e(x) - U_0^e(x), \quad x \in \Gamma^e, \tag{2.10c}$$

where L_2^e is an extension of the operator $L_2 = L_{2(2.2c)}$.

Let the data of the problem (2.2), (2.4) satisfy the condition a_{sk}, b_s, c, c^0, $f \in C^{l+\alpha}(\overline{D})$, where $l \geq k^0$, $k^0 \geq 2$, $\alpha \in (0, 1)$, and let the extensions of these data to the set $\overline{D}^e$ belong to the class $C^{l+\alpha}(\overline{D}^e)$. Then for the function $v_1^e(x)$, one has $v_1^e \in C^{l+\alpha}(\overline{D}^e)$, and also the estimate

$$\left| v_1^e(x) \right| \leq M \varepsilon^2, \quad x \in \overline{D}^e,$$

that is established using the maximum principle. In the boundary value problem for the function $v_1^e(x)$, $x \in \overline{D}^e$, we pass to the variables $\xi_s = \xi_s(x) =$

$\varepsilon^{-1} x_s$, where $s = 1, \ldots, n$. In these variables, the differential equation on the set $\overline{\widetilde{D}}$ becomes regular, where $\overline{\widetilde{D}}^e = \{\xi = \xi(x) : \ x \in \overline{D}^e\}$. Taking into account *interior a priori estimates* [69], we find the following estimates on $\overline{\widetilde{D}}$

$$\left| \frac{\partial^k}{\partial \xi_1^{k_1} \ldots \partial \xi_n^{k_n}} \widetilde{v}_1^e(\xi) \right| \le M \varepsilon^2, \quad 0 \le k \le k^0, \quad \xi \in \overline{\widetilde{D}},$$

where $\widetilde{v}_1^e(\xi) = v_1^e(x(\xi))$. Returning to the variables x, by virtue of the explicit form of the function $U_0^e(x)$, we obtain $U \in C^{l+\alpha}(\overline{D})$; moreover,

$$\left| \frac{\partial^k}{\partial x_1^{k_1} \ldots \partial x_n^{k_n}} U(x) \right| \le M \left[1 + \varepsilon^{2-k} \right], \quad x \in \overline{D}, \quad 0 \le k \le K, \tag{2.11}$$

where $K = k^0$. The regular component of the problem solution and its lowest derivatives (here they are no higher than second order) are bounded ε-uniformly on the set $\overline{D}$.

Using *majorant* functions of type

$$\exp\left(-m\varepsilon^{-1}(d_1^* - x_1)\right) + \exp\left(-m\varepsilon^{-1}(x_1 - d_{*1})\right), \quad x \in [d_{*1}, d_1^*],$$

we establish the inequality

$$|V(x)| \le M \exp\left(-m\varepsilon^{-1} r(x, \Gamma)\right), \quad x \in \overline{D}, \tag{2.12a}$$

where $r(x, \Gamma)$ is the distance from the point $x \in \overline{D}$ to the boundary Γ and m is an arbitrary constant in the interval $(0, m_0)$, the constant m_0 defined by the relation

$$m_0 = \min_{\overline{D}} \left[a_{11}^{-1}(x) c^0(x) \right]^{1/2}. \tag{2.12b}$$

For the data of the problem (2.2), (2.4), as well as the conditions for smoothness given above, let the condition $\varphi \in C^{l+\alpha}(\Gamma)$ be fulfilled also. Then $V \in C^{l+\alpha}(\overline{D})$. With regard to the estimates up to the boundary, in the variables $\xi = \xi(x)$ [69] one has

$$\left| \frac{\partial^k}{\partial \xi_1^{k_1} \ldots \partial \xi_n^{k_n}} \widetilde{V}(\xi) \right| \le M \exp\left(-m r(\xi, \widetilde{\Gamma})\right), \quad \xi \in \overline{\widetilde{D}}, \quad 0 \le k \le k^0,$$

where $\widetilde{V}(\xi) = V(x(\xi))$, $\xi \in \overline{\widetilde{D}}$, and we obtain

$$\left| \frac{\partial^k}{\partial x_1^{k_1} \ldots \partial x_n^{k_n}} V(x) \right| \le M \varepsilon^{-k} \exp\left(-m\varepsilon^{-1} r(x, \Gamma)\right), \quad x \in \overline{D}, \tag{2.13}$$

$$0 \le k \le k^0.$$

Let us refine the estimate (2.13) in the case of the problem (2.2), (2.4). The function $V(x)$ is the solution of the problem (2.9). We differentiate the equation and the boundary condition in x_j, where $j \neq 1$. In the left-hand side of the equation obtained, we leave a term that involves the function $V^{(1)}(x) \equiv (\partial/\partial x_j)V(x)$ and its derivatives. We transfer to the right-hand side of the equation those terms that contain the derivatives of the coefficients of (2.2a) as multipliers. We denote the right-hand side by $f^{(1)}(x)$, where $f^{(1)} \in C^{l-1+\alpha}(\overline{D}))$, and we have the inequality

$$|f^{(1)}(x)| \leq M \exp\left(-m\varepsilon^{-1} r(x, \Gamma)\right), \quad x \in \overline{D}.$$

Taking into account the inequality

$$\left| V^{(1)}(x) \right| = \left| \frac{\partial \varphi(x)}{\partial x_j} - \frac{\partial U(x)}{\partial x_j} \right| \leq M, \quad x \in \Gamma,$$

one can establish the estimate

$$\left| \frac{\partial^k}{\partial x_1^{k_1} \dots \partial x_{j-1}^{k_{j-1}} \partial x_j \, \partial x_{j+1}^{k_{j+1}} \dots \partial x_n^{k_n}} V(x) \right| \leq$$

$$\leq M \left[\varepsilon^{-k_1} + \varepsilon^{1-k} \right] \exp\left(-m\varepsilon^{-1} r(x, \Gamma)\right),$$

$$0 \leq k \leq k^0, \quad j > 1, \quad k = k_1 + \dots + k_{j-1} + 1 + k_{j+1} + \dots + k_n.$$

In a similar way, we obtain the estimate

$$\left| \frac{\partial^k}{\partial x_1^{k_1} \dots \partial x_n^{k_n}} V(x) \right| \leq M \left[\varepsilon^{-k_1} + \varepsilon^{2-k} \right] \exp\left(-m\varepsilon^{-1} r(x, \Gamma)\right), \qquad (2.14)$$

$$x \in \overline{D}, \quad 0 \leq k \leq K,$$

where m is an arbitrary constant in the interval $(0, m_0)$ with $m_0 = m_{0(2.12)}$ in the case of the problem (2.2), (2.4), for $K = k^0$. In the estimates (2.13), (2.14), one can choose the constant m as an arbitrary number in the interval $(0, m_0)$.

The singular component of the problem solution decays rapidly when moving away from the set Γ (i.e., exponentially decreases as $\varepsilon^{-1} r(x, \Gamma)$ grows); however, its derivatives along the normal to the boundary Γ in the neighborhood of Γ defined by $\varepsilon^{-1} r(x, \Gamma) = \mathcal{O}(1)$ grow without bounds as $\varepsilon \to 0$ and $\varepsilon^{-1} r(x, \Gamma) \to \infty$.

Thus, the following theorem is established.

Theorem 2.2.1 *Let the condition* a_{sk}, b_s, c, c^0, $f \in C^{l+\alpha}(\overline{D})$, *for* $s, k = 1, \dots, n$, $\varphi \in C^{l+\alpha}(\Gamma)$, $l \geq k^0$, $k^0 \geq 2$, $\alpha \in (0, 1)$ *hold. Then for the components* $U(x)$ *and* $V(x)$ *of the solution* $u(x)$ *to the problem* (2.2), (2.4) *in the representation* (2.7), *the estimates* (2.11), (2.14) *hold.*

Let us discuss estimates of the problem (2.2), (2.1) in the case when the domain D has sufficiently smooth *curvilinear boundary* Γ. Here it is convenient to consider the problem solution in the space $C^{l+\alpha}(D)$ (a definition can be found in Chapter 3, Section 3.2). Let $\Gamma_0 \subseteq \Gamma$ be a smooth part of the boundary Γ, i.e., $\Gamma_0 = \Gamma_0^g$, and let $\Gamma_0 \in C^{l+\alpha}$, where $l \geq k^0$, $k^0 \geq 2$ and $\alpha \in (0,1)$. Thus, in a neighborhood of each point of the boundary Γ, there exists a local representation of the boundary Γ_0 in the form

$$x_i = \eta_i\left(x_1,\ldots,x_{i-1},x_{i+1},\ldots,x_n\right);$$

furthermore, the function η_i belongs to the class $C^{l+\alpha}$. In the case when $\varphi \in C^{l+\alpha}(\Gamma_0)$, then the function $\varphi(x)$, $x \in \Gamma_0$ (as a function of the local variables $x_1,\ldots,x_{i-1},x_{i+1},\ldots,x_n$), belongs to the regarded class $C^{l+\alpha}$. In the case when a_{sk}, b_s, c, c^0, $f \in C^{l_0+\alpha}(\overline{D})$, $\varphi \in C^{l+\alpha}(\Gamma_0)$ for $l_0 = l - 2$, for the problem solution the estimate (similar to (2.6) and (3.4))

$$\left| \frac{\partial^k}{\partial x_1^{k_1}\ldots\partial x_n^{k_n}} u(x) \right| \leq M \left[\varepsilon^{-k} + r^{-k}(x, \Gamma\setminus\Gamma_0)\right], \quad x \in \overline{D}, \ 0 \leq k \leq K,$$

holds up to the smooth part of the boundary Γ_0, where $K = l$.

Write the problem solution as the sum of the functions

$$u(x) = U(x) + v_u(x), \quad x \in \overline{D},$$

where $U(x)$, $x \in \overline{D}$, is the restriction to $\overline{D}$ of the solution to the problem (2.8). For $l_0 = l$ the function $U(x)$ satisfies the estimate (2.11). For the function $v_u(x)$, one has $v_u \in C^{l+\alpha}(\overline{D})$, and also the estimate

$$|v_u(x)| \leq \begin{cases} M \exp\left(-m\varepsilon^{-1} r(x,\Gamma)\right), & r(x,\Gamma) \leq M_1\,\sigma, \\ M\varepsilon^2, & r(x,\Gamma) \geq m_1\,\sigma, \ x \in \overline{D}, \end{cases}$$

holds. Here m is an arbitrary value in the interval $(0, m_0)$, σ is sufficiently small value independent of ε; $m_0 = \left(a_0^{-1} c^0\right)^{1/2}$, $a_0 = a_{0(2.3)}$, $c^0 = c_{(2.3)}^0$; $M_1\,\sigma$ is less than the radius of curvature of the boundary Γ. Taking into account interior *a priori* estimates of the derivatives, we find the estimate of the derivatives outside an σ-neighborhood of the boundary Γ:

$$\left| \frac{\partial^k}{\partial x_1^{k_1}\ldots\partial x_n^{k_n}} v_u(x) \right| \leq M\varepsilon^{2-k}, \quad x \in D, \quad r(x,\Gamma) \geq \sigma.$$

For the problem solution outside an σ-neighborhood of the boundary Γ, we obtain the estimate

$$\left| \frac{\partial^k}{\partial x_1^{k_1}\ldots\partial x_n^{k_n}} u(x) \right| \leq M\left[1 + \varepsilon^{2-k}\right], \quad x \in \overline{D}, \tag{2.15}$$

$$r(x, \Gamma) \geq \sigma, \quad 0 \leq k \leq K,$$

where $K = k^0$.

In an σ-neighborhood of the boundary Γ, we write the problem solution as the sum

$$u(x) = U(x) + V(x), \quad x \in \overline{D}, \quad r(x, \Gamma) \leq \sigma, \tag{2.16}$$

where $U(x)$ and $V(x)$ are the regular and singular components of the problem solution, here $U(x) = U_{(2.8)}^e(x)$, $x \in \overline{D}$, $r(x, \Gamma) \leq \sigma$. For the function $U(x)$ we have the estimate

$$\left| \frac{\partial^k}{\partial x_1^{k_1} \dots \partial x_n^{k_n}} U(x) \right| \leq M \left[1 + \varepsilon^{2-k} \right], \quad x \in \overline{D}, \tag{2.17}$$

$$r(x, \Gamma) \leq \sigma, \quad 0 \leq k \leq K,$$

where $K = k^0$. In an σ-neighborhood of the boundary Γ we pass to a new coordinate system $x_s' = x_s'(x)$, for $s = 1, \dots, n$ such that some part of the boundary Γ belongs to the plane $x_1' = \text{const}$, where $x_1' = x_1'(x)$; $r(x, \Gamma) \leq \sigma$ is the distance from the point $x \in \overline{D}$ to the boundary Γ. Similar to constructions on the slab, we estimate the problem solution in a neighborhood of this part of the boundary. Returning to the original coordinate system, we find that for the regular and singular components of the problem solution in an σ-neighborhood of the boundary Γ, the estimates similar to (2.11), (2.14) hold.

Thus, assuming that the conditions of Theorem 2.2.1 are fulfilled and also the condition $\Gamma \in C^{l+\alpha}$ holds, one obtains that outside some sufficiently small neighborhood of the boundary Γ (the size of the neighborhood is independent of the parameter), the solution of the boundary value problem satisfies the estimate (2.15). In a neighborhood of the boundary, the problem solution has the representation (2.16). For the regular part of the problem solution, the estimate (2.17) is valid, and for the singular part, one has the estimate

$$\left| \frac{\partial^k}{\partial x_1'^{k_1} \dots \partial x_n'^{k_n}} V_{x'}(x') \right| \leq M \left[\varepsilon^{-k_1} + \varepsilon^{2-k} \right] \exp \left(-m \, \varepsilon^{-1} r(x', \Gamma_{x'}) \right),$$

$$\tag{2.18}$$

$$x' \in \overline{D}_{x'} \cap \{ r(x', \Gamma_{x'}) \leq \sigma \}, \quad 0 \leq k \leq K,$$

where $K = k^0$, $\overline{D}_{x'} \cap \{ r(x', \Gamma_{x'}) \leq \sigma \}$ and $\Gamma_{x'}$ are the set $\overline{D} \cap \{ r(x, \Gamma) \leq \sigma \}$ and the boundary Γ in the variables x', $V_{x'}(x') = V(x(x'))$; the derivative in x_1' is the derivative along the normal to the boundary Γ, and the derivatives in x_s', where $s = 2, \dots, n$, are the derivatives in the direction orthogonal to the normal to the boundary; $m \in (0, m_0)$, the constant m_0 in the case of the estimate (2.18), is defined by the relation

$$m_0 = \left(a_0^{-1} c^0 \right)^{1/2}, \quad a_0 = a_{0(2.3)}, \quad c^0 = c_{(2.3)}^0.$$

Theorem 2.2.2 *For the data of the boundary value problem (2.2), (2.1) in the domain D with the curvilinear boundary Γ, let the assumptions of Theorem 2.2.1 be fulfilled, and let $\Gamma \in C^{l+\alpha}$. Then for the problem solution $u(x)$ and for the components $U(x)$ and $V(x)$ in the representation (2.16), the estimates (2.15), (2.17), (2.18) hold.*

2.3 Conditions ensuring ε-uniform convergence of difference schemes for the problem on a slab

2.3.1 Sufficient conditions for ε-uniform convergence of difference schemes

When deriving sufficient conditions for ε-uniform convergence of a difference scheme, we use a standard algorithm (principle) that is used for convergence proofs in the case of regular problems: "convergence of a finite difference scheme follows from an approximation of the boundary value problem by a stable finite difference scheme" (see, e.g., [80, 109, 106]). This principle allows us to establish convergence of the stable finite difference scheme that approximates the boundary value problem.

We consider the boundary value problem (2.2) on a convex set

$$\overline{D}, \tag{2.19}$$

whose faces are orthogonal to coordinate axes.
    Let

$$\overline{D}_h \tag{2.20}$$

be a rectangular mesh on the set $\overline{D}$, and let $\overline{N} = N(D)$ be the number of nodes in the mesh $\overline{D}_h$, if the set $\overline{D}$ is bounded, and $\overline{N} = N(D)$ be the minimal number of nodes in the mesh $\overline{D}_h$ on the unit n-dimensional cube whose centre belongs to the set $\overline{D}$, if the set $\overline{D}$ is unbounded. On the mesh $\overline{D}_h$, consider a finite difference scheme that corresponds to the problem (2.2), (2.19) (see, e.g., schemes [79, 108], in particular, the scheme (2.47), (2.26) for the problem (2.2), (2.4)):

$$\Lambda z(x) = f^h(x), \quad x \in D_h, \quad z(x) = \varphi^h(x), \quad x \in \Gamma_h. \tag{2.21}$$

Here the functions $f^h(x)$, $x \in D_h$, and $\varphi^h(x)$, $x \in \Gamma_h$, as well as the coefficients multiplied by the difference derivatives in the operator Λ, are functionals of the data of the problem (2.2) and of the coefficients in the differential equation.
    We say that:

1) the finite difference *scheme* (2.21), (2.20) *approximates the boundary value problem* (2.2), (2.19) *ε-uniformly* (in the discrete maximum norm) if the following inequalities are fulfilled:

$$|(L - \Lambda) u(x)| \leq M \lambda(N(D)), \quad x \in D_h; \qquad (2.22)$$

$$\left.\begin{array}{l} |f(x) - f^h(x)| \leq M\lambda(N(D)), \quad x \in D_h \\ |\varphi(x) - \varphi^h(x)| \leq M\lambda(N(D)), \quad x \in \Gamma_h \end{array}\right\}, \qquad (2.23)$$

where $\lambda(N(D)) \underset{\varepsilon}{\rightarrow} 0$ as $N \rightarrow 0$ (the notation $f(\xi) \underset{\varepsilon}{\rightarrow} 0$ as $\xi \rightarrow \xi_0$ means that $f(\xi) \rightarrow 0$ as $\xi \rightarrow \xi_0$ ε-uniformly), and $M \lambda(N(D))$ is the ε-uniform error bound of the approximation of the boundary value problem by the finite difference scheme (2.21), (2.20);

2) the finite difference *scheme* (2.21), (2.20) is *ε-uniformly stable* (in the discrete maximum norm) if the following inequality is valid:

$$|z(x)| \leq M \max\left[\max_{D_h}|f^h(x)|, \ \max_{\Gamma_h}|\varphi^h(x)|\right], \quad x \in \overline{D}_h;$$

3) the finite difference *scheme* (2.21), (2.20) *converges ε-uniformly at the rate* of $\mathcal{O}(\lambda(N(D)))$ if the following inequality holds:

$$|u(x) - z(x)| \leq M \lambda(N(D)), \quad x \in \overline{D}_h.$$

Here it is convenient to give a formulation of the maximum principle for the grid problem (2.21), (2.20) [108] (see also [214]).

The difference equation (2.21) in the nodes $x \in D_h$ can be written in *the canonical form*

$$\Lambda z(x) \equiv -A(x) z(x) + \sum_{y \in S(x)} B(x, y) z(y) = f^h(x), \quad x \in D_h. \qquad (2.24a)$$

Here $A(x)$ and $B(x, y)$ are coefficients of the discrete equation. *The stencil of the finite difference scheme*, i.e., the grid equation (2.24a), is the set of nodes $x \bigcup S(x)$ on which the grid functions are involved in the grid equation written at the point x. The node x is the stencil centre and the set $S(x)$ is a stencil neighborhood of the node x. Let the coefficients $A(x)$ and $B(x, y)$ of the difference operator $\Lambda_{(2.21)}$ on the grid (2.20) satisfy the conditions [108]

$$A(x) > 0, \quad B(x, y) > 0, \quad C(x) = A(x) - \sum_{y \in S(x)} B(x, y) \geq 0, \qquad (2.24b)$$

$$x \in D_h, \quad y \in S(x), \quad \varepsilon \in (0, 1].$$

In this case, we say that *the scheme* (2.21), (2.20) (*the operator* $\Lambda_{(2.21)}$) is *ε-uniformly positive*.

In the case when the inequality $z(x) \geq 0$, $x \in \overline{D}_h$, for $\varepsilon \in (0, 1]$, follows from the condition

$$\Lambda z(x) \leq 0, \quad x \in D_h, \qquad z(x) \geq 0, \quad x \in \Gamma_h,$$

we say that the finite difference *scheme* (2.21), (2.20) (*the operator* $\Lambda_{(2.21)}$) is ε-*uniformly monotone.* The condition (2.24b) implies the monotonicity of the finite difference scheme (2.21), (2.20) (the operator $\Lambda_{(2.21)}$), but the converse is false. We call this *the monotonicity condition of the finite difference scheme (the operator).* We say that the condition (2.24b) is *the condition of strong* ε-*uniform monotonicity of the scheme* (2.21), (2.20) (*the operator* $\Lambda_{(2.21)}$) if in (2.24b) one has $C(x) > 0$ for $x \in D_h$, and $\varepsilon \in (0, 1]$.

The maximum principle for the discrete Dirichlet problem is stated in the following way.

Lemma 2.3.1 *Let the coefficients of the operator* Λ *satisfy the condition* (2.24). *Then from the condition*

$$\Lambda z(x) \geq 0 \quad x \in D_h, \quad z(x) \not\equiv const, \quad x \in \overline{D}_h,$$

it follows that $z(x)$ *can not take its largest positive value on the nodes of* D_h.

The maximum principle is often applied in the following form of *a comparison lemma.*

Lemma 2.3.2 *Let the coefficients of the operator* Λ *satisfy the condition* (2.24), *and for the functions* $v(x)$ *and* $w(x)$, $x \in \overline{D}_h$, *let the condition*

$$|\Lambda v(x)| \leq -\Lambda w(x) \quad x \in D_h, \quad |v(x)| \leq w(x), \quad x \in \Gamma_h,$$

be satisfied. Then $|v(x)| \leq w(x)$, $x \in \overline{D}_h$.

Here $w(x)$, $x \in \overline{D}_h$ is a *majorant (barrier) function* for $v(x)$, $x \in \overline{D}_h$.
The following convergence theorem holds.

Theorem 2.3.1 *Let the finite difference scheme* (2.21), (2.20) *approximate the boundary value problem* (2.2), (2.19) ε-*uniformly with the estimates* (2.22), (2.23), *and let this scheme be* ε-*uniformly stable. Then the difference scheme* (2.21), (2.20) *converges at the rate* $\mathcal{O}\bigl(\lambda(N(D))\bigr)$ ε-*uniformly as* $N(D) \to \infty$:

$$|u(x) - z(x)| \leq M\,\lambda(N(D)), \quad x \in \overline{D}_h.$$

In the hypotheses of Theorem 2.3.1, the ε-uniform monotonicity of the scheme (2.21), (2.20) is not assumed.

The solution of the boundary value problem (2.2), (2.19) is ε-uniformly stable, and it satisfies the estimate (2.5).

The difference scheme (2.21), (2.20) is also ε-uniformly stable if the operator Λ is ε-uniformly monotone and approximates the operator L ε-uniformly (in the maximum norm) on the smooth functions:

$$\sup_{\varepsilon,\, D_h} |(L - \Lambda)\, v(x)| \le M\, \lambda_1(N(D)). \qquad (2.25a)$$

Here $\lambda_1(N(D)) \underset{\varepsilon}{\to} 0$ as $N \to \infty$ and $v(x)$ is a sufficiently smooth function satisfying the condition

$$\left| \frac{\partial^k}{\partial x_1^{k_1} \dots \partial x_n^{k_n}} v(x) \right| \le M, \quad x \in \overline{D}, \quad 0 \le k \le 3, \quad k = k_1 + \dots + k_n. \quad (2.25b)$$

Inequality (2.25) is established using the maximum principle, where one takes into account the strong negativity of the operator L on the majorant function $v(x) = M \max \left[\max_{D_h} |f^h(x)|, \max_{\Gamma_h} |\varphi^h(x)| \right]$, $x \in \overline{D}$. Thus, the following statement is valid.

Lemma 2.3.3 *Let the operator $\Lambda_{(2.21)}$ be ε-uniformly monotone and approximate the operator L ε-uniformly on sufficiently smooth functions satisfying the condition (2.25). Then the difference scheme (2.21), (2.20) is ε-uniformly stable:*

$$|z(x)| \le M \max \left[\max_{D_h} |f^h(x)|, \max_{\Gamma_h} |\varphi^h(x)| \right], \quad x \in \overline{D}_h.$$

A corollary of Theorem 2.3.1 and Lemma 2.3.3 is the following result.

Theorem 2.3.2 *The difference scheme (2.21), (2.20) converges ε-uniformly at the rate $\mathcal{O}(\lambda(N(D)))$ if it approximates the boundary value problem (2.2), (2.19) ε-uniformly with the estimates (2.22), (2.23) and the operator Λ is ε-uniformly monotone.*

2.3.2 Sufficient conditions for ε-uniform approximation of the boundary value problem

We now discuss sufficient conditions for ε-uniform approximation of the boundary value problem (2.2) on the slab (2.4) by a difference scheme. We consider classical difference approximations on rectangular grids. Assume that the estimates of Theorem 2.2.1 hold for the solution of the boundary value problem (2.2).

The simplest *classical difference schemes* are obtained by changing derivatives in the differential equations into appropriate difference derivatives. In certain difference schemes, the coefficients in the differential and the difference equations (2.2) and (2.21), in general, do not coincide, but this discrepancy tends to zero as the maximal mesh step-size tends to zero; furthermore, the coefficients in the difference equations are independent of the solutions of the differential equations; such schemes are also classical. For boundary value problems, the construction of grid approximations including difference schemes can be found in, e.g., [79, 108]; for the approximation of mixed derivatives see also Subsections 2.4.1–2.4.3. Note that we do not consider schemes of fitted operator type.

On the set $\overline{D} = \overline{D}_{(2.4)}^{(0)}$ we introduce the grid

$$\overline{D}_h = \overline{\omega}_1 \times \omega_2 \times \ldots \times \omega_n. \tag{2.26}$$

Here $\overline{\omega}_1, \omega_2, \ldots, \omega_n$ are meshes with an arbitrary distribution of nodes,

$$\overline{\omega}_1 = \left\{ x_1^i : \ d_{*1} \le x_1^i \le d_1^*, \ \ i = 0, 1, \ldots, N_1, \ \ x_1^0 = d_{*1}, \ \ x_1^{N_1} = d_1^* \right\},$$

$$\omega_s = \{ x_s^i : \ |x_s^i| < \infty, \ \ i = 0, \pm 1, \pm 2, \ldots \}, \quad s = 2, \ldots, n.$$

Set $h = \max_s h_s$, where $h_s = \max_i h_s^i$ with $h_s^i = x_s^{i+1} - x_s^i$ for $s = 1, \ldots, n$. Let $N_1 + 1$ be the number of nodes in the mesh $\overline{\omega}_1$ and let $N_s + 1$ be the minimal number of nodes in the mesh ω_s, $s = 2, \ldots, n$, per unit length; $N = \min_s N_s$, $s = 1, \ldots, n$. Assume that the condition $h \le MN^{-1}$ is fulfilled. Set $D_h = D \cap \overline{D}_h$, $\Gamma_h = \Gamma \cap \overline{D}_h$.

Write the operator $L_{(2.2)}$ as the sum of operators

$$L_{(2.2)} = L^* + L^{**}, \tag{2.27a}$$

where

$$L^* \equiv \varepsilon^2 \, a_{11}(x) \, \frac{\partial^2}{\partial x_1^2}, \tag{2.27b}$$

$$L^{**} \equiv \varepsilon^2 \left\{ \sum_{\substack{s, k=1 \\ s+k>2}}^{n} a_{sk}(x) \frac{\partial^2}{\partial x_s \partial x_k} + \sum_{s=1}^{n} b_s(x) \frac{\partial}{\partial x_s} - c(x) \right\} - c^0(x). \tag{2.27c}$$

It is not difficult to show that in the class of functions $v(x)$, $x \in \overline{D}$, that satisfy the condition

$$\varepsilon^2 \left| \frac{\partial^k}{\partial x_s^{k_s} \partial x_r^{k_r}} v(x) \right| \le M, \quad x \in \overline{D}, \ \ 0 \le k \le K, \ \ 0 \le k_1 \le 2,$$

where $k = k_s + k_r$, $s, r = 1, \ldots, n$, $K \ge 3$, *the operator L^{**} is approximated ε-uniformly* by the classical finite difference *operators*

$$\Lambda^{**} \tag{2.28}$$

on the grids $\overline{D}_{h(2.24)}$ with an arbitrary distribution of the nodes x_s^i in the meshes ω_s for $s = 1, \ldots, n$:

$$|(L^{**} - \Lambda^{**}) v(x)| \le M \lambda_2(N), \quad x \in D_h, \tag{2.29}$$

$\lambda_2(N) \underset{\varepsilon}{\to} 0$ as $N \to \infty$. The same *operators* Λ^{**}, by virtue of the bounds of Theorem 2.2.1, *approximate ε-uniformly the operator* $L^{**}_{(2.27)}$ *on the solution of the boundary value problem:*

$$|(L^{**}_{(2.27)} - \Lambda^{**})\, u(x)| \le M\, \lambda_2(N), \quad x \in D_h.$$

However, already the difference operator $\varepsilon^2\, \delta_{\overline{x1}\,\widehat{x1}}$ does not approximate ε-uniformly the operator $\varepsilon^2\, \partial^2/\partial x_1^2$ (on the problem solution) on the grid $\overline{D}_{h(2.26)}$, where $\overline{\omega}_1$ is a mesh with an arbitrary distribution of nodes. Here $\delta_{\overline{x1}\,\widehat{x1}}$ is the *second-order difference derivative on a nonuniform mesh* [108]:

$$\delta_{\overline{x1}\,\widehat{x1}}\, u(x^i) = 2\left(h_1^i + h_1^{i-1}\right)^{-1}\left[\delta_{x1}\, u(x^i) - \delta_{\overline{x1}}\, u(x^i)\right],$$

$\delta_{x1}\, u(x^i)$ and $\delta_{\overline{x1}}\, u(x^i)$ are the first-order (forward and backward) difference derivatives

$$\delta_{x1}\, u(x^i) = \left(h_1^i\right)^{-1}\left[u(x^{i+1}) - u(x^i)\right],$$

$$\delta_{\overline{x1}}\, u(x^i) = \left(h_1^{i-1}\right)^{-1}\left[u(x^i) - u(x^{i-1})\right],$$

where $x^i,\ x^{i-1},\ x^{i+1} \in \overline{D}_h,\quad x^i = (x_1^i, x_2, \ldots, x_n)$.

Let us consider an error of an approximation of the operator L^* by the operator

$$\Lambda^* \equiv \varepsilon^2\, a_{11}^h(x)\, \delta_{\overline{x1}\,\widehat{x1}} \tag{2.30}$$

on the problem solution.

The *local error of the approximation of the operator* $\varepsilon^2\,(\partial^2/\partial x_1^2)$ *by the operator* $\varepsilon^2\, \delta_{\overline{x1}\,\widehat{x1}}$ *on the problem solution is*

$$\varepsilon^2\left|\left(\frac{\partial^2}{\partial x_1^2} - \delta_{\overline{x1}\,\widehat{x1}}\right) u(x^i)\right| \equiv \psi_{11}(x^i; u(\cdot)) = \psi_{11}(x^i), \quad x^i \in D_h. \tag{2.31}$$

This quantity is specified by local *a priori estimates* of derivatives of the problem solution and the stencil of the scheme:

$$\psi_{11}(x^i) \le M\, \varepsilon^2 \min\left\{ \left(h_1^i + h_1^{i-1}\right) \max_{\substack{\xi_1 \in \left[x_1^{i-1},\, x_1^{i+1}\right] \\ \xi_s = x_s,\ s=2,\ldots,n}} \left|\frac{\partial^3}{\partial x_1^3}\, u(\xi)\right|; \right.$$

$$\left. \max_{\substack{\xi_1 \in \left[x_1^{i-1},\, x_1^{i+1}\right] \\ \xi_s = x_s,\ s=2,\ldots,n}} \left|\frac{\partial^2}{\partial x_1^2}\, u(\xi)\right| \right\}, \quad x^i = (x_1^i, x_2, \ldots, x_n) \in D_h.$$

Taking into account the representation (2.7) of the solution to the boundary value problem and also the estimates (2.11), (2.14) for its components, we obtain the following estimate of the value $\psi_{11(2.31)}(x^i)$:

$$\psi_{11(2.31)}(x^i) \le M\left[\beta(x^i) + \lambda_3(N)\right], \tag{2.32}$$

where $\beta(x^i)$ is the main term (up to a constant multiplier) in the estimate of the local error of the approximation $\psi_{11(2.31)}(x^i)$:

$$\beta(x^i) = \beta(x_1^i) = \beta(x_1^i; \, \varepsilon, \overline{\omega}_1) \equiv \min\left[\, \eta(x_1^i), \, (h_1^i + h_1^{i-1})\, \varepsilon^{-1}\, \eta(x_1^i)\,\right]; \quad (2.33)$$

$$\eta(x_1^i) = \exp\left(-\, m\, \varepsilon^{-1}\, r(x_1^{i+1}, \{d_{*1}, d_1^*\})\right) + \exp\left(-\, m\, \varepsilon^{-1}\, r(x_1^{i-1}, \{d_{*1}, d_1^*\})\right);$$

$$\lambda_3(N) = N_1^{-1}; \quad x^i = (x_1^i, x_2, \ldots, x_n) \in D_h, \quad m < m_{0(2.12)}.$$

Let the grid function $a_{11}^h(x)$, $x \in \overline{D}_h$, from (2.30) approximate the function $a_{11}(x)$ ε-uniformly:

$$|a_{11}(x) - a_{11}^h(x)| \le M\, \lambda_4(N), \quad x \in \overline{D}_h, \tag{2.34}$$

where $\lambda_4(N) \underset{\varepsilon}{\to} 0$ as $N \to \infty$. The products

$$\varepsilon^2 \frac{\partial^2}{\partial x_1^2} u(x), \quad x \in \overline{D}, \quad \varepsilon^2 \delta_{\overline{x1}\,\widehat{x1}}\, u(x), \quad x \in \overline{D}_h$$

(we call them *normalized derivatives*) are bounded ε-uniformly (see the estimates (2.11), (2.14)). Hence for the error of the approximation of the operator L^* by the operator Λ^* on $u(x)$, we obtain the estimate

$$|(L^* - \Lambda^*)\, u(x)| \le M\left[\, \max_{x_1^i \in \omega_1} \beta(x_1^i) + \lambda_3(N) + \lambda_4(N)\,\right], \quad x \in D_h,$$

and for the error of the approximation of the operator L by the operator Λ, one has the estimate

$$|(L - \Lambda)\, u(x)| \le M\left[\, \max_{x_1^i \in \omega_1} \beta(x_1^i) + \sum_{n=2}^{4} \lambda_n(N)\,\right], \quad x \in D_h.$$

Here

$$\Lambda = \Lambda_{(2.30)}^* + \Lambda_{(2.28)}^{**}. \tag{2.35}$$

According to the statement of Theorem 2.3.2, for the boundary value problem (2.2), (2.4) the construction of the ε-uniformly convergent difference scheme $\{(2.21), (2.35), (2.26)\}$ in the case when the difference scheme is monotone amounts to the construction of a distribution of nodes in the mesh $\overline{\omega}_1$ (on the interval $[d_{*1}, d_1^*]$), under which the condition

$$\max_i \beta(x_1^i) \underset{\varepsilon}{\to} 0 \quad as \quad N_1 \to \infty \tag{2.36}$$

holds. This guarantees the ε-uniform approximation of the operator $\varepsilon^2 \partial^2 / \partial x_1^2$ by the operator $\varepsilon^2 \delta_{\overline{x1}\,\widehat{x1}}$ (in the case of the conditions (2.29), (2.34), one also has ε-uniform approximation of the operator $L_{(2.2)}$ by the operator $\Lambda_{(2.35)}$) on the solution of the boundary value problem.

Theorem 2.3.3 *For the solution of the boundary value problem* (2.2), (2.4), *let the estimates* (2.11), (2.14) *hold for* $K = 3$. *Then the condition* (2.36) *is necessary for the ε-uniform approximation of the boundary value problem by the difference scheme* $\{(2.21), (2.35), (2.26)\}$ *satisfying the conditions* (2.29), (2.30), (2.32), (2.34).

Theorem 2.3.4 *For the solution of the boundary value problem* (2.2), (2.4), *let the estimates* (2.11), (2.14) *hold for* $K = 3$. *In order that the scheme* $\{(2.21), (2.35), (2.26)\}$ *satisfying the conditions* (2.29), (2.30), (2.32), (2.34) *be ε-uniformly convergent at the rate* $\mathcal{O}\left(\max_i \beta(x_1^i) + \sum_{n=2}^{4} \lambda_n(N)\right)$, *it is sufficient for the operator* Λ *to be ε-uniformly monotone and for the condition* (2.36) *to hold.*

There exist meshes

$$\overline{\omega}_1^* = \overline{\omega}_1^*(d_{*1}, d_1^*), \tag{2.37}$$

that ensure the fulfillment of the condition (2.36) (and the ε-uniform convergence of the scheme $\{(2.21), (2.35), (2.26)\}$. For their construction, one can use, e.g., the technique of [15]; see also the following subsections.

2.3.3 Necessary conditions for distribution of mesh points for ε-uniform convergence of difference schemes. Construction of condensing meshes

For classical difference schemes, we now impose conditions on the distribution of grid nodes that are necessary for the ε-uniform approximation of the operator $L_{(2.2)}$ by the operator $\Lambda_{(2.35)}$ on the problem solution and that are sufficient for the ε-uniform convergence of the difference schemes. We construct piecewise-uniform meshes along the normal to the domain boundary that ensure the ε-uniform approximation of the operator L by the operator Λ. Note that, by virtue of the estimate (2.14), the k-th derivative along the normal to the boundary in an $M\varepsilon$-neighborhood of the boundary is of order $\mathcal{O}\left(\varepsilon^{-k}\right)$.

The following theorem defines *a condition* on the distribution of grid nodes in a boundary layer that is *necessary for the ε-uniform approximation of the boundary value problem* by the difference scheme.

Theorem 2.3.5 *Let a difference scheme be constructed by a classical approximation of the boundary value problem* (2.2), (2.4) *on the grid* $\overline{D}_{h(2.26)}$. *Then the condition*

$$\left.\begin{array}{c} \varepsilon^{-1} \max_{d_{*1} < x_1^i \leq d_{*1} + M\varepsilon} h_1^{i-1} \\ \varepsilon^{-1} \max_{d_1^* - M\varepsilon \leq x_1^i < d_1^*} h_1^i \end{array}\right\} \xrightarrow{\varepsilon} 0 \quad as \quad N_1 \to \infty, \quad x_1^i \in \omega_1, \tag{2.38}$$

is a necessary condition on the distribution of nodes in the mesh $\overline{\omega}_1$ for the ε-uniform approximation of the boundary value problem by the difference scheme.

To prove the theorem, it is sufficient to estimate $|(L - \Lambda)u(x)|$, $x \in D_h$ for $x_1 \leq d_{*1} + M\varepsilon$, i.e., the error in the approximation of the operator L by the operator Λ on the solution of the problem

$$L\,u(x) \equiv \varepsilon^2 \,\Delta u(x) - u(x) = 0, \quad x \in D^{(0)}, \quad u(x) = 1, \quad x \in \Gamma, \qquad (2.39)$$

where the operator Λ is constructed by a classical approximation of the operator L.

In the case of a mesh $\overline{\omega}_1$ that is uniform in $(M\varepsilon)$-neighborhoods of the endpoints of the interval $[d_{*1}, d_1^*]$, the condition (2.38) takes the form

$$\varepsilon^{-1} h_1^*, \, \varepsilon^{-1} h_2^* \to 0 \quad as \quad N_1 \to \infty,$$

where h_1^* and h_2^* are step-sizes in the mesh $\overline{\omega}_1$ on the intervals $[d_{*1}, d_{*1} + M\varepsilon]$ and $[d_1^* - M\varepsilon, d_1^*]$, respectively.

Let us consider the convergence of difference schemes on piecewise-uniform meshes, i.e., the *simplest meshes* condensing in boundary layers. Let

$$\overline{\omega}_1 = \overline{\omega}_1(d_1, d_2; \sigma, N_1) \qquad (2.40a)$$

be a mesh with a piecewise-constant step-size on the interval $[d_1, d_2]$ in the x_1-axis, the step-size in the mesh $\overline{\omega}_{1(2.40a)}$ on the intervals $[d_1, d_1 + \sigma]$ and $[d_2 - \sigma, d_2]$ equals $h_{(1)}$ while on the interval $[d_1 + \sigma, d_2 - \sigma]$ equals $h_{(2)}$; here $0 \leq \sigma \leq 2^{-1}(d_2 - d_1)$ and $N_1 + 1$ is the number of nodes in the mesh $\overline{\omega}_1$. Here the parameter σ (*the transition parameter*) defines the point of transition from the step-size $h_{(1)}$ to the step-size $h_{(2)}$ in the mesh $\overline{\omega}_1$. Using the meshes $\overline{\omega}_{1(2.40a)}$, one can construct the *simplest meshes* that condense in the boundary layer.

Lemma 2.3.4 *Let the difference scheme (2.21), (2.26) be constructed by a classical approximation of the boundary value problem (2.2), (2.4) on the grid*

$$\overline{D}_{h(2.26)}, \quad where \quad \overline{\omega}_1 = \overline{\omega}_{1(2.40a)}(d_{*1}, d_1^*; \sigma, N_1). \qquad (2.41)$$

For $\sigma = \sigma(\varepsilon)$ (σ is independent of N_1), there exist no meshes on which the difference scheme converges ε-uniformly.

To prove the lemma, it is sufficient to consider the problem (2.39), (2.4) and the scheme (2.21), (2.35) for

$$a_{11}^h(x) = c^{0h}(x) \equiv 1, \quad f^h(x) \equiv 0, \quad x \in \overline{D}_h, \quad \varphi^h(x) \equiv 1, \quad x \in \Gamma_h.$$

The statement of the lemma follows from a comparison of explicit solutions of the differential and the difference problems on the interval $[d_1 + \sigma, d_2 - \sigma]$

for $\sigma \leq M\varepsilon$ and on the interval $[d_1, d_1 + \sigma]$ for $\sigma \geq m\varepsilon\psi(\varepsilon)$, where $\psi(\varepsilon) \to \infty$ as $\varepsilon \to 0$.

The following theorem establishes *conditions* on the value $\sigma_{(2.40)}$ that are *necessary for the ε-uniform convergence of the scheme on simple meshes.*

Theorem 2.3.6 *Consider the difference scheme* (2.21), (2.35) *which is constructed by the classical approximation of the boundary value problem* (2.2), (2.4) *on the grid* (2.41), (2.40a). *In order that this scheme converge ε-uniformly, it is necessary for the value σ to depend on both ε and N_1:*

$$\sigma = \sigma_0(\varepsilon, N_1). \tag{2.40b}$$

In the class of functions

$$\sigma = \sigma_0(\varepsilon, N_1) = \min[m_1, \ \sigma^*(\varepsilon, N_1)]; \quad \sigma^*(\varepsilon, N_1) = \varepsilon\,\sigma_1(N_1), \tag{2.40c}$$

where m_1 is an arbitrary constant in the interval $(0, \ 2^{-1}(d_2 - d_1))$, for the ε-uniform convergence of the difference scheme it is necessary to satisfy the conditions

$$\sigma_1(N_1) \to \infty \quad as \quad N_1 \to \infty \tag{2.40d}$$

and

$$N_1^{-1}\sigma_1(N_1) \to 0 \quad as \quad N_1 \to \infty. \tag{2.40e}$$

We now consider the meshes

$$\overline{\omega}_1 = \overline{\omega}_{1(2.40a-e)}(d_1, d_2; \sigma, N_1), \tag{2.40f}$$

whose step-sizes $h_{(1)}$ and $h_{(2)}$ satisfy the conditions

$$h_{(1)} \leq M_1 \sigma N_1^{-1}, \quad h_{(2)} \leq M_2(d_2 - d_1) N_1^{-1}, \quad h_{(1)} \leq h_{(2)}.$$

Note that the condition (2.36) (the *sufficient condition for the ε-uniform convergence* of the monotone difference scheme $\{(2.21), (2.35), (2.26)\}$) holds on the grid (2.26), where $\overline{\omega}_1 = \overline{\omega}_{1(2.40)}(d_{*1}, d_1^*)$. Then we have the estimate

$$\max_i \beta(x_1^i) \leq M \left[N_1^{-1}\sigma_1(N_1) + \exp(-m\,\sigma_1(N_1)) \right], \quad m < m_{0(2.12)},$$

where $\beta(x_1^i) = \beta_{(2.33)}(x_1^i)$.

Theorem 2.3.7 *Let the solution of the problem* (2.2), (2.4) *satisfy the estimates* (2.11), (2.14) *for $K = 3$, and for the difference scheme* (2.21), (2.35) *on the grid $\overline{D}_{h(2.26)}$, let the following conditions be fulfilled:*
  a) *the functions $f^h(x)$ and $\varphi^h(x)$ satisfy the conditions* (2.23);
  b) *the operator Λ^{**} satisfies the condition* (2.29) *and the function $a_{11}^h(x)$ satisfies the relation* (2.34);
  c) *the mesh $\overline{\omega}_1$ in* (2.26) *is the mesh $\overline{\omega}_1 = \overline{\omega}_{1(2.40)}(d_{*1}, d_1^*)$;*

d) *the operator Λ is ε-uniformly monotone.*
Then the difference scheme $\{(2.21), (2.35), (2.26)\}$ converges ε-uniformly as $N \to \infty$, i.e.,

$$|u(x) - z(x)| \le M\left[\sum_{n=2}^{4} \lambda_n(N) + N_1^{-1}\sigma_1(N_1) + \exp(-m\,\sigma_1(N_1))\right], \quad x \in \overline{D}_h.$$

Remark 2.3.1 To justify Theorem 2.3.7, we use *the technique of majorant functions* [108]. For example, for the difference scheme $\{(2.21), (2.35), (2.26)\}$, let the conditions b) and c) of Theorem 2.3.7 be fulfilled. Consider the function

$$w(x) = M_1 + \exp\left(-m\varepsilon^{-1}r(x, \{x_1 = d_{*1}\})\right) +$$
$$+ \exp\left(-m\varepsilon^{-1}r(x, \{x_1 = d_1^*\})\right), \quad x \in \overline{D}.$$

Then for $m < m_{0(2.12)}$ and sufficiently large M_1, the inequality

$$\Lambda w(x) \le -m_1 < 0, \quad x \in D_h,$$

holds. The function $w(x)$ is used for the estimation of $u(x) - z(x)$, $x \in \overline{D}_h$. ∎

In the case of the boundary value problem (2.2), (2.4) we are interested in numerical methods whose solutions converge ε-uniformly in the maximum discrete norm. However, in the case of singularly perturbed problems the ε-uniform convergence of the discrete solution $z(x)$ at the nodes of the grid $\overline{D}_h$ in the maximum discrete norm is, in general, inadequate to describe the ε-uniform convergence of the approximation designed on the entire domain $\overline{D}$. For example, the solution of the difference scheme constructed by the classical approximation of the boundary value problem (2.2), (2.4) on the uniform grid $\overline{D}_{h(2.26)} = \overline{D}_h^u$ converges on the grid $\overline{D}_h$ as $\varepsilon^{-1}h_1 \to \infty$ for $h \to 0$, when the characteristic width of the boundary layer defined by the parameter ε is much less than the step-size of the mesh in x_1. Nevertheless, already the simplest *piecewise-linear interpolant*

$$\overline{z}(x) = \overline{z}\big(x;\ z(\cdot),\ \overline{D}_h\big), \quad x \in \overline{D}, \tag{2.42}$$

constructed from the values of the function $z(x)$, $x \in \overline{D}_h$ (which is linear on each $(n+1$-vertex) polyhedral element, i.e., partitions of elementary n-dimensional rectangular parallelepipeds with vertices at the nodes of the grid $\overline{D}_h$), does not converge on $\overline{D}$. Under the conditions given above for the grid $\overline{D}_{h(2.26)}$, the interpolant

$$\overline{u}^h(x) = \overline{z}_{(2.42)}\big(x;\ u^h(\cdot),\ \overline{D}_h\big), \quad x \in \overline{D},$$

constructed from the grid function $u^h(x) = u(x)$, $x \in \overline{D}_h$, where $u(x)$ is the solution of the problem (2.2), (2.4), also fails to converge on $\overline{D}$.

In the case when the interpolant $\bar{z}_{(2.42)}(x)$, $x \in \overline{D}$, converges on $\overline{D}$, we say that the difference *scheme resolves the boundary value problem* (for some values of the parameter ε); otherwise, we say that the difference scheme does not resolve the boundary value problem. If the interpolant $\bar{u}^h(x)$, $x \in \overline{D}$, converges on $\overline{D}$ (for some values of the parameter ε), we say that the *grid $\overline{D}_h$ is informative for the solution of the boundary value problem*; otherwise, we say that the grid $\overline{D}_h$ is not informative for the solution.

In the case when the interpolant $\bar{z}(x)$, $x \in \overline{D}$ (the interpolant $\bar{u}^h(x)$, $x \in \overline{D}$), converges ε-uniformly on $\overline{D}$, we say that *the difference scheme (the grid $\overline{D}_h$) resolves the boundary value problem (is informative for the solution of the boundary value problem) ε-uniformly.*

In the case when the solution of a difference scheme (or, briefly, the difference scheme itself) converges in the maximum norm on the grid $\overline{D}_h$ and the difference scheme resolves the boundary value problem, we say that *the solution (the difference scheme) converges.* But if the solution (the difference scheme) converges only on the grid $\overline{D}_h$, however, the solvability of the boundary value problem is, in general, not assumed; we say that *the solution (or, the scheme) converges on the grid $\overline{D}_h$.*

We are interested in difference schemes that converge ε-uniformly, i.e., difference schemes that converge in the discrete maximum norm and resolve boundary value problems ε-uniformly.

Remark 2.3.2 Let the condition (2.38) be violated. Then

(i) the difference scheme constructed by the classical approximation of the boundary value problem (2.2), (2.4) on the grid $\overline{D}_{h(2.26)}$ does not resolve the boundary value problem ε-uniformly;

(ii) the grid $\overline{D}_{h(2.26)}$ is not informative ε-uniformly.

Thus, *the condition (2.38) is necessary for the ε-uniform solvability of the boundary value problem* by the classical difference scheme.

Under the hypotheses of Theorem 2.3.7, the difference scheme {(2.21), (2.35), (2.26)} resolves the boundary value problem (2.2), (2.4) ε-uniformly, and the rate of ε-uniform convergence of the solution $z(x)$ to the difference scheme on the grid $\overline{D}_h$ is preserved for the interpolant $\bar{z}(x)$, $x \in \overline{D}$:

$$|u(x) - \bar{z}(x)| \leq M \left[\sum_{n=2}^{4} \lambda_n(N) + N_1^{-1} \sigma_1(N_1) + \exp(-m\,\sigma_1(N_1)) \right], \quad x \in \overline{D}.$$

The grid $\overline{D}_{h(2.26)}$, where $\bar{\omega}_1 = \bar{\omega}_{1(2.40)}(d_{*1}, d_1^*)$, is informative ε-uniformly for the solution to the boundary value problem (2.2), (2.4). ∎

On the interval $[d_1, d_2]$ we construct the *special piecewise-uniform mesh* $\bar{\omega}^{(1)}$ that we shall use later on:

$$\bar{\omega}^{(1)} = \bar{\omega}^{(1)}(d_1, d_2; m^0) \equiv \bar{\omega}_{1(2.40)}(d_1, d_2; \sigma, N_1), \qquad (2.43a)$$

where

$$\sigma = \sigma_0 = \sigma_0(\varepsilon, N_1, m^0; m^1); \qquad (2.43b)$$

here $\sigma_0(\varepsilon, N_1, m^0; m^1) = \min\left[\sigma_1^*, m^1\right]$ and $\sigma_1^* = (m^0)^{-1}\varepsilon \ln N_1$; the constants m^0 and m^1 take arbitrary values in the intervals $(0, m_{0(2.12)})$ and $(0, 2^{-1}(d_2 - d_1))$, respectively. Set $h_{(1)} = \sigma_0 N_{(*)}^{-1}$ and $h_{(2)} = (d_2 - d_1 - 2\sigma)(N_1 - 2N_{(*)})^{-1}$, where $N_{(*)} + 1$ is the number of nodes in the mesh $\overline{\omega}_{(2.43)}^{(1)}$ on the interval $[d_1, d_1 + \sigma]$ (or on $[d_2 - \sigma, d_2]$), $N_{(*)} = m^1(d_2 - d_1)^{-1}N_1$; furthermore, for $\sigma_1^* = m^1$ the mesh $\overline{\omega}^{(1)}$ is uniform. The constant $m_{(2.43)}^0$ satisfies the same condition as the constant m in the estimates (2.12), (2.14).

Let

$$\overline{\omega}_1 = \overline{\omega}_{1(2.44)}(d_{*1}, d_1^*; m) \equiv \overline{\omega}_{(2.43)}^{(1)}(d_{*1}, d_1^*; m) \qquad (2.44a)$$

be a mesh, where m is an arbitrary constant. For $m = m_{(2.43)}^0$, the following relations are fulfilled:

$$h_1 \leq M N_1^{-1}, \quad \varepsilon^{-1} h_{(1)1} \leq M N_1^{-1} \ln N_1, \qquad (2.44b)$$

where $h_1 = \max_i h_1^i$, $h_1 = h_{(2)1}$, $h_{(k)1} = h_{(k)(2.43)}$ for $k = 1, 2$; on the mesh $\overline{\omega}_{1(2.44)}$ we have the estimate

$$\max_i \beta(x_1^i) \leq M N_1^{-1} \ln N_1, \quad x_1^i \in \overline{\omega}_{1(2.44)}, \qquad (2.45)$$

where $\beta(x_1^i) = \beta_{(2.33)}(x_1^i)$.

Under the fulfillment of the conditions (2.29), (2.34), the mesh $\overline{\omega}_{1(2.44)}$ guarantees ε-uniform approximation of the operator $L_{(2.2)}$ by the operator $\Lambda_{(2.35)}$ on the solution of the boundary value problem; moreover, for the value $\beta(x_1^i)$ the estimate (2.45) holds.

2.4   Monotone finite difference approximations of the boundary value problem on a slab. $\mathcal{E}$-uniformly convergent difference schemes

2.4.1   Problems on uniform meshes

For the differential operator $L_{(2.2)}$ on the slab (2.4), we construct ε-*uniform monotone difference approximations* on uniform meshes and systems of difference equations (on these meshes) for which conditions of ε-uniform monotonicity are fulfilled.

The discrete operator

$$\Lambda = \Lambda(L) \equiv \varepsilon^2 \left\{ \sum_{s=1}^{n} a_{ss}(x)\delta_{\overline{xs}\,\widehat{xs}} + 2^{-1} \sum_{\substack{s,\,k=1 \\ s \neq k}}^{n} \left[a_{sk}^+(x)(\delta_{xs\,xk} + \delta_{\overline{xs}\,\overline{xk}}) + \right. \right. \quad (2.46)$$

$$\left. + a_{sk}^-(x)(\delta_{xs\,\overline{xk}} + \delta_{\overline{xs}\,xk}) \right] + \sum_{s=1}^{n} [b_s^+(x)\delta_{xs} + b_s^-(x)\delta_{\overline{xs}}] - c(x) \Bigg\} - c^0(x)$$

corresponds to the operator $L_{(2.2)}$ on the grid $\overline{D}_{h(2.26)}$. Here $\delta_{xs\,xk}$, $\delta_{\overline{xs}\,\overline{xk}}$, $\delta_{xs\,\overline{xk}}$, $\delta_{\overline{xs}\,xk}$ for $s \neq k$ are mixed *difference derivatives*; for example, for $s = 1$, $k = 2$, one has [108]

$$\delta_{x1\,x2}\, z(x) = (h_1^i)^{-1} [\delta_{x2}\, z(x^{i+1}) - \delta_{x2}\, z(x^i)],$$

$$\delta_{x1\,\overline{x2}}\, z(x) = (h_1^i)^{-1} [\delta_{\overline{x2}}\, z(x^{i+1}) - \delta_{\overline{x2}}\, z(x^i)],$$

$$\delta_{\overline{x1}\,x2}\, z(x) = (h_1^{i-1})^{-1} [\delta_{x2}\, z(x^i) - \delta_{x2}\, z(x^{i-1})],$$

$$\delta_{\overline{x1}\,\overline{x2}}\, z(x) = (h_1^{i-1})^{-1} [\delta_{\overline{x2}}\, z(x^i) - \delta_{\overline{x2}}\, z(x^{i-1})],$$

for $s = k = 1$, we have

$$\delta_{\overline{x1}\,\widehat{x1}}\, z(x) = 2\,(h_1^i + h_1^{i-1})^{-1} [\delta_{x1}\, z(x^i) - \delta_{\overline{x1}}\, z(x^i)],$$

$$\delta_{x1}\, z(x) = (h_1^i)^{-1} [\, z(x^{i+1}) - z(x^i)\,],$$

$$\delta_{\overline{x1}}\, z(x) = (h_1^{i-1})^{-1} [\, z(x^i) - z(x^{i-1})\,],$$

where

$$x = x^i = (x_1^i, x_2, \ldots, x_n), \quad x^{i \pm 1} = (x_1^{i \pm 1}, x_2, \ldots, x_n), \quad x^i, x^{i-1}, x^{i+1} \in \overline{D}_h,$$

while $\delta_{x1}\, z(x)$, $\delta_{\overline{x1}}\, z(x)$, $\delta_{x2}\, z(x)$, $\delta_{\overline{x2}}\, z(x)$ are the first-order upwind difference derivatives and $v^+(x) = 2^{-1} [v(x) + |v(x)|]$, $v^-(x) = 2^{-1} [v(x) - |v(x)|]$.

For the scheme

$$\Lambda\, z(x) = f(x), \quad x \in D_h; \quad z(x) = \varphi(x), \quad x \in \Gamma_h, \quad (2.47)$$

where $\Lambda = \Lambda_{(2.46)}(L)$ and $\overline{D}_h = \overline{D}_{h(2.26)}$, in the case of the piecewise-uniform meshes in x_1

$$\overline{\omega}_1 = \overline{\omega}_{1(2.40)}(d_{*1}, d_1^*) = \overline{\omega}_{1(2.40)}(d_{*1}, d_1^*; \sigma, N_1), \quad (2.48)$$

all hypotheses of Theorem 2.3.7 are valid, except for the monotonicity of the operator $\Lambda_{(2.46)}$.

Under the condition

$$a_{sk}(x) \geq 0, \quad x \in \overline{D}, \quad s, k = 1, \ldots, n, \quad s \neq k,$$

the monotonicity of the operator $\Lambda_{(2.46)}$ on the grid $\overline{D}_{h(2.26)}$ is violated, e.g., in the case when the following condition is violated:

$$
\min_{h_s^{(i_s)},k,D_h} \left[a_{kk}(x)\, 2 \left(h_k^{i_k} + h_k^{i_k-1} \right)^{-1} - \sum_{\substack{s=1 \\ s\neq k}}^{n} a_{sk}(x) \left(h_s^{(i_s)} \right)^{-1} \right] \geq 0,
$$

$$
x = (x_1^{i_1}, \dots, x_n^{i_n}) \in D_h,
$$

where $h_s^{(i_s)} = h_s^{i_s}$, or $h_s^{(i_s)} = h_s^{i_s-1}$, for $s = 1, \dots, n$.

Theorem 2.4.1 *In the case of the boundary value problem (2.2), (2.4), suppose that the equation (2.2a) does not contain mixed derivatives, i.e., let the following condition*

$$
a_{sk}(x) \equiv 0, \quad x \in \overline{D}, \quad s, k = 1, \dots, n, \quad s \neq k, \tag{2.49}
$$

be valid. Then the difference scheme (2.47), (2.46) is ε-uniform monotone. Furthermore, under the condition that the solution of the boundary value problem satisfies the estimates (2.11), (2.14) for $K = 3$, the difference scheme (2.47), (2.46) converges ε-uniformly on the grid $\overline{D}_{h(2.26)}$ with $\overline{\omega}_1 = \overline{\omega}_{1(2.48)}$.

Lemma 2.4.1 *On the grid $\overline{D}_{h(2.26)}$, where ω_s, for $s = 1, \dots, n$, are meshes with an arbitrary distribution of nodes, the operator $\Lambda_{(2.46)}(L)$ is not ε-uniform monotone if the operator $L_{2(2.2)}$ contains mixed derivatives.*

When the conditions of Lemma 2.4.1 and Theorem 2.3.7 are satisfied, except that condition (d) is omitted, then ε-uniform convergence of the difference scheme $\{(2.47), (2.46), (2.26), (2.48)\}$ is not implied.

The monotonicity condition for the operator $\Lambda_{(2.46)}$ is violated when approximating the operator

$$
L^*_{(2)} = L^*_{(2)}(L_{(2.2)}) \equiv \sum_{s,k=1}^{n} a_{sk}(x) \frac{\partial^2}{\partial x_s \partial x_k}, \tag{2.50}
$$

i.e., *the elliptic part of the operator $L_{(2.2)}$.* Let us define conditions that allow us to construct *grids $\overline{D}_h$ with a special distribution of nodes* on which the difference operators $\Lambda_{(2.46)}$ and

$$
\Lambda^*_{(2)} = \Lambda^*_{(2)}\left(L^*_{(2)}\right) \equiv \sum_{s=1}^{n} a_{ss}(x)\delta_{\overline{x_s}\,\widehat{x_s}} + \tag{2.51}
$$

$$
+2^{-1} \sum_{\substack{s,k=1 \\ s\neq k}}^{n} \left[a_{sk}^+(x)\left(\delta_{x_s\,x_k} + \delta_{\overline{x_s}\,\overline{x_k}}\right) + a_{sk}^-(x)\left(\delta_{x_s\,\overline{x_k}} + \delta_{\overline{x_s}\,x_k}\right) \right],
$$

and, hence, *the difference scheme* {(2.47), (2.46), (2.26), (2.48)}, *are ε-uniform monotone.* These conditions will be given below.

First, we discuss an approximation of the operator $L^*_{(2)}$ on uniform grids. On the set $\overline{D}$ we introduce the grid

$$\overline{D}_h = \overline{\omega}_1 \times \omega_2 \times \ldots \times \omega_n, \tag{2.52}$$

where ω_s for $s = 1,\ldots,n$, are uniform meshes with step-size h_s. Let Q be a convex subdomain of D. On the set $\overline{Q}$ we introduce the uniform grid

$$\overline{D}_h(Q) = \{\overline{Q}\}_h = \overline{Q} \cap \overline{D}_{h(2.52)}. \tag{2.53}$$

The difference operator $\Lambda^*_{(2)(2.51)}$ corresponds to the operator $L^*_{(2)(2.50)}$ on the grid $\overline{D}_{h(2.53)}(Q)$.

The elliptic operator $L^*_{(2)(2.50)}$, considered at the point x_0, can be brought to canonical form by an appropriate linear transformation. In the transformed operator $L^*_{(2)X}$, the maxima of the coefficients of the mixed derivatives at the point X depend on the distance to the point X_0, where $X_0 = X(x_0)$, and these maxima, in general, increase as this distance increases.

Let $L^*_{(2)(2.50)}$ be already in canonical form at the point x_0. Then in a small neighborhood of x_0, the operator $L^*_{(2)}$ has "almost canonical" form, i.e., the coefficients of the mixed derivatives are small compared with the coefficients of the second-order derivatives in each variable.

We shall assume that for each point x that is sufficiently close to x_0, one can find n pairs of values ρ_{*s} and ρ^*_s for $s = 1,\ldots,n$, where $\rho_{*s}, \rho^*_s > 0$, $\rho_{*s} \le \rho^*_s$, ρ_{*s} and ρ^*_s depend on a_{sk} for $s, k = 1,\ldots,n$, such that for any h_s satisfying the condition

$$h_s \in [\rho_{*s} h^*, \rho^*_s h^*], \quad s = 1,\ldots,n, \tag{2.54a}$$

where $h^* > 0$ is an arbitrary number, the inequality

$$a_{kk} h_k^{-1} - \sum_{\substack{s=1 \\ s \ne k}}^{n} |a_{sk}| h_s^{-1} > 0, \quad k = 1,\ldots,n, \tag{2.54b}$$

holds. This is equivalent to the inequality

$$\sum_{\substack{s=1 \\ s \ne k}}^{n} |a_{sk}| \xi_s \xi_k < a_{kk} \xi_k^2, \quad k = 1,\ldots,n, \quad \xi_k = h_{k(2.54a)}^{-1}.$$

At the point x_0 where the operator $L^*_{(2)}$ is canonical, such inequality is valid for all nonnegative values ξ_s and also in a small neighborhood of x_0 in the case when the values $\xi_s \xi_k^{-1}$ are bounded. In the case of the condition (2.54a), one has $\rho_{*s} \rho_k^{*-1} \le \xi_s \xi_k^{-1} \le \rho^*_s \rho_{*k}^{-1}$.

Recall that

$$\rho_{*s} = \rho_{*s}\left(L^*_{(2)}\right), \quad \rho^*_s = \rho^*_s\left(L^*_{(2)}\right), \quad s = 1, \ldots, n. \tag{2.54c}$$

The values ρ_{*s} and ρ^*_s can be chosen to satisfy the relations

$$\rho_{*s} \to 0, \quad \rho^*_s \to \infty \quad as \quad \max_{s,\ k,\ s\neq k}|a_{sk}| \to 0;$$

the values ρ_{*s} and ρ^*_s, as well as the coefficients a_{sk}, depend in general on x, i.e., $\rho_{*s} = \rho_{*s}(x)$, $\rho^*_s = \rho^*_s(x)$, and $a_{sk} = a_{sk}(x)$.

We say that the operator $L^*_{(2)}$ at the *point* x has *almost canonical* form in the variables $x_1, \ldots, x_n$ with *the condition of almost canonicity* (2.54) (or, briefly, has almost canonical form) if its coefficients satisfy the relations (2.54).

In the relations (2.54), the values ρ_{*s}, ρ^*_s, $h_s = \xi_s^{-1}$ for $s = 1, \ldots, n$ and h^* appear. The value h_s in the definition of almost canonicity of the operator $L^*_{(2)}$ in (2.54a), (2.54b) and the value h_s in the definition of the uniform grid (2.52) are, in general, not related to each other.

Note that the fulfillment of the condition (2.54b) at the point x guarantees the monotonicity of the operator $\Lambda^*_{(2)(2.51)}$ at this point in the case of the uniform grid $\overline{D}_{h(2.52)}$, where $h_s = h_{s(2.54a)}$. Thus, in the case of the difference scheme $\{(2.47),\ (2.46),\ (2.52)\}$ the condition (2.54) is the *condition of strong ε-uniform monotonicity of the operators* $\Lambda^*_{(2)(2.51)}$, $\Lambda_{(2.46)}$ *and the scheme at the point x itself*.

In general, there exist no values h_s for $s = 1, \ldots, n$, satisfying the condition (2.54a) such that the inequality (2.54b) is valid for all x in $\overline{D}$. But if it turns out that the relation

$$\max_{\overline{D}} \rho_{*s}(x) < \min_{\overline{D}} \rho^*_s(x), \quad s = 1, \ldots, n,$$

holds, then it is possible to find values h_s under which the inequality (2.54b) is valid for all x in $\overline{D}$.

We now generalize the relations (2.54). We shall assume that for an arbitrary point $x^* \in \overline{D}$, one can find (i) a sufficiently small neighborhood $Q(x^*)$ (we shall say, a neighborhood of the point x^*) with $x^* \in \overline{Q}(x^*)$, $\overline{Q}(x^*) \subseteq \overline{D}$ and (ii) functions $\rho_{*s}(x)$ and $\rho^*_s(x)$, $x \in \overline{Q}(x^*)$ such that the relation

$$\max_{\overline{Q}(x^*)} \rho_{*s}(x) \leq \min_{\overline{Q}(x^*)} \rho^*_s(x), \quad s = 1, \ldots, n, \tag{2.55a}$$

holds and also the following relation (similar to (2.54b)):

$$\min_{\overline{Q}(x^*)} a_{kk}(x)h_k^{-1} - \sum_{\substack{s=1 \\ s\neq k}}^{n} \max_{\overline{Q}(x^*)}|a_{sk}(x)|\, h_s^{-1} > 0, \quad k = 1, \ldots, n. \tag{2.55b}$$

Here the values h_s satisfy the condition

$$h_s \in \left[\max_{\overline{Q}(x^*)} \rho_{*s}(x)\, h^*, \ \min_{\overline{Q}(x^*)} \rho_s^*(x)\, h^* \right], \quad s = 1, \ldots, n, \tag{2.55c}$$

and $h^* > 0$ is an arbitrary number.

We emphasize that the functions $\rho_{*s}(x)$ and $\rho_s^*(x)$ are defined by the coefficients of the operator $L_{(2)(2.50)}^*$:

$$\rho_{*s}(x) = \rho_{*s}\big(x;\, L_{(2)}^*\big), \quad \rho_s^*(x) = \rho_s^*\big(x;\, L_{(2)}^*\big), \quad s = 1, \ldots, n. \tag{2.55d}$$

The values $\rho_{*s}(x)$ and $\rho_s^*(x)$ can be chosen to satisfy the relations

$$\max_{\overline{Q}(x^*)} \rho_{*s}(x) \to 0, \ \min_{\overline{Q}(x^*)} \rho_s^*(x) \to \infty \quad \textit{for} \quad \max_{s,k,s\neq k,\overline{Q}(x^*)} |a_{sk}(x)| \to 0. \tag{2.55e}$$

In the general case, one can ensure fulfillment of the relations (2.55) if one brings the operator $L_{(2)(2.50)}^*$ to the canonical form at the point x^* (or close to the canonical form) and chooses the neighborhood $Q(x^*)$ to be sufficiently small.

We say that the operator $L_{(2)}^*$ on the set $\overline{D}$ has *local almost canonical form* (in the variables $x_1, \ldots, x_n$), if for an arbitrary point $x^* \in \overline{D}$, one can find a neighborhood $Q(x^*)$ and n pairs of the functions $\rho_{*s}(x)$ and $\rho_s^*(x)$, $x \in \overline{Q}(x^*)$, such that the relations (2.55) hold for the coefficients of the operator $L_{(2)}^*$.

If on a set $\overline{D}^1 \subseteq \overline{D}$, the relations (2.55), where $Q(x^*) = \overline{D}^1$, hold for the coefficients of the operator $L_{(2)}^*$, we say that *the operator $L_{(2)}^*$ has almost canonical form* (or, briefly, is almost canonical) *on the set $\overline{D}^1$*. In the case when the coefficients of the operator $L_{(2)}^*$ allow us to construct the functions $\rho_{*s}(x)$, and $\rho_s^*(x)$ for $s = 1, \ldots, n$, such that the relations (2.55), where $\overline{Q}(x^*) = \overline{D}^1$, are valid, we say that *the operator $L_{(2)}^*$ permits the transformation to the almost canonical form* on the set $\overline{D}^1$.

Note that when defining the local almost canonical operator, in the relations (2.55) we use the values h_s for $s = 1, \ldots, n$, which, in general, have no relations to grids on the set $\overline{Q}(x^*)$; grids on the set $\overline{Q}(x^*)$ will be considered below.

For the set $\overline{Q} \cap \overline{D}_h$ and the operator Λ *we consider the node x_0 as strongly interior*, or, briefly, as *interior*, if all nodes of the stencil (for which the point x_0 is centre) of the operator Λ belong to the set $\overline{Q} \cap \overline{D}_h$.

The fulfillment of the condition (2.55b) under the condition (2.55c) ensures the monotonicity of the operator $\Lambda_{(2)(2.51)}^*$ on the interior nodes of the uniform grid $\overline{D}_{h(2.53)}(Q(x^*))$. In the case of the difference scheme $\{(2.47), (2.46), (2.52)\}$ the condition (2.55) is *the condition of strong ε-uniform monotonicity of the operators $\Lambda_{(2)}^*$, Λ and of the difference scheme* on the interior nodes of the set $\overline{D}_{h(2.53)}(Q(x^*))$.

We say that the grids $\overline{D}_h$, with a special distribution of nodes in the meshes $\omega_r, \ldots, \omega_k$ that ensure the ε-uniform monotonicity of the operator Λ on the interior nodes of the set $\overline{Q} \cap \overline{D}_h$, are *consistent grids on the set $\overline{Q}$ in the variables* $x_r, \ldots, x_k$ with the monotonicity condition for the operator Λ; restrictions on the meshes ω_s in the other variables are not imposed. For brevity, we shall call such meshes *the consistent meshes* (without indicating the variables) *guaranteeing the monotonicity of the operator Λ on the set $\overline{Q} \cap \overline{D}_h$.*

Theorem 2.4.2 *Let the grid $\overline{D}_h$ be uniform on the set $\overline{Q}(x^*)$ in each of the variables, i.e.,*

$$\overline{D}_h = \overline{D}_h^u \big(Q(x^*), \ L_{(2)}^* \big) = \overline{D}_{h^*} \big(Q(x^*), \ L_{(2)}^*, \ x_1, \ldots, x_n \big) = \qquad (2.56)$$

$$= \Big\{ \{ \overline{Q}(x^*) \}_{h(2.53)} \ \text{under the condition (2.55c) while assuming (2.55)} \Big\}.$$

Then the operators $\Lambda_{(2.46)}(L)$ and $\Lambda_{(2)(2.51)}^*\big(L_{(2)}^*\big)$ are ε-uniformly monotone on the set $\overline{D}_{h(2.56)}$ (in the case when the set of interior nodes is not empty).

Thus, the uniform grid $\overline{D}_{h(2.56)}$ is consistent on $\overline{Q}(x^*)$ in $x_1, \ldots, x_n$ with respect to the monotonicity condition for the operators $\Lambda_{(2)(2.51)}^*$ and $\Lambda_{(2.46)}$.

The lower index h^* in the notation of the grid $\overline{D}_{h^*}$ is used to show that the distribution of nodes in the grid satisfies a condition (the condition (2.55c)) controlled by the parameter h^*. In the list of "arguments" of the grid $\overline{D}_h$ in (2.56), the variables $x_1, \ldots, x_n$ in which the grid is consistent are written.

The conditions (2.55) that are imposed on the coefficients $a_{sk}(x)$ and on the relations between the values $h_1, \ldots, h_n$ allow us to construct uniform consistent grids on the set $\overline{Q}(x^*)$.

The difference scheme (2.47), (2.46) on the uniform grid

$$\overline{D}_{h(2.56)} = \overline{D}_{h^*(2.56)} \big(Q(x^*), \ L_{(2)}^* \big) \ \text{with} \ \overline{Q}(x^*) = \overline{D}$$

is monotone, but this scheme does not satisfy the condition (2.36) (sufficient for ε-uniform convergence of a difference scheme) nor the condition (2.40b) (necessary for ε-uniform convergence of a difference scheme on the family of the grids $\overline{D}_{h(2.41)}$ when the grid $\overline{D}_{h(2.56)} \subset \overline{D}_{h(2.41)}$).

2.4.2   Problems on piecewise-uniform meshes

We are interested in monotone difference approximations of the boundary value problem (2.2), (2.4).

The following statements are valid.

Lemma 2.4.2 *For the mesh $\overline{\omega}_1$ that defines the grid $\overline{D}_{h(2.26)}$, let the condition (2.38) be violated. Then the operator $\Lambda_{(2.46)}(L)$ does not approximate the operator $L_{(2.2)}$ ε-uniformly on the problem solution.*

This lemma follows from Theorem 2.3.5. The condition (2.38) is necessary for the ε-uniform approximation of the operator L by the operator $\Lambda_{(2.46)}$ on the problem solution.

Theorem 2.4.3 *Suppose that the coefficients $a_{1s}(x)$ for $s = 2, \ldots, n$, where $a_{1s}(x) = a_{s1}(x)$, do not satisfy the condition*

$$a_{1s}(x) = 0, \quad x \in \Gamma, \quad s = 2, \ldots, n. \tag{2.57}$$

Then the difference scheme (2.47), (2.46) on the grid $\overline{D}_{h(2.26)}$ under the condition (2.38) is not ε-uniformly monotone for any distribution of nodes in the meshes ω_s, for $s = 1, \ldots, n$. In particular, one does not obtain ε-uniform monotonicity on the grid

$$\overline{D}_h = \overline{D}_h(\overline{\omega}_{1(2.48)}) = \overline{\omega}_1 \times \omega_2 \times \ldots \times \omega_n, \tag{2.58}$$

where $\overline{\omega}_1 = \overline{\omega}_{1(2.48)}$ is the piecewise-uniform mesh; and ω_s, for $s = 2, \ldots, n$, are uniform meshes (with step-size h_s); $h_1 = \max\limits_i h_1^i$.

The condition (2.57) is necessary for the ε-uniform monotonicity of the operator $\Lambda_{(2.46)}$ on the grid $\overline{D}_{h(2.58)}$ satisfying the condition (2.38). Thus, the condition (2.57) is *the necessary condition for ε-uniform monotonicity* of the operator $\Lambda_{(2.46)}$ under the condition that the difference *scheme approximates the boundary value problem.*

We shall consider the difference scheme (2.47), (2.46) on the grid (2.58).

The case $a_{1s}(x) \equiv 0$ for $x \in \overline{D}$

We consider the boundary value problem in the case when the equation (2.2a) does not contain mixed derivatives involving differentiation in x_1, i.e., under the condition

$$a_{1s}(x) \equiv 0, \quad x \in \overline{D}, \quad s = 2, \ldots, n. \tag{2.59}$$

In the case of the condition

$$a_{sk}(x) = 0, \quad x \in \overline{D}^1, \quad s = k+1, \ldots, n, \quad k = 1, \ldots, q,$$

where

$$\overline{D}^1 \subseteq \overline{D}, \quad 1 \leq q \leq n-1, \quad s, k = 1, \ldots, n, \quad s \neq k,$$

we say that the operator $L^*_{(2)(2.50)}$ on the set $\overline{D}^1$ is *canonical in the variables* $x_1, \ldots, x_q$ for $q < n-1$ and canonical in the variables $x_1, \ldots, x_n$ for $q = n-1$. Note that $a_{sk}(x) = a_{ks}(x)$, $x \in \overline{D}$ for $s, k = 1, \ldots, n$. In a similar way, one can define the canonical form of the operator $L^*_{(2)}$ in the variables $x_q, \ldots, x_n$, where $1 \leq q \leq n$.

In the case of the conditions (2.57) and (2.59), the operator $L^*_{(2)(2.50)}$ has canonical form in the variable x_1 on Γ and on $\overline{D}$, respectively.

We say that the condition (2.57) (the condition (2.59)) is *the canonicity condition in the variable x_1 of the elliptic part of the differential equation (of the elliptic part of the operator L) on the boundary Γ (respectively, on the set $\overline{D}$), or, briefly, the canonicity condition of the boundary value problem in x_1 on the boundary Γ (respectively, on the set $\overline{D}$).*

On the set $\overline{Q} \subseteq \overline{D}$ we introduce the grid

$$\overline{D}_h(Q) = \{\overline{Q}\}_h = \overline{Q} \cap \overline{D}_{h(2.58)} \tag{2.60}$$

that is piecewise-uniform in x_1 and give conditions that guarantee the ε-uniform monotonicity of the operator $\Lambda_{(2.46)}$ on this grid.

In the case of the condition (2.59), using an appropriate linear transformation of the variables $x_2, \ldots, x_n$ (the transformation $X = X(x)$, $X_1 = x_1$, $x \in \overline{D}$), the operator $L^*_{(2)(2.50)}$ can be brought to canonical form at the point $X_0 \in \overline{D}_X$, $X_0 = X(x_0)$. In the canonical form, the coefficients of the mixed derivatives not involving differentiation in X_1 (here $X_1 = x_1$) are sufficiently small in a small neighborhood of the point X_0. In the new variables, the coefficients of the mixed derivatives involving X_1 are equal to zero on the whole domain.

Next, we shall assume that the coefficients of the operator $L^*_{(2)(2.50)}$ already satisfy both the condition (2.59) and the following condition:

> *For an arbitrary point $x^* \in \overline{D}$, one can find a sufficiently small neighborhood $Q(x^*)$, where $\overline{Q}(x^*) \subseteq \overline{D}$, and $n-1$ pairs of functions $\rho_{*s}(x)$ and $\rho^*_s(x)$, for $s = 2, \ldots, n$, such that the following relations are fulfilled:*

$$\max_{\overline{Q}(x^*)} \rho_{*s}(x) \le \min_{\overline{Q}(x^*)} \rho^*_s(x), \quad s = 2, \ldots, n; \tag{2.61a}$$

$$\min_{\overline{Q}(x^*)} a_{kk}(x) h_k^{-1} - \sum_{\substack{s=2 \\ s \ne k}}^{n} \max_{\overline{Q}(x^*)} |a_{ks}(x)| \, h_s^{-1} > 0, \quad k = 2, \ldots, n, \tag{2.61b}$$

where

$$h_s \in \left[\max_{\overline{Q}(x^*)} \rho_{*s}(x) \, h^*, \ \min_{\overline{Q}(x^*)} \rho^*_s(x) \, h^* \right], \quad s = 2, \ldots, n, \tag{2.61c}$$

$h^* > 0$ *is an arbitrary number.*

In the case of the conditions (2.59) and (2.61), we say that the operator $L^*_{(2)}$ on the set $\overline{D}$ is *canonical in x_1* (because the condition (2.59) holds) and *locally almost canonical in $x_2, \ldots, x_n$* (because, besides (2.59), for an arbitrary point $x^* \in \overline{D}$, one can find a neighborhood $Q(x^*)$ and functions $\rho_{*s}(x)$ and $\rho^*_s(x)$, for $s = 2, \ldots, n$, such that the relations (2.61) hold). Note that the conditions (2.61) involve only the coefficients $a_{sk}(x)$, for $s, k = 2, \ldots, n$, of the *truncated operator*

$$L_{(2)}^{*[1]} = L_{(2)}^{*[1]}(L_{(2)}^*), \qquad (2.62)$$

i.e., the operator involving differentiation in the variables $x_2, \ldots, x_n$; thus, the truncated operator $L_{(2)}^{*[1]}$ is local almost canonical.

In the case of the condition (2.59), the locally almost canonicity of the operator $L_{(2)}^{*[1]}$ allows us to construct piecewise-uniform meshes in x_1 on which the operator Λ is monotone.

Theorem 2.4.4 *Consider the grid* $\overline{D}_h$ *(constructed on the set* $\overline{Q}(x^*)$*), which is piecewise-uniform in* x_1 *and uniform in* $x_2, \ldots, x_n$*:*

$$\overline{D}_h = \overline{D}_{h^*}\left(Q(x^*), L_{(2)}^*\right) = \qquad (2.63)$$

$$= \overline{D}_{h^*}\left(Q(x^*), L_{(2)}^*, \overline{\omega}_1 = \overline{\omega}_{1(2.48)}, \omega_s = \omega_s^u, s = 2, \ldots, n, x_2, \ldots, x_n\right) =$$

$$= \left\{\{\overline{Q}(x^*)\}_{h(2.60)} \text{ under the condition (2.61c) subject to (2.61)}\right\}.$$

In the case of the conditions (2.59), (2.61), the operator $\Lambda_{(2.46)} = \Lambda_{(2.46)}(L)$ *is* ε*-uniformly monotone.*

Thus, the grid $\overline{D}_h = \overline{D}_{h^*(2.63)}(Q(x^*), L_{(2)}^*)$ is consistent in the variables $x_2, \ldots, x_n$ on $\overline{Q}(x^*) \subseteq \overline{D}$ for the operators $\Lambda_{(2)(2.51)}^*(L_{(2)}^*)$ and $\Lambda_{(2.46)}(L)$ under the condition (2.59).

The case $a_{1s}(x) = 0$ for $x \in \Gamma$

We consider the boundary value problem in the case when the coefficients $a_{1s}(x)$ vanish on the boundary Γ (i.e., the condition (2.57) holds) and, furthermore, for these coefficients one has

$$a_{1s} \in C^1(\overline{D}), \quad s = 2, \ldots, n. \qquad (2.64)$$

We construct a grid that is consistent in $x_1, \ldots, x_n$, uniform in $x_2, \ldots, x_n$, and condenses in the boundary layer in x_1.

In the case of condition (2.57), it is appropriate to use the piecewise-uniform meshes (2.48).

We shall assume that the coefficients of the operator $L_{(2)(2.50)}^*$, as well as the condition (2.57) (i.e., the operator $L_{(2)}^*$ on the set Γ is canonical in x_1), satisfy the following condition. For an arbitrary point x^* in $\overline{D}$, one can find sufficiently small values m^* and $\nu > 0$, a neighborhood $Q(x^*)$ and n pairs of functions $\rho_{*s}(x)$ and $\rho_s^*(x)$, for $s = 1, \ldots, n$, such that the relations (2.55) are valid for the coefficients $a_{sk}(x)$, for $s, k = 1, \ldots, n$, of the operator $L_{(2)}^*$ (i.e., the operator $L_{(2)}^*$ is locally almost canonical in $x_1, \ldots, x_n$ on the set $\overline{D}$); the values m^* and $\nu > 0$ do not appear in the relations (2.55). Furthermore, in a neighborhood of Γ, the coefficients $a_{sk}(x)$, for $s, k = 2, \ldots, n$, of the operator $L_{(2)(2.62)}^{*[1]}$ satisfy the relations (2.55), where $s, k = 2, \ldots, n$, as well as the additional relation $\big($similar to (2.61b)$\big)$

$$
\left[\min_{\substack{\overline{Q}(x^*) \\ r(x,\Gamma) \leq m^*}} a_{kk}(x) - \nu \right] h_k^{-1} - \sum_{\substack{s=2 \\ s \neq k}}^{n} \max_{\substack{\overline{Q}(x^*) \\ r(x,\Gamma) \leq m^*}} |a_{sk}(x)| \, h_s^{-1} > 0, \qquad (2.65)
$$

$$
k = 2, \ldots, n.
$$

In the case when the relations (2.55) and (2.65) are both fulfilled with the same "arguments", we shall use the notation $\{(2.55), (2.65)\}$.

The fulfillment of $\{(2.55), (2.65)\}$ in the case of the condition (2.57) can be ensured by a linear transformation of the variables $x_2, \ldots, x_n$ that brings the operator $L_{(2)}^*$ to the canonical form at the point $x^* \in \Gamma$ (or to a form close to canonical), and by choosing the neighborhood $Q(x^*)$ and the values m^* and ν to be sufficiently small.

Note that the relations (2.65), in general, follow from the relation (2.55b). Because the values $h_2, \ldots, h_n$ are commensurable, one can subtract from the left part of the inequality (2.55b), without violating it, the value $\nu \, h_k^{-1}$, choosing the value ν and the neighborhood $Q(x^*)$ sufficiently small. This way of obtaining the relations (2.65) is less preferable because the smallness of the values ν and m^* leads to strong restrictions imposed on the parameters of the piecewise-uniform mesh $\overline{\omega}_1$ constructed later (see the condition (2.67) below).

In the case when the relations $\{(2.55), (2.65)\}$ hold for the coefficients of the operator $L_{(2)}^{*[1]}$ for $s, k \geq 2$, we say that the operator $L_{(2)}^{*[1]}$ is *strongly locally almost canonical* in $x_2, \ldots, x_n$ in a neighborhood of the set Γ.

If on a set $\overline{D}^1 \subseteq \overline{D}$ the relations $\{(2.55), (2.65)\}$ hold for the coefficients of the operator $L_{(2)(2.62)}^{*[1]}$ for $s, k = 2, \ldots, n$ and $Q(x^*) = \overline{D}^1$, we say that the operator $L_{(2)}^{*[1]}$ is *strongly almost canonical* in $x_2, \ldots, x_n$ *on the subset* $\overline{D}^1$ in a neighborhood of the set Γ.

In an m^*-neighborhood of Γ, the condition (2.65) is more restrictive than the condition (2.61b); note that the conditions (2.55a), (2.55c) for $s, k = 2, \ldots, n$ and the conditions (2.61a), (2.61c) are similar.

The condition (2.57) and the condition

$$
\begin{array}{l}
\textit{the coefficients } a_{sk}(x) \textit{ of the operator } L_{(2)}^* \textit{ on the set } \overline{Q}(x^*) \subseteq \overline{D} \\[4pt]
\textit{satisfy the relations } \{(2.55), (2.65)\}
\end{array}
\qquad (2.66)
$$

together allow us to construct on the set $\overline{Q}(x^*)$ piecewise-uniform (in the x_1-axis) consistent grids and thereby ensure the monotonicity of the operators $\Lambda^*_{(2)(2.51)}$ and $\Lambda_{(2.46)}$. Let us construct such a mesh.

Let the conditions (2.57) and (2.66) hold. We now impose conditions on the parameters of the grid (2.60) under which the operator $\Lambda^*_{(2)(2.51)}$ is monotone.

We consider the operator $\Lambda^*_{(2)(2.51)}$ on the grid $\overline{D}_{h(2.58)}$ that is uniform in x_s, for $s = 2, \ldots, n$, and piecewise-uniform in x_1. Let the step-sizes h_s in the meshes ω_s, for $s = 2, \ldots, n$, and the step-size h_1 (i.e., the maximal step-size in the mesh $\overline{\omega}_{1(2.48)}$) satisfy the condition (2.55c). Note that $h_1 = h_{(2)}$ in the mesh (2.48). Let the value $\sigma = \sigma_{(2.48)}$ satisfy the condition

$$\sigma \le m^*_{(2.65)}, \tag{2.67a}$$

and let the condition (2.64) hold. By virtue of the conditions (2.57), (2.64), one has $|a_{1s}(x)| \le M_1 \sigma$ for $r(x, \Gamma) \le \sigma$, where $M_1 = \max\limits_{s, s \ne 1, \overline{D}} |(\partial/\partial x_1) a_{1s}(x)|$. Therefore $|a_{1k}(x)| h_{(1)}^{-1} \le M_1 h_{(1)}^{-1} \sigma$, for $k \ne 1$. The inequality

$$\nu h_k^{-1} - \max\limits_{\substack{\overline{Q}(x^*) \\ x_1 \notin (d_{*1} + \sigma,\, d_1^* - \sigma)}} |a_{1k}(x)| h_{(1)}^{-1} > 0, \quad k = 2, \ldots, n$$

holds if $M_1 h_{(1)}^{-1} \sigma < \nu \min\limits_{k=2,\ldots,n} [h_k^{-1}]$, i.e., under the condition

$$N_{*1} \equiv \sigma h_{(1)}^{-1} < M_1^{-1} \nu \min\limits_{k=2,\ldots,n} [h_k^{-1}], \quad N_1 \le M N_{*1}, \tag{2.67b}$$

where $N_{*1} + 1$ is the number of nodes in the mesh $\overline{\omega}_{1(2.48)}$ in the interval $[d_{*1}, d_{*1} + \sigma]$ or on $[d_1^* - \sigma, d_1^*]$.

Under the above conditions on the values σ and N_{*1}, the operators $\Lambda^*_{(2)(2.51)}$ and $\Lambda_{(2.46)}$ are ε-uniformly monotone on the grid $\overline{D}_{h(2.60)}(Q)$, where Q is $\overline{Q}(x^*)$ from $\{(2.55), (2.65)\}$. Thus in the case of (2.57), (2.66), (2.64), the condition (2.67) is a *sufficient condition for the local ε-uniform monotonicity of the operators $\Lambda^*_{(2)}$ and Λ on the piecewise-uniform mesh $\overline{D}_{h(2.60)}(Q)$.*

In the case of the condition (2.66), when constructing consistent meshes based on the grid

$$\overline{D}_h = \overline{D}_{h^*}(Q) = \overline{D}_{h^*}(Q(x^*)) = \Big\{ \overline{D}_{h(2.58)} \cap \overline{Q}(x^*),$$
$$\text{under the condition (2.55c) subject to } \{(2.55), (2.65)\} \Big\}, \tag{2.68}$$

we use the meshes

$$\overline{\omega}_1 = \Big\{ \overline{\omega}_{1(2.48)} \text{ under the condition (2.67)} \Big\}. \tag{2.69}$$

Under the conditions (2.57), (2.66), (2.64), (2.67) the grid

$$\overline{D}_h = \overline{D}_{h^*}\left(Q(x^*),\, L^*_{(2)}\right) = \overline{D}_{h^*}\left(Q(x^*),\, L^*_{(2)},\, \overline{\omega}_{1(2.69)},\, x_1,\ldots,x_n\right) =$$

$$= \left\{\overline{D}_{h^*(2.68)},\; \overline{\omega}_1 = \overline{\omega}_{1(2.69)},\; \omega_s = \omega^u_s,\; s = 2,\ldots,n\right\} =$$

$$\tag{2.70}$$

$$= \left\{\overline{D}_{h(2.58)} \cap \overline{Q}(x^*),\; \overline{\omega}_1 = \overline{\omega}_{1(2.48)}\; subject\; to\; (2.55c)\; and\; (2.67)\right.$$

$$\left. in\; the\; case\; of\; the\; relations\; (2.57),\; \{(2.55),\, (2.65)\}\right\}$$

is consistent on the set $\overline{Q}(x^*)$ in the variables $x_1,\ldots,x_n$ with the monotonicity condition for the operators $\Lambda^*_{(2)(2.51)}$ and $\Lambda_{(2.46)}$.

Thus, when constructing consistent meshes, the passage from the uniform grids $\overline{D}_{h(2.56)}$ to the piecewise-uniform grids $\overline{D}_{h(2.60)}$ requires us to impose (besides the condition (2.55) for the coefficients $a_{sk}(x)$ and the parameters h_s of the meshes $\overline{\omega}_s$) additional conditions (2.65) and (2.67) on the coefficients $a_{sk}(x)$ and on the mesh $\overline{\omega}_{1(2.48)}$.

Note that the grid $\overline{D}_{h(2.70)}$ (the grid on the set $\overline{Q}(x^*)$) is generated by the meshes $\overline{\omega}_{1(2.58)}$ defined on $[d_{*1}, d^*_1]$, and by the meshes ω_s defined on the x_s-axis for $s = 2,\ldots,n$. However, any change of the distribution of nodes in the meshes $\overline{\omega}_1$, or ω_2, $\ldots$, ω_n outside the projection of the set $\overline{Q}(x^*)$ on the x_s-axis does not affect the distribution of nodes in the grid $\overline{D}_{h(2.70)} = \overline{D}_{h(2.70)}\left(Q(x^*),\, L^*_{(2)}\right)$.

Theorem 2.4.5 *Under the conditions (2.57), (2.66), (2.64), (2.67) the operator $\Lambda_{(2.46)} = \Lambda_{(2.46)}\,(L)$ on the grid $\overline{D}_h = \overline{D}_{h^*\,(2.70)}\left(Q(x^*),\, L^*_{(2)}\right)$ is ε-uniformly monotone.*

Remark 2.4.1 Let the coefficients of the mixed derivatives satisfy the condition

$$\max_{\overline{D}} |a_{sk}(x)| < (n-1)^{-1} \min_{\overline{D}} \left[\alpha_s\, \alpha^{-1}_k\, a_{ss}(x),\; \alpha_k\, \alpha^{-1}_s\, a_{kk}(x)\right], \tag{2.71}$$

$$s,k = 1,\ldots,n,\quad s \neq k,$$

where $\alpha_1,\ldots,\alpha_n$ are some positive numbers, i.e., we have *dominance of the diagonal elements in the matrix of the coefficients of the operator* $L^*_{(2)(2.50)}$ on the whole set $\overline{D}$. Then one can choose the value ν and the functions $\rho_{*s}(x)$ and $\rho^*_s(x)$, for $s = 1,\ldots,n$, such that the condition (2.66) is valid for

$$\overline{Q}(x^*) = \overline{D}. \tag{2.72}$$

The fulfillment of the condition (2.71) ensures existence of a transformation of the operator $L^*_{(2)}$ to the strongly almost canonical form (in the variables

$x_1, \ldots, x_n)$ on the set $\overline{D}$. Indeed, under the condition (2.71), one can choose the value ν to be sufficiently small such that the same inequality (2.71) will be satisfied if the coefficients $a_{pp}(x)$ in its right-hand side are changed to $a_{pp}(x) - \nu$, for $p = s, k$. Setting $\rho_{*s}(x) = \rho_s^*(x) = \rho_s = \alpha_1 \alpha_s^{-1}$, for $s = 1, \ldots, n$, we come to the condition (2.55) and also to the condition (similar to (2.65))

$$\left[\min_{\overline{D}} a_{kk}(x) - \nu \right] h_k^{-1} - \sum_{\substack{s=1 \\ s \neq k}}^{n} \max_{\overline{D}} |a_{sk}(x)| \, h_s^{-1} > 0, \quad k = 1, \ldots, n.$$

In the case of the condition (2.71), under the additional condition (2.57), using the grid $\overline{D}_{h(2.58)}$, where $\overline{\omega}_1 = \overline{\omega}_{1(2.69)}$ for $h_s = \rho_s h^*$, with $s = 1, \ldots, n$ (for appropriate $m_{1(2.40)}$, ν, $N_{*1} = N_{*1}(N_1)$), one can satisfy the conditions (2.66), (2.67) that ensure ε-uniform monotonicity of the operator $\Lambda_{(2.46)}$ for $\overline{Q}(x^*) = \overline{D}$ and any h^*.

In the case when the coefficients $a_{sk}(x)$, for $s, k > 1$, satisfy the condition

$$\max_{\overline{D}} |a_{sk}(x)| < (n-2)^{-1} \min_{\overline{D}} \left[\alpha_s \alpha_k^{-1} a_{ss}(x), \ \alpha_k \alpha_s^{-1} a_{kk}(x) \right], \quad (2.73)$$

$$s, k = 2, \ldots, n, \quad s \neq k,$$

where $\alpha_2, \ldots, \alpha_n$ are some positive numbers (i.e., we have *dominance of the diagonal elements in the matrix of the coefficients of the operator* $L_{(2)(2.62)}^{*[1]}$), under the condition $\rho_{*s}(x) = \rho_s^*(x) = \rho_s = \alpha_2 \alpha_s^{-1}$, for $s = 2, \ldots, n$, the relations (2.61) hold for

$$\overline{Q}(x^*) = \overline{D}. \quad (2.74)$$

Conditions (2.73) guarantee existence of a transformation of the operator $L_{(2)}^{*[1]}$ to almost canonical form (in the variables $x_2, \ldots, x_n$) on the set $\overline{D}$. In the case of conditions (2.59) and (2.73) on the grid $D_{h(2.63)}$, which is uniform in $x_2, \ldots, x_n$ for $h_s = \rho_s h^*$, with $s = 2, \ldots, n$, condition (2.61b) holds and ensures the ε-uniform monotonicity of the operator $\Lambda_{(2.46)}$ for $\overline{Q}(x^*) = \overline{D}$ and any h^* specifying the meshes $\overline{\omega}_s$, for $s = 2, \ldots, n$; here $\overline{\omega}_1 = \overline{\omega}_{1(2.48)}$.   ∎

2.4.3   Consistent grids on subdomains

For the construction of difference schemes of the domain decomposition method on overlapping subdomains, in particular, for boundary value problems in domains with curvilinear boundaries, we need piecewise-uniform consistent meshes that condense in a neighborhood of only one of the side-faces adjoining to the boundary. Let us discuss the construction of such consistent meshes.

In the mesh $\overline{\omega}_{1(2.69)} = \overline{\omega}_{1(2.69)}\big(\overline{\omega}_{1(2.48)}\big)$ from $\overline{D}_{h(2.70)}\big(Q(x^*), L_{(2)}^*\big)$, the use of the uniform mesh $\overline{\omega}_1^u\big(d_{*1}, d_1^*\big)$ whose step-size h_1 equals the step-size

$h_{(2)}$ in the mesh $\overline{\omega}_{1(2.48)}$ instead of the piecewise-uniform mesh $\overline{\omega}_{1(2.48)} = \overline{\omega}_{1(2.48)}(d_{*1}, d_1^*)$ leads to the uniform grid

$$\overline{D}_h = \overline{D}_h^u\Big(Q(x^*),\ L_{(2)}^*\Big) = \overline{D}_{h(2.70)}\Big(Q(x^*),\ L_{(2)}^*,\ \overline{\omega}_{1(2.69)}(\overline{\omega}_1^u)\Big) = \quad (2.75a)$$

$$= \overline{D}_{h(2.70)}\Big(Q(x^*),\ L_{(2)}^*,\ \overline{\omega}_1 = \overline{\omega}_1^u\ under\ the\ condition\ (2.55c)$$

in the case of the relations $\{(2.55), (2.65)\}\Big),$

which is consistent (in the variables $x_1, \ldots, x_n$) on the set $\overline{Q}(x^*)$. The consistency of the grid $\overline{D}_{h(2.70)}(Q(x^*),\ L_{(2)}^*)$ on $\overline{Q}(x^*)$ is preserving also in the case when instead of the mesh $\overline{\omega}_{1(2.48)}$ that condenses in neighborhoods of both endpoints of the interval $\big[d_{*1},\ d_1^*\big]$, we use the piecewise-uniform mesh either $\overline{\omega}_1^l(d_{*1},\ d_1^*)$ or $\overline{\omega}_1^r(d_{*1},\ d_1^*)$ that condenses in a neighborhood either of the left or the right endpoint of the interval $\big[d_{*1},\ d_1^*\big]$, i.e., in the case of the grids

$$\overline{D}_h = \overline{D}_h^l\Big(Q(x^*),\ L_{(2)}^*\Big) = \overline{D}_{h(2.70)}\Big(Q(x^*),\ L_{(2)}^*,\ \overline{\omega}_{1(2.69)}(\overline{\omega}_1^l)\Big), \quad (2.75b)$$

$$\overline{D}_h = \overline{D}_h^r\Big(Q(x^*),\ L_{(2)}^*\Big) = \overline{D}_{h(2.70)}\Big(Q(x^*),\ L_{(2)}^*,\ \overline{\omega}_{1(2.69)}(\overline{\omega}_1^r)\Big). \quad (2.75c)$$

For the consistency of the grids $\overline{D}_h^l$ and $\overline{D}_h^r$ on $\overline{Q}(x^*)$, it is required that the parameters σ, $h_{(1)}$, and $h_{(2)}$ specifying the meshes $\overline{\omega}_1^l(d_{*1},\ d_1^*)$ and $\overline{\omega}_1^r(d_{*1},\ d_1^*)$ would be the same as in the mesh $\overline{\omega}_{1(2.48)}(d_{*1},\ d_1^*)$.

Note that the consistent uniform grid $\overline{D}_{h(2.75a)}^u(Q(x^*),\ L_{(2)}^*)$ is, in general, different from the consistent uniform grid $\overline{D}_{h(2.56)}(Q(x^*),\ L_{(2)}^*)$. Produced requirements to the coefficients $a_{sk}(x)$ and generating meshes $\overline{\omega}_1$ and ω_s, for $s = 2, \ldots, n$, in the case of the grid $\overline{D}_{h(2.75a)}^u$, turn out to be more restricted compared with the grid $\overline{D}_{h(2.56)}$.

We now give sufficiently natural conditions imposed on the coefficients $a_{sk}(x)$ and the meshes $\overline{\omega}_1, \omega_2, \ldots, \omega_n$ that guarantee the consistency of the grids $\overline{D}_h$ constructed based on *the elementary meshes in the* x_1*-axis*, i.e., $\overline{\omega}_1^u$, $\overline{\omega}_1^l$, $\overline{\omega}_1^r$, or $\overline{\omega}_1^b$, where $\overline{\omega}_1^b$ is a mesh that condenses in neighborhoods of both endpoints of the interval $\big[d_{*1},\ d_1^*\big]$.

Let

$$\Gamma = \cup_j \Gamma_j, \quad j = 1, 2, \quad (2.76)$$

where Γ_1 and Γ_2 are the left and right parts of the boundary Γ. Let the piecewise-uniform (in x_1) grid $\overline{D}_h$ condense in a neighborhood either of the side Γ_1 and Γ_2, or the grid $\overline{D}_h$ is uniform. We denote by Γ^e those parts of the boundary Γ in whose neighborhoods it is required to condense the mesh $\overline{\omega}_1$. Let J^*

$$J^* = J^*(D) = J^*(D; \Gamma^e) \quad (2.77)$$

be the set of indexes j related to Γ_j in whose neighborhood the mesh $\overline{\omega}_1$ condenses; here $\Gamma_j \subseteq \Gamma^e$. For example, $J^* = \emptyset$ in the case of the uniform mesh $\overline{\omega}_1$; but if the mesh $\overline{\omega}_1$ is $\overline{\omega}_1^l$ then $J^* = \{j = 1\}$; here $J^*(D; \Gamma^e = \emptyset) = \emptyset$. Thus, the set of the indexes J^* is defined by the set Γ^e of the parts of the boundary Γ.

Suppose that on the boundary Γ the coefficients $a_{sk}(x)$ of the operator $L_{(2)}^*$ satisfy the condition

$$a_{sk}(x) = 0, \ x \in \Gamma_j, \ j \in J^*, \ s = 1 \text{ or } k = 1, \ s \neq k, \text{ if } J^* \neq \emptyset; \qquad (2.78)$$

restrictions on $a_{sk}(x)$ on the set Γ are not imposed if $J^* = \emptyset$,

i.e., the operator $L_{(2)}^*$ is canonical on the side Γ_j for $j \in J^*$, where $J^* \neq \emptyset$.

Assume that on the set $\overline{D}$ the coefficients $a_{sk}(x)$ satisfy the conditions (2.55). In the case when $J^* \neq \emptyset$, we assume that the following additional condition similar to (2.65) holds:

$$\left[\min_{\substack{\overline{Q}(x^*) \\ r(x, \Gamma_j) \leq m^* \\ j \in J^*}} a_{kk}(x) - \nu \right] h_k^{-1} - \sum_{\substack{s=2 \\ s \neq k}}^{n} \max_{\substack{\overline{Q}(x^*) \\ r(x, \Gamma_j) \leq m^* \\ j \in J^*}} |a_{sk}(x)| h_s^{-1} > 0, \qquad (2.79)$$

$$k = 2, \ldots, n, \quad J^* \neq \emptyset.$$

In the case of condition (2.78) together with the relations $\{(2.55), (2.79)\}$, the operator $L_{(2)}^*$ on the set $\overline{D}$ has *local almost canonical form* in $x_1, \ldots, x_n$ and the operator $L_{(2)(2.62)}^{*[1]}$ has *strongly local almost canonical form* in $x_2, \ldots, x_n$ in a neighborhood of the set Γ_j, where $j \in J^*$.

The condition

the coefficients $a_{sk}(x)$ of the operator $L_{(2)}^*$ on the set $\overline{D}$ $\qquad (2.80)$

satisfy the relations (2.78), $\{(2.55), (2.79)\}$ if $J^* \neq \emptyset$,

and the relations (2.55) if $J^* = \emptyset$, where $\overline{Q}(x^*) \subseteq \overline{D}$, $J^* = J_{(2.77)}^*(D)$,

allows us on the set $\overline{Q}(x^*)$ to construct piecewise-uniform (controlled by the set $J^*(D)$) consistent grids that ensure the monotonicity of the operators $\Lambda_{(2)(2.51)}^*$ and $\Lambda_{(2.46)}$. We now construct such meshes.

For $J^* = \emptyset$ the uniform grid

$$\overline{D}_h = \overline{D}_{h(2.81)}\left(Q(x^*), L_{(2)}^*\right), \ \overline{\omega}_1 = \overline{\omega}_1^u, \ \omega_s = \omega_s^u, \ s = 2, \ldots, n$$

under the condition (2.55c) in the case of the relations (2.55)) $\qquad (2.81)$

(this is the grid $\overline{D}_h = \overline{D}_{h(2.56)}\left(Q(x^*), L_{(2)}^*\right)$) is consistent on the set $\overline{Q}(x^*)$.

Let the condition (2.80) hold. In the case of the condition $J^* \neq \emptyset$ (i.e., when the mesh $\overline{\omega}_1$ is piecewise uniform), we assume that the step-sizes $h_1 = h_{(2)1}$

and h_s in the meshes $\overline{\omega}_1$ and ω_s, for $s = 2, \ldots, n$, satisfy the condition (2.55c) (on the set $\overline{Q}(x^*)$) and, moreover, for the mesh $\overline{\omega}_1$, one has the condition (similar to (2.67a)):

$$\begin{aligned}
\sigma = \sigma^l \leq m^*_{(2.79)}, \quad \text{for } j = 1, \quad j \in J^*, \\
\sigma = \sigma^r \leq m^*_{(2.79)}, \quad \text{for } j = 2, \quad j \in J^*;
\end{aligned} \tag{2.82a}$$

and also the following condition (similar to (2.67b)):

$$N_{*1} \equiv \sigma \, h^{-1}_{(1)} < M_1^{-1} \nu \min_{k=2,\ldots,n} [h_k^{-1}], \quad j \in J^*, \tag{2.82b}$$

where $N_{*1} + 1$ is the number of nodes on the interval $[d_{*1}, d_{*1} + \sigma]$ for $\{j = 1\} \in J^*$ and on the interval $[d_1^* - \sigma, d_1^*]$ for $\{j = 2\} \in J^*$; $M_1 = M_{1(2.67)}$. In the case when $J^* = \{j = 1, 2\}$ (i.e., the mesh $\overline{\omega}_1$ condenses in neighborhoods of both endpoints of the interval $[d_{*1}, d_1^*]$), set $\sigma^l = \sigma^r$.

Furthermore, we assume that the condition (2.80) is fulfilled. Along the x_1-axis, we use the piecewise-uniform or uniform meshes

$$\overline{\omega}_1 = \begin{cases}
\overline{\omega}_1^l & \text{if } J^* = \{j = 1\} \\
\overline{\omega}_1^r & \text{if } J^* = \{j = 2\} \\
\overline{\omega}_1^b & \text{if } J^* = \{j = 1, 2\} \\
\overline{\omega}_1^u & \text{under the condition (2.55c) if } J^* = \emptyset;
\end{cases} \quad \begin{array}{l} \text{under the conditions} \\ (2.55c), \ (2.82) \\ \text{if } J^* \neq \emptyset, \end{array} \tag{2.83a}$$

the mesh $\overline{\omega}_{1(2.83)}$ condenses only in a neighborhood of the sets Γ_j for which $j \in J^*$, where $J^* \neq \emptyset$.

We expose the mesh

$$\overline{\omega}_{1(2.83)}(d_{*1}, d_1^*) = \overline{\omega}_1(d_{*1}, d_1^*; \overline{\omega}_{1(2.48)}), \tag{2.83b}$$

that is constructed based on the mesh $\overline{\omega}_{1(2.48)}(d_{*1}, d_1^*)$. Set

$$\overline{\omega}_1 = \overline{\omega}_1^b(d_{*1}, d_1^*) = \overline{\omega}_{1(2.48)}(d_{*1}, d_1^*) \quad \text{if } J^* = \{j = 1, 2\}. \tag{2.83c}$$

For $J^* = \{j = 1\}$ we define the mesh $\overline{\omega}_1$ by the relation

$$\overline{\omega}_1 = \overline{\omega}_1^l(d_{*1}, d_1^*) = \overline{\omega}_{1(2.48)}(d_{*1}, d_1^{(*)}) \cap [d_{*1}, d_1^*], \quad J^* = \{j = 1\}. \tag{2.83d}$$

Here $\overline{\omega}_{1(2.48)}(d_{*1}, d_1^{(*)})$ is the mesh on the interval $[d_{*1}, d_1^{(*)}]$ with $2N_1 + 1$ nodes; moreover, $x_1 = d_1^*$ is the midpoint of the interval $[d_{*1}, d_1^{(*)}]$; $d_1^* = 2^{-1}(d_{*1} + d_1^{(*)})$, $N_1 + 1$ is the number of nodes in the mesh $\overline{\omega}_{1(2.83)}(d_{*1}, d_1^*)$. In a similar way, for $J^* = \{j = 2\}$ the mesh $\overline{\omega}_1 = \overline{\omega}_1^r(d_{*1}, d_1^*)$ is constructed. The parameters σ, $h_{(1)1}$, and $h_{(2)1}$ of the mesh $\overline{\omega}_{1(2.48)}(d_{*1}, d_1^{(*)})$ are the same as those in the meshes $\overline{\omega}_1^l, \overline{\omega}_1^r$.

Theorem 2.4.6 *In the case of the conditions* (2.80), (2.64), (2.82), *the grid*

$$\overline{D}_h = \overline{D}_h\big(Q(x^*), L^*_{(2)}; J^*\big) = \tag{2.84}$$

$$= \overline{D}_h\Big(Q(x^*), L^*_{(2)}, \overline{\omega}_1 = \overline{\omega}_{1(2.83)}(\overline{\omega}_{1(2.48)})\Big),$$

$$\omega_s = \omega_s^u, \ s = 2, \dots, n \ \textit{under the conditions} \ (2.55c) \ \textit{and} \ (2.82)$$

in the case of the relations (2.78), {(2.55), (2.79)} *if* $J^* \neq \emptyset$,

and under the condition (2.55c)

in the case of the relations (2.55) *if* $J^* = \emptyset$; $\quad J^* = J^*_{(2.77)}(D)\Big)$,

is consistent on the set $\overline{Q}(x^*)$.

But if in (2.84) $\overline{Q}(x^*) = \overline{D}^i$, where $\overline{D}^i$ is a subset from $\overline{D}$, the grid $\overline{D}_{h(2.84)}$ is consistent on the set $\overline{D}^i$. In the case when

$$\overline{D}^i = D^i \bigcup \Gamma^i, \quad D^i = \{x : \ d^i_{*s} < x_s < d^{i*}_s, \ s = 1, \dots, n\} \tag{2.85}$$

is a rectangular parallelepiped whose faces Γ^i_j belong to coordinate planes ($\Gamma^i = \bigcup_j \Gamma^i_j$, for $j = 1, \dots, 2n$), it is convenient to construct a consistent grid directly on $\overline{D}^i$ based on the meshes $\overline{\omega}^i_s$, for $s = 1, \dots, n$, introduced on the intervals $[d^i_{*s}, d^{i*}_s]$. We show such grids.

By j^* we denote the index of the face $\Gamma_j \subseteq \Gamma^e$, for which the condition $\Gamma^i \cap \Gamma_j \neq \emptyset$ holds (i.e., the set $\overline{D}^i$ adjoins to the face Γ_{j*}); $\Gamma_j = \Gamma_{j(2.76)}$, $\Gamma^e = \Gamma^e_{(2.77)}$, $\Gamma^i = \Gamma(D^i)$. We denote by J^{i*}, where J^{i*}

$$J^{i*} = J^{i*}(D^i) = J^{i*}(D^i; \Gamma^e), \tag{2.86}$$

the set of all indexes j^*.

We assume that the following condition (similar to (2.80)) holds:

the coefficients $a_{sk}(x)$ *of the operator* $L^*_{(2)}$ *on the set* $\overline{D}^i \subseteq \overline{D}$ (2.87)

satisfy the relations (2.78), {(2.55), (2.79)} *if* $J^{i*} \neq \emptyset$,

and the relations (2.55) *if* $J^{i*} = \emptyset$, *where* $\overline{Q}(x^*) = \overline{D}^i$, $J^{i*} = J^{i*}_{(2.86)}(D^i)$.

On the intervals $[d^i_{*s}, d^{i*}_s]$, we construct the meshes $\overline{\omega}^i_s$, for $s = 1, \dots, n$. The mesh $\overline{\omega}^i_1$ is uniform, i.e., $\overline{\omega}^{iu}_1$ if $J^{i*} = \emptyset$, or is piecewise uniform, i.e., $\overline{\omega}^{il}_1$, $\overline{\omega}^{ir}_1$, or $\overline{\omega}^{ib}_1$ if J^{i*} is the corresponding set $\{j = 1\}$, $\{j = 2\}$, or $\{j = 1, 2\}$. Let

$$\overline{\omega}^i_1 = \begin{cases} \overline{\omega}^{il}_1 \text{ if } J^* = \{j = 1\} \\ \overline{\omega}^{ir}_1 \text{ if } J^* = \{j = 2\} \\ \overline{\omega}^{ib}_1 \text{ if } J^* = \{j = 1, 2\} \\ \overline{\omega}^{iu}_1 \text{ under the condition (2.55c)} \end{cases} \begin{array}{l} \text{under the conditions} \\ (2.55c), (2.82) \\ \text{if } J^* \neq \emptyset, \\ \\ \text{if } J^* = \emptyset; \end{array} \qquad (2.88a)$$

$$J^* = J^{i*} = J^{i*}_{(2.86)}(D^i; \Gamma^e),$$

and let the meshes $\overline{\omega}^i_s$, for $s = 2, \dots, n$, be uniform. The mesh

$$\overline{\omega}^i_1 = \overline{\omega}^i_{1(2.88)}(d^i_{*1}, d^{i*}_1) = \overline{\omega}^i_1(d^i_{*1}, d^{i*}_1; \overline{\omega}_{1(2.48)}), \qquad (2.88b)$$

constructed on $[d^i_{*1}, d^{i*}_1]$ (from the set $[d_{*1}, d^*_1]$) based on the mesh $\overline{\omega}_{1(2.48)}$, is defined by the relation

$$\overline{\omega}^i_1 = \overline{\omega}^i_{1(2.88)}(d^i_{*1}, d^{i*}_1; \overline{\omega}_{1(2.48)}) = \overline{\omega}_{1(2.83)}(d^i_{*1}, d^{i*}_1; \overline{\omega}_{1(2.48)}). \qquad (2.88c)$$

Assume that the parameters of the meshes $\overline{\omega}^i_s = \overline{\omega}^i_s(d^i_{*s}, d^{i*}_s)$, for $s = 1, \dots, n$, satisfy the conditions (2.55c), (2.82), where $\overline{Q}(x^*) = \overline{D}^i$ and $J^* = J^{i*}_{(2.86)} \neq \emptyset$; for $J^* = J^{i*} = \emptyset$ the parameters of the meshes $\overline{\omega}^i_s$ satisfy the conditions (2.55c).

Theorem 2.4.7 *In the case of the conditions* (2.64), (2.87), (2.82) *under the condition that in* (2.78), (2.79), (2.82) J^* *is* $J^{i*}_{(2.86)}$, *the grid (similar to* (2.84)*):*

$$\overline{D}^i_h = \overline{D}^i_h(D^i, L^*_{(2)}; J^{i*}) = \overline{D}^i_h(Q(x^*) = D^i_{(2.85)}, L^*_{(2)}, \qquad (2.89)$$

$$\overline{\omega}^i_1 = \overline{\omega}^i_{1(2.88)}(\overline{\omega}_{1(2.48)}), \quad \overline{\omega}^i_s = \overline{\omega}^{iu}_s, \quad for \ s = 2, \dots, n,$$

under the conditions (2.55c) *and* (2.82)

subject to (2.78), {(2.55), (2.79)} *if* $J^{i*} \neq \emptyset$,

and under the condition (2.55c) *subject to* (2.55) *if* $J^{i*} = \emptyset$,

where $\overline{Q}(x^*) = \overline{D}^i$, $J^* = J^{i*} = J^{i*}_{(2.86)}(D^i)$,

is consistent on the set $\overline{D}^i$.

For $J^{i*} = \emptyset$ (i.e., for either $\overline{D}^i \cap \Gamma^e = \emptyset$ and $\Gamma^e \neq \emptyset$ or $\Gamma^e = \emptyset$) the grid $\overline{D}^i_{h(2.89)}$ is uniform.

Note that the grid $\overline{D}_{h(2.89)}$ is constructed directly on the set $\overline{D}^i$ at that time as the grid $\overline{D}_h = \overline{D}_{h(2.84)}(Q(x^*)$ is constructed on $\overline{D}^i$ for $\overline{Q}(x^*) = \overline{D}^i$ based on meshes introduced on the set $\overline{D}$.

Grids similar to (2.84), (2.89) are used for the construction of domain-decomposition-based difference schemes, in particular, of a difference scheme for boundary value problems in domains with curvilinear boundaries; see Subsections 2.5.1, 2.5.2 in this section.

2.4.4 ε-uniformly convergent difference schemes

We now consider difference schemes for the boundary value problem (2.2) on the slab $\overline{D}_{(2.4)}^{(0)}$ in the case when the operator $L_{(2)}^*$ is almost canonical on the set $\overline{D}$. This property of the operator $L_{(2)}^*$ allows us by means of a choice of a consistent mesh on $\overline{D}^{(0)}$ to construct a monotone difference scheme that approximates the boundary value problem ε-uniformly for different type restrictions imposed on coefficients of the mixed derivatives, in particular, on the coefficients $a_{1s}(x)$.

A corollary of Theorems 2.3.7, 2.4.4, and 2.4.5 is the following theorems on the convergence of the scheme (2.47), (2.46).

Theorem 2.4.8 *Let the conditions $a_{sk} \in C^1(\overline{D})$, for $s \neq k$, and a_{ss}, b_s, c, c^0, $f \in C(\overline{D})$, for $s, k = 1, \ldots, n$, hold and the coefficients $a_{sk}(x)$ satisfy the conditions (2.57), (2.71), and also (2.66), (2.67), where $\overline{Q} = \overline{D}$. For the solution of the problem (2.2), (2.4), let the estimates of Theorem 2.2.1 be satisfied for $K = 3$. Then the difference scheme (2.47), (2.46) on the grid $\overline{D}_{h(2.70)}$, where $\overline{Q} = \overline{D}$, $\overline{\omega}_{1(2.48)} = \overline{\omega}_{1(2.44)}$, converges ε-uniformly at the rate $\mathcal{O}(N^{-1} \ln N)$ as $N \to \infty$:*

$$|u(x) - z(x)| \leq M\,N^{-1} \ln N, \quad x \in \overline{D}_h.$$

Theorem 2.4.9 *Let a_{sk}, b_s, c, c^0, $f \in C(\overline{D})$, for $s, k = 1, \ldots, n$, the coefficients $a_{sk}(x)$ satisfy the conditions (2.59), (2.73), and also (2.61), where $\overline{Q} = \overline{D}$, and let for the solution of the problem (2.2), (2.4), the estimates of Theorem 2.2.1 be satisfied for $K = 3$. Then the difference scheme (2.47), (2.46) on the grid $\overline{D}_{h(2.63)}$, where $\overline{Q} = \overline{D}$, $\overline{\omega}_{1(2.48)} = \overline{\omega}_{1(2.44)}$, converges ε-uniformly at the rate $\mathcal{O}(N^{-1} \ln N)$ as $N \to \infty$.*

Theorem 2.4.10 *Let a_{ss}, b_s, c, c^0, $f \in C(\overline{D})$, for $s = 1, \ldots, n$, the coefficients $a_{sk}(x)$ satisfy the condition (2.49), and let for the solution of the problem (2.2), (2.4), the estimates of Theorem 2.2.1 be satisfied for $K = 3$. Then the difference scheme (2.47), (2.46) on the grid $D_{h(2.58)}$, where $\overline{\omega}_{1(2.48)} = \overline{\omega}_{1(2.44)}$, converges ε-uniformly at the rate $\mathcal{O}(N^{-1} \ln N)$ as $N \to \infty$.*

Under the hypotheses of Theorems 2.4.8, 2.4.9, and 2.4.10, for the interpolant $\overline{z}(x) = \overline{z}_{(2.42)}(x; z(\cdot), \overline{D}_h)$, $x \in \overline{D}$, where $z(x)$, $x \in \overline{D}_h$ is the solution of the difference scheme (2.47), (2.46), respectively, on the grids (2.70), (2.63), where $\overline{Q} = \overline{D}$, and (2.58), one has the estimate

$$|u(x) - \overline{z}(x)| \leq M\,N^{-1} \ln N, \quad x \in \overline{D}.$$

2.5   Boundary value problems in domains with curvilinear boundaries

2.5.1   A domain-decomposition-based difference scheme for the boundary value problem on a slab

Theorem 2.4.8 establishes ε-uniform convergence of the difference scheme $\{(2.47), (2.46), (2.70), (2.72)\}$ constructed in Subsections 2.4.1 and 2.4.2 for the problem on the slab (2.4) in the case of the conditions (2.57), (2.71). When considering the boundary value problem in a domain with a curvilinear boundary, it is appropriate to pass to a new coordinate system in which the new boundary of the domain belongs to the coordinate plane and for the transformed problem to construct ε-uniformly convergent scheme. Keeping in mind such an idea, we construct a finite difference scheme for the problem (2.2) on the slab $\overline{D}^{(0)}_{(2.4)}$ in the case when the coefficients $a_{1s}(x)$ satisfy only the condition (2.57) and fulfillment of the condition (2.71) is not assumed. Emphasize that the condition (2.72), where $\overline{Q}(x^*)$ is a neighborhood from the relations $\{(2.55), (2.65)\}$, in general, is violated if the condition (2.71) is not satisfied. In this case, to approximate the boundary value problem we use a scheme of the domain decomposition method. Accuracy of the solution of an iterative scheme to the domain decomposition method depends on the width of overlapping to subdomains covering $\overline{D}^{(0)}$, the number of nodes in the grid domain and the number of iterations used for the solution of the difference scheme and also on the value of the parameter ε. We are interested in methods whose errors of solutions are independent of the value of the parameter ε and are defined only by the number of nodes in the mesh and the number of iterations used for the solution of the boundary value problem. For simplicity, when constructing the scheme, we assume that the boundary value problem is periodic. An approach that we use in this subsection for the construction of the difference scheme, we shall apply in Subsection 2.5.2 for a problem in a domain with a curvilinear boundary.

In the case of a periodic problem in the domain $D^{(0)}$, one can cover this domain by a finite system of overlapping bounded simply connected sets $Q(x^{*i}) = D^i$, for $i = 1, \ldots, I$ (neighborhoods of the points x^{*i}), such that $\overline{D}^{(1)} \subset \bigcup_i \overline{D}^i$ and $x^{*i} \in D^{(1)}$, for $i = 1, \ldots, I$, where

$$D^{(1)} = \{x : \; d_{*1} < x_1 < d_1^*, \; 0 < x_s < 2\pi, \; s = 2, \ldots, n\}, \qquad (2.90)$$

is *a domain of periodicity to the boundary value problem.*

We assume that the sets

$$D^i, \quad i = 1, \ldots, I, \qquad (2.91a)$$

are 2π-periodic in x_s, for $s = 2, \ldots, n$.

To solve the boundary value problem (2.2), (2.4), we use a discrete analogue of the *Schwartz method*. Let us first describe this method for the *differential problem.*

On the set $\overline{D}_{(2.4)}^{(0)}$ we construct a sequence of functions $u^{[k]}(x)$, where $k = 1, 2, 3, \ldots$ is the iteration number, in the following way. We assume that

$$u^{[0]}(x) = 0, \quad x \in \overline{D}^{(0)}. \tag{2.91b}$$

Let the function $u^{[k-1]}(x)$ be already constructed for $k \geq 1$. We construct $u^{[k]}(x)$. For this we consider the auxiliary functions

$$u_i^{[k]}(x) = u_i^{[k]}(x; u_i^{(k)}(\cdot), u_{i-1}^{[k]}(\cdot)) \equiv$$

$$\equiv \left\{ \begin{array}{ll} u_i^{(k)}(x), & x \in \overline{D}^i \\ u_{i-1}^{[k]}(x), & x \in \overline{D}^{(1)} \setminus \overline{D}^i \end{array} \right\}, \quad x \in \overline{D}^{(0)}; \tag{2.91c}$$

$$u_0^{[k]}(x) = u^{[k-1]}(x), \quad x \in \overline{D}^{(0)}, \quad i = 1, \ldots, I;$$

the values i and k define the number of interior and exterior iterations. Here $u_i^{(k)}(x)$, $x \in \overline{D}^i$, is the solution of the boundary value problem on the set $\overline{D}^i$

$$L u_i^{(k)}(x) = f(x), \quad x \in D^i, \tag{2.91d}$$

$$u_i^{(k)}(x) = \left\{ \begin{array}{ll} \varphi(x), & x \in \Gamma^i \cap \Gamma \\ u_{i-1}^{[k]}(x), & x \in \Gamma^i \setminus \Gamma \end{array} \right\}, \quad x \in \Gamma^i,$$

where $\Gamma^i = \overline{D}^i \setminus D^i$. Thus,

$$u_i^{[k]}(x) = u_i^{[k]}\big(x; \{u_l^{(k)}(\cdot)\}, 1 \leq l \leq i, u^{[k-1]}(\cdot)\big), \quad x \in \overline{D}^{(0)}, \tag{2.91e}$$

$$i = 1, \ldots, I.$$

The functions $u_i^{(k)}(x)$, $x \in \overline{D}^i$, and $u_i^{[k]}(x)$, $x \in \overline{D}^{(0)}$, are 2π-periodic in x_s, for $s = 2, \ldots, n$. The function $u^{[k]}(x)$, $x \in \overline{D}^{(0)}$, is determined by the relation

$$u^{[k]}(x) = u_I^{[k]}(x), \quad x \in \overline{D}^{(0)}. \tag{2.91f}$$

We say that the function $u_{(2.91f)}^{[k]}(x)$, $x \in \overline{D}^{(0)}$, is the solution on the k-th iteration *of the continual Schwartz method* (2.91), where $k \geq 1$.

We assume that the following *condition for the width of overlapping to subdomains* holds. Let D^i be the set from (2.91a), where $i = 1, \ldots, I$, and let $\Gamma^{i,(1)}$ be the part of the boundary to the set D^i that belongs to $D^{(1)}$. Let $D^{\{i\}} = \bigcup_{i_1} D^{i_1}$, where $i_1 = 1, \ldots, I$ and $i_1 \neq i$, be an union of the sets from (2.91a) which do not contain D^i, and let $\Gamma^{\{i\},(1)}$ be the part of the boundary to the set $D^{\{i\}}$ that belongs to $D^{(1)}$. The minimum of the minimal distance between the sets $\Gamma^{i,(1)}$ and $\Gamma^{\{i\},(1)}$ for all i, where $i = 1, \ldots, I$ (their infimum), is *the minimal overlapping of the subdomains* $\{D^i \cap D^{(1)}\}$, *covering* $D^{(1)}$. Assume that

the minimal overlapping of subdomains,
covering $D^{(1)}$, *is independent of* ε.

This condition ensures a convergence rate of *the iterative Schwartz method* (as the number of iteration grows) which is independent of the value of the parameter ε. About the iterative method with such convergence rate, we say that *the method* (with growth of the iteration number) *converges* ε-*uniformly*.

Theorem 2.5.1 *The function* $u^{[k]}_{(2.91)}(x)$, *i.e., the solution of the continual Schwartz method* (2.91) *for the boundary value problem* (2.2), (2.4), *converges to the solution of the boundary value problem* ε-*uniformly at the rate of a geometric progression as* $k \to \infty$:

$$|u(x) - u^{[k]}(x)| \le M\,q^k, \quad x \in \overline{D}^{(0)}, \quad q \le 1 - m. \qquad (2.92)$$

In (2.92) and in what follows, the constant M is independent of k, and the value q is independent of ε.

For the Schwartz method (2.91) we construct a difference scheme and formulate conditions which ensure ε-uniform convergence of this scheme.

For simplicity, we assume that the sets D^i are chosen in the form of rectangular parallelepipeds, whose faces are formed by the coordinate planes $x_s = $ const, for $s = 1, \ldots, n$. Under the condition that $\overline{D}^i \cap \Gamma = \emptyset$, the distance between the sets D^i and the boundary Γ is independent of ε.

May happen that the inequality (2.71) is violated on $\overline{D}$, but a similar inequality is satisfied pointwisely:

$$|a_{sk}(x)| < (n-1)^{-1} \left[\alpha_s(x)\,\alpha_k^{-1}(x)\,a_{ss}(x), \ \alpha_k(x)\,\alpha_s^{-1}(x)\,a_{kk}(x) \right],$$

$$x \in \overline{D}, \quad s,k = 1, \ldots, n, \quad s \ne k, \qquad (2.93)$$

moreover,

$$\max_{s,k,\overline{D}} \left[\alpha_s(x)\,\alpha_k^{-1}(x) \right] \le M, \quad s,k = 1,\ldots,n, \quad s \ne k,$$

i.e., we have *pointwise dominance of the diagonal elements* in the matrix of the coefficients of the operator $L^*_{(2)}$ (or, *the diagonal elements* of the elliptic part of the differential equation).

In this case, one can choose the diameter of the sets $\overline{D}^i$ sufficiently small such that the following inequality takes place on $\overline{D}^i$

$$\max_{\overline{D}^i} |a_{sk}(x)| < (n-1)^{-1} \min_{\overline{D}^i} \left[\alpha_s^i\,(\alpha_k^i)^{-1}\,a_{ss}(x), \ \alpha_k^i\,(\alpha_s^i)^{-1}\,a_{kk}(x) \right],$$

$$s,k = 1,\ldots,n, \quad s \ne k, \qquad (2.94)$$

that ensures a transformation of the operator $L^*_{(2)}$ to almost canonical form on $\overline{D}^i$ and also a transformation of the operator $L^{*[1]}_{(2)}$ to strongly almost canonical form on $\overline{D}^i$ if $\overline{D}^i \cap \Gamma \neq \emptyset$ and the coefficients $a_{1s}(x)$ on the boundary Γ satisfy the condition (2.57).

Otherwise, for sufficiently small diameter (independent of ε) of the set $\overline{D}^i$, it is possible to transfer to new variables $X^i = X^i(x)$, in which the elliptic operator containing the second-order derivatives (we denote it by $L^*_{X^i(2)}$) will be almost canonical form on the set $\overline{D}^i_{X^i}$, i.e., the set $\overline{D}^i$ in the variables X^i. But if $\overline{D}^i \cap \Gamma \neq \emptyset$, then new variables can be chosen such that the truncated operator $L^{*[1]}_{X^i(2)} = L^{*[1]}_{X^i(2)}(L^*_{X^i(2)})$ on the face $\Gamma^{iL}_{X^i}$, where $\Gamma^{iL} = \Gamma \cap \Gamma^i$, $\Gamma^i = \overline{D}^i \setminus D^i$, is canonical with respect to a variable orthogonal to the face $\Gamma^{iL}_{X^i}$ and, besides, on the set $\overline{D}^i_{X^i}$ is strongly almost canonical with respect to variables orthogonal to the face $\Gamma^{iL}_{X^i}$. The coefficients of the operator $L^*_{X^i(2)}$ on the set $\overline{D}^i_{X^i}$ satisfy relations similar to (2.55) for $\overline{D}^i \cap \Gamma = \emptyset$ and satisfy the relations $\{(2.55), (2.65)\}$ for $\overline{D}^i \cap \Gamma \neq \emptyset$; moreover, in (2.55) and $\{(2.55), (2.65)\}$ the set $\overline{Q}(x^*)$ is $\overline{D}^i_{X^i}$. Such approach is considered in Subsection 2.5.2.

Here we assume that the coefficients $a_{1s}(x)$ satisfy the condition (2.57). We consider that the operator $L^*_{(2)}$ is locally almost canonical on $\overline{D}$ and almost canonical on the subdomains $\overline{D}^i$, and the operator $L^{*[1]}_{(2)}$ under the condition $\overline{D}^i \cap \Gamma \neq \emptyset$ is strongly almost canonical. We also suppose that the condition (2.64), (2.67) holds.

Furthermore, we assume to be fulfilled the condition

> *the coefficients $a_{sk}(x)$ of the operator $L^*_{(2)}$ on the set $\overline{D}^i$* $\qquad$ (2.95)
>
> *satisfy the relations*
>
> *(2.94), (2.57), $\{(2.55), (2.65)\}$, (2.64), (2.67) if $\overline{D}^i \cap \Gamma \neq \emptyset$,*
>
> *and the relations (2.55) if $\overline{D}^i \cap \Gamma = \emptyset$,*
>
> *where $\overline{Q}(x^*) = \overline{D}^i$, $i = 1, \ldots, I$,*

that admits the construction of consistent grids in the variables $x_1, \ldots, x_n$ with the monotonicity condition for the difference operator $\Lambda^*_{(2)(2.51)}(L^*_{(2)})$ on $\overline{D}^i$. Note that the *grids* defined on the subdomains $\overline{D}^i$ and $\overline{D}^j$, *on the intersection of the subdomains $\overline{D}^i \cap \overline{D}^j$, for $i \neq j$, in general, do not coincide.* That requires a special organization of interaction to the grid solutions on the boundaries of subdomains.

Taking account of the condition (2.95) on the set $\overline{D}^i$ we construct the consistent grid

$$\overline{D}_h^i = \begin{cases} \overline{D}_{h(2.56)}\left(Q(x^*) = D^i\right) & \text{if } \overline{D}^i \cap \Gamma = \emptyset, \\ \overline{D}_{h(2.70)}\left(Q(x^*) = D^i\right) & \text{if } \overline{D}^i \cap \Gamma \neq \emptyset, \quad i = 1,\dots,I, \end{cases} \tag{2.96}$$

that ensures the monotonicity of the operators $\Lambda_{(2)}^* = \Lambda_{(2)(2.51)}^*(L_{(2)}^*)$ and $\Lambda = \Lambda_{(2.46)}(L)$ on $\overline{D}_h^i$. The nodes of the grid $\overline{D}_h^i$ belong to $\overline{D}^i \cap \Gamma$ if this set is not empty. We assume that the sets $\overline{D}^i$ and the grids $\overline{D}_h^i$ are 2π-periodic in x_s, for $s = 2,\dots,n$.

The *grid* $\overline{D}_{h(2.96)}^i$ on the set $\overline{D}^i$ is constructed *based on the grid* $\overline{D}_{h(2.56)}$ or $\overline{D}_{h(2.70)}$ *(introduced on the set* $\overline{D}_{(2.4)}^{(0)}$*)* and on the basis, respectively, the relations (2.55) or {(2.55), (2.65)}, (2.67) for $\overline{Q}(x^*) = \overline{D}^i$. Note that any change of the distribution of nodes in the meshes $\overline{\omega}_s$ (generating $\overline{D}_{h(2.96)}^i$) outside the projection of the set $\overline{D}^i$ on the x_s-axis does not affect the distribution of nodes in the grid $\overline{D}_h^i$. In general, $\overline{D}_h^i \cap \{\overline{D}^i \cap \overline{D}^j\} \neq \overline{D}_h^j \cap \{\overline{D}^i \cap \overline{D}^j\}$ for $\overline{D}^i \cap \overline{D}^j = \emptyset$, where $i \neq j$.

On the slab $\overline{D} = \overline{D}_{(2.4)}^{(0)}$ on the basis of the grids $\overline{D}_h^i = \overline{D}_{h(2.96)}^i$, for $i = 1,\dots,I$, we construct the grid

$$\overline{D}_h = \overline{D}_h^{\{0\}}, \tag{2.97a}$$

where we set

$$\overline{D}_{(1)h} = \overline{D}_h^1, \quad \overline{D}_{(i)h} = \overline{D}_h^i \cup \left\{\overline{D}_{(i-1)h} \setminus \overline{D}^i\right\}, \quad i = 2,\dots,I, \tag{2.97b}$$

$$\overline{D}_h^{\{0\}} = \overline{D}_{(I)h}.$$

On the slab $\overline{D}^{(0)}$ we also introduce a set of the grids $\overline{D}_h^{\{i\}}$, for $i = 1,\dots,I$, i.e.,

$$\overline{D}_h^{\{i\}} = \overline{D}_h^i \cup \{\overline{D}_h^{\{i-1\}} \setminus \overline{D}^i\}, \quad i = 1,\dots,I. \tag{2.97c}$$

It is not difficult to see that $\overline{D}_h^{\{0\}} = \overline{D}_h^{\{I\}}$.

On the grid $\overline{D}_{h(2.97a)}$ we construct a sequence of grid functions $z^{[k]}(x)$, for $k = 1,2,3,\dots$, in the following way. Set $z^{[0]}(x) \equiv 0$, $x \in \overline{D}_h$. We construct $z^{[k]}(x)$ assuming that the function $z^{[k-1]}(x)$, $x \in \overline{D}_h$, for $k \geq 1$, has already been constructed. Let us consider the auxiliary functions $z_i^{[k]}(x)$ defined on the grids $\overline{D}_{h(2.97c)}^{\{i\}}$:

$$z_i^{[k]}(x) = z_i^{[k]}\big(x; \, z_i^{(k)}(\cdot), \, z_{i-1}^{[k]}(\cdot)\big) \equiv$$

$$\equiv \left\{ \begin{array}{ll} z_i^{(k)}(x), & x \in \overline{D}_h^i \\ z_{i-1}^{[k]}(x), & x \in \overline{D}_h^{\{i-1\}} \setminus \overline{D}^i \end{array} \right\}, \quad x \in \overline{D}_h^{\{i\}}, \qquad (2.98a)$$

$$z_0^{[k]}(x) = z^{[k-1]}(x), \quad x \in \overline{D}_h, \quad i = 1, \ldots, I;$$

here the values i and k define the number of interior and exterior iterations. In these relations $z_i^{(k)}(x)$, $x \in \overline{D}_h^i$, is the solution of the discrete boundary value problem on the set $\overline{D}_h^i$

$$\Lambda_i \, z_i^{(k)}(x) = f(x), \quad x \in D_h^i; \qquad (2.98b)$$

$$z_i^{(k)}(x) = \left\{ \begin{array}{ll} \varphi(x), & x \in \Gamma_h^i \cap \Gamma \\ \check{z}_{i-1}^{[k]}(x), & x \in \Gamma_h^i \setminus \Gamma \end{array} \right\}, \quad x \in \Gamma_h^i, \quad i = 1, \ldots, I.$$

Here Λ_i is $\Lambda = \Lambda_{(2.46)}(L)$ on the grid $\overline{D}_h^i$,

$$\check{z}_0^{[k]}(x) = \check{z}^{[k-1]}(x), \quad x \in \Gamma_h^1 \setminus \Gamma; \qquad (2.98c)$$

the functions $\check{z}_{i-1}^{[k]}(x)$, $x \in \overline{D}$, and $\check{z}^{[k-1]}(x)$, $x \in \overline{D}$, are constructed as the interpolation of the functions $z_{i-1}^{[k]}(x)$, $x \in \overline{D}_h^{\{i-1\}}$, and $z^{[k-1]}(x)$, $x \in \overline{D}_h$, for $i = 1, \ldots, I$. Thus,

$$z_i^{[k]}(x) = z_i^{[k]}\big(x; \, \{z_l^{(k)}(\cdot)\}, \, 1 \leq l \leq i, \, z^{[k-1]}(\cdot)\big), \quad x \in \overline{D}_h^{(i)}, \qquad (2.98d)$$

$$i = 1, \ldots, I.$$

The function $z^{[k]}(x)$, $x \in \overline{D}_h$, is determined by the relation

$$z^{[k]}(x) = z_I^{[k]}(x), \quad x \in \overline{D}_h. \qquad (2.98e)$$

We say that the function $z^{[k]}(x)$, $x \in \overline{D}_h$, where $\overline{D}_h = \overline{D}_{h(2.97)}$, is k-th iteration *of the discrete Schwartz method* $\{(2.98), (2.97), (2.96)\}$, where $k \geq 1$.

The difference scheme $\{(2.98), (2.97), (2.96)\}$ is *the iterative scheme of the domain decomposition method on the overlapping subdomains.*

The functions $\check{z}_i^{[k]}(x)$ and $\check{z}^{[k]}(x)$, $x \in \overline{D}$, i.e., the interpolants used in (2.98b), (2.98c), are defined by the relations

$$\check{z}_i^{[k]}(x) = \left\{ \begin{array}{ll} \overline{z}_i^{(k)}(x), & x \in \overline{D}^i \\ \check{z}_{i-1}^{[k]}(x), & x \in \overline{D} \setminus \overline{D}^i \end{array} \right\}, \quad x \in \overline{D}, \quad i = 1, \ldots, I, \qquad (2.98f)$$

$$\check{z}_0^{[k]}(x) = \check{z}^{[k-1]}(x), \quad \check{z}^{[k]}(x) = \check{z}_I^{[k]}(x), \quad x \in \overline{D}, \ k \geq 1;$$

$$\check{z}^{[0]}(x) = 0, \quad x \in \overline{D}.$$

Here the interpolant $\overline{z}_i^{(k)}(x)$, $x \in \overline{D}^i$ is defined by the relation

$$\overline{z}_i^{(k)}(x) = \overline{z}_{i(2.42)}^{(k)}(x; \ z_i^{(k)}(\cdot), \overline{D}_h^i), \quad x \in \overline{D}^i.$$

Such interpolation preserves accuracy of the solution of the discrete problem (2.98b) that approximates the problem (2.91d).

The iterative difference scheme $\{(2.98), (2.97), (2.96)\}$ is defined by the *canonical elements* of the difference scheme of the domain decomposition method, i.e., the operator $\Lambda_{(2.46)}$ and the grids $\overline{D}_{h(2.56)}(\overline{\omega}_1^u)$ and $\overline{D}_{h(2.70)}$ $(\overline{\omega}_{1(2.48)})$ on the basis of which the difference scheme is constructed. Set $N^i = \min\limits_s N_s^i$ and $N = \min\limits_i N^i$, where $N_s^i + 1$ is the number of nodes in the grid $\overline{D}_h^i$ along the x_s-axis, for $s = 1, \dots, n$.

We now show a *reduced problem for the iterative difference scheme* $\{(2.98), (2.97), (2.96)\}$.

For $k \to \infty$ the discrete problem (2.98b) transforms into the following problem on the set $\overline{D}_h^i$

$$\Lambda_i \, z_i^0(x) = f(x), \quad x \in D_h^i; \tag{2.99a}$$

$$z_i^0(x) = \begin{cases} \varphi(x), & x \in \Gamma_h^i \cap \Gamma \\ \check{z}_{i-1}(x), & x \in \Gamma_h^i \setminus \Gamma \end{cases}, \quad x \in \Gamma_h^i, \ i = 1, \dots, I.$$

The function $z_i(x)$, $x \in \overline{D}_h^{\{i\}}$ that corresponds to the function $z_i^{[k]}(x)$ from (2.98a) is defined by the relation

$$z_i(x) = z_i\big(x; \ z_i^0(\cdot), z_{i-1}(\cdot)\big) \equiv$$

$$\equiv \begin{cases} z_i^0(x), & x \in \overline{D}_h^i \\ z_{i-1}(x), & x \in \overline{D}_h^{\{i-1\}} \setminus \overline{D}^i \end{cases}, \quad x \in \overline{D}_h^{\{i\}}, \ i = 1, \dots, I, \tag{2.99b}$$

$$z_0(x) = z_I(x), \quad x \in \overline{D}_h.$$

The interpolant $\check{z}_i(x)$, $x \in \overline{D}$, in (2.99a) is defined by the relation

$$\check{z}_i(x) = \begin{cases} \overline{z}_i^0(x), & x \in \overline{D}^i \\ \check{z}_{i-1}(x), & x \in \overline{D} \setminus \overline{D}^i \end{cases}, \quad x \in \overline{D}, \ i = 1, \dots, I, \tag{2.99c}$$

$$\check{z}_0(x) = \check{z}_I(x), \quad x \in \overline{D}.$$

Here $\bar{z}_i^0(x) = \bar{z}_{i(2.42)}^0(x; z_i^0(\cdot), \overline{D}_h^i)$, $x \in \overline{D}^i$, is an interpolant of the function $z_i^0(x)$, $x \in \overline{D}_h^i$. Thus,

$$z_i(x) = z_i(x; \{z_l(\cdot)\}, l = 1, \ldots, n, l \neq i), \quad x \in \overline{D}_h^{\{i\}}, \tag{2.99d}$$

$$i = 1, \ldots, I.$$

We define the functions $z(x)$, $x \in \overline{D}_h$, by the relation

$$z(x) = z_I(x), \quad x \in \overline{D}_h; \tag{2.99e}$$

the interpolant of the discrete function $z(x)$, $x \in \overline{D}_h$, is defined by the relation

$$\check{z}(x) = \check{z}_I(x), \quad x \in \overline{D}. \tag{2.99f}$$

The function $z(x)$, $x \in \overline{D}_h$, is the solution of the difference scheme $\{(2.99),$ $(2.97), (2.96)\}$, i.e., *the noniterative difference scheme of the domain decomposition method* which is reduced for the scheme $\{(2.98), (2.97), (2.96)\}$.

The noniterative difference scheme $\{(2.99), (2.97), (2.96)\}$ is defined by canonical elements, i.e., the operator $\Lambda_{(2.46)}$ and the grids $\overline{D}_{h(2.56)}(\overline{\omega}_1^u)$ and $\overline{D}_{h(2.70)}(\overline{\omega}_{1(2.48)})$.

The schemes $(2.98), (2.97), (2.96)$ and $(2.99), (2.97), (2.96)$ are ε-uniformly monotone and approximate the boundary value problem $(2.2), (2.4)$ ε-uniformly.

We assume that the minimal overlapping of the subdomains $\{D^i \cap D^{(1)}\}$ is independent of ε. For $k \to \infty$ *the solution of the iterative scheme* $\{(2.98),$ $(2.97), (2.96)\}$ *converges to the solution of the noniterative scheme* $\{(2.99),$ $(2.97), (2.96)\}$ ε-*uniformly* with the estimate

$$|z(x) - z^{[k]}(x)| \leq M q^k, \quad x \in \overline{D}_h, \quad q \leq 1 - m. \tag{2.100}$$

Here and below the value q is independent of ε and N, and the constant M is independent of q.

For the solution of the boundary value problem $(2.2), (2.4)$, let the estimates of Theorem 2.2.1 be fulfilled for $K = 3$. Then the difference scheme $\{(2.99), (2.97), (2.96)\}$, where $\overline{D}_{h(2.70)} = \overline{D}_{h(2.70)}(\overline{\omega}_{1(2.48)} = \overline{\omega}_{1(2.44)})$ converges ε-uniformly with the estimate

$$|u(x) - z(x)| \leq M N^{-1} \ln N, \quad x \in \overline{D}_h, \tag{2.101}$$

as $N \to \infty$.

We have the following theorems on the convergence of the noniterative and iterative difference schemes of the domain decomposition method.

Theorem 2.5.2 *Let* $a_{sk}, b_s, c, c^0, f \in C(\overline{D})$, $s, k = 1, \ldots, n$, *the coefficients* $a_{sk}(x)$ *satisfy the conditions* $(2.93), (2.95)$, *and let for the solution of the problem* $(2.2), (2.4)$, *the estimates of Theorem 2.2.1 be satisfied for*

K = 3. Then the iterative difference scheme $\{(2.98), (2.97), (2.96)\}$ on the grid $\overline{D}_{h(2.97a,2.96)}$, where $\overline{\omega}_{1(2.48)}$ is $\overline{\omega}_{1(2.44)}$, converges ε-uniformly at the rate $\mathcal{O}\left(N^{-1}\ln N + q^k\right)$ as $N, k \to \infty$:

$$|u(x) - z^{[k]}(x)| \leq M[N^{-1}\ln N + q^k], \quad x \in \overline{D}_h.$$

Theorem 2.5.3 *The solution of the iterative difference scheme $\{(2.98), (2.97), (2.96)\}$, for $k \to \infty$, converges to the solution of the noniterative difference scheme $\{(2.99), (2.97), (2.96)\}$ ε-uniformly. In the case when the hypotheses of Theorem 2.5.2 hold, the solution of the noniterative difference scheme $\{(2.99), (2.97), (2.96)\}$ on the grid $\overline{D}_{h(2.97a,2.96)}$, where $\overline{\omega}_{1(2.48)}$ is $\overline{\omega}_{1(2.44)}$ converges ε-uniformly to the solution of the problem (2.2), (2.4) as $N \to \infty$. The discrete solutions satisfy the estimates (2.100), (2.101).*

The interpolants $\check{z}_{(2.99f)}(x)$ and $\check{z}^{[k]}_{(2.98f)}(x)$ of the solutions of the noniterative and iterative difference schemes converge ε-uniformly with the estimates

$$|u(x) - \check{z}(x)| \leq M\,N^{-1}\ln N, \quad x \in \overline{D},$$

$$|u(x) - \check{z}^{[k]}(x)| \leq M[N^{-1}\ln N + q^k], \quad x \in \overline{D}, \quad q \leq 1 - m.$$

Remark 2.5.1 Let k^f be *the number of iterations required for the solution of the iterative difference scheme of the decomposition method $\{(2.98), (2.97), (2.96)\}$ (where $\overline{\omega}_{1(2.48)}$ is $\overline{\omega}_{1(2.44)}$), i.e., the number of iterations on which the estimate*

$$|u(x) - z^{[k^f]}(x)| \leq M\,N^{-1}\ln N, \quad x \in \overline{D}_h,$$

holds. The number of iterations k^f is ε-uniformly bounded. The value k^f satisfies the estimate

$$k^f \leq M\ln N.$$

Remark 2.5.2 When constructing a difference scheme for the problem (2.2), (2.4), along with the grids (2.96) (constructed based on the grid $\overline{D}_{h(2.56)}$, $\overline{D}_{h(2.70)}$ introduced on the set $\overline{D}$) it is convenient to apply also consistent grids *(that use the grids $\overline{D}_{h(2.89)}$, constructed on $\overline{D}^i$)*

$$\overline{D}_h^i = \overline{D}_h^i(D^i, L^*_{(2)}; J^{i*}) = \tag{2.102}$$

$$= \overline{D}_{h(2.89)}\left(D^i, L^*_{(2)}; \overline{\omega}_1^i = \overline{\omega}_{1(2.88)}^i(\overline{\omega}_{1(2.48)}), J^{i*}_{(2.86)}\right), \quad i = 1, \ldots, I,$$

where $\overline{\omega}_{1(2.48)}$ is $\overline{\omega}_{1(2.44)}$. The statement of Theorem 2.5.2 is preserved in the case when, instead of the grid $\overline{D}_{h(2.96)}^i$, we use the grid $\overline{D}_{h(2.102)}^i$. ∎

2.5.2 A difference scheme for the boundary value problem in a domain with curvilinear boundary

We construct a difference scheme for *the boundary value problem* (2.2) in the case when $\overline{D}$ is *a domain with a curvilinear boundary*. The fulfillment of the condition of a dominance of the diagonal elements of the elliptic operator on whole set $\overline{D}$, i.e., the condition (2.71), as well as restrictions on the coefficients of the mixed derivatives on the domain boundary similar to the condition (2.57), is not assumed.

We consider the *problem* (2.2) *in a doubly-connected bounded domain*

$$\overline{D} = D \cup \Gamma, \quad \Gamma = \Gamma_1 \cup \Gamma_2, \quad \Gamma_1 \cap \Gamma_2 = \emptyset; \tag{2.103}$$

the boundaries Γ_1 and Γ_2 are assumed to be sufficiently smooth. The problem (2.2), (2.103) is equivalent to a periodical boundary value problem on the slab with a curvilinear boundary.

We construct a difference scheme using the *domain decomposition method*. First, we made some preliminary constructions.

Let $x^* \in \Gamma$, and let on the closure $\overline{Q}(x^*)$, i.e., a neighborhood $Q(x^*)$ of the point x^*, the boundary Γ be given by the equation $x_1 = x_1(x_2, \ldots, x_n)$, $x \in \overline{Q}(x^*)$. By transforming to the variables $X = X(x)$:

$$X_1 = x_1 - x_1(x_2, \ldots, x_n), \quad X_s = x_s, \quad s = 2, \ldots, n, \tag{2.104}$$

we make the boundary Γ "plane" in a neighborhood of the point x^* (in a neighborhood of the part to the boundary Γ, the tangent plane for which is not parallel to x_1-axis).

In the case when the neighborhood $Q(x^*)$ is sufficiently small, the system (2.104) can be uniquely solved with respect to $x_1, \ldots, x_n$ for $x \in \overline{Q}(x^*)$. We denote by $X^{-1}(X)$ the map inverse to $X(x)$. We assume that $\overline{D}^* = \overline{D} \cap \overline{Q}(x^*)$. For the functions $v(x)$, $W(X)$ and subdomains $D^0 \subseteq \overline{D}^*$ we will use the notation

$$v(x(X)) = v_X(X) = \{v(x)\}_X, \quad W(X(x)) = W_{X^{-1}}(x),$$
$$D_X^0 = \{D^0\}_X = X(D^0) = \{X : X^{-1}(X) \in D^0\}.$$

We assume that

$$\widetilde{D}_{X^{-1}} = X^{-1}(\widetilde{D}) = \{x : X(x) \in \widetilde{D}\},$$

where $\widetilde{D}$ is a subset of the set $\overline{D}_X^* = \{\overline{D}^*\}_X$.

Let the face $\{\Gamma \cap \overline{D}^*\}_X$ be belong to the plane $X_1 = 0$, and let the set $\overline{D}_X^*$ belong to the half-space $X_1 \geq 0$. We choose the neighborhood $Q(x^*)$ such that the set $\overline{D}_X^i = \{\overline{D}^i\}_X$ (which belongs to $\overline{D}_X^*$) is a rectangular parallelepiped whose faces are formed by the coordinate planes $X_s = \text{const}$, for $s = 1, \ldots, n$; here $\overline{D}^i \cap \Gamma \neq \emptyset$. We denote the face $\{\Gamma \cap \overline{D}^i\}_X$ by Γ_X^{iL}; $\Gamma^{iL} = \Gamma \cap \Gamma^i$, $\Gamma^i = \overline{D}^i \setminus D^i$.

In the new variables, the problem (2.2), (2.103) on the set $D_X^i \cup \Gamma_X^{iL}$ is transformed into the subproblem

$$L_X U(X) = F(X), \quad X \in D_X^i, \tag{2.105a}$$

$$U(X) = \Phi(X), \quad X \in \Gamma_X^{iL}. \tag{2.105b}$$

Here

$$L_X \equiv \varepsilon^2 L_{2X} + L_{0X}, \quad L_{0X} = -C^0(X),$$

$$L_{2X} \equiv \sum_{s,k=1}^n A_{sk}(X) \frac{\partial^2}{\partial X_s \partial X_k} + \sum_{s=1}^n B_s(X) \frac{\partial}{\partial X_s} - C(X);$$

set

$$L_{X(2)}^* \equiv \sum_{s,k=1}^n A_{sk}(X) \frac{\partial^2}{\partial X_s \partial X_k}. \tag{2.105c}$$

The coefficients A_{sk} and B_s are determined from the formulae

$$A_{sk}(X) = \left\{ \sum_{r,p=1}^n a_{rp}(x) \frac{\partial}{\partial x_r} X_s \frac{\partial}{\partial x_p} X_k \right\}_X, \quad s, k = 1, \ldots, n;$$

$$B_s(X) = \left\{ \sum_{r=1}^n b_r(x) \frac{\partial}{\partial x_r} X_s + \sum_{r,p=1}^n a_{rp}(x) \frac{\partial^2}{\partial x_r \partial x_p} X_s \right\}_X, \quad s = 1, \ldots, n.$$

The functions U, C, C^0, F, and Φ are defined by the relation $V(X) = v_X(X)$, where $v(x)$ is one of the functions $u(x)$, $c(x)$, $c^0(x)$, $f(x)$, $\varphi(x)$. Note that $A_{sk} \in C^1(\overline{D}_X^i)$, $B_s \in C(\overline{D}_X^i)$ under the condition $a_{sk} \in C^1(\overline{D}^i)$, $b_s \in C(\overline{D}^i)$, $\Gamma \in C^2$.

The coefficients A_{sk} of the operator $L_{X(2.105)}$ satisfy the condition of strong ellipticity

$$A_0 \sum_{s=1}^n \xi_s^2 \leq \sum_{s,k=1}^n A_{sk}(X) \xi_s \xi_k \leq A^0 \sum_{s=1}^n \xi_s^2, \quad X \in \overline{D}_X^i, \quad A_0 > 0. \tag{2.106}$$

We assume the fulfillment of the canonicity condition on the boundaries Γ_X^{iL}

$$A_{1s}(X) = 0, \quad X \in \Gamma_X^{iL}, \quad s = 2, \ldots, n. \tag{2.107}$$

If the condition (2.107) is not fulfilled we pass to the new variables X^1, $X^1 = X^1(X)$:

$$X_1^1(X) = X_1, \quad X_s^1(X) = X_s - A_{11}^{-1}(0, X_2, \ldots, X_n) A_{1s}(0, X_2, \ldots, X_n) X_1,$$

$$s = 2, \ldots, n$$

and then consider the problem (2.2) on the set $\{D^i \cup \Gamma^{iL}\}_{XX^1}$ in these new variables. For the coefficients $A^1_{1s}(X^1)$ of the operator L_{XX^1}, one has the condition

$$A^1_{1s}(X^1) = 0, \quad X^1 \in \Gamma^{iL}_{XX^1}, \quad s = 2, \ldots, n.$$

In this case we choose the neighborhood $Q(x^*)$ such that the set $\{D^i\}_{XX^1}$ is a rectangular parallelepiped whose faces are formed by the coordinate planes $X^1_s = \text{const}$, where $s = 1, \ldots, n$.

Assume that under the condition $\Gamma \cap \overline{D}^i \neq \emptyset$ the operator $L^*_{X(2)(2.105)}$ defined on the set $\overline{D}^i_X$ has almost canonical form in $X_1, \ldots, X_n$ on $\overline{D}^i_X$ (for sufficiently small diameter of the set $\overline{D}^i$), and the operator $L^{*[1]}_{X(2)} = L^{*[1]}_{X(2)}(L^*_{X(2)})$ (i.e., the truncated operator involving differentiation in the variables $X_2, \ldots, X_n$; see (2.62)) has *strongly almost canonical form* in $X_2, \ldots, X_n$ in a neighborhood of Γ^{iL}_X. Let the coefficients of the operator $L^*_{X(2)}$ defined on $\overline{D}^i_X = \{X : d^i_{*s} \leq X_s \leq d^{i*}_s, \; s = 1, \ldots, n\}$ satisfy the condition (similar to (2.87)):

the coefficients $A_{sk}(X)$ of the operator $L^*_{X(2)}$ on the set $\overline{D}^i_X$

satisfy the relations (2.78), $\{(2.55), (2.79)\}$ if $J^{i*} \neq \emptyset$,

and the relations (2.55) if $J^{i*} = \emptyset$, (2.108)

where $a_{sk}(x), \ldots, \overline{Q}(x^*)$ is $A_{sk}(X), \ldots, \{\overline{Q}(x^*)\}_X$,

under the condition $\{\overline{Q}(x^*)\}_X = \overline{D}^i_X, \quad J^{i*} = J^{i*}(D^i_X) = \{j = 1\}.$

Here the condition $J^{i*}(D^i_X) = \{j = 1\}$ means that when constructing piecewise-uniform grids $\overline{D}^i_{Xh}$ on the set $\overline{D}^i_X$, the piecewise-uniform mesh $\overline{\omega}_{X1}$ in the X_1-axis (orthogonal to the face Γ^{iL}_X) condenses (for small ε) in a neighborhood of the left endpoint of the interval $[d^i_{*1}, d^{i*}_1]$.

Note that on the face Γ^{iL}_X the condition (2.107) holds (i.e., the condition (2.78), where the coefficients $a_{1k}(x)$ on the set Γ_j, for $j \in J^*$, are the coefficients $A_{1k}(X)$ on the set Γ^{iL}_X). In the case of the operator $L^*_{X(2)}$, the condition (similar to (2.55c) for $\overline{Q}(x^*) = \overline{D}^i$, where $\overline{D}^i \subseteq \overline{D}$) imposed on the parameters h_s takes the form

$$h_s \in \left[\max_{\overline{D}^i_X} \rho_{*s}(X)\, h^*, \; \min_{\overline{D}^i_X} \rho^*_s(X)\, h^* \right], \quad s = 1, \ldots, n. \quad (2.109)$$

The fulfillment of the condition (2.108) allows us to construct on the set $\overline{D}^i_X$ a piecewise-uniform grid that ensures the monotonicity of the operators $\Lambda^*_{X(2)} = \Lambda^*_{X(2)(2.51)}\left(L^*_{X(2)(2.105c)}\right)$ and $\Lambda_X = \Lambda_{X(2.46)}\left(L_{X(2.105a)}\right)$. The operators $\Lambda^*_{X(2)(2.51)}$ and $\Lambda_{X(2.46)}$ are constructed using the operators $L^*_{X(2)}$ and L_X, respectively.

Let the conditions $a_{sk} \in C^1(\overline{D})$ and $\Gamma \in C^2$ hold, and let also the conditions (2.108) and (2.82), where $J^* = J^{i*}(D_X^i) = \{j = 1\}$, imposed on the parameters of the piecewise-uniform mesh $\overline{\omega}_{X1} = \overline{\omega}_{X1}^l$ and on the parameters of uniform meshes $\overline{\omega}_{Xs} = \overline{\omega}_{Xs}^u$, for $s = 2, \ldots, n$, where $\overline{\omega}_{Xs} = \overline{\omega}_{Xs}(d_{*s}^i, d_s^{i*})$, for $s = 1, \ldots, n$, be fulfilled. Then the following piecewise-uniform grid (the grid (2.89), where $J^* = \{j = 1\}$),

$$\overline{D}_{Xh}^i = \overline{D}_{Xh}^i(\{Q(x^*)\}_X = D_X^i, L_{X(2)}^*; J^{i*}) = \tag{2.110}$$

$$= \overline{D}_{Xh(2.89)}^i\Big(\{Q(x^*)\}_X = D_X^i, L_{X(2)}^*; \overline{\omega}_{X1}^i = \overline{\omega}_{X1}^l, \overline{\omega}_{Xs}^i = \overline{\omega}_{Xs}^u,$$

for $s = 2, \ldots, n$, *under the conditions* (2.109), (2.82),

$$J^{i*} = \{j = 1\} \text{ in the case of the condition (2.108)}\Big)$$

is consistent (in the variables $X_1, \ldots, X_n$) on the set $\overline{D}_X^i$. The grid $\overline{D}_{Xh(2.110)}^i$ is generated by the meshes $\overline{\omega}_{X1}^l$ and $\overline{\omega}_{Xs}^u$, for $s = 2, \ldots, n$,

$$\overline{D}_{Xh(2.110)}^i = \overline{\omega}_{X1}^l \times \overline{\omega}_{X2}^u \times \ldots, \overline{\omega}_{Xn}^u.$$

For the subproblem (2.105) we construct a grid approximation, i.e., the monotone difference scheme

$$\Lambda_X Z(X) = F(X), \quad X \in D_{Xh}^i; \quad Z(X) = \Phi(X), \quad X \in \Gamma_{Xh}^{iL}. \tag{2.111}$$

Here $\Lambda_X = \Lambda_{X(2.46)}(L_X)$, and $\overline{D}_{Xh}^i = \overline{D}_{Xh(2.110)}^i$.

Returning to the variables $x = (x_1, \ldots, x_n)$, we obtain the monotone difference scheme

$$\Lambda^* z^*(x) = f(x), \quad x \in D_{Xh\,X^{-1}}^i; \quad z^*(x) = \varphi(x), \quad x \in \Gamma_{Xh\,X^{-1}}^{iL}. \tag{2.112}$$

Here
$$D_{Xh\,X^{-1}}^i = D^i \cap \overline{D}_{Xh\,X^{-1}}^i, \quad \Gamma_{Xh\,X^{-1}}^{iL} = \Gamma \cap \overline{D}_{Xh\,X^{-1}}^i,$$

$$\overline{D}_{Xh\,X^{-1}}^i = \{\overline{D}_{Xh}^i\}_{X^{-1}}; \quad z^*(x) = Z(X(x)), \quad x \in \overline{D}_{Xh\,X^{-1}}^i.$$

Note that *the grid* $\overline{D}_{Xh\,X^{-1}}^i$ *is*, in general, *not rectangular*.

We now give a finite difference scheme for the problem (2.2), (2.103).

Let the subdomains $\overline{D}^i$, for $i = 1, \ldots, I$, form a covering of the set $\overline{D}$; moreover, the subdomains $\overline{D}^i$, for $i = 1, \ldots, I_1$, adjoin to the boundary Γ, and the set $\overline{D}^0$ formed by the union $\cup_i \overline{D}^i$, for $i = I_1 + 1, \ldots, I$, has no common points with the boundary Γ.

Let the distance between the sets $\overline{D}^0$ and Γ be independent of ε.

We assume that *the condition for the width of overlapping to subdomains* holds, i.e., for all i, where $i = 1, \ldots, I$, the minimum of the minimal distance between the boundary of the set D^i and the boundary of the union of all sets which do not contain D^i (the minimal overlapping of subdomains that cover D) is independent of the value of the parameter ε.

The sets $\overline{D}^i$, for $i = I_1 + 1, \ldots, I$, which have no common points with the boundary Γ, are chosen as rectangular parallelepipeds formed by coordinate planes in the variables $x_1, \ldots, x_n$. The sets $\overline{D}^i$, for $i = 1, \ldots, I_1$, adjoining to the boundary Γ, in new variables $X^i = X^i(x)$ are the sets $\overline{D}^i_{X^i}$ which are rectangular parallelepipeds formed by coordinate planes in the variables $X^i_1, \ldots, X^i_n$.

We assume that the coefficients of the operator L_{X^i}, for $i = 1, \ldots, I_1$, on the set $\overline{D}^i_{X^i}$ satisfy the condition (2.108). We also consider that on the subdomains $\overline{D}^i$, for $i = I_1 + 1, \ldots, I$, the operator L is almost canonical; on these subdomains, let the following condition (similar to (2.87), where $J^* = \emptyset$) hold:

> the coefficients $a_{sk}(x)$ of the operator $L^*_{(2)}$ on the set $\overline{D}^i$
>
> satisfy the relations (2.55), $\hspace{3cm}$ (2.113)
>
> where $\overline{Q}(x^*) = \overline{D}^i$, for $i = I_1 + 1, \ldots, I$.

But if the operator L is not almost canonical on $\overline{D}^i$, for $i = I_1 + 1, \ldots, I$, then on the set $\overline{D}^i$ we transfer to new variables $X = X(x)$, in which the operator $L^*_{X(2)}$ has almost canonical form on $\overline{D}^i_X$ (for sufficiently small diameter $\overline{D}^i_X$, the coefficients of the operator $L^*_{X(2)}$ satisfy relations similar to (2.55) under the condition $\{\overline{Q}(x^*)\}_X = \overline{D}^1_X$ in the new variables).

These assumptions admit the construction of the grids $\overline{D}^i_{X^i h}$ and $\overline{D}^i_h$ that are consistent in the variables $X^i_1, \ldots, X^i_n$ and $x_1, \ldots, x_n$ with the monotonicity condition for the corresponding grid operators $\Lambda^*_{X^i(2)(2.51)}(L^*_{X^i(2)})$ and $\Lambda^*_{(2)(2.51)}(L^*_{(2)})$.

On the sets $\overline{D}^i$, for $i = 1, \ldots, I_1$, on the basis of the consistent grids $\overline{D}^i_{X^i h(2.110)}$, we construct the grids

$$\overline{D}^i_h = \overline{D}^i_{X^i h \{X^i\}^{-1}}, \quad i = 1, \ldots, I_1, \hspace{2cm} (2.114a)$$

and the monotone difference operators

$$\Lambda_i = \Lambda^*_{(2.112)} \quad \text{for} \quad x \in \overline{D}^i_{X^i h \{X^i\}^{-1}}, \quad i = 1, \ldots, I_1. \hspace{1cm} (2.114b)$$

Here $\overline{D}^i_{X^i h}$ are grids that are piecewise-uniform in the variable orthogonal to the boundary $\Gamma^{iL}_{X^i} = \{\Gamma \cap \overline{D}^i\}_{X^i}$ and uniform in other variables. On the sets $\overline{D}^i$, for $i = I_1 + 1, \ldots, I$, we introduce consistent grids that are uniform in all variables (similar to the grid (2.89), where $J^* = \emptyset$):

$$\overline{D}_h^i = \overline{D}_h^i\left(Q(x^*) = D^i, L_{(2)}^*; J^{i*}(D^i)\right) = \tag{2.114c}$$

$$= \overline{D}_{h(2.89)}^i\left(Q(x^*) = D^i, L_{(2)}^*, \ \overline{\omega}_s = \overline{\omega}_s^u, \ for \ s = 1,\ldots,n,\right.$$

$$\left.under \ the \ condition \ (2.55c); \ J^{i*} = \emptyset\right), \quad i = I_1 + 1,\ldots,I,$$

and the monotone difference operators

$$\Lambda_i = \Lambda_{(2.46)}(L_{(2.2a)}) \ for \ x \in D_h^i, \quad i = I_1 + 1,\ldots,I. \tag{2.114d}$$

By $N_s^i + 1$ we denote the number of nodes in the grid $\overline{D}_{X^ih}^i$, for $i = 1,\ldots,I_1$ (the grid $\overline{D}_h^i$, for $i = I_1 + 1,\ldots,I$), on the X_s^i-axis (respectively, on the x_s-axis), for $s = 1,\ldots,n$. Set $N^i = \min_s N_s^i$ and $N = \min_i N^i$.

To solve the problem (2.2), (2.103) in the domain with curvilinear boundary, we use the difference scheme (2.98), (2.97), where

$$\overline{D}_h^i = \begin{cases} \overline{D}_{h(2.114a)}^i & for \ i = 1,\ldots,I_1, \\ \overline{D}_{h(2.114c)}^i & for \ i = I_1 + 1,\ldots,I; \end{cases} \tag{2.114e}$$

$$\Lambda_i = \begin{cases} \Lambda_{i(2.114b)} & for \ i = 1,\ldots,I_1, \\ \Lambda_{i(2.114d)} & for \ i = I_1 + 1,\ldots,I. \end{cases} \tag{2.114f}$$

In this difference scheme unlike the scheme {(2.98), (2.97), (2.96)} for the problem (2.2), (2.103), the functions $\check{z}_i^{[k]}(x)$, $x \in \overline{D}$ for $k \geq 1$, used in (2.98b), now are defined by relations (different from (2.98f))

$$\check{z}_i^{[k]}(x) = \left\{ \begin{cases} \{\overline{Z}_i^{(k)}(X^i)\}_{\{X^i\}^{-1}}, & i = 1,\ldots,I_1 \\ \check{z}_i^{(k)}(x), & i = I_1 + 1,\ldots,I \\ \check{z}_{i-1}^{[k]}(x), & \end{cases} \begin{matrix} x \in \overline{D}^i \\ \\ x \in \overline{D} \setminus \overline{D}^i \end{matrix} \right\}, \tag{2.114g}$$

$$x \in \overline{D}, \quad i = 1,\ldots,I,$$

$$\check{z}_0^{[k]}(x) = \check{z}^{[k-1]}(x), \quad \check{z}^{[k]}(x) = \check{z}_I^{[k]}(x), \quad x \in \overline{D}, \quad k \geq 1;$$

$$\check{z}^{[0]}(x) = 0, \quad x \in \overline{D}.$$

Here the interpolant $\overline{Z}_i^{(k)}(X^i)$, $X^i \in \overline{D}_{X^i}^i$, is defined by the relation

$$\overline{Z}_i^{(k)}(X^i) = \overline{Z}_{i(2.42)}^{(k)}(X^i; z_i^{(k)}(\cdot), \overline{D}_{X^ih}^i), \quad X^i \in \overline{D}_{X^i}^i;$$

this *interpolant is piecewise-linear in the variables* X_s^i, for $s = 1,\ldots,n$, at the $(n+1\text{-vertex})$ polyhedral elements, i.e., partitions of elementary n-dimensional rectangular parallelepipeds formed by the nodes of the grid $\overline{D}_{X^ih}^i$.

The difference scheme of the domain decomposition method $\{(2.98), (2.97), (2.114)\}$ is determined by canonical elements, i.e., the operators $\Lambda_{(2.46)}$ (L) and $\Lambda_{X^i(2.46)}(L_{X^i})$ and the consistent uniform and piecewise-uniform grids $\overline{D}^i_{h(2.114c)}$ and $\overline{D}^i_{X^i h(2.114a)}$.

The difference scheme $\{(2.98), (2.97), (2.114)\}$ is ε-uniformly monotone and approximates the boundary value problem (2.2), (2.103) ε-uniformly.

The following convergence theorem holds.

Theorem 2.5.4 *Let* $a_{sk} \in C^1(\overline{D})$, b_s, c, c^0, $f \in C(\overline{D})$, *for* $s, k = 1, \ldots, n$, *and* $\Gamma \in C^2$, *the coefficients* $a_{sk}(x)$, $x \in \overline{D}^i$, *and* $A_{sk}(X^i)$, $X^i \in \overline{D}^i_{X^i}$, *satisfy, respectively, the conditions* (2.93), (2.113) *for* $i = I_1 + 1, \ldots, I$ *and* (2.107), (2.108), *where* $X = X^i$, *for* $i = 1, \ldots, I_1$, *and let for the solution of the problem* (2.2), (2.103), *the estimates of Theorem 2.2.2 be satisfied for* $K = 3$. *Then, on the grid* $\overline{D}_{h(2.97a,2.114e)}$, *in which the mesh* $\overline{\omega}^i_{X^i 1}$ *in* $\overline{D}^i_{X^i h}$, *for* $i = 1, \ldots, I_1$, *is defined by the mesh* $\overline{\omega}^i_{1(2.44)}$, *the difference scheme* $\{(2.98), (2.97), (2.114)\}$ *converges* ε-*uniformly as* N, $k \to \infty$:

$$|u(x) - z^{[k]}(x)| \le M \left(N^{-1} \ln N + q^k \right), \quad x \in \overline{D}_h, \quad q \le 1 - m.$$

In the case of the scheme $\{(2.98), (2.97), (2.114)\}$, the interpolant $\check{z}^{[k]}(x)$, $x \in \overline{D}$, of the solution $z^{[k]}(x)$, $x \in \overline{D}_h$, to the difference scheme of the decomposition method is defined by the relation

$$\check{z}^{[k]}(x) = \check{z}^{[k]}_{I(2.114g)}(x), \quad x \in \overline{D}, \quad k \ge 1.$$

For the interpolant $\check{z}^{[k]}(x)$ we have the estimate

$$|u(x) - \check{z}^{[k]}(x)| \le M \left(N^{-1} \ln N + q^k \right), \quad x \in \overline{D}.$$

For the solution of the difference scheme $\{(2.98), (2.97), (2.114)\}$, the statement of Remark 2.5.1 to Theorem 2.5.2 holds.

Remark 2.5.3 When solving differential subproblems on the sets $\overline{D}^i$, for $i = 1, \ldots, I_1$, in the decomposition method it is convenient to use the difference equations

$$\Lambda_i Z_i^{(k)}(X^i) = F(X^i), \qquad\qquad\qquad X^i \in D^i_{X^i h},$$

$$Z_i^{(k)}(X^i) = \begin{cases} \Phi(X^i), & X^i \in \Gamma^i_{X^i h} \cap \Gamma_{X^i}, \\ \check{Z}^{[k]}(X^i), & i = 2, \ldots, I_1 \\ \{\check{z}^{[k-1]}(x)\}_{X^i}, & i = 1 \end{cases}, \quad X^i \in \Gamma^i_{X^i h} \setminus \Gamma_{X^i},$$

$$i = 1, \ldots, I_1$$

not transferring to the original variables x. $\blacksquare$

Chapter 3

Boundary value problems for elliptic reaction-diffusion
equations in domains with piecewise-smooth boundaries

In this chapter the boundary value problem is considered for an elliptic reaction-diffusion equation in a domain with a piecewise-smooth boundary. In this problem in a neighborhood of the corner points, edges, and smooth parts of the boundary, the derivatives of the problem solution in each direction orthogonal to the boundary grow without boundary as the parameter ε tends to zero. To construct finite difference schemes, classical difference approximations of the corresponding differential operators are used. Sufficient conditions for ε-uniform convergence of the finite difference schemes are derived. The fulfillment of these conditions is ensured by choosing special grids that condense near the boundary and have a consistent distribution of nodes.

3.1 Problem formulation. The aim of the research

On an n-dimensional domain D with a piecewise-smooth boundary Γ, we consider the Dirichlet problem for the elliptic *reaction-diffusion equation*

$$L\,u(x) = f(x), \quad x \in D; \tag{3.1a}$$

$$u(x) = \varphi(x), \quad x \in \Gamma. \tag{3.1b}$$

Here

$$L \equiv \varepsilon^2 L_2 + L_0,$$

the operators L_2 and L_0 are defined by the relations (2.2c). The coefficients satisfying the conditions (2.3) and the right-hand side f in the differential equation are assumed to be sufficiently smooth. Let Γ^c be the set of vertices and edges of the domain D and Γ_j, for $j = 1, \dots, J$, be its curvilinear faces; here $\Gamma = \bigcup_j \Gamma_j$ and $\Gamma_j = \overline{\Gamma}_j$. Assume that the neighboring faces are intersected at nonzero angles and the domain D has not entering corners and edges. The function $\varphi(x)$ is sufficiently smooth on each of the sides Γ_j, for $j = 1, \dots, J$ and continuous on Γ. The fulfillment of other compatibility conditions on the set Γ^c is not assumed (under these assumptions, in general, one has $u \notin C^2(\overline{D})$).

We shall consider the boundary value problem either on a rectangular parallelepiped

$$D = \{x : d_{*s} < x_s < d_s^*, \quad s = 1, \ldots, n\}, \tag{3.2}$$

or in a domain with a piecewise-smooth curvilinear boundary.

Our **aim** for the boundary value problem (3.1) in a domain with a piecewise-smooth boundary is to construct a finite difference scheme that converges ε-uniformly.

3.2   Estimates of solutions and derivatives

We estimate the solution of the problem (3.1) on the rectangular parallelepiped $\overline{D}_{(3.2)}$. The estimates are established using *a priori* estimates [69, 37].

Let the functions $u(x)$, $x \in \overline{D}$ on the boundary Γ take values of $\varphi(x)$, moreover, on the set Γ^g the function $\varphi(x)$ is "good", i.e., $\varphi \in C^{l+\alpha}(\Gamma^g)$ and $\Gamma^g \in C^{l+\alpha}$, $\Gamma^g = \overline{\Gamma}^g$. The set Γ^g can coincide with Γ, or can be empty. We denote by $C^{l+\alpha}(D)$ the Banach space whose elements are continuous functions in $D \bigcup \Gamma^g$ taking "good" boundary values on the set Γ^g and having in $D \bigcup \Gamma^g$ continuous derivatives of order up to l which are in the sense of Hölder continuous with exponent α. For these elements a finite value is taken by the norm

$$|u|_{l+\alpha}^g = |u|_l^g + \sum_{\substack{k_1+\ldots+k_n=k \\ k=l}} H_\alpha^g \left(d^k \frac{\partial^k}{\partial x_1^{k_1} \ldots \partial x_n^{k_n}} u \right).$$

Here

$$|u|_l^g = \sum_{k=0}^{l} \sum_{k_1+\ldots+k_n=k} \left| d^k \frac{\partial^k}{\partial x_1^{k_1} \ldots \partial x_n^{k_n}} u \right|_0^g, \quad |u|_0^g = \sup_D |u(x)|,$$

$$H_\alpha^g(d^p u) = \sup_{x,\,x' \in D} d_{x,\,x'}^{p+\alpha} \frac{|u(x) - u(x')|}{|x - x'|^\alpha}, \quad |d^p v|_0^g = \sup_D |d_x^p v(x)|,$$

$$d_x = d_x^g = r(x, \Gamma \setminus \Gamma^g), \quad d_{x,\,x'} = d_{x,\,x'}^g = \min(d_x, d_{x'}).$$

We assume that the coefficients in the equation (3.1a) and the functions $f(x)$ and $\varphi(x)$ are sufficiently smooth $\overline{D}$ and Γ_j, and $\varphi(x)$ is continuous on Γ. Under these conditions, the problem solution is sufficiently smooth on D, however, $u \notin C^2(\overline{D})$ (see, e.g., [212, 69, 38]).

Using a maximum principle we establish the ε-uniform stability of the problem solution

$$|u(x)| \le M \max \left[\max_{\overline{D}} |f(x)|, \max_{\Gamma} |\varphi(x)| \right], \quad x \in \overline{D}. \tag{3.3}$$

Under the condition a_{sk}, b_s, c, c^0, $f \in C^{l+\alpha}(\overline{D})$, $\varphi \in C^1(\Gamma_j)$, for $j = 1, \ldots, 2n$, we have $u \in C^{\alpha_1}(\overline{D}) \cap C^{l+2+\alpha}(D)$, where $l \geq 0$ is an integer number and α, $\alpha_1 \in (0,1)$. From interior *a priori* estimates and estimates up to the smooth parts of the boundary Γ (derived in the variables ξ for the function $\widetilde{u}(\xi)$, see [69], where $\widetilde{u}(\xi) = u(x(\xi))$ is the solution of a regular equation in the variables ξ, $\xi = \xi(x)$ and $\xi_s = \varepsilon^{-1} x_s$, for $s = 1, \ldots, n$), it follows the estimate

$$\left| \frac{\partial^k}{\partial x_1^{k_1} \ldots \partial x_n^{k_n}} u(x) \right| \leq M \left[\varepsilon^{-k} + r^{-k}(x, \Gamma) \right], \tag{3.4}$$

$$x \in \overline{D}, \quad x \notin \Gamma \ \text{ for } k > 0, \quad 0 \leq k \leq K,$$

where $K = l + 2$. Furthermore, under the condition $\varphi \in C^{l_1}(\Gamma)$, for $l_1 \geq 2$, using a maximum principle, we establish the following estimate for the function $u(x)$:

$$|u(x) - \varphi(x^*(x))| \leq M\varepsilon^{-1} r(x, \Gamma), \quad x \in \overline{D}, \tag{3.5}$$

where $x^*(x)$ is a point on the boundary Γ nearest to the point $x \in \overline{D}$.

We shall study the behaviour of the solution in a neighborhood of the boundary layer and outside it. For convenience of presentation and formulation of the results obtained, we introduce some sets on which we shall consider auxiliary problems.

We assume that the faces

$$\Gamma_j = \Gamma_j(D), \quad j = 1, \ldots, 2n \tag{3.6a}$$

of the parallelepiped D for $j = s$ and $j = n + s$, where $s = 1, \ldots, n$, are orthogonal to the x_s-axis and the face Γ_s contains the vertex $(d_{*1}, \ldots, d_{*n})$; here $\Gamma_j = \overline{\Gamma}_j$, for $j = 1, \ldots, 2n$. By $D_{(j)}$ we denote the half-space $\{x : d_{*j} < x_j < \infty\}$ for $j \leq n$ and the half-space $\{x : -\infty < x_{j-n} < d^*_{j-n}\}$ for $j > n$. We set

$$D_{(j \ldots r)} = \bigcap_q D_{(q)}, \quad q = j, \ldots, r, \quad 1 \leq j, \ldots, r \leq 2n. \tag{3.6b}$$

The set $\Gamma_{(j \ldots r)} = \Gamma(D_{(j \ldots r)})$, i.e., the boundary of the set $D_{(j \ldots r)}$, is formed by the faces $\Gamma_q(D_{(j \ldots r)})$, where $\Gamma(D_{(j \ldots r)}) = \bigcup_q \Gamma_q(D_{(j \ldots r)})$, for $q = j, \ldots, r$; the faces $\Gamma_q(D_{(j \ldots r)})$ are defined similar to those $\Gamma_q = \Gamma_q(D)$, moreover, $\Gamma_q \subseteq \Gamma_q(D_{(j \ldots r)})$, for $q = j, \ldots, r$. Note that $D = D^{(1)}$ and $\Gamma_q = \Gamma_q(D^{(1)})$, for $q = 1, \ldots, 2n$, in the case when $D^{(1)} = D_{(1 \ldots 2n)}$.

The problem solution on the set $\overline{D}$ is decomposed into the sum

$$u(x) = U(x) + V(x), \quad x \in \overline{D}, \tag{3.7}$$

where $U(x)$ and $V(x)$ are the regular and singular components of the solution. The function $U(x)$, $x \in \overline{D}$, is the restriction of the function $U^e(x)$, $x \in D^e$, to the set $\overline{D}$, i.e., $U(x) = U^e(x)$, $x \in \overline{D}$; here, for simplicity, as the set D^e we choose an n-dimensional space $\mathbb{R}^n$. The function $U^e(x)$ is the bounded solution of the following problem on the unbounded domain D^e:

$$L^e U^e(x) = f^e(x), \quad x \in D^e. \tag{3.8}$$

The domain D^e is an extension of the domain D beyond the boundary Γ; the coefficients and right-hand side f^e of the equation (3.8) are smooth continuations of those in the equation (3.1a) with preserving their properties. Let $L^e = \varepsilon^2 \triangle - c$, and let the function $f^e(x)$ be vanished beyond an m-neighborhood of the set D. The function $V(x)$, $x \in \overline{D}$, is the solution of the problem

$$L V(x) = 0, \quad x \in D; \quad V(x) = \varphi_V(x), \quad x \in \Gamma,$$

where $\varphi_V(x) = \varphi(x) - U(x)$, $x \in \Gamma$.

Let us estimate the functions $U(x)$, $V(x)$ and their derivatives.

For the data of the problem (3.1), (3.2), let the condition a_{sk}, b_s, c, c^0, $f \in C^{l+\alpha}(\overline{D})$ be fulfilled for $l \geq 2$ and $\alpha \in (0,1)$. Then $U \in C^{l+\alpha}(\overline{D})$. Furthermore, let the condition $\varphi \in C^{l+\alpha}(\Gamma_j)$ hold for $j = 1, \ldots, 2n$. Then $V \in C^{\alpha_1}(\overline{D}) \cap C^{l+\alpha}(D)$, $\alpha_1 \in (0,1)$, moreover, $\varphi_V \in C^{l+\alpha}(\Gamma_j)$, i.e., $\Gamma_j = \Gamma_j^g$, for $j = 1, \ldots, 2n$, in the case of the function $V(x)$, $x \in \overline{D}$.

For the function $U(x)$, the estimate (2.11) holds:

$$\left| \frac{\partial^k}{\partial x_1^{k_1} \ldots \partial x_n^{k_n}} U(x) \right| \leq M \left[1 + \varepsilon^{2-k} \right], \quad x \in \overline{D}, \quad 0 \leq k \leq K. \tag{3.9}$$

We write the singular part $V(x)$ as the sum of functions

$$V(x) = V_0(x) + v(x), \quad x \in \overline{D}. \tag{3.10}$$

Here the functions $V_0(x)$ and $v(x)$, i.e., the main term of the singular part of the solution and the remainder term, are solutions of homogeneous equations.

We represent the function $V_0(x)$ as the sum of functions of type to regular and corner boundary layers:

$$V_0(x) = \sum_{j=1,\ldots,2n} V_{(j)}(x) + \sum_{\substack{j,\ldots,r=1,\ldots,2n \\ j<\ldots<r, \, 1<|j\ldots r|\leq n \\ \Gamma_j \cap \ldots \cap \Gamma_r \neq \emptyset}} V_{(j\ldots r)}(x), \quad x \in \overline{D}, \tag{3.11}$$

where $V_{(j)}(x)$ and $V_{(j\ldots r)}(x)$ are the regular and corner boundary layers; $|j\ldots r|$ is the number of components $j, \ldots, r$, which define the dimension of the layers (corner if $|j\ldots r| > 1$ and regular if $|j\ldots r| = 1$), here $1 \leq |j\ldots r| \leq n$, moreover, $\Gamma_j \cap \ldots \cap \Gamma_r \neq \emptyset$ under the condition $|j\ldots r| > 1$. The function $V_{(j\ldots r)}(x)$ is a restriction to $\overline{D}$ of the function $V_{(j\ldots r)}^e(x)$, $x \in \overline{D}^{(1)}$, where $D^{(1)} = D_{(j\ldots r)}$. The function $V_{(j\ldots r)}^e(x)$ is the solution of the problem

$$L^e V_{(j\ldots r)}^e(x) = 0, \qquad x \in D^{(1)}; \tag{3.12}$$

$$V_{(j\ldots r)}^e(x) = \varphi_{(j\ldots r)}^e(x), \quad x \in \Gamma(D^{(1)}),$$

where $L^e = L^e_{(3.8)}$, $j, \ldots, r = 1, \ldots, 2n$, $j < \ldots < r$, $1 \leq |j \ldots r| \leq n$ and $\Gamma_j \cap \ldots \cap \Gamma_r \neq \emptyset$; the functions $V^e_{(j \ldots r)}(x)$ exponentially decrease when they move away from the set $\Gamma(D^{(1)})$ for $|j \ldots r| = 1$ and from the set $\cap_q \{\Gamma_q(D^{(1)})\}$, $q = j, \ldots, r$ for $|j \ldots r| > 1$. The functions $\varphi^e_{(j \ldots r)}(x)$ are sufficiently smooth on the faces $D^{(1)}$, and on the faces $\Gamma_q \subset \Gamma$ they satisfy the condition

$$U(x) + \varphi^e_{(j)}(x) = \varphi(x), \quad x \in \Gamma_q, \quad q = j, \quad |j \ldots r| = 1;$$

$$U(x) + \sum_{q=j,\ldots,r} V_{(q)}(x) + \sum_{\substack{q,\ldots,k=j,\ldots,r \\ q<\ldots<k, \\ 1<|q\ldots k|\leq|j\ldots r|-1 \\ \Gamma_q\cap\ldots\cap\Gamma_k\neq\emptyset}} V_{(q\ldots k)}(x) + \varphi^e_{(j\ldots r)}(x) = \varphi(x),$$

$$x \in \cup_q \Gamma_q, \quad q = j, \ldots, r, \quad |j \ldots r| \geq 2.$$

For the functions $V(x)$, $V_{(j\ldots r)}(x)$ and $v(x)$ from (3.11) and (3.10) one has the estimates

$$|V(x)| \leq M \exp\left(-m\varepsilon^{-1} r(x, \Gamma)\right), \quad x \in \overline{D}; \tag{3.13a}$$

$$|V_{(j\ldots r)}(x)| \leq M \exp\left(-m\varepsilon^{-1} r(x, \cap_{q=j,\ldots,r}\Gamma_q)\right), \quad x \in \overline{D},$$

$$j, \ldots, r = 1, \ldots, 2n, \quad 1 \leq |j \ldots r| \leq n;$$

$$|v(x)| \leq M\varepsilon^2, \quad x \in \overline{D}.$$

The constant m is chosen arbitrarily from the interval $(0, m_0)$, where

$$m_0 = \left(a_0^{-1} c^0\right)^{1/2}, \tag{3.13b}$$

with $a_0 = a_{0(2.3)}$ and $c^0 = c^0_{0(2.3)}$.

We now find the estimates for the derivatives of $V_{(j\ldots r)}(x)$.

Taking into account interior *a priori* estimates and also the estimates of the functions $V^e_{(j\ldots r)}(x)$ similar to (3.13), we sequentially, for $|j \ldots r| = 1, 2, \ldots, n$, find the estimate of the derivatives for the functions $V^e_{(j\ldots r)}(x)$, $x \in \overline{D}_{(j\ldots r)}$,

$$\left|\frac{\partial^k}{\partial x_1^{k_1} \ldots \partial x_n^{k_n}} V^e_{(j\ldots r)}(x)\right| \leq M\left[\varepsilon^{-k} + r^{-k}(x, \Gamma(D_{(j\ldots r)}))\right] \times \tag{3.14}$$

$$\times \exp\left(-m\varepsilon^{-1} r(x, \cap_{q=j,\ldots,r}\Gamma_q(D_{(j\ldots r)}))\right),$$

$$x \in \overline{D}_{(j\ldots r)}, \quad x \notin \Gamma(D_{(j\ldots r)}) \text{ for } k > 0, \quad 0 \leq k \leq K,$$

where $K = l$.

Let us refine the estimate (3.14). We write the function $V^e_{(j\ldots r)}(x)$ as the decomposition

$$V^e_{(j\ldots r)}(x) = V^e_{(j\ldots r)0}(x) + v_{V_{(j\ldots r)}}(x), \quad x \in \overline{D}_{(j\ldots r)},$$

where $V^e_{(j...r)0}(x)$ and $v_{V_{(j...r)}}(x)$ are the main and remainder terms. The function $V^e_{(j...r)0}(x)$ is the solution of the problem

$$L^e_{(j...r)} V^e_{(j...r)0}(x) = 0, \qquad x \in D_{(j...r)}, \tag{3.15a}$$

$$V^e_{(j...r)0}(x) = \varphi^e_{(j...r)}(x), \qquad x \in \Gamma\big(D_{(j...r)}\big),$$

where

$$L^e_{(j...r)} \equiv \varepsilon^2 \left[\sum_{s,k} a^e_{sk}(x) \frac{\partial^2}{\partial x_s\, \partial x_k} + \sum_s b^e_s(x) \frac{\partial}{\partial x_s} - c^e(x) \right] - c^{0e}(x), \tag{3.15b}$$

$$s, k = \bar{j}, \dots, \bar{r}; \quad \bar{j} = \bar{j}(j),$$

$\bar{j} = j$ for $j \le n$ and $\bar{j} = j - n$ for $j > n$; here the variables x_s for $s \ne j, \dots, r$ appear as parameters. For the derivatives of $V^e_{(j...r)0}(x)$, taking into account (3.13), we obtain the estimate

$$\left| \frac{\partial^k}{\partial x_1^{k_1} \dots \partial x_n^{k_n}} V^e_{(j...r)0}(x) \right| \le M \left[\varepsilon^{-(\bar{k}_j + \dots + \bar{k}_r)} + \right. \tag{3.16}$$

$$\left. + r^{-(\bar{k}_j + \dots + \bar{k}_r)}\big(x,\, \Gamma(D_{(j...r)})\big) \right] \exp\left(-m\varepsilon^{-1} r\big(x,\, \cap_{q=j,\dots,r} \Gamma_q(D_{(j...r)})\big) \right),$$

$$x \in \overline{D}_{(j...r)}, \quad x \notin \Gamma(D_{(j...r)}) \quad \text{for } k > 0, \ k \le l.$$

Here the values $\bar{k}_p = \bar{k}_p(k_s)$ for $p = j, \dots, r$, $s = 1, \dots, n$, where $j, \dots, r = 1, \dots, 2n$, $|j \dots r| \ge 1$, $k = \sum_{s=1}^n k_s$, are defined by the following relations. In the case when $k_s \ne 0$ and either $p = s$ or $p = n + s$, we set $\bar{k}_p(k_s) = k_s$; otherwise, we set $\bar{k}_p(k_s) = 0$. The value $\bar{k}_p = \bar{k}_p(k_s)$ is the order of the derivative along the x_s-axis orthogonal to the face Γ_p, $p = j, \dots, r$ for $|j \dots r| \ge 1$.

For the function $v_{V_{(j...r)}}(x)$, using majorant functions, we find

$$\left| v_{V_{(j...r)}}(x) \right| \le M \varepsilon \exp\left(-m\varepsilon^{-1} r\big(x,\, \cap_{q=j,\dots,r} \Gamma_q(D_{(j...r)})\big) \right), \qquad x \in \overline{D}_{(j...r)};$$

for the derivatives we have the estimate

$$\left| \frac{\partial^k}{\partial x_1^{k_1} \dots \partial x_n^{k_n}} v_{V_{(j...r)}}(x) \right| \le M \varepsilon \left[\varepsilon^{-k} + r^{-k}\big(x,\, \Gamma(D_{(j...r)})\big) \right] \times$$

$$\times \exp\left(-m\varepsilon^{-1} r\big(x,\, \cap_{q=j,\dots,r} \Gamma_q(D_{(j...r)})\big) \right),$$

$$x \in \overline{D}_{(j...r)}, \quad x \notin \Gamma(D_{(j...r)}) \quad \text{for } k > 0, \ 0 \le k \le K, \ K = l.$$

Taking into account the estimates of the derivatives of the functions $V_{(j\ldots r)0}^e(x)$ and $v_{V_{(j\ldots r)}}(x)$, we obtain

$$\left| \frac{\partial^k}{\partial x_1^{k_1} \ldots \partial x_n^{k_n}} V_{(j\ldots r)}^e(x) \right| \le M \left\{ \varepsilon^{-(\overline{k}_j + \ldots + \overline{k}_r)} + r^{-(\overline{k}_j + \ldots + \overline{k}_r)}\left(x, \Gamma(D_{(j\ldots r)})\right) + \right.$$

$$\left. + \varepsilon \left[\varepsilon^{-k} + r^{-k}(x, \Gamma(D_{(j\ldots r)}))\right] \right\} \exp\left(- m\varepsilon^{-1} r\left(x, \cap_{q=j,\ldots,r} \Gamma_q(D_{(j\ldots r)})\right) \right),$$

$$x \in \overline{D}_{(j\ldots r)}, \quad x \notin \Gamma(D_{(j\ldots r)}) \quad \text{for } k > 0, \quad 0 \le k \le K, \tag{3.17}$$

where $K = l$ and $\overline{k}_p = \overline{k}_{p(3.16)}(k_s)$ with $p = j, \ldots, r$.

For the function $v(x)$ from (3.10) one has the estimate

$$\left| \frac{\partial^k}{\partial x_1^{k_1} \ldots \partial x_n^{k_n}} v(x) \right| \le M \varepsilon^2 \left[\varepsilon^{-k} + r^{-k}(x, \Gamma)\right], \tag{3.18}$$

$$x \in \overline{D}, \quad x \notin \Gamma \quad \text{for } k > 0, \quad 0 \le k \le K.$$

Next, we consider the problem solution $u(x)$ as the decomposition

$$u(x) = U_{(0)}(x) + V_0(x), \quad x \in \overline{D}, \tag{3.19}$$

where

$$U_{(0)}(x) = U(x) + v(x).$$

For the function $u(x)$ and its components $U_{(0)}(x)$ and $V_{(j\ldots r)}(x)$ in the representations (3.19), (3.11) we have the estimates

$$\left| \frac{\partial^k}{\partial x_1^{k_1} \ldots \partial x_n^{k_n}} u(x) \right| \le M \left[\varepsilon^{-k} + r^{-k}(x, \Gamma)\right], \tag{3.20}$$

$$\left| \frac{\partial^k}{\partial x_1^{k_1} \ldots \partial x_n^{k_n}} U_{(0)}(x) \right| \le M \left[1 + \varepsilon^{2-k} + \varepsilon^2 r^{-k}(x, \Gamma)\right],$$

$$\left| \frac{\partial^k}{\partial x_1^{k_1} \ldots \partial x_n^{k_n}} V_{(j\ldots r)}(x) \right| \le M \left\{ \varepsilon^{-(\overline{k}_j + \ldots + \overline{k}_r)} + r^{-(\overline{k}_j + \ldots + \overline{k}_r)}(x, \Gamma) + \right.$$

$$\left. + \varepsilon \left[\varepsilon^{-k} + r^{-k}(x, \Gamma)\right] \right\} \exp\left(- m\varepsilon^{-1} r(x, \cap_{q=j,\ldots,r} \Gamma_q) \right),$$

$$x \in \overline{D}, \quad x \notin \Gamma \quad \text{for } k > 0,$$

$$|u(x) - \varphi(x^*(x))| \le M \varepsilon^{-1} r(x, \Gamma), \quad x \in \overline{D};$$

$$j, \ldots, r = 1, \ldots, 2n, \quad 1 \le |j \ldots r| \le n, \quad 0 \le k \le K,$$

where $x^*(x) = x_{(3.5)}^*(x)$. In (3.20), $\overline{k}_p = \overline{k}_{p(3.16)}$, $p = j, \ldots, r$, $m = m_{(3.13)}$.
The following theorem holds.

Theorem 3.2.1 *Let a_{sk}, b_s, c, c^0, $f \in C^{l+\alpha}(\overline{D})$, for s, $k = 1, \ldots, n$, $\varphi \in C(\Gamma)$, $\varphi \in C^{l+\alpha}(\Gamma_j)$ for $j = 1, \ldots, 2n$, $l \geq K$, $K \geq 2$, $\alpha \in (0, 1)$. Then for the function $u(x)$ and its components $U_{(0)}(x)$ and $V_{(j \ldots r)}(x)$ in the representations (3.19), (3.11) of the solution to the boundary value problem (3.1), (3.2), the estimates (3.20) hold.*

In some problems for the function $u(x)$ one has the inclusion

$$u \in C^{K+\alpha}(\overline{D}), \quad \alpha \in (0, 1), \tag{3.21}$$

if the data of the problem (3.1), (3.2) on the set $\Gamma^c(D)$ satisfy special conditions. *Compatibility conditions* on the set $\Gamma^c(D)$ are special conditions imposed on the coefficients in the equation, its right-hand side and the boundary conditions in a neighborhood of the set $\Gamma^c(D)$ that guarantee membership of the solution to a class of functions having required smoothness in the neighborhood of the set $\Gamma^c(D)$, (see, e.g., [212, 213, 66]). The membership (3.21) allows us to obtain estimates of derivatives for the components in (3.19), (3.11) on the set $\overline{D}$. The smoothness of the data of the problem on the parallelepiped does not, in general, imply the smoothness of the solution on the set $\overline{D}$ (see, e.g., [38]).

In [212, 213] for Poisson equation $\Delta u = f(x)$ at $n = 2, 3$, imposed compatibility conditions are given for the functions $f(x)$ and $\varphi(x)$ $\big($here $u(x) = \varphi(x)$, $x \in \Gamma\big)$ on the set Γ^c that ensure the smoothness $u \in C^{4+\alpha}(\overline{D})$. For the problem (3.1), where

$$L \equiv \varepsilon^2 \triangle - c^0(x), \quad x \in \overline{D}, \quad r(x, \Gamma^c) \leq m\varepsilon, \tag{3.22}$$

on the basis of results from [212, 213], it is not difficult to write down compatibility conditions on the set $\Gamma^c(D)$ under which one has (3.21) (in the case $n = 2$ see, e.g., [180]). Applying *a priori* estimates of the solution up to the smooth parts of the boundary $\big($for $r\big(x, \Gamma^c(D_{(j \ldots r)})\big) \geq m_1 \varepsilon\big)$ and estimates of [212, 213] in a neighborhood of the set $\Gamma^c(D_{(j \ldots r)})$ $\big($for $r\big(x, \Gamma^c(D_{(j \ldots r)})\big) \leq m_2 \varepsilon$, for $m_2 > m_1\big)$, we justify that for the functions $V^e_{(j \ldots r)}(x)$ and $v(x)$ for $n = 2, 3$, the following estimates are valid:

$$\left| \frac{\partial^k}{\partial x_1^{k_1} \ldots \partial x_n^{k_n}} V^e_{(j \ldots r)}(x) \right| \leq M \left[\varepsilon^{-(\overline{k}_j + \ldots + \overline{k}_r)} + \varepsilon^{2-k} \right] \times \tag{3.23}$$

$$\times \exp\big(-m\varepsilon^{-1} r(x, \cap_{q=j, \ldots, r} \Gamma_q) \big), \quad x \in \overline{D}_{(j \ldots r)};$$

$$\left| \frac{\partial^k}{\partial x_1^{k_1} \ldots \partial x_n^{k_n}} v(x) \right| \leq M \varepsilon^{2-k}, \quad x \in \overline{D}, \quad 0 \leq k \leq K. \tag{3.24}$$

For $n \leq 3$ in the case of the condition (3.22) if (3.21) holds, then the functions $u(x)$, $U_{(0)}(x)$, and $V_{(j...r)}(x)$ satisfy the estimates

$$\left| \frac{\partial^k}{\partial x_1^{k_1} \dots \partial x_n^{k_n}} u(x) \right| \le M\varepsilon^{-k}, \tag{3.25}$$

$$\left| \frac{\partial^k}{\partial x_1^{k_1} \dots \partial x_n^{k_n}} U_{(0)}(x) \right| \le M \left[1 + \varepsilon^{2-k} \right],$$

$$\left| \frac{\partial^k}{\partial x_1^{k_1} \dots \partial x_n^{k_n}} V_{(j...r)}(x) \right| \le M \left[\varepsilon^{-(\bar{k}_j + \dots + \bar{k}_r)} + \varepsilon^{2-k} \right] \times$$

$$\times \exp \left(-m\varepsilon^{-1} r(x, \cap_{q=j,\dots,r} \Gamma_q) \right), \quad x \in \overline{D};$$

$$j, \dots, r = 1, \dots, 2n, \quad 1 \le |j \dots r| \le n, \quad 0 \le k \le K.$$

In (3.25) one has $\bar{k}_p = \bar{k}_{p(3.16)}$, where $p = j, \dots, r$, and $m = m_{(3.13)}$.
The following theorem holds.

Theorem 3.2.2 *Let $n \le 3$, the hypotheses of Theorem 3.2.1 hold, and also the condition (3.22) and the membership (3.21) be satisfied. Then for the function $u(x)$ and its components $U_{(0)}(x)$ and $V_{(j...r)}(x)$ in the representations (3.19), (3.11) of the solution to the boundary value problem (3.1), (3.2), the estimates (3.25) hold.*

Let us discuss estimates in the case of the problem (3.1) in a domain with a *piecewise-smooth boundary* having curvilinear faces. It is convenient to write the estimates of the solution in the case of the problem (3.1), (3.2) in a form different from the estimates (3.20), (3.25).
We write the solution $u(x)$ of the problem (3.1), (3.2) in an σ-neighborhood of the boundary Γ, where σ is sufficiently smooth, as the sum

$$u(x) = U_{(0)}(x) + V_0(x), \quad x \in \overline{D}, \quad r(x, \Gamma) \le \sigma, \tag{3.26}$$

where

$$U_{(0)}(x) = U_0(x) + v(x).$$

For the function $u(x)$ considered outside the σ-neighborhood of the boundary Γ and its components $U_{(0)}(x)$ and $V_{(j...r)}(x)$ in the representations (3.19), (3.11) considered only in the σ-neighborhood of the boundary Γ, by virtue of (3.20), we have the estimates

$$\left| \frac{\partial^k}{\partial x_1^{k_1} \dots \partial x_n^{k_n}} u(x) \right| \le M \left[1 + \varepsilon^{2-k} + \varepsilon^2 \sigma^{-k} \right], \quad r(x, \Gamma) \ge \sigma; \tag{3.27a}$$

$$\left| \frac{\partial^k}{\partial x_1^{k_1} \dots \partial x_n^{k_n}} U_{(0)}(x) \right| \le M \left[1 + \varepsilon^{2-k} + \varepsilon^2 r^{-k}(x, \Gamma) \right], \tag{3.27b}$$

$$\left| \frac{\partial^k}{\partial x_1^{k_1} \dots \partial x_n^{k_n}} V_{(j\dots r)}(x) \right| \le M \Big\{ \varepsilon^{-(\overline{k}_j + \dots + \overline{k}_r)} + \tag{3.27c}$$

$$+ r^{-(\overline{k}_j + \dots + \overline{k}_r)}(x, \Gamma) + \varepsilon \left[\varepsilon^{-k} + r^{-k}(x, \Gamma) \right] \Big\} \times$$

$$\times \exp \Big(- m\varepsilon^{-1} r \big(x, \cap_{q=j,\dots,r} \Gamma_q \big) \Big), \qquad x \notin \Gamma \ \text{for} \ k > 0,$$

$$|u(x) - \varphi(x^*(x))| \le M \varepsilon^{-1} r(x, \Gamma), \quad r(x, \Gamma) \le \sigma; \tag{3.27d}$$

$$x \in \overline{D}, \quad j, \dots, r = 1, \dots, 2n, \quad 1 \le |j \dots r| \le n, \quad 0 \le k \le K.$$

For $n \le 3$ in the case of the condition (3.22), by virtue of (3.25), for the function $u(x)$, $U_{(0)}(x)$ and $V_{(j\dots r)}(x)$ the estimates

$$\left| \frac{\partial^k}{\partial x_1^{k_1} \dots \partial x_n^{k_n}} u(x) \right| \le M \left[1 + \varepsilon^{2-k} \right], \quad r(x, \Gamma) \ge \sigma; \tag{3.28a}$$

$$\left| \frac{\partial^k}{\partial x_1^{k_1} \dots \partial x_n^{k_n}} U_{(0)}(x) \right| \le M \left[1 + \varepsilon^{2-k} \right], \tag{3.28b}$$

$$\left| \frac{\partial^k}{\partial x_1^{k_1} \dots \partial x_n^{k_n}} V_{(j\dots r)}(x) \right| \le M \left[\varepsilon^{-(\overline{k}_j + \dots + \overline{k}_r)} + \varepsilon^{1-k} \right] \times \tag{3.28c}$$

$$\times \left[\exp \Big(- m\varepsilon^{-1} r \big(x, \cap_{q=j,\dots,r} \Gamma_q \big) \Big) \right], \quad r(x, \Gamma) \le \sigma;$$

$$x \in \overline{D}, \quad j, \dots, r = 1, \dots, 2n, \quad 1 \le |j \dots r| \le n, \quad 0 \le k \le K$$

are valid. In (3.27), (3.28) one has $\overline{k}_p = \overline{k}_{p(3.16)}$, where $p = j, \dots, r$, and $m = m_{(3.13)}$.

Estimates of type (3.27), (3.28) take place also in the case of the problem (3.1) in a domain with a piecewise-smooth boundary, in particular, in the domain $\overline{D}$, which is an n-dimensional parallelepiped with curvilinear faces, when the hypotheses of Theorem 3.2.1, 3.2.2 are fulfilled if $\Gamma_j \in C^{l+\alpha}$, for $j = 1, \dots, 2n$. Outside the σ-neighborhood of the boundary Γ, the estimates (3.27a) and (3.28a) hold for the solutions of the boundary value problem. We cover the σ-neighborhood of the boundary Γ by finite set of subdomains. On each such subdomain we introduce a new coordinate system $x'_s = x'_s(x)$, for $s = 1, \dots, n$, in which the curvilinear faces become pieces of planes orthogonal to the coordinate axes. In a neighborhood of the set $\Gamma^{(j\dots r)}$, the variables $x'_s = x'_s(x)$, for $s = j, \dots, r$ are the distances from the point $x \in \overline{D}$, with $r(x, \Gamma) \le \sigma$, to the faces $\Gamma_{j_0}, \dots, \Gamma_{r_0}$ generating $\Gamma^{(j\dots r)}$. Similar to the constructions on the rectangular parallelepiped, we shall derive an estimate of the problem solution in variables x', $x' = (x'_1, \dots, x'_n)$ on subdomains from the σ-neighborhood of the boundary Γ. In a neighborhood of the boundary, the problem solution has a representation similar to (3.22). Let in the original variables the membership (3.21) hold and in the new variables the condition (3.22) be satisfied. Now we shall preserve previous notations on the partition subdomains from the σ-neighborhood of the domain boundary for the new variables and also for the domain and its boundary, for the solution of the boundary value problem and its regular and singular components.

Let the functions $U_{(0)}(x)$ and $V_{(j...r)}(x)$, which are the regular and singular parts of the solution, be determined as above. For these functions, the estimates (3.28b) and (3.28c), respectively, are valid in the case of additional conditions (3.21), (3.22), or the estimates (3.27b) and (3.27c) when these conditions are violated, moreover, the estimate (3.27d) is fulfilled for the solution of the boundary value problem. In (3.27c), (3.28c) in the case of the function $V_{(j...r)}(x)$ the value $\bar{k}_p = \bar{k}_p(k_s)$ is the order of the derivative along the x_s-axis orthogonal to the face Γ_p, for $p = j, \ldots, r$, from the σ-neighborhood of the domain boundary. The constant m in these estimates is chosen to satisfy the condition $m < m'_0$, where

$$m'_0 = \left[(a'_0)^{-1} c^{0\,\prime} \right]^{1/2}, \tag{3.29}$$

a'_0 and $c^{0\,\prime}$ are the constant from the ellipticity condition and the constant from the estimate of the reaction term in the differential operator written in the new variables in the σ-neighborhood of the boundary Γ.

Theorem 3.2.3 *For the data of the boundary value problem (3.1) in the domain $\overline{D}$, i.e., an n-dimensional parallelepiped with curvilinear faces, let the hypotheses of Theorem 3.2.1 be fulfilled and also the condition $\Gamma_j \in C^{l+\alpha}$, for $j = 1, \ldots 2n$. Then for the solution $u(x)$ and its components in the representation (3.26), the estimates (3.27), (3.29) hold; in the representation (3.26) and in the estimates (3.27b, c, d) the variable x is new. But if for $n \leq 3$ under the additional condition (3.21), the condition (3.22) holds, then the estimates (3.28), (3.29) are satisfied; in the relations (3.26), (3.22) and in the estimates (3.28b, c) the variable x is new.*

3.3 Sufficient conditions for ε-uniform convergence of a difference scheme for the problem on a parallelepiped

The solution of the boundary value problem (3.1), (3.2) does not, in general, belong to $C^2(\overline{D})$ and that does not allow us immediately to use results of Chapter 2 for the construction and justification of ε-uniformly convergent finite difference schemes. We shall study finite difference schemes on grids condensing in a neighborhood of the boundary layers and find sufficient conditions for ε-uniform convergence of the difference schemes. Assume that the solution of the problem (3.1) satisfies the condition

$$u \in C(\overline{D}) \cap C^3(D) \tag{3.30}$$

and the estimates (3.20) of Theorem 3.2.1.

On the parallelepiped $\overline{D}_{(3.2)}$ we introduce the rectangular grid

$$\overline{D}_h = \overline{\omega}_1 \times \ldots \times \overline{\omega}_n, \tag{3.31}$$

where $\overline{\omega}_s$, for $s = 1, \ldots, n$ is a mesh on the interval $[d_{*s}, d_s^*]$ at the x_s-axis. We set $h = \max_s h_s$, $h_s = \max_i h_s^i$, and $N = \min_s N_s$, where $N_s + 1$ is the number of nodes in the mesh $\overline{\omega}_s$, for $s = 1, \ldots, n$; assume that the condition $h \leq MN^{-1}$ holds. On the grid $\overline{D}_{h(3.31)}$ for the problem (3.1) we consider a difference scheme

$$\Lambda z(x) = f^h(x), \quad x \in D_h, \quad z(x) = \varphi^h(x), \quad x \in \Gamma_h, \tag{3.32}$$

and find sufficient conditions for its ε-uniform convergence.

In the case when the problem solution is sufficiently smooth on $\overline{D}$ for each fixed value of the parameter, the conditions sufficient for the ε-uniform convergence of the scheme (3.32), (3.31) (and for its stability) are given by Theorem 2.3.2 that follows from Theorem 2.3.1 and Lemma 2.3.3 (Lemma 2.3.3 provides the stability conditions).

If, however, $u \in C^{2+\alpha}(D)$, but $u \notin C^2(\overline{D})$ (ε-uniformly), then the difference operator Λ from the scheme (3.32), (3.31) approximating the operator L on the problem solution at each point $x \in D_h$

$$|(L - \Lambda)u(x)| \leq M\,\mu\,(h(x),\,x), \quad x \in D_h,$$

in general, does not approximate the operator L uniformly on $\overline{D}$. Here $\mu(h(x),\,x) \to 0$ as $h(x) \to 0$; $h(x)$ is the maximal mesh step in the stencil of the operator Λ centered at the point x (ε is fixed). Therefore, Theorems 2.3.1 and 2.3.2 and Lemma 2.3.3 that establish the ε-uniform convergence of the scheme turn out to be directly inapplicable for study of the ε-uniform convergence of the scheme (3.32), (3.31). Thus, a problem arises to justify ε-uniform convergence of the schemes in the case of boundary value problems whose solutions belong to the class $C^{2+\alpha}(D)$ but not belong to $C^2(\overline{D})$.

Let us discuss some concepts on the basis of which we shall formulate conditions that allow us to justify ε-uniform convergence of the schemes constructed.

Let $u \notin C^2(\overline{D})$, and let for the solution of the problem the estimates (3.20) hold. In that case, the estimates of the problem solution considered on the set $\overline{D}(\rho)$ from D, where

$$\overline{D}(\rho) = D(\rho) \cup \Gamma(D(\rho)), \quad D(\rho) = \{x : r(x, \Gamma) > \rho\}, \tag{3.33}$$

are the same as the estimates (3.25) in the case when $u \in C^3(\overline{D})$ up to a multiplying factor (dependent on ρ). For example, by virtue of (3.20), for the derivative $(\partial^3/\partial x_1^3)\, u(x)$ on $\overline{D}(\rho)$, we have

$$\left| \frac{\partial^3}{\partial x_1^3}\, u(x) \right| \le M(\widetilde{\rho})\, \varepsilon^{-3}, \quad x \in \overline{D}(\rho),$$

where $M(\widetilde{\rho}) = M_{(3.20)}(1+\widetilde{\rho}^3)$ and $\widetilde{\rho} = \varepsilon^{-1}\rho$ is the normalized distance to the boundary. If on the boundary $\Gamma(D(\rho))$ for not too small ρ the solution of the difference problem (3.32), (3.31) for $N \to \infty$ approximates the solution of the problem (3.1) ε-uniformly, one would use the conditions of Theorems 2.3.1 and 2.3.2 and Lemma 2.3.3 as sufficient conditions for ε-uniform convergence of the difference scheme (3.32), (3.31) on $\overline{D}(\rho)$.

To estimate an error of the solution on the set $\overline{D} \setminus D(\rho)$, we shall use the following. Assume that the solution of the problem (3.32), (3.31) satisfies the estimate similar to (3.5):

$$|z(x) - z(x^*(x))| \le M\,\varepsilon^{-1}\, r(x, \Gamma), \quad x \in \overline{D}_h,$$

where $x^*(x) = x^*_{(3.5)}(x)$ (in the case of monotone schemes the estimate is derived similar to (3.5)). Then for the error in the problem solution we have

$$|u(x) - z(x)| \le M\,\varepsilon^{-1}\, r(x, \Gamma), \quad x \in \overline{D}_h,$$

and, on the set $\overline{D} \setminus D(\rho)$ that is the ρ-neighborhood of the boundary Γ and also on the boundary $\Gamma(D(\rho))$, we obtain the estimate

$$|u(x) - z(x)| \le M\,\widetilde{\rho}, \quad x \in \overline{D} \setminus D(\rho), \quad x \in \overline{D}_h. \tag{3.34}$$

By choosing the value $\widetilde{\rho}$ (the value ρ) that satisfies the condition $\widetilde{\rho} = \widetilde{\rho}(N)$, where $\widetilde{\rho}(N) \to 0$ as $N \to \infty$, it is possible to satisfy the conditions of convergence of the scheme on the set $\overline{D}(\rho)$ that follow from Theorems 2.3.1 and 2.3.2 and Lemma 2.3.3 and, furthermore, to ensure the ε-uniform convergence of the scheme on $\overline{D} \setminus D(\rho)$ by virtue of the estimate (3.34) that leads to the ε-uniform convergence on $\overline{D}_h$.

Thus, the ε-uniform convergence principle:

> *ε-uniform convergence of a finite difference scheme follows from*
> *ε-uniform approximation of the boundary value problem by an*
> *ε-uniformly stable finite difference scheme*

transforms into the principle:

> *ε-uniform convergence of a finite difference scheme follows from*
>
> *(i) continuity in the variable $\widetilde{\rho}$ of the solutions to the boundary value problem and to the difference scheme on the domain boundary*

and

(ii) ε-uniform approximation and ε-uniform stability of the diffe-rence scheme outside an ρ-neighborhood of the boundary with con-stants in the error bounds of the approximation and stability, which depend on $\widetilde{\rho}$.

Keeping in mind such an approach, we obtain conditions under which the finite difference scheme converges ε-uniformly.

Let *the functions* $f^h(x)$, $x \in \overline{D}_h$, *and* $\varphi^h(x)$, $x \in \Gamma_h$, *approximate the functions* $f(x)$, $x \in \overline{D}$, *and* $\varphi(x)$, $x \in \Gamma$, ε-*uniformly on the sets* $\overline{D}_h$ *and* $\overline{\Gamma}_h$, respectively:

$$|f(x) - f^h(x)| \leq M\,\lambda(N), \quad x \in \overline{D}_h; \tag{3.35}$$

$$|\varphi(x) - \varphi^h(x)| \leq M\,\lambda(N), \quad x \in \Gamma_h, \tag{3.36}$$

where $\lambda(N) \underset{\varepsilon}{\to} 0$ as $N \to \infty$.

We say that *the operator* Λ *from the scheme* (3.32), (3.31) *acting on the solution of the problem* (3.1) *approximates the operator* $L_{(3.1a)}$ ε-*uniformly on the set* $D(\rho)$ $\big($*or* ε-*uniformly inside the set* $D\big)$ if the following inequality holds:

$$|(L - \Lambda)\,u(x)| \leq M\,\psi(\widetilde{\rho})\,\lambda(N), \quad x \in D_h \cap D(\rho), \tag{3.37}$$

where $M\,\psi(\widetilde{\rho})$ is a constant in the error bound of the approximation; in gen-eral, $\psi(\widetilde{\rho}) \to \infty$ as $\widetilde{\rho} \to 0$, and ψ is independent of ε.

Let us define *stability* of the finite difference scheme (3.32), (3.31) inside the set D.

Let D_h^1 be a subset of $\overline{D}_h$. We call *a point* $x \in \overline{D}_h$ *the nearest node to the set* D_h^1 if $x \notin D_h^1$ and this point x is a mesh node in the stencil of the difference scheme (3.32), (3.31) with the centre at any point ξ belonging to D_h^1. We denote the set of these nodes by

$$S(D_h^1), \tag{3.38}$$

and we call it *the boundary of the subset* D_h^1. We say that the difference scheme *or, the scheme* (3.32), (3.31) ε-*uniformly stable on the set* $D_{(3.33)}(\rho)$ *uniformly in* ρ $\big($or,ε-*uniformly stable inside the set* $D\big)$ if for the function $w(x)$, $x \in \{D_h \cap D(\rho)\} \cup S(D_h \cap D(\rho))$ that satisfies the relations

$$\Lambda w(x) = f_w^h(x), \ x \in D_h \cap D(\rho),$$
$$w(x) = \varphi_w^h(x), \ x \in S(D_h \cap D(\rho)),$$

the estimate holds

$$|w(x)| \leq M \max\left[\max_{D_h \cap D(\rho)} |f_w^h(x)|, \ \max_{S(D_h \cap D(\rho))} |\varphi_w^h(x)|\right], \tag{3.39}$$

$$x \in \Big\{D_h \cap D(\rho)\Big\} \cup S\Big(D_h \cap D(\rho)\Big),$$

where M is independent of ρ.

We assume that the error bound of the difference scheme satisfies the following *condition of convergence in the nearest neighborhood of the boundary*

$$|u(x) - z(x)| \leq M\,\lambda_0 \left(\varepsilon^{-1} r(x, \Gamma) \right), \quad x \in \overline{D}_h, \qquad (3.40)$$

where $\lambda_0(\xi) \underset{\varepsilon}{\to} 0$ as $\xi \to 0$, i.e., the error bound of the difference scheme for $\varepsilon^{-1} r(x, \Gamma) \to 0$ converges to zero ε-uniformly with the estimate (3.40).

The following statements are valid.

Theorem 3.3.1 *For the difference scheme (3.32), (3.31) let the following conditions be fulfilled: a) the functions $f^h(x)$ and $\varphi^h(x)$ approximate the functions $f(x)$ and $\varphi(x)$ ε-uniformly (with the estimates (3.35), (3.36)); the operator $\Lambda_{(3.32)}$ acting on the solution of the boundary value problem approximates the operator $L_{(3.1a)}$ ε-uniformly inside the set D (with the estimate (3.37)); b) the difference scheme (3.32), (3.31) is ε-uniformly stable inside the set D (with the estimate (3.39)); c)the error bound of the solution to the difference scheme satisfies the condition (3.40). Then the difference scheme (3.32), (3.31) converges ε-uniformly at the rate $\mathcal{O}\left(\lambda_0(\rho_1(\lambda(N)))\right)$ as $N \to \infty$:*

$$|u(x) - z(x)| \leq M\,\lambda_0 \Big(\rho_1\big(\lambda(N)\big) \Big), \quad x \in \overline{D}_h,$$

where $\rho_1(\lambda)$ is the solution of the equation $\lambda\,\psi(\rho) = \lambda_0(\rho)$.

Lemma 3.3.1 *Let the operator Λ from the difference scheme (3.32), (3.31) be monotone and approximate the operator L ε-uniformly on smooth functions (satisfying the condition (2.25)). Then the scheme (3.32), (3.31) is ε-uniformly stable inside the set D (and on the set $\overline{D}$).*

Theorem 3.3.2 *The scheme (3.32), (3.31) converges ε-uniformly at the rate $\mathcal{O}\left(\lambda_0(\rho_1(\lambda(N)))\right)$ if the conditions a), c) of Theorem 3.3.1 and the hypotheses of Lemma 3.3.1 are fulfilled.*

3.4 A difference scheme for the boundary value problem on a parallelepiped

We construct a difference scheme for the problem (3.1) on the parallelepiped $\overline{D}_{(3.2)}$. We assume that the membership (3.30) and either the estimates (3.25) or (3.20) are fulfilled.

Unlike problem (2.2), classical difference approximations to each of the terms of the equation (3.1a) that contain the second-order derivatives do not approximate these terms ε-uniformly everywhere in the domain if at least one of the meshes $\overline{\omega}_s$, for $s = 1, \ldots, n$ (generating the grid $\overline{D}_{h(3.29)}$), is a mesh with an arbitrary distribution of its nodes.

The behaviour of the solution of the boundary value problem (3.1) in neighborhoods of the smooth parts of the boundary Γ is similar to the behaviour of the solution of (2.2). Therefore, when constructing difference schemes for (3.1), we will use special meshes condensing in a neighborhood of the boundary layer.

Let us consider the error of approximation of the operators

$$L^{(sk)} \equiv \varepsilon^2 \frac{\partial^2}{\partial x_s \partial x_k}, \quad s, k = 1, \ldots, n$$

by difference operators on the problem solution. We assume that the estimates (3.25) hold for the problem solution and its components in the representations (3.19), (3.11).

In the case of the operator $L^{(ss)}$, we write the problem solution on the parallelepiped as the sum of two functions

$$u(x) = u_1^{(s)}(x) + u_2^{(s)}(x), \quad x \in \overline{D}, \quad s = 1, \ldots, n,$$

which satisfy the estimates

$$\left| \frac{\partial^{k_s}}{\partial x_s^{k_s}} u_1^{(s)}(x) \right| \leq M \left[1 + \varepsilon^{1-k_s} \right],$$

$$\left| \frac{\partial^{k_s}}{\partial x_s^{k_s}} u_2^{(s)}(x) \right| \leq M \varepsilon^{-k_s} \left[\exp\left(-m \varepsilon^{-1} r(x, \Gamma_s \cup \Gamma_{n+s}) \right) \right],$$

$$x \in \overline{D}, \quad 0 \leq k_s \leq 3,$$

similar to the estimates (2.11), (2.14). According to the results of Section 2.1, for the ε-uniform approximation of the operator $L^{(ss)}$ by the operator

$$\Lambda^{(ss)} \equiv \Lambda^{(ss)}(L^{(ss)}) \equiv \varepsilon^2 \delta_{\overline{x_s}\,\widehat{x_s}}, \quad s = 1, \ldots, n, \tag{3.41}$$

on the problem solution, it suffices to use the mesh $\overline{\omega}_s$ that ensures the fulfillment of the condition

$$\max_i \beta^{(s)}(x_s^i) \underset{\varepsilon}{\to} 0 \quad for \quad N_s \to \infty, \quad s = 1, \ldots, n. \tag{3.42}$$

Here

$$\beta^{(s)}(x_s^i) = \beta_{(2.33)}(x_s^i; \varepsilon, \overline{\omega}_s), \quad x_s^i \in \omega_s, \quad s = 1, \ldots, n.$$

To approximate the operator $L^{(sr)}$, for $s, r = 1, \ldots, n$, $s \neq r$, by the operator

$$\Lambda^{(sr)} = \Lambda^{(sr)}(L^{(sr)}) = \begin{cases} either & 2^{-1}\varepsilon^2 (\delta_{xs\,xr} + \delta_{\overline{xs}\,\overline{xr}}) \\ or & 2^{-1}\varepsilon^2 (\delta_{xs\,\overline{xr}} + \delta_{\overline{xs}\,xr}) \end{cases}, \tag{3.43}$$

$$s, r = 1, \ldots, n, \quad s \neq r,$$

we write the problem solution as the sum of the functions

$$u(x) = u_1^{(sr)}(x) + u_2^{(sr)}(x), \quad x \in \overline{D}, \quad s, r = 1, \ldots, n, \quad s \neq r,$$

which satisfy the estimates

$$\left| \frac{\partial^{k_s + k_r}}{\partial x_s^{k_s} \partial x_r^{k_r}} u_1^{(sr)}(x) \right| \leq M \left[1 + \varepsilon^{1 - (k_s + k_r)} \right],$$

$$\left| \frac{\partial^{k_s + k_r}}{\partial x_s^{k_s} \partial x_r^{k_r}} u_2^{(sr)}(x) \right| \leq M \varepsilon^{-(k_s + k_r)} \min_{q = s, r} \left[\exp \left(-m \varepsilon^{-1} r \left(x, \Gamma_q \cup \Gamma_{n+q} \right) \right) \right],$$

$$x \in \overline{D}, \quad 0 \leq k_s + k_r \leq 3, \quad k_s, k_r \leq 2.$$

On the function $u_1^{(sr)}(x)$, the operators $(\partial / \partial x_q) L^{(sr)}$, for $q = s, r$, are ε-uniformly bounded. On the function $u_2^{(sr)}(x)$, the operators $(\varepsilon^p \partial^p / \partial x_q^p) L^{(sr)}$, for $q = s, r$ and $p = 0, 1$, are ε-uniformly bounded and when moving away from the set $\Gamma_{q_1} \cap \Gamma_{q_2}$, where $q_1, q_2 = s, r, n+s, n+r$ with $q_1 \neq q_2$ and $s \neq r$, they exponentially decrease as $\varepsilon^{-1} r \left(x, \Gamma_{q_1} \cap \Gamma_{q_2} \right) \to \infty$. For the ε-uniform approximation of the operator $L^{(sr)}$ by the operator $\Lambda^{(sr)}$ on the problem solution, it suffices to use the meshes $\overline{\omega}_s$ and $\overline{\omega}_r$ that ensure the fulfillment of the conditions (3.42) (where s is r in the case of the mesh $\overline{\omega}_r$).

Lemma 3.4.1 *For the solution of the problem and its components in the representations (3.19), (3.11), let the estimates (3.25) hold. Then the operators* $\Lambda_{(3.41;3.43)}^{(sk)}$, *for* $s, k = 1, \ldots, n$, *defined on the grid* $\overline{D}_{h(3.31)}$, *where* $\overline{\omega}_s$, *for* $s = 1, \ldots n$, *are meshes satisfying the condition (3.42), approximate the operators* $L^{(sk)}$ *on the problem solution* ε-*uniformly.*

Since, in the case of estimates (3.25) the products

$$\varepsilon^2 \frac{\partial^k}{\partial x_1^{k_1} \ldots \partial x_n^{k_n}} u(x), \quad x \in \overline{D}, \quad 0 \leq k \leq 2$$

are ε-uniformly bounded, then the products $\varepsilon^2 (\partial / \partial x_s) u(x)$, $x \in \overline{D}$, for $s = 1, \ldots, n$, are approximated ε-uniformly by the products of the parameter ε^2 and corresponding classical difference derivatives on the grids $\overline{D}_{h(3.29)}$ with an arbitrary distribution of nodes.

Thus, we approximate the operator $L_{(3.1a)}$ on the grid $\overline{D}_{h(3.31)}$ by the operator $\Lambda = \Lambda_{(2.46)}(L)$.

For the scheme

$$\Lambda z(x) = f(x), \quad x \in D_h, \quad z(x) = \varphi(x), \quad x \in \Gamma_h, \tag{3.44a}$$

where $\Lambda_{(3.44)} = \Lambda_{(2.46)}(L)$, on the piecewise-uniform grid

$$\overline{D}_h = \left\{ \overline{D}_{h(3.31)} \text{ for } \overline{\omega}_s = \overline{\omega}_{1(2.48)}(d_{*s}, d_s^*), \quad s = 1, \ldots, n \right\}, \tag{3.45}$$

in the case of the estimates (3.25), the conditions similar to those of Theorem 2.3.7 are fulfilled, except for the ε-uniform monotonicity of the operator $\Lambda_{(3.44)}$. Note that

$$
\begin{aligned}
L^*_{(2)} &= L^*_{(2)}(L_{(3.1)}) = L^*_{(2)(2.50)}(L_{(2.2)}), \\
\Lambda^*_{(2)} &= \Lambda^*_{(2)}(\Lambda_{(3.44)}) = \Lambda^*_{(2)(2.51)}(\Lambda_{(2.46)}).
\end{aligned}
\tag{3.44b}
$$

Lemma 3.4.2 *Assume that the equation (3.1a) does not contain mixed derivatives, i.e., the following condition*

$$
a_{sk}(x) \equiv 0, \quad x \in \overline{D}, \quad s, k = 1, \ldots, n, \quad s \neq k,
\tag{3.46}
$$

is valid. Then the difference scheme (3.44) is ε-uniform monotone on the grid $\overline{D}_{h(3.31)}$ and converges ε-uniformly on the grid $\overline{D}_{h(3.45)}$ in the case of the estimates (3.25).

Lemma 3.4.3 *The condition*

$$
\left.
\begin{aligned}
\varepsilon^{-1} &\max_{d_{*s} < x^i_s \leq d_{*s} + M\varepsilon} h^{i-1}_s, \\
\varepsilon^{-1} &\max_{d^*_s - M\varepsilon \leq x^i_s < d^*_s} h^i_s,
\end{aligned}
\right\}
\xrightarrow[\varepsilon]{} 0 \text{ as } N \to \infty, \ x^i_s \in \overline{\omega}_s, \ s = 1, \ldots, n,
\tag{3.47}
$$

that is imposed on the distribution of nodes in the meshes $\overline{\omega}_s$, is necessary for the ε-uniform approximation of the operator L by the operator $\Lambda_{(3.44)}$ on the problem.

The condition (3.47) is necessary for the ε-uniform convergence of the difference scheme (3.44), (3.31) and also for the ε-uniform informativity of the grid $\overline{D}_{h(3.31)}$ for the solution of the boundary value problem.

Note that the condition (3.47) is valid for the grid $\overline{D}_{h(3.45)}$.

Lemma 3.4.4 *Let for the grid $\overline{D}_{h(3.31)}$ the condition (3.47) hold. Then the condition that is imposed on the coefficients of the mixed derivatives of the operator L, i.e.,*

$$
a_{sk}(x) = 0, \quad x \in \Gamma^{(p)}, \ p = s \text{ or } p = k, \quad s, k, p = 1, \ldots, n, \quad s \neq k,
\tag{3.48}
$$

where $\Gamma^{(p)} = \Gamma_p \cup \Gamma_{n+p}$, is necessary for the ε-uniform monotonicity of the operators $\Lambda^*_{(2)}(\Lambda_{(3.44)})$ and $\Lambda_{(3.44)}$.

Thus, the conditions (3.47) and (3.48) are necessary for the ε-uniform approximation of the boundary value problem (3.1), (3.2) by the difference scheme (3.44), (3.31) (the condition (3.47)) and for the ε-uniform monotonicity of this scheme (the conditions (3.47), (3.48)).

Under the fulfillment of the condition (3.48), for the system of difference equations (3.44) it is required to construct a grid that is consistent in the variables $x_1, \ldots, x_n$ and condenses in a neighborhood of the boundary Γ. When constructing the consistent grid, we use the mesh (3.31), where $\overline{\omega}_s = \overline{\omega}_{1(2.48)}(d_{*s}, d_s^*; \sigma_s, N_s)$, $s = 1, \ldots, n$.

Next, we impose conditions on both the operator $L_{(2)(2.50)}^*$ and on the grids $\overline{D}_h$, which are piecewise uniform in all variables, that are sufficient for the existence of consistent grids guaranteeing the monotonicity of the operators $\Lambda_{(2)(2.51)}^*$ and $\Lambda_{(2.46)}$.

In the case of condition (3.48), the operator $L_{(2)}^* = L_{(2)(2.50)}^*(L_{(3.1a)})$ has canonical form in the variables x_p on the set $\Gamma^{(p)}$ (the sum of pair of the faces Γ_p and Γ_{n+p}), for $p = 1, \ldots, n$, and also canonical form in the variables x_s, where $s = j, \ldots, r$ for $j, \ldots, r = 1, \ldots, n$, on the set

$$\Gamma^{(j\ldots r)} = \cap_p \Gamma^{(p)}, \quad p = j, \ldots, r \tag{3.49}$$

that is the intersection of pairs of the faces $\Gamma^{(p)}$.

We say that the condition $\{(3.48), (3.49)\}$ is *the canonicity condition in the variables x_s, where $s = j, \ldots, r$, for $j, \ldots, r = 1, \ldots, n$, of the elliptic part of the differential equation (of the differential operator) on the boundary set $\Gamma^{(j\ldots r)}$*, or, briefly, the canonicity condition in the variables x_s, where $s = j, \ldots, r$, for $j, \ldots, r = 1, \ldots, n$, on the set $\Gamma^{(j\ldots r)}$ (in the variables orthogonal to the faces Γ_j, ..., Γ_r, Γ_{j+n}, ..., Γ_{r+n}).

The operator $L_{(2)}^*$ at the point $x_0 \in \overline{D}$ in a neighborhood of the set $\Gamma^{(p)}$, by an appropriate linear transformation of the variables $x_1, \ldots, x_{p-1}$, $x_{p+1}, \ldots, x_n$ (written as $X = X(x)$, where $X_p = x_p$) can be brought to a canonical form in a part of the variables, i.e., to canonical form in the variables $X_1, \ldots, X_{p-1}, X_{p+1}, \ldots, X_n$, where the coefficients of the mixed derivatives not involving differentiation in X_p vanish at the point X_0, $X_0 = X(x_0)$. The coefficients of the mixed derivatives at the point $X \in \overline{D}_X$ depend on the distance to the point X_0 and, in general, increase as this distance increases. Note that the coefficients of the mixed derivatives involving differentiation in X_p are equal to zero on $\Gamma_X^{(p)}$, i.e., on the faces orthogonal to the X_p-axis.

At the point $x_0 \in \overline{D}$ in a neighborhood of the set $\Gamma^{(j\ldots r)}$ (when $|j \ldots r| \le n - 2$), using a transformation of the variables x_s, for $s = 1, \ldots, n$, and $s \ne j, \ldots, r$ (written as $X = X(x)$, where $X_s = x_s$, with $s = j, \ldots, r$), one can bring the operator $L_{(2)}^*$ to canonical form in the variables X_s, for $s = 1, \ldots, n$, with $s \ne j, \ldots, r$, where the coefficients of the mixed derivatives not involving differentiation in X_s, for $s = j, \ldots, r$, vanish at the point X_0. The coefficients of the mixed derivatives involving differentiation in X_s, for $s = j, \ldots, r$, are equal to zero on $\Gamma_X^{(j\ldots r)}$.

Assume that the operator $L_{(2)}^*$ has the above canonical form at the point $x_0 \in \overline{D}$, which depends on the relative position of the point x_0 and the faces. If a neighborhood of the point x_0 is sufficiently small then in this neighborhood the operator $L_{(2)}^*$ has almost canonical form (in all variables), i.e., the coefficients of the mixed derivatives $\partial^2/\partial x_s \partial x_k$, for $s, k = 1, \ldots, n$, with $s \neq k$, are small compared with the coefficients of the second-order derivatives in each direction $\partial^2/\partial x_s^2$, for $s = 1, \ldots, n$.

Under the condition (3.48), let the operator $L^*_{(2)(2.50)}$ have *locally almost canonical form* in $x_1, \ldots, x_n$ in the set $\overline{D}$ and *local almost canonical form* in $x_1, \ldots, x_{p-1}, x_{p+1}, \ldots, x_n$ in a *neighborhood* of $\Gamma^{(p)}$, and in x_s, for $s = 1, \ldots, n$, with $s \neq j, \ldots, r$ in a *neighborhood* of $\Gamma^{(j \cdots r)}$. That is, in the case of the condition (3.48), for an arbitrary point $x^* \in \overline{D}$, one can find sufficiently small values m^* and $\nu > 0$, a sufficiently small neighborhood $Q(x^*)$ and n pairs of the functions $\rho_{*s}(x)$ and $\rho_s^*(x)$, for $s = 1, \ldots, n$, for which the relation (2.55) holds. Furthermore, let one have the following condition (similar to (2.65))

$$
\left[\min_{\substack{\overline{Q}(x^*) \\ r(x, \Gamma^{(j \cdots r)}) \leq m^*}} a_{kk}(x) - \nu \right] h_k^{-1} - \sum_{\substack{s=1, s \neq k \\ s \neq j, \ldots, r}}^{n} \max_{\substack{\overline{Q}(x^*) \\ r(x, \Gamma^{(j \cdots r)}) \leq m^*}} |a_{sk}(x)| \, h_s^{-1} > 0,
$$

$$
k, j, \ldots, r = 1, \ldots, n, \quad k \neq j, \ldots, r, \quad j < \ldots < r, \quad 1 \leq |j \ldots r| \leq n - 2.
\tag{3.50}
$$

If on a set $\overline{D}^1 \subset \overline{D}$, where $\Gamma^{(j \cdots r)} \cap \overline{D}^1 \neq \emptyset$, the coefficients of the operator $L^*_{(2)}$ satisfy the relations $\{(2.55), (3.50)\}$, where $\overline{Q}(x^*) = \overline{D}^1$, we say that the operator $L^*_{(2)}$ has *almost canonical form* $\left(\text{in } x_1, \ldots, x_n\right)$ on $\overline{D}^1$ (by virtue of (2.55)), and *strongly almost canonical form* $\left(\text{in } x_s, \text{ for } s = 1, \ldots, n, \text{ with } s \neq j, \ldots, r\right)$ in an m^*-neighborhood of $\Gamma^{(j \cdots r)} \cap \overline{D}^1$ for $1 \leq |j \ldots r| \leq n-2$ (by virtue of (3.50)), or, briefly, strongly almost canonical form in a neighborhood of $\Gamma \cap \overline{D}^1$.

The condition (3.48) and the condition

the coefficients $a_{sk}(x)$ of the operator $L^*_{(2)}$ on the set $\overline{Q}(x^*) \subseteq \overline{D}$ satisfy the relations $\{(2.55), (3.50)\}$ (3.51)

together allow us to construct on the set $\overline{Q}(x^*)$ piecewise-uniform consistent grids and thereby ensure the monotonicity of the operators $\Lambda^*_{(2)(2.51)}$ and $\Lambda_{(2.46)}$. We now give such a mesh.

Let the conditions (3.48), (3.51) be fulfilled. We now impose conditions on the parameters of the piecewise-uniform grid $\overline{D}_{h(3.45)}$ (piecewise-uniform in all variables) under which the operator $\Lambda^*_{(2)(2.51)}$ is monotone.

Let the step-sizes h_s, i.e., maximal step-sizes in the meshes

$$
\overline{\omega}_s = \overline{\omega}_{1(2.48)}(d_{*s}, d_s^*; \sigma_s, N_s), \quad s = 1, \ldots, n
$$

satisfy the conditions $\{(2.55), (3.50)\}$. Note that $h_s = h_{(2)s}$, where $h_{(1)s}$ and $h_{(2)s}$ are the step-sizes in the mesh $\overline{\omega}_s$ in neighborhoods of the endpoints of the interval $[d_{*s}, d_s^*]$ and outside it, respectively. Assume that the value σ_s that defines the transition points in the mesh $\overline{\omega}_s$ satisfies the condition

$$
\sigma_s \leq m^*_{(3.50)}, \quad s = 1, \ldots, n,
\tag{3.52a}
$$

and that
$$a_{sk} \in C^1(\overline{D}), \quad s, k = 1, \ldots, n, \quad s \neq k.$$

By virtue of the condition (3.48), one has
$$\left| a_{sk}(x) \right| \leq M_1 \sigma_s \quad \text{for} \ r(x, \Gamma_s \cup \Gamma_{s+n}) \leq \sigma_s, \quad s \neq k,$$

where
$$M_1 = \max_{s,k,r, \, s \neq k, \, \overline{D}} \left| \frac{\partial}{\partial x_r} a_{sk}(x) \right|.$$

Then
$$\left| a_{sk}(x) \right| h_{(1)s}^{-1} \leq M_1 h_{(1)s}^{-1} \sigma_s \quad \text{for} \ r(x, \Gamma_s \cup \Gamma_{s+n}) \leq \sigma_s, \quad s \neq k.$$

The inequality
$$\nu h_k^{-1} - \sum_{s=j,\ldots,r} \max_{\substack{\overline{Q}(x^*) \\ x_s \notin (d_{*s}+\sigma_s, \, d_s^* - \sigma_s)}} \left| a_{sk}(x) \right| h_{(1)s}^{-1} > 0,$$

$$k, j, \ldots, r = 1, \ldots, n, \quad k \neq j, \ldots, r \quad \text{for} \ r(x, \Gamma^{(j \ldots r)}) \leq m^*,$$

where $Q(x^*)$ is taken from $\{(2.55), (3.50)\}$, is satisfied if
$$M_1 \sum_{s=1}^{n} \left[\sigma_s h_{(1)s}^{-1} \right] < \nu \min_{s=1,\ldots,n} \left[h_s^{-1} \right],$$

i.e., when one has the condition
$$\sum_{s=1}^{n} N_{*s} < M_1^{-1} \nu \min_k \left[h_k^{-1} \right], \quad k = 1, \ldots, n. \tag{3.52b}$$

Here $N_{*s} + 1$ is the number of nodes in the mesh $\overline{\omega}_s$ on each of the intervals $[d_{*s}, d_{*s} + \sigma_s]$ and $[d_s^* - \sigma_s, d_s^*]$; $h_{(1)s} = \sigma_s N_{*s}^{-1}$ and $h_s = h_{(2)s} < (d_s^* - d_{*s})(N_s - 2N_{*s})^{-1}$, with $N_s \leq M N_{*s}$.

Let the parameters of the meshes $\overline{\omega}_s$, for $s = 1, \ldots, n$, be chosen to satisfy the conditions (2.55c), (3.52), where $h_s = h_{(2)s}$. Write these meshes $\overline{\omega}_s$ in the following form:

$$\overline{\omega}_s = \{\overline{\omega}_s = \overline{\omega}_{1(2.48)}(d_s^*, d_s^*) \ \text{under the conditions} \ (2.55c), (3.52) \}, \tag{3.53}$$
$$s = 1, \ldots, n.$$

In the case of

(i) the condition (3.48) on the coefficients $a_{sk}(x)$ on the boundary Γ,

(ii) the condition (3.51) on the coefficients of the operator $L_{(2)}^*$ on the set $\overline{D}$,

(iii) the conditions (2.55c), (3.52) on the meshes $\overline{\omega}_s$ that generate the piecewise-uniform grid $\overline{G}_{h(5.45)}$, the operators $\Lambda^*_{(2)(3.44)}$ and $\Lambda_{(3.44)}$ are ε-uniformly monotone on the grid $\overline{D}_{h(3.45)} \cap \overline{Q}$, where $Q = Q(x^*)$ is taken from $\{(2.55),\ (3.50)\}$. Thus, in the case of the conditions (3.48), (3.51), the grid

$$\overline{D}_h = \overline{D}_h \left(Q(x^*),\ L^*_{(2)},\ \overline{\omega}_{s(3.53)},\ x_1, \ldots, x_n \right) = \tag{3.54}$$

$$= \left\{ \overline{Q}(x^*) \cap \overline{D}_{h(3.45)} \quad \text{under the conditions (2.55c), (3.52)} \right.$$

$$\left. \text{in the case of the relations } \{(2.55),\ (3.50)\} \right\}$$

is consistent on the set $\overline{Q}(x^*)$ in the variables $x_1, \ldots, x_n$ with the monotonicity condition on the operators $\Lambda^*_{(2)(3.44)}$ and $\Lambda_{(3.44)}$.

We consider the difference scheme (3.44) on the grid

$$\overline{D}_h = \overline{D}_{h(3.54)} \left(Q(x^*),\ L^*_{(2)},\ \overline{\omega}_{s(3.53)},\ x_1, \ldots, x_n \right), \tag{3.55}$$

where $\overline{\omega}_{1(2.48)}$ in the mesh $\overline{\omega}_{s(3.53)}(\overline{\omega}_{1(2.48)})$ is $\overline{\omega}_{1(2.44)}(d_{*s},\ d^*_s,\ m)$, and m is the constant from the estimates (3.13), (3.20), (3.25) of Theorems 3.2.1 and 3.2.2. In the case of the condition (2.71), one can choose the set Q in (3.55) as $\overline{Q} = \overline{D}$.

The following theorems on the convergence are a corollary of Theorem 3.3.2.

Theorem 3.4.1 *Let the following conditions be satisfied: a)* $a_{sk} \in C^1(\overline{D})$, $b_s,\ c,\ c^0,\ f \in C(\overline{D})$, *for* $s,\ k = 1, \ldots, n;$ *b) the coefficients* $a_{sk}(x)$ *satisfy the conditions (2.71), (3.48), and* $\{(2.55),\ (3.50)\}$, *where* $\overline{Q} = \overline{D};$ *c) for the solution of the problem (3.1) on* $\overline{D}_{(3.2)}$ *the estimates of Theorem 3.2.1 (Theorem 3.2.2) hold for* $K = 3$. *Then the difference scheme (3.44) on the grid* $\overline{D}_{h(3.55)}$, *where* $\overline{Q} = \overline{D}$, *converges* ε-*uniformly as* $N \to \infty$

$$|u(x) - z(x)| \le M \left(N^{-1} \ln N \right)^\nu, \quad x \in \overline{D}_h, \tag{3.56}$$

where $\nu = 1$ *in the case of the estimates (3.25) and* $\nu = 4^{-1}$ *in the case of (3.20).*

Theorem 3.4.2 *Let the conditions a), c) of Theorem 3.4.1, and also the condition (3.46) be satisfied. Then the difference scheme (3.44) on the grid* $\overline{D}_{h(3.31)}$, *where* $\overline{\omega}_s = \overline{\omega}_{1(2.44)}(d_{*s},\ d^*_s;\ m)$ *and* $m = m_{(3.13)}$, *converges* ε-*uniformly as* $N \to \infty$ *with the estimate (3.56).*

Under the hypotheses of Theorems 3.4.1 and 3.4.2, let $\overline{z}(x) = \overline{z}_{(2.42)}(x;\ z(\cdot),\ \overline{D}_h)$, $x \in \overline{D}$, be the interpolant of the function $z(x)$, $x \in \overline{D}_h$, which is the solution of the difference scheme (3.44) on the grids (3.55) and (3.31), respectively. Then one has the estimate

$$|u(x) - \overline{z}(x)| \le M \left(N^{-1} \ln N \right)^\nu, \quad x \in \overline{D}, \quad \nu = \nu_{(3.56)}.$$

3.5 Consistent grids on subdomains

When constructing a difference scheme for a boundary value problem in a domain with piecewise-uniform curvilinear boundary in Section 3.6, we shall need piecewise-uniform consistent grids on subdomains, namely, on parallelepipeds; such grids condense in neighborhoods of only some faces of the subdomains-parallelepipeds. Let us now discuss the construction of such consistent grids.

Let

$$\overline{D}_h = \overline{D}_h \left(Q(x^*), \ L^*_{(2)}, \ \overline{\omega}_s, \ x_1, \ldots, x_n \right)$$

be a grid of type $\overline{D}_{h(3.54)}$; however, this grid differs from the last one such that in $\overline{D}_{h(3.54)}$ the piecewise-uniform meshes $\overline{\omega}_{s(3.53)}\left(d_{*s}, d_s^*\right)$ along with some of the x_s-axis (or in all directions) are replaced by uniform meshes $\overline{\omega}_s^u = \overline{\omega}_s^u(d_{*s}, d_s^*)$, whose step-size h_s equals $h_{(2)s}$ in the mesh $\overline{\omega}_{s(3.53)}\left(d_{*s}, d_s^*\right)$; this grid is consistent in $x_1, \ldots, x_n$ on $\overline{Q}(x^*)$ with the monotonicity condition for the operators $\Lambda^*_{(2)(3.44)}$ and $\Lambda_{(3.44)}$. The consistency of the grid on $\overline{Q}(x^*)$ is preserving also in that case when in $\overline{D}_{h(3.54)}$ some meshes $\overline{\omega}_{s(3.53)}\left(d_{*s}, d_s^*\right)$ that condense in neighborhoods of both ends of the interval $\left[d_{*s}, d_s^*\right]$ are replaced by the piecewise uniform meshes $\overline{\omega}_s^l\left(d_{*s}, d_s^*\right)$ or $\overline{\omega}_s^r\left(d_{*s}, d_s^*\right)$ that condense in a neighborhood, respectively, of the left or right end of the interval $\left[d_{*s}, d_s^*\right]$; or by the uniform meshes $\overline{\omega}_s^u(d_{*s}, d_s^*)$. In this case, it is required that the parameters σ_s, $h_{(1)s}$, and $h_{(2)s}$ that define the meshes $\overline{\omega}_s^l\left(d_{*s}, d_s^*\right)$ and $\overline{\omega}_s^r\left(d_{*s}, d_s^*\right)$ (just as the parameter $h_{(2)s} = h_s$) be the same as those in the mesh $\overline{\omega}_{s(3.53)}\left(d_{*s}, d_s^*\right)$.

When constructing the consistent grids based on the grid $\overline{D}_{h(3.54)}$, imposed requirements to the coefficients $a_{sk}(x)$ and meshes $\overline{\omega}_s$, for $s = 1, \ldots, n$, turn out to be more strong than it is necessary. For example, in the case of a consistent uniform grid

$$\overline{D}_h = \overline{D}_h^u \left(Q(x^*), \ L^*_{(2)} \right) = \overline{D}_h \left(Q(x^*), \ L^*_{(2)}, \ \overline{\omega}_s^u, \ x_1, \ldots, x_n \right),$$

which is the grid (3.54), where $\overline{\omega}_s$ is $\overline{\omega}_s^u$ in all directions, it was required that the conditions (3.48) and (3.50) be satisfied. However, for a uniform mesh in order to be consistent, restrictions for $a_{sk}(x)$ on the boundary Γ (i.e., the condition (3.48)) and for the parameters of the grid in a neighborhood of the boundary Γ (i.e., the condition (3.50)) are not required.

In the case of the domain $\overline{D}_{(3.2)}$, we give sufficiently natural conditions imposed on the coefficients $a_{sk}(x)$ and on the meshes $\overline{\omega}_s$ that guarantee the consistency of the grids $\overline{D}_h$ that are constructed based on the "elementary meshes", i.e., $\overline{\omega}^u$, $\overline{\omega}_s^l$, $\overline{\omega}_s^r$, and $\overline{\omega}_s^b$; the mesh $\overline{\omega}_s^b$ condenses in neighborhoods of both endpoints to the interval $\left[d_{*s}, d_s^*\right]$.

In the case of the grid $\overline{D}_h$, let the piecewise-uniform mesh on the x_s-axis condense in a neighborhood of the face Γ_s and/or Γ_{s+n}, or this mesh is uniform, for $s = 1, \ldots, n$. We denote by Γ^e those faces of the boundary Γ in whose neighborhoods the meshes $\overline{\omega}_s$, for $s = 1, \ldots, n$, condense; here $\overline{\omega}_s = \overline{\omega}_s(d_{*s}, d_s^*)$. We denote by J^*, where

$$J^* = J^*(D) = J^*(D; \Gamma^e) \tag{3.57}$$

the set of indexes j, related to Γ_j in whose neighborhoods the grid $\overline{D}_h$ condenses in the direction orthogonal to the face Γ_j; here $\Gamma_j \subseteq \Gamma^e$. For example, in the case of the uniform grid $\overline{D}_h$ we have $J^* = \emptyset$; but if $\overline{\omega}_s = \overline{\omega}_s^l$, for $s = 1, \ldots, n$, then $J^* = \{j = 1, \ldots, n\}$. We denote by $J_s^* = J_s^*(D; \Gamma^e)$, with $s = 1, \ldots, 2n$, the set of indexes j from $J^*(D)$, related to the faces $\Gamma_j \subseteq \Gamma^e$ orthogonal to the x_s-axis.

We suppose that the coefficients $a_{sk}(x)$ of the operator $L_{(2)}^*$ on the boundary Γ satisfy the following condition, which is similar to (3.48):

$$a_{sk}(x) = 0, \ x \in \Gamma_j, \ j \in J^*, \ \ s = j \text{ or } k = j, \ s \neq k, \ if J^* \neq \emptyset; \tag{3.58}$$

restrictions on $a_{sk}(x)$ on the boundary Γ are not imposed if $J^* = \emptyset$.

We also assume that the coefficients $a_{sk}(x)$, which define the operator $L_{(2)}^*$, satisfy the condition (2.55), independent of the structure of the set J^*. Furthermore, if $J^* \neq \emptyset$, we assume that the following additional condition holds, which is similar to (3.50):

$$\left[\min_{\substack{\overline{Q}(x^*) \\ r(x, \cap_{j_i} \Gamma_{j_i}) \leq m^* \\ j_i \in J^*, \ i=1,\ldots,r}} a_{kk}(x) - \nu \right] h_k^{-1} - \sum_{s=1}^{n} \max_{\substack{\overline{Q}(x^*) \\ s \neq k, p_{j_i} \ r(x, \cap_{j_i} \Gamma_{j_i}) \leq m^* \\ j_i \in J^*, \ i=1,\ldots,r}} |a_{sk}(x)| \, h_s^{-1} > 0,$$

$$k, p_{j_1}, \ldots, p_{j_r} = 1, \ldots, n, \ \ k \neq p_{j_1}, \ldots, p_{j_r}, \ \ 1 \leq r \leq n-2, \tag{3.59}$$

$$\cap_{j_i} \Gamma_{j_i} \neq \emptyset, \ J^* \neq \emptyset,$$

where $p_j = p_j(j, n)$, where $j = j_i \in J^*$; $p_j = j$ for $j \leq n$, and $p_j = j - n$ for $j > n$; m^* and ν are sufficiently small values.

When the relations $\{(2.55), (3.59)\}$ are fulfilled, we say that *the operator $L_{(2)}^*$ has local almost canonical form in $x_1, \ldots, x_n$ on the set $\overline{D}$, and the operator $L_{(2)}^{*[p_{j_1}, \ldots, p_{j_r}]}$ has strongly local almost canonical form in x_s, for $s = 1, \ldots, n$, with $s \neq p_{j_1}, \ldots, p_{j_r}$, in a neighborhood of the set $\cap_{j_i} \Gamma_{j_i}$, for $i = 1, \ldots, r$, with $j_i \in J^*$*. Here $L_{(2)}^{*[p_{j_1}, \ldots, p_{j_r}]}$ is *the truncated operator*, i.e., the operator $L_{(2)}^*$ that does not contain derivatives in the variables x_s, where $s = p_{j_1}, \ldots, p_{j_r}$.

The condition

the coefficients $a_{sk}(x)$ of the operator $L^*_{(2)}$ on the set $\overline{D}$

satisfy the relations (3.58), {(2.55), (3.59)} if $J^* \neq \emptyset$, $\qquad$ (3.60)

and the relations (2.55) if $J^* = \emptyset$, where $\overline{Q}(x^*) \subseteq \overline{D}$, $J^* = J^*_{(3.57)}(D)$,

allows us on the set $\overline{Q}(x^*)$ to construct piecewise-uniform (controlled by the set $J^*(D)$) consistent grids that guarantee the monotonicity of the operators $\Lambda^*_{(2)(3.44)}$ and $\Lambda_{(3.44)}$. Let us give such grids.

On the set $\overline{D}$ we construct consistent grids, which are defined by the structure of the set J^*. Let the condition (3.60) hold.

For $J^* = \emptyset$, the uniform grid $\overline{D}_h$ (generated by the meshes $\overline{\omega}_s^u$, for $s = 1, \ldots, n$, whose step-size satisfies the condition (2.55c))

$$\overline{D}_h = \overline{D}_h(Q(x^*), L^*_{(2)}), \ \overline{\omega}_s = \overline{\omega}_s^u \ \text{ under the condition } (2.55c)$$

in the case of the relations (2.55) for $J^* = \emptyset$) $\qquad$ (3.61)

is consistent on the set $\overline{Q}(x^*)$.

In the case of the condition $J^* \neq \emptyset$ (i.e., at least one of the meshes $\overline{\omega}_s$, for $s = 1, \ldots, n$, generating $\overline{D}_h$, is piecewise uniform), we suppose that for the step-sizes $h_{(2)s}$ of the piecewise uniform meshes $\overline{\omega}_s$ the condition (2.55c) holds, and furthermore, the meshes $\overline{\omega}_s$ satisfy the condition (which is similar to (3.52a)):

$$\begin{aligned}\sigma_s = \sigma_s^l \leq m^*_{(3.59)} \ \text{ for } s = p_j \ \text{ if } p_j = j, \ \ j \in J^*; \\ \sigma_s = \sigma_s^r \leq m^*_{(3.59)} \ \text{ for } s = p_j \ \text{ if } p_j \neq j, \ \ j \in J^*,\end{aligned} \qquad (3.62a)$$

and also the following condition (which is similar to (3.52b)):

$$\sum_{s=p_j} N_{*s} < M_1^{-1} \nu \min_k [h_k^{-1}], \ \ k = 1, \ldots, n, \ \ j \in J^*. \qquad (3.62b)$$

Here $p_j = p_{j(3.59)}$, for $j \in J^*$, and $N_{*s} + 1$ is the number of nodes in the mesh $\overline{\omega}_s$ on the interval $[d_{*s}, d_{*s} + \sigma_s]$ if the mesh $\overline{\omega}_s$ condenses in a neighborhood of the left endpoint of the interval $[d_{*s}, d_s^*]$, and on the interval $[d_s^* - \sigma_s, d_s^*]$ if the mesh $\overline{\omega}_s$ condenses in a neighborhood of the right endpoint of the interval, and $M_1 = M_{1(3.52)}$. In the case when the mesh $\overline{\omega}_s$ condenses in neighborhoods of both endpoints of the interval $[d_{*s}, d_s^*]$, we set

$$\sigma_s^l = \sigma_s^r = \sigma_s \ \text{ for } \ s = p_{j_1} = p_{j_2}, \ \ j_1, j_2 \in J^*, \ \ j_1 \neq j_2. \qquad (3.62c)$$

In the grid $\overline{D}_h$, we apply piecewise-uniform or uniform meshes along the x_s-axis (depending on the structure of the set J^*)

$$
\overline{\omega}_s = \begin{cases}
\overline{\omega}_s^l \ \text{if}\ J_s^* = \{j = s\} \\
\overline{\omega}_s^r \ \text{if}\ J_s^* = \{j = s+n\} \\
\overline{\omega}_s^b \ \text{if}\ J_s^* = \{j = s, s+n\} \\
\qquad\qquad\qquad\qquad \text{if}\ J^* \neq \emptyset, \\
\overline{\omega}_s^u \ \text{under the condition (2.55c) if}\ J_s^* = \emptyset,\ J^* \neq \emptyset, \\
\qquad\qquad\qquad\qquad\qquad\quad \text{or}\quad \text{if}\ J^* = \emptyset;
\end{cases}
$$

with the conditions $(2.55c), (3.62)$ (3.63)

$$
J^* = J_{(3.57)}^*(D; \Gamma^e);
$$

the mesh $\overline{\omega}_{s(3.63)}$ is piecewise uniform on at least one of the x_s-axes for $J^* \neq \emptyset$. The mesh $\overline{\omega}_{s(3.63)}$ is constructed based on the mesh $\overline{\omega}_{1(2.48)}$ similar to $\overline{\omega}_{1(2.83)}$

$$
\overline{\omega}_s = \overline{\omega}_{1(2.83)}\big(d_{*s}, d_s^*; \overline{\omega}_{1(2.48)}\big).
$$

Theorem 3.5.1 *Let the condition $a_{sk} \in C^1(\overline{D})$ and also the condition (3.60) be fulfilled. Then the grid*

$$
\overline{D}_h = \overline{D}_h\big(Q(x^*),\ L_{(2)}^*;\ J^*\big) = \overline{D}_h\big(Q(x^*),\ L_{(2)}^*,\ \overline{\omega}_s = \overline{\omega}_{s(3.63)}, \tag{3.64}
$$

for $s = 1,\ldots,n$, under the conditions $(2.55c)$ and (3.62)

subject to $\{(2.55), (3.59)\}$ if $J^* \neq \emptyset$,

and under the condition $(2.55c)$ subject to (2.55) if $J^* = \emptyset$;

$$
J^* = J_{(3.57)}^*(D))
$$

is consistent on the set $\overline{Q}(x^*)$ with the monotonicity condition for the operators $\Lambda_{(2)(3.44)}^*$ and $\Lambda_{(3.44)}$.

The grid $\overline{D}_{h(3.64)}$ is the grid $\overline{D}_{h(3.61)}$ for $J^* = \emptyset$, and the grid $\overline{D}_{h(3.55)}$ for $J^* = \{j = 1,\ldots,2n\}$.

If in (3.64) one has $\overline{Q}(x^*) = \overline{D}^i$, where $\overline{D}^i$ is a subset in $\overline{D}$, then the grid $\overline{D}_{h(3.64)}$ is consistent on the set $\overline{D}^i$. In the case when

$$
\overline{D}^i = D^i \cup \Gamma^i,\quad D^i = \{x:\ d_{*s}^i < x_s < d_s^{i*},\ s = 1,\ldots,n\} \tag{3.65}
$$

is a parallelepiped whose faces belong to the coordinate planes $\big(\Gamma^i = \cup_j \Gamma_j^i,$ for $j = 1,\ldots,2n\big)$, it is convenient to construct a consistent grid on $\overline{D}^i$ based on the meshes $\overline{\omega}_s^i$, for $s = 1,\ldots,n$, introduced on the intervals $[d_{*s}^i, d_s^{i*}]$. We show such grids (the construction of the grid is similar to that of (2.89)).

For the set $\overline{D}^i \subseteq \overline{D}$, where $\overline{D}^i = \overline{D}^i_{(3.65)}$, by j^* we denote the index of the face $\Gamma_j \subseteq \Gamma^e$, for which the condition $\Gamma^i \cap \Gamma_j \neq \emptyset$ holds, with $\Gamma_j = \Gamma_j(D)$, $\Gamma^i = \Gamma(D^i)$, and $\Gamma^e = \Gamma^e_{(3.57)}$. We denote by J^{i*}, where J^{i*}

$$J^{i*} = J^{i*}(D^i) = J^{i*}(D^i; \Gamma^e),\tag{3.66}$$

the set of all indexes j^*.

We denote by $J^{i*}_s = J^{i*}_s(D^i; \Gamma^e)$, for $s = 1, \ldots, 2n$, the set of the indexes j from $J^{i*}(D^i)$, related to the face $\Gamma_j \subseteq \Gamma^e$ orthogonal to the x_s-axis.

We assume that the following condition (similar to (3.60)) holds:

the coefficients $a_{sk}(x)$ of the operator $L^*_{(2)}$ on the set $\overline{D}^i \subseteq \overline{D}$

satisfy the relations (3.58), {(2.55), (3.59)} if $J^* \neq \emptyset$, (3.67)

and the relations (2.55) if $J^* = \emptyset$, where $\overline{Q}(x^*) = \overline{D}^i$, $J^* = J^{i*}_{(3.66)}(D^i)$.

On the intervals $[d^i_{*s}, d^i_s]$, we construct the meshes $\overline{\omega}^i_s$, $s = 1, \ldots, n$. The mesh $\overline{\omega}^i_s$ is uniform, i.e., $\overline{\omega}^{iu}_s$ if $s, s+n \notin J^{i*}$, or is piecewise uniform, i.e., $\overline{\omega}^{il}_s, \overline{\omega}^{ir}_s$, or $\overline{\omega}^{ib}_s$ if, respectively, $s \in J^{i*}$, $s+n \in J^{i*}$, or $s, s+n \in J^{i*}$. Let

$$\overline{\omega}^i_s = \begin{cases} \overline{\omega}^{il}_s \ \text{if} \ J^*_s = \{j = s\} & \text{under the conditions} \\ \overline{\omega}^{ir}_s \ \text{if} \ J^*_s = \{j = s+n\} & (2.55c),\ (3.62) \\ \overline{\omega}^{ib}_s \ \text{if} \ J^*_s = \{j = s, s+n\} & \text{if} \ J^* \neq \emptyset, \\ \overline{\omega}^{iu}_s \ \text{under the condition} \ (2.55c) \ \text{if} \ J^*_s = \emptyset, \ J^* \neq \emptyset, \\ \qquad \qquad or \quad \text{if} \ J^* = \emptyset; \end{cases}\tag{3.68a}$$

$$J^* = J^{i*} = J^{i*}_{(3.66)}(D; \Gamma^e).$$

In the case when the mesh $\overline{\omega}^i_{s(3.68)}(d^i_{*s}, d^i_s)$ is constructed based on the mesh $\overline{\omega}_{1(2.48)}(d^i_{*s}, d^i_s)$, the mesh $\overline{\omega}^i_s$ is defined by the relation

$$\overline{\omega}^i_s \equiv \overline{\omega}^i_s(d^i_{*s}, d^{i*}_s) = \overline{\omega}_{1(2.88)}(d^i_{*s}, d^{i*}_s; \overline{\omega}_{1(2.48)}).\tag{3.68b}$$

Assume that the parameters of the meshes $\overline{\omega}^i_s = \overline{\omega}^i_s(d^i_{*s}, d^{i*}_s)$, for $s = 1, \ldots, n$, satisfy the conditions (2.55c), (3.62), where $\overline{Q}(x^*) = \overline{D}^i$, $J^* = J^{i*}_{(3.66)}$.

Theorem 3.5.2 *In the case of the condition (3.67), the grid (see (3.64))*

$$\overline{D}^i_h = \overline{D}^i_h(D^i, L^*_{(2)};\ J^{i*}) = \overline{D}^i_h(Q(x^*) = D^i_{(3.65)},\ L^*_{(2)},\ \overline{\omega}^i_s = \overline{\omega}^i_{s(3.68)},$$

$$s = 1, \ldots, n, \quad \text{under the conditions} \ (2.55c),\ (3.62)\tag{3.69}$$

$$\text{subject to} \ \{(2.55),\ (3.59)\} \ \text{if} \ J^* \neq \emptyset,$$

and under the condition (2.55c) subject to (2.55) if $J^* = \emptyset$,

$$\text{where} \ \ \overline{Q}(x^*) = \overline{D}^i, \quad J^* = J^{i*}_{(3.66)}(D^i)),$$

is consistent on the set $\overline{D}^i$ with the monotonicity condition for the operators $\Lambda^*_{(2)(3.44)}$ *and* $\Lambda_{(3.44)}$.

For $J^{i*} = \emptyset$ (i.e., for $J^*_{(3.57)}(D) = \emptyset$, or $\overline{D}^i \cap \Gamma^e = \emptyset$), the grid $\overline{D}_{h(3.69)}$ is uniform.

Grids similar to (3.64) and (3.69) are used for the construction of difference schemes for boundary value problems in domains with piecewise-uniform boundaries (see, e.g., Subsection 3.6).

3.6 A difference scheme for the boundary value problem in a domain with piecewise-uniform boundary

We construct a difference scheme for the problem (3.1) in the domain

$$\overline{D} = D \bigcup \Gamma \qquad (3.70)$$

with a piecewise-uniform boundary Γ; the faces Γ_j of the set $\overline{D}$ are, in general, curvilinear. For simplicity, let $\overline{D}$ be an n-dimensional parallelepiped. Assume that on the set $\overline{D}$ the coefficients of the elliptic part of the differential operator satisfy the condition of pointwise dominance of the diagonal elements (the condition (2.93)). Suppose also that in parts of the domain adjoining to the boundary, the boundary value problem is reduced to problems (in a local coordinate system) on subdomains which are rectangular parallelepipeds. Furthermore, on the boundaries of these subdomains, which are common with the domain boundary, the canonicity condition (similar to (3.48)) of the elliptic part of the operator in the variables orthogonal to the smooth parts of the domain boundary is fulfilled. Boundary value problems having such properties arise, e.g., in the case of diffusion (heat) processes acting in isotropic mediums when parts of the domain boundary in the local coordinate system are pieces of the coordinate planes [19].

To solve the boundary value problem (3.1), (3.70), we use the grid Schwartz method constructed for the problem (2.2), (2.4).

The continual Schwartz method (2.91) in the case of the (3.1), (3.70) converges ε-uniformly as the number of iteration grows.

Theorem 3.6.1 *The function* $u^{[k]}_{(2.91)}(x)$, *i.e., the solution of the continual Schwartz method (2.91) for the boundary value problem (3.1), (3.70), converges ε-uniformly to the solution of the boundary value problem at the rate of a geometric progression as* $k \to \infty$:

$$|u(x) - u^{[k]}(x)| \leq M\, q^k, \quad x \in \overline{D}, \quad q < 1 - m. \qquad (3.71)$$

For the problem $\{(2.91), (3.1), (3.70)\}$ of the continual Schwartz method, we construct a difference scheme of the decomposition method similar to the one $\{(2.98), (2.97), (2.114)\}$.

Subdomains $D^i \subset D$ in the case of the condition $\overline{D}^i \cap \Gamma = \emptyset$ are chosen as rectangular parallelepipeds whose faces are formed by the planes $x_s = \mathrm{const}$, for $s = 1, \ldots, n$. Distance between the sets D^i and the boundary Γ is independent of ε.

Let $x^* \in \Gamma$, and let $x^* \in \Gamma_{(3.49)}^{(j \ldots r)}$, where $1 \leq |j \ldots r| \leq n$. On the set $\overline{D}^i = \overline{Q}(x^*)$, i.e., a closure of the neighborhood $Q(x^*)$, we pass to new variables $X = X(x)$. We choose the set $\overline{D}_X^i$ as a rectangular parallelepiped whose faces are formed by the coordinate planes $X_s = \mathrm{const}$, for $s = 1, \ldots, n$; set $\Gamma^{iL} = \overline{D}^i \cap \Gamma$, with $\Gamma^{iL} \neq \emptyset$.

In the new variables, the boundary value problem (3.1), (3.70) considered on the set $D^i \cup \Gamma^{iL}$ is transformed into the problem (similar to (2.105))

$$L_X U(X) = F(X), \quad X \in D_X^i; \tag{3.72}$$

$$U(X) = \Phi(X), \quad X \in \Gamma_X^{iL}.$$

Here

$$L_X \equiv \varepsilon^2 L_{2X} + L_{0X}; \quad L_{0X} = -C^0(X);$$

$$L_{2X} \equiv \sum_{s,k=1}^{n} A_{sk}(X) \frac{\partial^2}{\partial X_s \partial X_k} + \sum_{s=1}^{n} B_s(X) \frac{\partial}{\partial X_s} - C(X);$$

set

$$L_{X(2)}^* \equiv \sum_{s,k=1}^{n} A_{sk}(X) \frac{\partial^2}{\partial X_s \partial X_k}.$$

In the case when *the operator* $L_{X(2)}^*$ *is canonical in the variables* X_s, *for* $s = j, \ldots, r$ *on the boundary* $\Gamma_X^{iL} \cap \Gamma_X^{(j \ldots r)}$, this operator, by an appropriate linear transformation $X^1 = X^1(X)$, $X_s^1 = X_s$, for $s = j, \ldots, r$, can be brought to *almost canonical form in the variables* X_s^1, for $s = 1, \ldots, n$, with $s \neq j, \ldots, r$ on the set $\overline{D}_{XX^1}^i$ (choosing (it is possible) the sizes of the set $\overline{D}^i$ sufficiently small). The coefficients of the second-order derivatives involving differentiation in X_s^1, for $s = j, \ldots, r$, are equal to zero on $\{\Gamma^{iL} \cap \Gamma^{(j \ldots r)}\}_{X X^1}$, i.e., the operator $L_{XX^1(2)}^*$ *remains to be canonical in the variables* X_s^1, *for* $s = j, \ldots, r$ *on the boundary* $\Gamma_{XX^1}^{iL} \cap \Gamma_{XX^1}^{(j \ldots r)}$.

Let the subdomains $\overline{D}^i$, for $i = 1, \ldots, I$, form a covering of the set $\overline{D}$, moreover, the subdomains $\overline{D}^i$, for $i = 1, \ldots, I_1$, adjoin to the boundary Γ, and the set $\overline{D}^0 = \cup_i \overline{D}^i$, for $i = I_1 + 1, \ldots, I$, has no common points with the boundary Γ. Let the distance between the sets $\overline{D}^0$ and Γ be independent of ε. We assume that the condition for the width of overlapping to the subdomains $\overline{D}^i$, for $i = 1, \ldots, I$, holds, i.e., for all i, where $i = 1, \ldots, I$, the minimum of the minimal distance between the boundary of the set D^i and the boundary of the union of all sets which do not contain D^i (the minimal overlapping of subdomains that cover D) is independent of the value of the parameter ε.

We suppose that the sets $\overline{D}^i$, for $i = I_1 + 1, \ldots, I$, which have no common points with the boundary Γ, are rectangular parallelepipeds formed by coordinate planes in the variables $x_1, \ldots, x_n$. The sets $\overline{D}^i$, for $i = 1, \ldots, I_1$, adjoining to the boundary Γ, in new variables $X^i = X^i(x)$ are the sets $\overline{D}^i_{X^i}$; these sets are rectangular parallelepipeds formed by coordinate planes in the variables $X_1^i, \ldots, X_n^i$.

We assume that the coefficients of the operator $L_{(2)}^*$ on the set $\overline{D}$ satisfy the condition of pointwise dominance of the diagonal elements (see (2.93)):

$$|a_{sk}(x)| < (n-1)^{-1} \left[\alpha_s(x)\, \alpha_k^{-1}(x)\, a_{ss}(x), \; \alpha_k(x)\, \alpha_s^{-1}(x)\, a_{kk}(x) \right],$$

$$x \in \overline{D}, \quad s, k = 1, \ldots, n, \quad s \neq k, \tag{3.73}$$

moreover,

$$\max_{s,k,\overline{D}} \left[\alpha_s(x)\, \alpha_k^{-1}(x) \right] \leq M, \quad s, k = 1, \ldots, n, \quad s \neq k.$$

Let the diameters of the sets $\overline{D}^i$, for $i = I_1 + 1, \ldots, I$, be sufficiently small such that the following inequality holds on $\overline{D}^i$ (see (2.94)):

$$\max_{\overline{D}^i} |a_{sk}(x)| < (n-1)^{-1} \min_{\overline{D}^i} \left[\alpha_s^i\, (\alpha_k^i)^{-1}\, a_{ss}(x), \; \alpha_k^i\, (\alpha_s^i)^{-1}\, a_{kk}(x) \right],$$

$$s, k = 1, \ldots, n, \quad s \neq k, \quad i = I_1 + 1, \ldots, I,$$

that ensures a transformation of the operator $L_{(2)}^*$ to almost canonical form on $\overline{D}^i$. Next, we consider to be fulfilled the condition (see (2.113)):

the coefficients $a_{sk}(x)$ of the operator $L_{(2)}^*$ on the set $\overline{D}^i$ satisfy
the relations (2.55), where $\overline{Q}(x^*) = \overline{D}^i$, for $i = I_1 + 1, \ldots, I$. $\qquad$ (3.74)

We assume that the operators $L_{X^i(2)}^*$ satisfy the canonicity condition on the boundaries $\Gamma_{X^i}^{iL} \cap \Gamma_{X^i}^{(j\ldots r)}$, for $i = 1, \ldots, I_1$ (see (2.107)):

the coefficients $A_{sk}(X^i)$ of the operator $L_{X^i(2)}^*$ $\qquad\qquad\qquad$ (3.75)

on the boundaries $\Gamma_{X^i}^{iL} \cap \Gamma_{X^i}^{(j\ldots r)}$ satisfy the condition $A_{sk}(X^i) = 0$,

where $X^i \in \{\Gamma_{X^i}^{iL} \cap \Gamma_{X^i}^{(j\ldots r)}\}$, for $s, k = j, \ldots, r$, with $s \neq k$,

and $j, \ldots, r = 1, \ldots, n$, for $\Gamma^{iL} \cap \Gamma^{(j\ldots r)} \neq \emptyset$, with $i = 1, \ldots, I_1$.

By virtue of Lemma 3.4.4, this condition is necessary for the ε-uniform monotonicity of the operator $\Lambda_{X^i} = \Lambda_{(3.44)}(L_{X^i})$ on grids, which are ε-uniformly informative for the solution of the subproblem (3.72) on the set $\left\{D^i \bigcup \Gamma^{iL}\right\}_{X^i}$ under the condition $\Gamma^{iL} \bigcap \Gamma^{(j \ldots r)} \neq \emptyset$.

Set

$$J^{i*} = J^{i*}(D^i) = J^{i*}_{(3.66)}(D^i; \Gamma), \quad i = 1, \ldots, I; \tag{3.76}$$

$J^{i*} = \emptyset$ for $i = I_1 + 1, \ldots, I$, and $J^{i*} \neq \emptyset$ for $i = 1, \ldots, I_1$.

We assume that the following condition holds that is similar to (3.67) (see also (2.108))

the coefficients $A_{sk}(X^i)$ of the operator $L^*_{X^i(2)}$ on the set $\overline{D}^i_{X^i}$

satisfy the relations (3.75), $\{(2.55), (3.59)\}$ if $J^{i*} \neq \emptyset$; $\tag{3.77}$

in $\{(2.55), (3.59)\}$ $a_{sk}(x), \ldots, \overline{Q}(x^*)$ is $A_{sk}(X^i), \ldots, \{\overline{Q}(x^*)\}_{X^i}$

under the condition $\{\overline{Q}(x^*)\}_{X^i} = \overline{D}^i_{X^i}, \quad J^{i*} = J^{i*}_{(3.76)}(D^i_{X^i}), \quad i = 1, \ldots, I_1.$

Conditions (3.74) and (3.77) together allow us to construct on the sets $\overline{D}^i$ and $\overline{D}^i_{X^i}$ consistent grids and thereby ensure the monotonicity of the operators $\Lambda = \Lambda_{(3.44)}(L)$ and $\Lambda_{X^i} = \Lambda_{(3.44)}(L_{X^i})$, respectively.

On the set $\overline{D}^i$, for $i = I_1 + 1, \ldots, I$, we construct uniform grids similar to (3.69) for $J^{i*} = \emptyset$ (see (2.114c)):

$$\overline{D}^i_h = \overline{D}^i_h(D^i, L^*_{(2)}; J^{i*}) = \overline{D}^i_h(Q(x^*) = D^i, L^*_{(2)}), \tag{3.78}$$

$$\overline{\omega}^i_s = \overline{\omega}^{iu}_{s(3.68)}, \; for \; s = 1, \ldots, n \; under \; the \; condition \; (2.55c)$$

in the case of the relations (2.55), where $\overline{Q}(x^*) = \overline{D}^i$,

$$J^{i*} = J^{i*}(D^i) = \emptyset), \quad i = I_1 + 1, \ldots, I.$$

On the sets $\overline{D}^i_{X^i}$ we construct piecewise-uniform grids similar to (3.69) for $J^{i*} \neq \emptyset$ (see (2.110)):

$$\overline{D}^i_{X^i h} = \overline{D}^i_{X^i h}\left(D^i_{X^i}, L^*_{X^i(2)}; J^{i*}\right) = \tag{3.79}$$

$$= \overline{D}^i_{X^i h}\left(\{Q(x^*)\}_{X^i} = D^i_{X^i}, L^*_{X^i(2)};\right.$$

$$\overline{\omega}^i_{X^i \, s} = \overline{\omega}^i_{X^i \, s(3.68)}, \; for \; s = 1, \ldots, n, \; under \; the \; conditions$$

$$(2.55c), \; (3.62) \; subject \; to \; \{(2.55), (3.59)\}, \; where$$

$$a_{sk}(x), \ldots, \overline{Q}(x^*), \; J^* \; is \; A_{sk}(X^i), \ldots, \{\overline{Q}(x^*)\}^i_{X^i}, \; J^{i*}$$

under the condition $\{\overline{Q}(x^*)\}^i_{X^i} = \overline{D}^i_{X^i},$

$$J^{i*} = J^{i*}_{(3.76)}\left(D^i_{X^i}; \Gamma_{X^i}\right), \; J^{i*} \neq \emptyset\right), \quad i = 1, \ldots, I_1.$$

For the subproblem (3.72) we construct a monotone difference scheme (see the scheme (2.111))

$$\Lambda_{X^i} Z(X^i) = F(X^i), \quad X^i \in D^i_{X^i h},$$ (3.80)

$$Z(X^i) = \Phi(X^i), \quad X^i \in \Gamma^{iL}_{X^i h}, \quad i = 1, \dots, I_1.$$

Here $\Lambda_{X^i} = \Lambda_{(2.46)}(L_{X^i})$ and $\overline{D}^i_{X^i h} = \overline{D}^i_{X^i h(3.79)}$.

Returning to the original variables, we obtain the monotone difference scheme (see (2.112)):

$$\Lambda_i z^*(x) \equiv \Lambda^* z^*(x) = f(x), \quad x \in D^i_{X^i h \{X^i\}^{-1}},$$ (3.81)

$$z^*(x) = \varphi(x), \quad x \in \Gamma^{iL}_{X^i h \{X^i\}^{-1}}, \quad i = 1, \dots, I_1.$$

Here

$$D^i_{X^i h \{X^i\}^{-1}} = D^i \cap \overline{D}^i_{X^i h \{X^i\}^{-1}}, \quad \Gamma^{iL}_{X^i h \{X^i\}^{-1}} = \Gamma^{iL} \cap \overline{D}^i_{X^i h \{X^i\}^{-1}},$$

$$\overline{D}^i_{X^i h \{X^i\}^{-1}} = \{\overline{D}^i_{X^i h}\}_{\{X^i\}^{-1}}; \quad z^*(x) = Z(X^i(x)), \quad x \in \overline{D}^i_{X^i h \{X^i\}^{-1}}.$$

We approximate the equation (3.1a) on the set D^i, for $i = I_1 + 1, \dots, I$, by the monotone difference scheme

$$\Lambda_i z(x) \equiv \Lambda z(x) = f(x), \quad x \in D^i_h, \quad i = I_1 + 1, \dots, I,$$ (3.82)

where $D^i_h = D^i_{h(3.78)}$ and $\Lambda = \Lambda_{(2.46)}(L)$.

To solve the problem (3.1), (3.70), we use the iterative scheme of the domain decomposition method $\{(2.98), (2.97), (2.114)\}$, where (see (2.114e), (2.114f)):

$$\overline{D}^i_h = \begin{cases} \overline{D}^i_{X^i h \{X^i\}^{-1}(3.81)} & for \quad i = 1, \dots, I_1, \\ \overline{D}^i_{h(3.78)} & for \quad i = I_1 + 1, \dots, I; \end{cases}$$ (3.83)

$$\Lambda_i = \begin{cases} \Lambda_{i(3.81)} & for \quad i = 1, \dots, I_1, \\ \Lambda_{i(3.82)} & for \quad i = I_1 + 1, \dots, I. \end{cases}$$ (3.84)

Set $N = \min_{i, s} N^i_s$, for $s = 1, \dots, n$, and $i = 1, \dots, I$.

The difference scheme $\{(2.98), (2.97), (2.114), (3.84), (3.83)\}$ is determined by *canonical elements*, i.e., the operators $\Lambda_{(2.46)}(L)$ and $\Lambda_{(2.46)}(L_{X^i})$ and the uniform and piecewise-uniform grids $\overline{D}^i_{h(3.78)}$ and $\overline{D}^i_{X^i h(3.79)}$, respectively.

The following theorem holds.

Theorem 3.6.2 *For the boundary value problem* (3.1), (3.70), *let the condition* $a_{sk} \in C^1(\overline{D})$, b_s, c, c^0, $f \in C(\overline{D})$, *for* $s, k = 1, 2, 3$, $\Gamma_j \in C^2$, *hold. Let* the coefficients $a_{sk}(x)$, $x \in \overline{D}$, and also $a_{sk}(x)$, $x \in \overline{D}^i$, for $i = I_1 + 1, \ldots, I$, and $A_{sk}^i(X^i)$, $X^i \in \overline{D}_{X^i}^i$, for $i = 1, \ldots, I_1$, satisfy, respectively, the conditions (3.73), and also (3.74) and (3.75), (3.77); and let the solution of the problem satisfy the estimates of Theorem 3.2.3 for $K = 3$. Then the solution $z^{[k]}(x)$ of the difference scheme $\{(2.98), (2.97), (2.114), (3.84), (3.83)\}$ converges to the solution of the boundary value problem $u(x)$ ε-uniformly as $N, k \to \infty$:

$$|u(x) - z^{[k]}(x)| \leq M \left[(N^{-1} \ln N)^\nu + q^k \right], \quad x \in \overline{D}_h, \quad q \leq 1 - m, \quad (3.85)$$

where $\nu = 1$ ($\nu = 4^{-1}$) in the case when the solution of the boundary value problem and its components satisfy the estimates (3.28) (the estimates (3.27)).

The interpolant $\breve{z}^{[k]}(x)$, $x \in \overline{D}$, of the solution of the iterative difference scheme, where

$$\breve{z}^{[k]}(x) = \breve{z}_{(2.114g)}^{[k]}(x), \quad x \in \overline{D}, \quad k \geq 1,$$

converges ε-uniformly

$$|u(x) - \breve{z}^{[k]}(x)| \leq M [(N^{-1} \ln N)^\nu + q^k], \quad x \in \overline{D}, \quad \nu = \nu_{(3.85)}.$$

The number of iterations k^f required for the solution of the iterative difference scheme with the estimate

$$|u(x) - z^{[k^f]}(x)| \leq M (N^{-1} \ln N)^\nu, \quad x \in \overline{D}_h, \quad \nu = \nu_{(3.85)},$$

is ε-uniformly bounded

$$k^f \leq M \ln N.$$

Chapter 4

Generalizations for elliptic reaction-diffusion equations

4.1 Monotonicity of continual and discrete Schwartz methods

Justification of Theorems 2.5.1–2.5.4 on the convergence of the continual and discrete Schwartz method is deduced using their ε-uniform monotonicity. First, we give the *monotonicity principle* for *the continual Schwartz method*.

Let on the sets $\overline{D}^{\,i} \subset \overline{D}^{\,(0)}$, for $i = 1, \ldots, I$, with $\overline{D}^{\,i} = \overline{D}^{\,i}_{(2.91)}$, that form a covering of the set $\overline{D}^{\,(1)}_{(2.90)}$, functions $w_i^{(k)}(x)$, $x \in \overline{D}^{\,i}$, where $i = 1, \ldots, I$ and $k = 1, 2, 3, \ldots$, be defined; these functions are 2π-periodic in x_s for $s = 2, \ldots, n$. Let $w^{[0]}(x)$, $x \in \overline{D}^{\,(0)}$, be a prescribed 2π-periodic function in x_s, for $s = 2, \ldots, n$; the functions $w_i^{[k]}(x)$, $w^{[k]}(x)$, $x \in \overline{D}^{\,(0)}$, are defined by the relations

$$w_i^{[k]}(x) = u_{i(2.91e)}^{[k]}\big(x; \{w_l^{(k)}(\cdot)\}, 1 \le l \le i, w^{[k-1]}(\cdot)\big), \quad i = 1, \ldots, I;$$

$$w^{[k]}(x) = w_I^{[k]}(x), \quad x \in \overline{D}^{\,(0)}, \quad k = 1, 2, 3, \ldots.$$

the functions $w_i^{[k]}(x)$, $x \in \overline{D}^{\,[k]}$, are 2π-periodic in x_s, for $s = 2, \ldots, n$.

The following lemma holds.

Lemma 4.1.1 *Let the functions* $w_i^{(k)}(x)$, $x \in \overline{D}^{\,i}$, *for* $i = 1, \ldots, I$ *and* $k = 1, 2, 3, \ldots$, *and also the functions* $w^{[0]}(x)$, $x \in \overline{D}^{\,(0)}$, *satisfy the relations*

$$Lw_i^{(k)}(x) \le 0, \quad x \in D^i, \quad w_i^{(k)}(x) \ge \left\{ \begin{array}{ll} 0, & x \in \Gamma^i \cap \Gamma \\ w_{i-1}^{[k]}(x), & x \in \Gamma^i \backslash \Gamma \end{array} \right\}, \quad x \in \Gamma^i,$$

$$w^{[0]}(x) \ge 0, \quad w_0^{[k]}(x) = w^{[k-1]}(x), \quad x \in \overline{D}^{\,(0)}, \quad i = 1, \ldots, I, \quad k = 1, 2, 3, \ldots.$$

Then the functions $w^{[k]}(x)$ *and* $w_i^{[k]}(x)$, $x \in \overline{D}^{\,(0)}$, *satisfy the inequalities*

$$w^{[k]}(x), \ w_i^{[k]}(x) \ge 0, \quad x \in \overline{D}^{\,(0)}, \quad i = 1, \ldots, I, \quad k = 1, 2, 3, \ldots.$$

The statement of the lemma is obtained by induction in both k and i for a fixed k.

To prove Theorem 2.5.1, we consider the functions

$$v_i^{(k)}(x) = u(x) - u_i^{(k)}(x), \quad x \in \overline{D}^i,$$

$$v_i^{[k]}(x) = u(x) - u_i^{[k]}(x), \quad v^{[k]}(x) = u(x) - u^{[k]}(x), \quad x \in \overline{D}^{(0)},$$

$$i = 1, \ldots, I, \quad k = 1, 2, 3, \ldots.$$

Here $u^{[k]}(x)$, $x \in \overline{D}^{(0)}$, is the solution of the problem (2.91); $u_i^{[k]}(x)$, $x \in \overline{D}^{(0)}$, and $u_i^{(k)}(x)$, $x \in \overline{D}^i$, are intermediate functions used for the construction of $u^{[k]}(x)$.

We construct majorant functions for the functions $v_i^{(k)}(x)$, $x \in \overline{D}^i$, $v_i^{[k]}(x)$, $v^{[k]}(x)$, $x \in \overline{D}^{(0)}$.

On each set $\overline{D}^i$, for $i = 1, \ldots, I$, one can choose a subset $\overline{D}^{i*}$ such that the set of all $\overline{D}^{i*}$, for $i = 1, \ldots, I$, covered $\overline{D}_{(2.90)}^{(1)}$, moreover, the sets $\overline{D}_\sigma^{i*} \subset \overline{D}^{(0)}$, where $\overline{D}_\sigma^{i*}$ are closed σ-neighborhoods of D^{i*}, are contained in $\overline{D}^i$, for $i = 1, \ldots, I$. The sets $\overline{D}^{i*}$ adjoin to the boundary Γ if $\overline{D}^i \cap \Gamma \neq \emptyset$. The value σ is chosen sufficiently small and independent of ε.

Let the functions $w_i(x)$, $x \in \overline{D}^i$, be solutions of the problems

$$L w_i(x) = 0, \quad x \in D^i, \quad w_i(x) = \begin{cases} 0, x \in \Gamma^i \cap \Gamma \\ 1, x \in \Gamma^i \setminus \Gamma \end{cases}, \quad x \in \Gamma^i, \quad i = 1, \ldots, I.$$

For the functions $w_i(x)$ one has the estimate

$$w_i(x) \leq q_0, \quad x \in \overline{D}^{i*}.$$

Here

$$q_0 = \min_\xi w_0(\xi), \quad \xi = (\xi_1, \ldots, \xi_n), \quad \sum_{i=1}^n \xi_i^2 \leq \sigma^2,$$

where

$$w_0(\xi) = 1 - m_1 + m_1 \sigma^{-2} \sum_{i=1}^n \xi_i^2, \quad m_1 = c^0 \left(c^0 + 2n\sigma^{-2} a^0 + 2n\sigma^{-1} b^0 \right)^{-1},$$

$$c^0 = c_{(2.3)}^0, \quad a^0 = a_{(2.3)}^0, \quad b^0 = \max_{s, \overline{D}^{(0)}} |b_{s(2.2)}(x)|.$$

The value q_0 satisfies the estimate

$$q_0 = 1 - m_1 < 1.$$

As majorants, we use the functions

$$w_i^{(k)}(x) = M q_0^k w_i(x), \quad x \in \overline{D}^i,$$

$$w^{[0]}(x) = M, \quad w_i^{[k]}(x) = M\, q_0^k, \quad x \in \overline{D}^{(0)}, \quad i = 1, \ldots, I, \quad k = 1, 2, 3, \ldots.$$

Applying Lemma 4.1.1 to the functions $w_i^{(k)}(x) \pm v_i^{(k)}(x)$, $x \in \overline{D}^i$, $w^{[0]}(x) \pm (u(x) - u^{[0]}(x))$, $x \in \overline{D}^{(0)}$, we verify that Theorem 2.5.1 holds where $q_{(2.92)} \leq q_0$.

Note that for the functions

$$w_i^*(x) = 1 - m_2 + m_2\, w_i(x), \quad x \in \overline{D}^i, \quad m_2 \in (0, 1)$$

one has the inequality

$$L\, w_i^*(x) \leq -(1 - m_2)\, c^0(x) \leq -m_3, \quad x \in \overline{D}^i, \quad i = 1, \ldots, I.$$

Next, we give the monotonicity principle for the discrete Schwartz method $\{(2.98), (2.97), (2.96)\}$ in the case of the periodic boundary value problem (2.2), (2.4).

Let on the sets $\overline{D}^i \subset \overline{D}^{(0)}$, $\overline{D}^i = \overline{D}_{(2.91)}^i$, for $i = 1, \ldots, I$, i.e., rectangular parallelepipeds, that form a covering of the set $\overline{D}_{(2.90)}^{(1)}$, rectangular grids $\overline{D}_{h(2.96)}^i$ be introduced, on which the operators $\Lambda_i = \Lambda_{i(2.46)}$ ε-uniformly monotone. Let on the grid sets $\overline{D}_h^i$ the functions $w_i^{(k)}(x)$, $x \in \overline{D}_h^i$, where $i = 1, \ldots, I$ and $k = 1, 2, 3, \ldots$, be defined, which are 2π-periodic in x_s, for $s = 2, \ldots, n$. Let $w^{[0]}(x)$, $x \in \overline{D}_h$, where $\overline{D}_h = \overline{D}_{h(2.97)}$, be an 2π-periodic function in x_s, for $s = 2, \ldots, n$. The functions $w_i^{[k]}(x)$, $x \in \overline{D}_h^{\{i\}}$, and $w^{[k]} \in \overline{D}_h$ are defined by the relations

$$w_i^{[k]}(x) = z_{i(2.98\mathrm{d})}^{[k]}(x)\big(x;\ \{w_l^{(k)}(\cdot)\},\ 1 \leq l \leq i,\ w^{[k-1]}(\cdot)\big) \quad x \in \overline{D}_h^{\{i\}},$$

$$w^{[k]}(x) = w_I^{[k]}(x), \quad x \in \overline{D}_h, \quad i = 1, \ldots, I, \quad k = 1, 2, 3, \ldots.$$

The functions $w^{[k]}(x)$, $x \in \overline{D}_h$, $w_i^{[k]}(x)$, $x \in \overline{D}^{\{i\}}$, are 2π-periodic in x_s, for $s = 2, \ldots, n$.

The following lemma holds.

Lemma 4.1.2 *In the case of grids $\overline{D}_h^i$, for $i = 1, \ldots, I$, let the operators Λ_i be ε-uniformly monotone, and let for the grid functions $w_i^{(k)}(x)$, $x \in \overline{D}_h^i$, where $i = 1, \ldots, I$ and $k = 1, 2, \ldots$, as well as for $w^{[0]}(x)$, $x \in \overline{D}_h$, the following relations be fulfilled:*

$$\Lambda_i\, w_i^{(k)} \leq 0, \quad x \in D_h^i,$$

$$w_i^{(k)}(x) \geq \left\{ \begin{array}{ll} 0, & x \in \Gamma_h^i \cap \Gamma \\ \tilde{w}_{i-1}^{[k]}(x), & x \in \Gamma_h^i \backslash \Gamma \end{array} \right\}, \quad x \in \Gamma_h^i,$$

$$w^{[0]}(x) \geq 0, \quad w_0^{[k]}(x) = w^{[k-1]}(x), \quad x \in \overline{D}_h, \quad i = 1, \ldots, I, \quad k = 1, 2, 3, \ldots.$$

Then the functions $w^{[k]}(x)$, $x \in \overline{D}_h$, $w_i^{[k]}(x)$, $x \in \overline{D}_h^{\{i\}}$, satisfy the inequalities

$$w^{[k]}(x) \geq 0, \quad x \in \overline{D}_h, \quad w_i^{[k]}(x) \geq 0, \quad x \in \overline{D}_h^{\{i\}}, \quad i = 1, \dots, I, \quad k = 1, 2, 3, \dots.$$

The estimate (2.100) is derived similar to (2.92).

Let us formulate a monotonicity principle for the noniterative difference scheme of the domain decomposition method $\{(2.99), (2.97), (2.96)\}$.

Let on the grid sets $\overline{D}_h^i = \overline{D}_{h(2.96)}^i$, the operators $\Lambda_i = \Lambda_{i(2.46)}$ be defined which are ε-uniformly monotone. Let on the grids $\overline{D}_h^i$ the functions $w_i^0(x)$, $x \in \overline{D}_h^i$, for $i = 1, \dots, I$, be defined, which are 2π-periodic function in x_s, for $s = 2, \dots, n$. The functions $w_i(x)$, $x \in \overline{D}_h^{\{i\}}$, and $w(x)$, $x \in \overline{D}_h$, are defined by the relations

$$w_i(x) = z_{i(2.99d)}\big(x; \{w_l^0(\cdot)\}, l = 1, \dots, n, l \neq i\big), \quad x \in \overline{D}_h^{\{i\}}, \quad i = 1, \dots, I,$$

$$w(x) = w_I(x), \quad x \in \overline{D}_h.$$

Lemma 4.1.3 *In the case of grids $\overline{D}_h^i$, for $i = 1, \dots, I$, let the operators Λ_i be ε-uniformly monotone, and let for the grid functions $w_i^0(x)$, $x \in \overline{D}_h^i$, for $i = 1, \dots, I$, as well as for the function $w(x)$, $x \in \overline{D}_h$, the following relations be fulfilled:*

$$\Lambda_i w_i^0(x) \leq 0, \quad x \in D_h^i;$$

$$w_i^0(x) \geq \left\{ \begin{array}{ll} 0, & x \in \Gamma_h^i \cap \Gamma \\ \breve{w}_{i-1}(x), & x \in \Gamma_h^i \backslash \Gamma \end{array} \right\}, \quad x \in \Gamma_h^i, \quad i = 1, \dots, I;$$

$$\breve{w}_0(x) = \breve{w}(x), \quad x \in \overline{D}_h.$$

Then the functions $w(x)$, $x \in \overline{D}_h$, $w_i(x)$, $x \in \overline{D}_h^{\{i\}}$, satisfy the inequalities

$$w(x) \geq 0, \quad x \in \overline{D}_h, \quad w_i(x) \geq 0, \quad x \in \overline{D}_h^{\{i\}}, \quad i = 1, \dots, I.$$

Using Lemma 4.1.3, and under the hypotheses of Theorem 2.5.2, we obtain the estimate (2.101).

Statement of Theorem 2.5.2 on the convergence of the difference scheme $\{(2.98), (2.97), (2.96)\}$ to the solution of the boundary value problem (2.2), (2.4) follows from the estimates (2.100), (2.101).

4.2   Approximation of the solution in a bounded subdomain for the problem on a strip

For the boundary value problem on the slab, i.e., the problem (2.2), (2.4) in Section 2.1, difference schemes are constructed that converge ε-uniformly (see, e.g., Theorems 2.4.8-2.4.10, 2.5.2). Under the construction of these schemes, we have used grids with infinite numbers of nodes. Schemes on grids with infinite numbers of nodes, i.e., *formal difference schemes*, are, in general, inapplicable for computations (for example, schemes for the problem (2.2), (2.4) which is not periodic). In the case of problems in unbounded domains, one needs numerical methods used grids with finite numbers of nodes, i.e., *constructive numerical methods* that are applicable for computations. In particular, for the nonperiodic problem (2.2), (2.4) we are interested in *constructive difference schemes* that converge ε-uniformly.

In this connection, it is suggested to consider the following computational problem, namely, for a boundary value problem on a slab, to find its solution on a prescribed bounded domain (say on a subdomain of our interest). For such a particular problem, it is required to construct grid approximations that use grids with finite numbers of nodes. In the case of a convection-diffusion problem, a perturbation of the solution, caused by a perturbation of the data of the problem on some distance from the subdomain of our interest, exponentially decreases as the distance grows. Perturbations of the data far from the subdomain of our interest hardly affect the problem solution on this subdomain. We use this property of the solution of the boundary value problem for the construction of a constructive difference scheme for the solution of the particular problem. It is required that the discrete solution on the prescribed subdomain would converge ε-uniformly with growth of the number of nodes in the meshes used.

For the problem (2.2) on the slab (2.4), it is required to find its solution on the prescribed bounded set

$$\overline{D}_0 \subset \overline{D}. \tag{4.1}$$

We assume that the coefficients of the operator L satisfy the conditions (2.57) and (2.71) (the diagonal elements in the matrix of the coefficients of the operator $L^*_{(2)(2.50)}$ dominate on the whole set $\overline{D}$). To construct continual approximations of the solution on the set $\overline{D}_0$, we consider auxiliary boundary value problems on extending bounded subdomains that contain the set $\overline{D}_0$. We denote by Q_0, where

$$Q_0 = Q_0(D_0; \eta)$$

a rectangular η-neighborhood of the set D_0 from R^n; Q_0 is a rectangular parallelepiped whose faces belong to the planes $x_s =$const, for $s = 1, \ldots, n$, moreover,

$$\min_{x \in \overline{D}} r(x, \Gamma(Q_0)) = \eta;$$

here $\Gamma(Q_0)$ is the boundary of the set Q_0. Set

$$\overline{D}^\eta = D^\eta \cup \Gamma^\eta, \quad D^\eta = D^\eta(D_0; \eta) = Q_0(D_0; \eta) \cap D_{(2.4)}; \tag{4.2}$$

D^η is a rectangular η-neighborhood (of the set D_0) from $D_{(2.4)}$.

On the set $\overline{D}^\eta$ we consider the boundary value problem

$$L\,u_0(x) = f(x), \qquad x \in D^\eta, \tag{4.3}$$

$$u_0(x) = \left\{ \begin{array}{ll} \varphi(x), & x \in \Gamma^\eta \cap \Gamma \\ \varphi^*(x), & x \in \Gamma^\eta \backslash \Gamma \end{array} \right\} \equiv \varphi_{u_0}(x), \quad x \in \Gamma,$$

where $L = L_{(2.2)}$, $\Gamma = \Gamma(D)$, and $\varphi^*(x)$ is a bounded function; $\varphi_{u_0}(x)$ is continuous on Γ^η.

The solution of the problem (4.3), (4.2) considered on $\overline{D}_0$ satisfies the estimate

$$|u(x) - u_0(x)| \le M \exp(-m\,\eta), \quad x \in \overline{D}_0, \tag{4.4}$$

where m is an arbitrary number in the interval $(0, m^0)$, with

$$m^0 = \left(a^{0-1} c^0 \right)^{1/2}, \quad a^0 = a^0_{(2.3)}, \quad c^0 = c^0_{(2.3)}.$$

Thus, the solution of the problem (4.3), (4.2) converges on the set $\overline{D}_0$ to the solution of the boundary value problem (2.2), (2.4) ε-uniformly as $\eta \to \infty$.

We consider a grid approximation of the problem (4.3), (4.2) that converges ε-uniformly.

On the set $\overline{D}^\eta$ we construct a rectangular grid

$$\overline{D}_h = \overline{D}_h(D^\eta) = \overline{\omega}_1 \times \ldots \overline{\omega}_n, \tag{4.5a}$$

where $\overline{\omega}_s$ is, in general, nonuniform mesh on the interval $[d_{*s}, d_s^*]$ that is a projection of the parallelepiped $\overline{D}^\eta$ on the x_s-axis, for $s = 1, \ldots, n$; let $N_s^\eta + 1$ be the number of nodes in the mesh $\overline{\omega}_s$, and $N^\eta = \min_s N_s^\eta$, moreover, $h_s \le M\,(N_s^\eta)^{-1}(d_s^* - d_{*s})$, where h_s is the maximal step-size in the mesh $\overline{\omega}_s$, for $s = 1, \ldots, n$. The value η that defines the set $\overline{D}^\eta$ satisfies the condition

$$\eta = m^{-1} \ln N_1^\eta, \quad m = m_{(4.4)}. \tag{4.5b}$$

For the problem (4.3), (4.2) we consider a difference scheme

$$\Lambda z(x) = f(x), \qquad x \in D_h, \tag{4.6}$$

$$z(x) = \left\{ \begin{array}{ll} \varphi(x), & x \in \Gamma_h \cap \Gamma, \\ \varphi^*(x), & x \in \Gamma_h \backslash \Gamma, \end{array} \right.$$

where $\Lambda = \Lambda_{(2.46)}$, $D_h = D^\eta \cap \overline{D}_h$, $\Gamma_h = \Gamma^\eta \cap \overline{D}_h$, and $\Gamma = \Gamma(D)$. As the grid (4.5) we shall use the piecewise-uniform grid

$$\overline{D}_h = \overline{D}_{h(2.58)} \cap \overline{D}^\eta. \tag{4.5c}$$

The fulfillment of the conditions (2.57) and (2.71) ensures a transformation of the operator $L_{(2)}^*$ to strongly almost canonical form on the set $\overline{D}^\eta$. On the set $\overline{D}^\eta$ we introduce the grid

$$\overline{D}_h = \overline{D}_{h(2.70)} \cap \overline{D}^\eta, \tag{4.7a}$$

where $\overline{D}_{h(2.70)} = \overline{D}_{h(2.70)}(Q = \overline{D}^{\eta})$, on which the operator $\Lambda_{(4.6)}$ is monotone.

Let the parameters of the grid (4.7) satisfy the relations

$$h_s \approx (N_1^{\eta})^{-1}, \quad N^{\eta} = \min_s N_s^{\eta} = N_1^{\eta}, \quad s = 1, \ldots, n; \tag{4.7b}$$

$$N_s^{\eta} \approx N_1^{\eta} \ln N_1^{\eta}, \quad \overline{N}^{\eta} = \min_s N_s^{\eta} \approx N_1^{\eta} \ln N_1^{\eta}, \quad s = 2, \ldots, n.$$

Taking into account *a priori* estimates of Theorem 2.2.1, for the solution of the difference scheme (4.6), (4.7) on the set $\overline{D}_{0h}$ we obtain the ε-uniform estimate

$$|u(x) - z(x)| \le M \, (N_1^{\eta})^{-1} \ln N_1^{\eta}, \quad x \in \overline{D}_{0h}, \tag{4.8}$$

where $\overline{D}_{0h} = \overline{D}_{h(4.7)} \cap \overline{D}_{0(4.1)}$.

The following theorem holds.

Theorem 4.2.1 *Let* $a_{sk} \in C^1(\overline{D})$, b_s, c, c^0, $f \in C(\overline{D})$, *for* $s, k = 1, \ldots, n$; *the coefficients* $a_{sk}(x)$ *satisfy the conditions* (2.57), (2.66), (2.71), *and let for the solution of the problem* (2.2), (2.4) *the estimates of Theorem 2.2.1 hold for* $K = 3$. *Then the solution of the difference scheme* (4.6) *on the grid* $\overline{D}_{h(4.7)}$, *where* $\overline{\omega}_{1(2.48)} = \overline{\omega}_{1(2.44)}$, *converges* ε-*uniformly on the set* $\overline{D}_{0h}$ *as* $N^{\eta} \to \infty$ *with the estimate* (4.8).

Remark 4.2.1 The number of nodes in the grid $\overline{D}_{h(4.7)}$, on which the discrete problem (4.6), (4.7) is solved, is a value of order $\mathcal{O}\big((N_1^{\eta})^n \ln^{n-1} N_1^{\eta}\big)$. The number of nodes on the set $\overline{D}_{0h}$ is a value of order $\mathcal{O}\big((N_1^{\eta})^n\big)$. Thus, the logarithmic multiplier $\ln^{n-1} N_1^{\eta}$ is an additional price for solving of the boundary value problem on the set $\overline{D}_0$ in the case when the boundary value problem is given on the unbounded domain, i.e., on the n-dimensional slab $\overline{D}_{(2.4)}^{(0)}$. ∎

For the problem (2.2), (2.4), it is required to find its solution on the set (4.1). We assume that the coefficients of the operator L satisfy the conditions (2.57) and (2.93) (the diagonal elements in the matrix of the coefficients of the operator $L_{(2)(2.50)}^*$ dominate on $\overline{D}$ pointwisely); periodicity of the problem (2.2), (2.4) is not assumed. When constructing a difference scheme for the problem (4.3), (4.2), we use the discrete Schwartz method $\{(2.98), (2.97), (2.96)\}$ adapted to the problem (4.3), (4.2)

We give the continual Schwartz method for the differential problem (4.3), (4.2).

Let the system of overlapping sets, i.e., rectangular parallelepiped,

$$D^i, \quad i = 1, \ldots, I, \tag{4.9a}$$

cover the set D^{η}, i.e., $\overline{D}^{\eta} = \bigcup_i \overline{D}_i$. We assume that the condition for the width of overlapping to the sets holds, i.e., the minimal distance from the boundary of the set D^i, for $i = 1, \ldots, I$, to the boundary of an union of the sets which do not contain D^i (i.e., the minimal overlapping of the subdomains D^i, for $i = 1, \ldots, I$, that cover D^η) is independent of ε. The distance between the sets D^i and the boundary Γ, under the condition that $\overline{D}^i \cap \Gamma = \emptyset$, is independent of ε.

On the set $\overline{D}^\eta$ we construct functions $u^{[k]}(x)$, for $k = 1, 2, 3, \ldots$, in the following way. Assume that

$$u^{[0]}(x) = 0, \quad x \in \overline{D}^\eta. \tag{4.9b}$$

Let for $k \geq 1$ the function $u^{[k-1]}(x)$ be already constructed. Construct $u^{[k]}(x)$. For this, we consider auxiliary functions

$$u_i^{[k]}(x) = \left\{ \begin{array}{ll} u_i^{(k)}(x), & x \in \overline{D}^i \\ u_{i-1}^{[k]}(x), & x \in \overline{D}^\eta \setminus \overline{D}^\eta \end{array} \right\}, \quad x \in \overline{D}^\eta; \tag{4.9c}$$

$$u_0^{[k]}(x) = u^{[k-1]}(x), \quad x \in \overline{D}^\eta, \quad i = 1, \ldots, I.$$

Here $u_i^{(k)}(x)$, $x \in \overline{D}^i$, is the solution of the boundary value problem on the set $\overline{D}^i$

$$L u_i^{(k)}(x) = f(x), \qquad x \in D^i, \tag{4.9d}$$

$$u_i^{(k)}(x) = \left\{ \begin{array}{ll} \varphi(x), & x \in \Gamma^i \cap \Gamma \\ \varphi^*(x), & x \in \Gamma^i \cap \{\Gamma^\eta \setminus \Gamma\} \\ u_{i-1}^{[k]}(x), & x \in \Gamma^i \setminus \Gamma^\eta \end{array} \right\}, \quad x \in \Gamma^i,$$

where $\Gamma^i = \overline{D}^i \setminus D^i$. The function $u^{[k]}(x)$, $x \in \overline{D}^\eta$, is defined by the relation

$$u^{[k]}(x) = u_I^{[k]}(x), \quad x \in \overline{D}^\eta. \tag{4.9e}$$

The function $u_{(4.9e)}^{[k]}(x)$, $x \in \overline{D}^\eta$, is k-th iteration *the continual Schwartz method* (2.91), $k \geq 1$ for the problem (4.3), (4.2).

Theorem 4.2.2 *The function $u_{(4.9)}^{[k]}(x)$, i.e., the solution of the continual Schwartz method (4.9) for the boundary value problem (4.3), (4.2), converges to the solution of the boundary value problem on the set $\overline{D}^\eta$ ε-uniformly as $k \to \infty$*

$$|u_0(x) - u^{[k]}(x)| \leq M q^k, \quad x \in \overline{D}^\eta, \quad q < 1. \tag{4.10}$$

For the Schwartz method (4.9), we construct a finite difference scheme and formulate conditions under which the scheme converges ε-uniformly.

In the case of conditions (2.57) and (2.93), one can choose the diameters of the sets $\overline{D}^i$ sufficiently small such that the following inequality holds on $\overline{D}^i$:

$$\max_{\overline{D}^i} |a_{sk}(x)| < (n-1)^{-1} \min_{\overline{D}^i} \left[\alpha_s^i \, (\alpha_k^i)^{-1} \, a_{ss}(x), \; \alpha_k^i \, (\alpha_s^i)^{-1} \, a_{kk}(x) \right],$$

$$s, k = 1, \ldots, n, \quad s \neq k, \quad i = 1, \ldots, I. \tag{4.11}$$

This ensures a transformation of the operator $L^*_{(2)(2.50)}$ to almost canonical form on $\overline{D}^i$ and also a transformation of the operator $L^{*[1]}_{(2)(2.62)}$ to strongly almost canonical form if $\overline{D}^i \cap \Gamma \neq \emptyset$.

We assume that the operator $L^*_{(2)}$ is locally almost canonical on $\overline{D}$ and almost canonical on the subdomains $\overline{D}^i$, and the operator $L^{*[1]}_{(2)}$ under the condition $\overline{D}^i \cap \Gamma \neq \emptyset$ is strongly almost canonical. We also suppose that the condition

the coefficients $a_{sk}(x)$ of the operator $L^*_{(2)}$ on the set $\overline{D}^i$

satisfy the relations $\{(2.55), (2.65)\}$, for $\overline{D}^i \cap \Gamma \neq \emptyset$, and $\qquad$ (4.12)

the relations (2.55), for $\overline{D}^i \cap \Gamma = \emptyset$, where $\overline{Q}(x^*) = \overline{D}^i$, $i = 1, \ldots, I$,

holds that admits the construction of consistent grids in the variables $x_1, \ldots, x_n$ with the monotonicity condition for the difference operator $\Lambda^*_{(2)(2.51)}$ $(L^*_{(2)})$ on $\overline{D}^i$.

On the set $\overline{D}^i$ we construct the consistent grid

$$\overline{D}^i_h = \overline{D}^i_h \left(\overline{D}_{h(2.56)}, \, \overline{D}_{h(2.70)} \right) \equiv \tag{4.13}$$

$$\equiv \left\{ \begin{array}{l} \overline{D}_{h(2.56)} \left(Q(x^*) = D^i \right) \; for \; \overline{D}^i \cap \Gamma = \emptyset \\ \overline{D}_{h(2.70)} \left(Q(x^*) = D^i \right) \; for \; \overline{D}^i \cap \Gamma \neq \emptyset \end{array} \right\}, \quad i = 1, \ldots, I,$$

that ensures the monotonicity of the operators $\Lambda^*_{(2)} = \Lambda^*_{(2)(2.51)}(L^*_{(2)})$ and $\Lambda = \Lambda_{(2.46)}(L)$ on $\overline{D}^i_h$. The *grid* $\overline{D}^i_{h(4.13)}$ on the set $\overline{D}^i$ is constructed *based on the grid* $\overline{D}_{h(2.56)}$, or $\overline{D}_{h(2.70)}$ (*introduced on the set* $\overline{D}^\eta_{(4.2)}$) and on the basis, respectively, the relations (2.55), or $\{(2.55), (2.65)\}$ for $\overline{Q}(x^*) = \overline{D}^i$.

On the set $\overline{D}^\eta$ we construct the grid

$$\overline{D}_h = \overline{D}_h^{\{0\}}, \tag{4.14a}$$

where we set

$$\overline{D}_{(1)h} = \overline{D}^1_h, \quad \overline{D}_{(i)h} = \overline{D}^i_h \cup \left\{ \overline{D}_{(i-1)h} \setminus \overline{D}^i \right\}, \quad i = 2, \ldots, I, \tag{4.14b}$$

$$\overline{D}_h^{\{0\}} = \overline{D}_{(I)h}.$$

On $\overline{D}^{\,\eta}$ we also introduce a set of the grids $\overline{D}_h^{\{i\}}$, for $i = 1, \dots, I$,

$$\overline{D}_h^{\{i\}} = \overline{D}_h^i \cup \{\overline{D}_h^{\{i-1\}} \setminus \overline{D}^i\}, \quad i = 1, \dots, I. \tag{4.14c}$$

On the grid $\overline{D}_{h(4.14a)}$ we construct a sequence of grid functions $z^{[k]}(x)$, for $k = 1, 2, \dots$. Set $z^{[0]}(x) \equiv 0$, $x \in \overline{D}_h$. We construct $z^{[k]}(x)$ assuming that the function $z^{[k-1]}(x)$ for $k \geq 1$ has already been constructed. Next, we introduce the auxiliary functions $z_i^{[k]}(x)$

$$z_i^{[k]}(x) = \begin{cases} z_i^{(k)}(x), & x \in \overline{D}_h^i \\ z_{i-1}^{[k]}(x), & x \in \overline{D}_h^{\{i-1\}} \setminus \overline{D}^i \end{cases}, \quad x \in \overline{D}_h^{\{i\}}, \tag{4.15a}$$

$$z_0^{[k]}(x) = z^{[k-1]}(x), \quad x \in \overline{D}_h, \quad i = 1, \dots, I.$$

In these relations $z_i^{(k)}(x)$, $x \in \overline{D}_h^i$, is the solution of the discrete boundary value problem on the set $\overline{D}_h^i$

$$\Lambda_i z_i^{(k)}(x) = f(x), \quad x \in D_h^i; \tag{4.15b}$$

$$z_i^{(k)}(x) = \begin{cases} \varphi(x), & x \in \Gamma_h^i \cap \Gamma \\ \varphi^*(x), & x \in \Gamma_h^i \cap \{\Gamma^\eta \setminus \Gamma\} \\ \check{z}_{i-1}^{[k]}(x), & x \in \Gamma_h^i \setminus \Gamma^\eta \end{cases}, \quad x \in \Gamma_h^i, \quad i = 1, \dots, I.$$

Here Λ_i is $\Lambda = \Lambda_{(2.46)}(L)$ on the grid $\overline{D}_h^i$,

$$\check{z}_0^{[k]}(x) = \check{z}^{[k-1]}(x), \quad x \in \Gamma_h^1 \setminus \Gamma; \tag{4.15c}$$

the functions $\check{z}_{i-1}^{[k]}(x)$, $x \in \overline{D}$ and $\check{z}^{[k-1]}(x)$, $x \in \overline{D}$, are constructed as the interpolation of the functions $z_{i-1}^{[k]}(x)$, $x \in \overline{D}_h^{\{i-1\}}$, and $z^{[k-1]}(x)$, $x \in \overline{D}_h$, for $i = 1, \dots, I$.

The function $z^{[k]}(x)$, $x \in \overline{D}_h$, is determined by the relation

$$z^{[k]}(x) = z_I^{[k]}(x), \quad x \in \overline{D}_h. \tag{4.15d}$$

The interpolants $\check{z}_i^{[k]}(x)$ and $\check{z}^{[k]}(x)$, $x \in \overline{D}$, are defined by the relation (2.98f).

The function $z^{[k]}(x)$, $x \in \overline{D}_h$, $\overline{D}_h = \overline{D}_{h(4.14)}$, for $k \geq 1$, is the solution of the difference scheme $\{(4.15), (4.14), (4.13)\}$, i.e., *the iterative scheme of the domain decomposition method on the overlapping subdomains.*

The iterative difference scheme $\{(4.15), (4.14), (4.13)\}$ is defined by *cano-nical elements* of the iterative scheme of the domain decomposition method, i.e., the operator $\Lambda_{(2.46)}$ and the grids $\overline{D}_{h(2.56)}(\overline{\omega}_1^u)$ and $\overline{D}_{h(2.70)}$ $(\overline{\omega}_{1(2.48)})$, on the basis of which the difference scheme is constructed. Set $N^i = \min_s N_s^i$ and

$N = \min\limits_{i} N^i$, where $N_s^i + 1$ is the number of nodes in the grid $\overline{D}_h^{\,i}$ along the x_s-axis, for $s = 1, \ldots, n$.

The parameters of the grids $\overline{D}_h^{\,i}$ satisfy the relations

$$d_s^{*i} - d_{*s}^i \approx 1, \quad N_s^i \approx N, \quad s = 1, \ldots, n, \quad i = 1, \ldots, I. \qquad (4.16\text{a})$$

The value η satisfies the condition

$$\eta = m^{-1} \ln N_1, \qquad (4.16\text{b})$$

where $N_1 = \min_i N_1^i$, for $i = 1, \ldots, I$, and $m = m_{(4.4)}$. Thus, the value I, i.e., the number of the subdomains $\overline{D}^{\,i}$ covering the set $\overline{D}^{\,\eta}$, depends on N

$$I = I(N) \qquad (4.16\text{c})$$

and increases unboundedly as N grows. For I one has the estimate

$$I = I(N) \le M \ln^{n-1} N. \qquad (4.17)$$

For the solution of the difference scheme $\{(4.15),\ (4.14),\ (4.13),\ (4.16)\}$ under the additional condition

$$\text{In } \overline{D}_{h(4.13)}^{\,i}\left(\overline{D}_{h(2.56)}(\overline{\omega}_1^u),\ \overline{D}_{h(2.70)}(\overline{\omega}_{1(2.48)})\right)$$
$$\text{the mesh } \overline{\omega}_{1(2.48)} \text{ is } \overline{\omega}_{1(2.44)} \qquad (4.16\text{d})$$

the following ε-uniform estimate on the set $\overline{D}_{0h}$ holds:

$$|z(x) - z^{[k]}(x)| \le M\,[N^{-1} \ln N + q^k], \quad x \in \overline{D}_{0h}, \qquad (4.18)$$

where $\overline{D}_{0h} = \overline{D}_{h(4.14)} \cap \overline{D}_{0(4.1)}$.

The number of iterations k^f required for the solution of the difference scheme $\{(4.15),\ (4.14),\ (4.13),\ (4.16)\}$ (where $\overline{\omega}_{1(2.48)}$ is $\overline{\omega}_{1(2.44)}$) with the estimate

$$|u(x) - z^{[k^f]}(x)| \le M\,N^{-1} \ln N, \quad x \in \overline{D}_{0h}, \qquad (4.19\text{a})$$

is ε-uniformly bounded.

$$k^f \le M \ln N. \qquad (4.19\text{b})$$

The following theorem on the convergence of the iterative difference scheme holds.

Theorem 4.2.3 *Let the conditions $a_{sk} \in C^1(\overline{D})$, b_s, c, c^0, $f \in C(\overline{D})$, for $s, k = 1, \ldots, n$, hold and the coefficients $a_{sk}(x)$ satisfy the conditions (2.57), (2.93), (2.95). For the solution of the problem (2.2), (2.4), let the estimates of Theorem 2.2.1 be satisfied for $K = 3$. Then the iterative difference scheme (4.15), (4.14) on the grids $\overline{D}_{h(4.13,4.16)}$ converges ε-uniformly on the set $\overline{D}_{0h}$ as $N, k \to \infty$ with the estimates (4.18), (4.19).*

The interpolants $\check{z}^{[k]}(x)$ of the solutions of the iterative difference scheme converge ε-uniformly on the set $\overline{D}_0$ with the estimates

$$|u(x) - \check{z}^{[k]}(x)| \le M \left[N^{-1} \ln N + q^k \right], \quad x \in \overline{D}_0,$$

$$|u(x) - \check{z}^{[k^f]}(x)| \le M N^{-1} \ln N, \quad x \in \overline{D}_0, \quad k^f \le M \ln N.$$

Remark 4.2.2 When computing the solutions of the iterative difference scheme $\{(4.15), (4.14), (4.13), (4.16)\}$ on (one) an iterative step, intermediate solutions are computed on grids with the common number of nodes of order $\mathcal{O}(N^n \ln^{n-1} N)$. Taking account of the number of iterations required for convergence of the iterative process, intermediate solutions are computed on grids with the common number of nodes of order $\mathcal{O}(N^n \ln^n N)$. At the same time, the number of nodes in the grid $\overline{D}_{0h}$, just as the number of the nodes in the grids $\overline{D}_h^i$ that belong to $\overline{D}_{0h}$, is $\mathcal{O}(N^n)$. According to the problem formulation, it is required to find approximate solutions of the problem (2.2), (2.4) only on the bounded grid set (the grid) $\overline{D}_{0h}$. Thus, the logarithmic multiplier $\ln^n N$ is *an additional price* for computation (with accuracy of $\mathcal{O}(N^{-1} \ln N)$) on the set $\overline{D}_0$ of the solution of the boundary value problem (2.2), (2.4) when we apply the iterative domain decomposition method. ∎

4.3 Difference schemes of improved accuracy for the problem on a slab

We now construct monotone difference schemes that converge ε-uniformly with an order of convergence close to two.

For the differential operator $L_{(2.2)}$ we consider monotone approximations of improved accuracy on grids that condense in the boundary layer.

To the operator $L_{(2.2)}$ on the grid $\overline{D}_h$ we associate the operator:

$$\Lambda = \Lambda(L) \equiv \varepsilon^2 \left\{ \sum_{s=1}^{n} a_{ss}(x) \delta_{\overline{xs}\,\widehat{xs}} + 2^{-1} \sum_{\substack{s,k=1 \\ s \ne k}}^{n} \left[a_{sk}^+(x) \left(\delta_{xs\,xk} + \delta_{\overline{xs}\,\overline{xk}} \right) + \right. \right.$$

$$\left. \left. + a_{sk}^-(x) (\delta_{xs\,\overline{xk}} + \delta_{\overline{xs}\,xk}) \right] + \sum_{s=1}^{n} b_s(x) \delta_{\widehat{xs}} - c(x) \right\} - c^0(x), \tag{4.20}$$

in which, unlike the operator $\Lambda_{(2.46)}$, the first-order derivatives are approximated by central difference derivatives.

The operator $\Lambda_{(4.20)}$ acting on smooth functions, in the case of uniform meshes, approximates the operator $L_{(2.2)}$ with the second order of accuracy.

On the grid (2.26) with an arbitrary distribution of nodes, under the condition (2.49) the operator $\Lambda_{(4.20)}$ is ε-uniformly monotone for

$$N \geq 2M \max_{s,\overline{D}} \left[a_{ss}^{-1}(x) \, |b_s(x)| \right], \quad M = M_{(2.26)}, \quad s = 1,\dots,n.$$

We consider an approximation of the operator L by the operator $\Lambda_{(4.20)}$ on the solution of the problem (2.2), (2.4) in the case of piecewise-uniform grids.

For the value $\psi_{11(2.31)}(x^i)$, i.e., the *local error* of the approximation of the operator $\varepsilon^2 \dfrac{\partial^2}{\partial x_1^2}$ by the operator $\varepsilon^2 \delta_{\overline{x1}\,\widehat{x1}}$ on the grid (2.26), the following estimate holds:

$$\psi_{11}(x^i) = \psi_{11}(x^i; u) \leq M\varepsilon^2 \min\left\{ (h_1^i + h_1^{i-1})^2 \max_\xi \left| \frac{\partial^4}{\partial x_1^4} u(\xi) \right|, \right.$$

$$\left. |h_1^i - h_1^{i-1}| \max_\xi \left| \frac{\partial^3}{\partial x_1^3} u(\xi) \right|, \ \max_\xi \left| \frac{\partial^2}{\partial x_1^2} u(\xi) \right| \right\}, \ x^i \in D_h, \ x^i = (x_1^i, x_2, \dots, x_n),$$

$$\xi = (\xi_1, \xi_2, \dots, \xi_n), \quad \xi_1 \in [x_1^{i-1}, x_1^{i+1}], \quad \xi_s = x_s, \quad s = 2, \dots, n.$$

For $\psi_{11}(x; u(\cdot))$ on the regular component $U(x)$ in (2.7) in the case of the piecewise-uniform grid $\overline{D}_{h(2.58)}$, we have

$$|\psi_{11}(x; U(\cdot))| \leq \begin{cases} M \, N_1^{-2}, & x_1 \neq d_{*1} + \sigma, \, d_1^* - \sigma, \\ M \varepsilon \, N_1^{-1}, & x_1 = d_{*1} + \sigma, \, d_1^* - \sigma, \ x \in D_h. \end{cases} \quad (4.21a)$$

On the singular component $V(x)$ in (2.7), one has

$$|\psi_{11}(x; V(\cdot))| \leq \begin{cases} M \, \varepsilon^{-2} h_{(1)}^2, & x_1 \notin [d_{*1} + \sigma, \, d_1^* - \sigma], \\ M \exp(-m\varepsilon^{-1}\sigma), & x_1 \in [d_{*1} + \sigma, \, d_1^* - \sigma], \ x \in D_h. \end{cases} \quad (4.21b)$$

For the operator $\Lambda_{(4.20)}$ on the components $U(x)$ and $V(x)$ in the case of the piecewise-uniform grid

$$\overline{D}_h = \overline{D}_{h(2.58)}, \quad (4.22a)$$

where

$$\overline{\omega}_1 = \overline{\omega}_{(2.43)}^{(1)}(d_{*1}, d_1^*; \ 2^{-1}m), \quad m < m_{0(2.12)}, \quad (4.22b)$$

by virtue of (4.21), one obtains the estimates

$$|(L - \Lambda) U(x)| \leq \begin{cases} M \, N^{-2}, & x_1 \neq d_{*1} + \sigma, \, d_1^* - \sigma, \\ M \, [\varepsilon \, N_1^{-1} + N^{-2}], & x_1 = d_{*1} + \sigma, \, d_1^* - \sigma, \ x \in D_h; \end{cases}$$

$$|(L - \Lambda) V(x)| \leq M \, N^{-2} \ln^2 N, \quad x \in D_h.$$

In the case when the difference scheme

$$\Lambda z(x) = f(x), \ x \in D_h, \quad z(x) = \varphi(x), \ x \in \Gamma_h, \quad (4.23)$$

where $\Lambda = \Lambda_{(4.20)}$, is ε-uniformly monotone on the grids (4.22), for its solution we obtain the estimate

$$|u(x) - z(x)| \leq M\,N^{-2}\,\ln^2 N, \quad x \in \overline{D}_h. \tag{4.24}$$

Here for the construction of the majorant we use the function

$$w(x) = 1 - \max\left[\exp\left(-m\,\varepsilon^{-1}\,r(x,\Gamma)\right),\ \exp\left(-m\,\varepsilon^{-1}\,\sigma)\right)\right], \quad x \in \overline{D}_h,$$

where $m < m_{0(2.14)}$.

Theorem 4.3.1 *Let for the solution of the boundary value problem (2.2), (2.4) the estimates of Theorem 2.2.1 hold for $K = 4$. Then the solution of the monotone difference scheme (4.23), (4.20), (4.22) satisfies the estimate (4.24).*

We now study conditions that ensure the monotonicity of the operator $\Lambda_{(4.20)}$; the fulfillment of the condition (2.49) is not assumed.

We write the operator $\Lambda_{(4.20)}$ as the sum of the operators

$$L = L^1 + L^2. \tag{4.25a}$$

Here

$$L^1 = L^1(\lambda) \equiv \varepsilon^2\left\{(1-\lambda)\sum_{s=1}^{n} a_{ss}(x)\frac{\partial^2}{\partial x_s^2} + \sum_{\substack{s,k=1 \\ s\neq k}}^{n} a_{sk}(x)\frac{\partial^2}{\partial x_s \partial x_k}\right\} \equiv \tag{4.25b}$$

$$\equiv \varepsilon^2 \sum_{s,k=1}^{n} a_{sk}^\lambda(x)\frac{\partial^2}{\partial x_s \partial x_k};$$

$$L^2 = L^2(\lambda) \equiv \varepsilon^2\left\{\lambda\sum_{s=1}^{n} a_{ss}(x)\frac{\partial^2}{\partial x_s^2} + \sum_{s=1}^{n} b_s(x)\frac{\partial}{\partial x_s} - c(x)\right\} - c^0(x), \tag{4.25c}$$

the parameter λ, which takes values in the interval $(0,1]$ and depends on the coefficients of the operator $L_{2(2.2)}$, is chosen below. The operator

$$L^1 = L^1(\lambda) = L^*_{(2)(2.50)}(L^1)$$

is strongly elliptic if the parameter λ satisfies the condition

$$\lambda \in [0, \lambda_0), \tag{4.26a}$$

where λ_0 is a constant in the interval $(0,1]$. For the value λ_0, the following upper estimate holds:

$$\lambda_0 \leq a_0\,(a^0)^{-1}, \tag{4.26b}$$

where a_0 and a^0 are the constants from the ellipticity condition (2.3). The value λ_0 can be chosen to satisfy the condition

$$\lambda_0 = \lambda_0(L) \to 1 \quad as \quad \max_{s,k,\overline{D}} |a_{sk}(x)| \to 0, \quad s,k = 1,\ldots,n, \ s \neq k. \qquad (4.26c)$$

In the case when the equation does not contain mixed derivatives we set $\lambda = 1$. The operator

$$\Lambda^1 = \Lambda^1\big(L^1_{(4.25,\,4.26)}\big) \equiv \sum_{s=1}^{n} a^{\lambda}_{ss}(x)\,\delta_{\overline{x_s x_s}} + \qquad (4.27)$$

$$+2^{-1}\sum_{\substack{s,k=1 \\ s \neq k}}^{n} \big[a^{\lambda +}_{sk}(x)\,(\delta_{x_s x k} + \delta_{\overline{x_s x k}}) + a^{\lambda -}_{sk}(x)\,(\delta_{x_s \overline{x k}} + \delta_{\overline{x_s} x k})\big],$$

$$a^{\lambda}_{sk}(x) = a^{\lambda}_{sk(4.25b)}(x), \quad s,k = 1,\ldots,n,$$

is defined only by the coefficients $a^{\lambda}_{sk}(x)$ of the operator $L^1_{(4.25,\,4.26)}$. The operator Λ^1 is ε-uniformly monotone in the following cases:

(i) on the uniform grid

$$\overline{D}_h = \overline{D}_h\big(Q(x^*),\ L^1_{(4.25,\,4.26)}(\lambda)\big) = \qquad (4.28)$$

$$= \overline{D}_{h(2.56)}\big(Q(x^*),\ L^*_{(2)(2.50)}(L^1) = L^1_{(4.25,\,4.26)}(\lambda),\ x_1,\ldots,x_n;$$

$$\textit{under the condition (2.55c) subject to (2.55)}$$

(the fulfillment of the conditions (2.49), (2.59), or (2.57) is not assumed);

(ii) and also on piecewise-uniform grids in the case of the conditions:

(ii_a) (2.57), on the grid

$$\overline{D}_h = \overline{D}_{h(2.70)}\big(Q(x^*),\ L^*_{(2)(2.50)}(L^1) = L^1_{(4.25,\,4.26)}(\lambda),\ x_1,\ldots,x_n;$$

$$\overline{\omega}_1 = \overline{\omega}_{1(4.22)} \textit{ under the conditions (2.67) and (2.55c)} \qquad (4.29)$$

$$\textit{subject to } \{(2.55),\ (2.65)\}\big);$$

(ii_b) (2.59), on the grid

$$\overline{D}_h = \overline{D}_{h(2.63)}\big(Q(x^*),\ L^*_{(2)(2.50)}(L^1) = L^1_{(4.25,\,4.26)}(\lambda),\ x_2,\ldots,x_n; \qquad (4.30)$$

$$\overline{\omega}_1 = \overline{\omega}_{1(4.22)} \textit{ under the condition (2.61c) subject to (2.61)}\big)$$

(ii_c) and (2.49), on the grid

$$\overline{D}_h = \overline{D}_{h(2.58)}\big(\overline{\omega}_{1(4.22)}\big). \qquad (4.31)$$

Consistent grids (4.28)–(4.31) are constructed depending on the coefficients $a_{sk}^{\lambda}(x)$, for $s, k = 1, \ldots, n$, of the operator $L_{(4.25)}^1$, at that time as consistent grids in Section 2.1 have been constructed depending on the coefficients $a_{sk}(x)$ of the operator $L_{(2)(2.50)}^*$.

The operator

$$\Lambda^2 = \Lambda^2\big(L_{(4.25,\,4.26)}^2(\lambda)\big) \equiv \Lambda\big(L_{(4.25,\,4.26)}^2(\lambda)\big) = \qquad (4.32)$$

$$= \varepsilon^2 \left\{ \lambda \sum_{s=1}^{n} a_{ss}(x)\,\delta_{\widehat{xs}\widehat{xs}} + \sum_{s=1}^{n} b_s(x)\,\delta_{\widehat{xs}} - c(x) \right\} - c^0(x), \quad \lambda = \lambda_{(4.26)}$$

on the grid $\overline{D}_{h(2.26)}$ with an arbitrary distribution of nodes is ε-uniformly monotone under the condition

$$N \geq 2\,M\,\lambda^{-1} \max_{s,\,\overline{D}} \big[a_{ss}^{-1}(x)\,|b_s(x)|\big], \quad M = M_{(2.26)}, \quad s = 1, \ldots, n, \qquad (4.33)$$

where $\lambda > 0$.

Under the condition (4.33) the operator

$$\Lambda_{(4.20)} = \Lambda^1 + \Lambda^2,$$

where $\Lambda^1 = \Lambda_{(4.27)}^1\big(L_{(4.25,\,4.26)}^1\big)$ and $\Lambda^2 = \Lambda_{(4.32)}^2\big(L_{(4.25,\,4.26)}^2\big)$, preserves the monotonicity property of the operator $\Lambda_{(4.27)}^1$ on the grids (4.28)–(4.31).

For the operator $\Lambda_{(4.20)}$ statements similar to that of Theorems 2.4.2, 2.4.4, 2.4.5 are valid.

Theorem 4.3.2 *The operator $\Lambda_{(4.20)}$ under the condition (4.33) is ε-uniformly monotone on the uniform grid (4.28) and also on the piecewise-uniform grids (4.29), (4.30), and (4.31) in the case of the conditions (2.57), (2.59), and (2.49), respectively.*

We associate the scheme (4.23) with the problem (2.2) on $\overline{D}_{(2.4)}^{(0)}$. Assume that the condition (4.33) holds. Depending on the fulfillment of the conditions (2.57), (2.59), and (2.49), we chose the grid, respectively, $\overline{D}_{h(4.29)}$, $\overline{D}_{h(4.30)}$, and $\overline{D}_{h(4.31)}$; otherwise, we chose the grid $\overline{D}_{h(4.28)}$. On each of these grids the operator $\Lambda_{(4.23)}$ is ε-uniformly monotone.

The following theorems similar to those of 2.4.8, 2.4.9, 2.4.10 are valid.

Theorem 4.3.3 *Let the conditions $a_{sk} \in C^1(\overline{D})$, b_s, c, c^0, $f \in C(\overline{D})$, for $s, k = 1, \ldots, n$, be fulfilled and the coefficients $a_{sk}(x)$ and $a_{sk(4.25)}^{\lambda}(x)$ satisfy, respectively, the conditions (2.57) and (2.66), (2.71) with $a_{sk(2.66,2.71)}(x) = a_{sk(4.25)}^{\lambda}(x)$ and $\overline{Q}(x^*) = \overline{D}$. For the solution of the problem (2.2), (2.4), let the estimates of Theorem 2.2.1 be satisfied for $K = 4$. Then the difference scheme (4.23) on the grid $\overline{D}_{h(4.29)}$, where $\overline{Q} = \overline{D}$, converges ε-uniformly at the rate $\mathcal{O}(N^{-2} \ln^2 N)$ as $N \to \infty$:*

$$|u(x) - z(x)| \leq M\,N^{-2} \ln^2 N, \quad x \in \overline{D}_h.$$

Theorem 4.3.4 *Let a_{sk}, b_s, c, c^0, $f \in C(\overline{D})$, for $s, k = 1, \ldots, n$, hold and the coefficients $a_{sk}(x)$ and $a^{\lambda}_{sk(4.25)}(x)$ satisfy, respectively, the conditions (2.59) and (2.61), (2.73) with $a_{sk(2.61,2.73)}(x) = a^{\lambda}_{sk(4.25)}(x)$, $\overline{Q}(x^*) = \overline{D}$. For the solution of the problem (2.2), (2.4), let the estimates of Theorem 2.2.1 be satisfied for $K = 4$. Then the difference scheme (4.23) on the grid $\overline{D}_{h(4.30)}$, where $\overline{Q} = \overline{D}$ converges ε-uniformly at the rate $\mathcal{O}(N^{-2} \ln^2 N)$ as $N \to \infty$.*

Theorem 4.3.5 *Let the coefficients $a_{sk}(x)$ satisfy the condition (2.49), and let a_{ss}, b_s, c, c^0, $f \in C(\overline{D})$, for $s = 1, \ldots, n$. For the solution of the problem (2.2), (2.4), let the estimates of Theorem 2.2.1 hold for $K = 4$. Then the difference scheme (4.23) on the grid $D_{h(4.31)}$ converges ε-uniformly at the rate $\mathcal{O}(N^{-2} \ln^2 N)$ as $N \to \infty$.*

Under the hypotheses of Theorems 4.3.3, 4.3.4, and 4.3.5, for the interpolant $\overline{z}(x) = \overline{z}_{(2.42)}(x; z(\cdot), \overline{D}_h)$, $x \in \overline{D}$, where $z(x)$, $x \in \overline{D}_h$, is the solution of the difference scheme (4.23), respectively, on the grids (4.29), (4.30), and (4.31), one has the estimate

$$|u(x) - \overline{z}(x)| \leq M N^{-2} \ln^2 N, \quad x \in \overline{D}.$$

4.4   Domain-decomposition method for improved iterative schemes

We now consider difference schemes of the domain-decomposition method that converge ε-uniformly with the convergence order close to two.

For the boundary value problem (2.2), (2.4), let the coefficients $a_{1s}(x)$ satisfy only the condition (2.57). The fulfillment of the condition (2.71) is not assumed, but the condition (2.93) holds. To solve the boundary value problem, we apply an iterative difference scheme similar to $\{(2.98), (2.97), (2.96)\}$.

In the case of condition (2.93), one can choose the diameter of the set $\overline{D}^i$ and the values λ^i sufficiently small such that the following inequality holds:

$$\max_{\overline{D}^i} |a^{\lambda^i}_{sk}(x)| < (n-1)^{-1} \min_{\overline{D}^i} \left[\alpha^i_s (\alpha^i_k)^{-1} a_{ss}(x), \; \alpha^i_k (\alpha^i_s)^{-1} a_{kk}(x) \right],$$

$$a^{\lambda^i}_{sk}(x) = a^{\lambda^i}_{sk(4.25)}(x), \quad s, k = 1, \ldots, n, \quad s \neq k.$$

This ensures a transformation of the regular operator $L^{*R}_{(2)}$, i.e.,

$$L^{*R}_{(2)} = L^{*R}_{(2)(2.50)} \left(\varepsilon^{-2} L^1_{(4.25)}(\lambda^i) \right) \equiv \varepsilon^{-2} L^1_{(4.25)}(\lambda^i), \tag{4.34a}$$

to almost canonical form on $\overline{D}^i$ and also a transformation of the operator $L_{(2)}^{*R[1]}$, i.e.,

$$L_{(2)}^{*R[1]} = L_{(2)(2.62)}^{*[1]}\left(L_{(2)(4.34a)}^{*R}\right), \tag{4.34b}$$

to strongly almost canonical form if $\overline{D}^i \cap \Gamma \neq \emptyset$ and the coefficients $a_{1s}(x)$ on the boundary Γ satisfy the condition (2.57).

Assume that the operator $L_{(2)(4.34a)}^{*R}$ is locally almost canonical on $\overline{D}$, moreover, on the subdomains $\overline{D}^i$ it is almost canonical, and under the condition $\overline{D}^i \cap \Gamma \neq \emptyset$, the operator $L_{(2)(4.34b)}^{*R[1]}$ is strongly almost canonical. Suppose also that the condition

> the coefficients $a_{sk}^{\lambda^i}(x)$ of the operator $L_{(2)(4.34a)}^{*R}$ on the set $\overline{D}^i$
>
> satisfy the relations $\{(2.55), (2.65)\}$ if $\overline{D}^i \cap \Gamma \neq \emptyset$,
>
> and the relations (2.55) if $\overline{D}^i \cap \Gamma = \emptyset$,   (4.35)
>
> where $\overline{Q}(x^*) = \overline{D}^i$, $a_{sk}(x) = a_{sk(4.25)}^{\lambda^i}(x)$, for $i = 1, \ldots, I$,

holds. Then the above conditions together with the condition (2.57) admit us to construct on $\overline{D}^i$ consistent grids in the variables $x_1, \ldots, x_n$ with respect to the monotonicity condition for the difference regular operator $\Lambda_{(2)}^{*R}$

$$\Lambda_{(2)}^{*R} = \Lambda_{(2)(2.51)}^*\left(L_{(2)(4.34a)}^{*R}\right) \equiv \varepsilon^{-2} \Lambda_{(4.27)}^1\left(L_{(4.25)}^1(\lambda^i)\right).$$

On the set $\overline{D}^i$ we construct the consistent grid

$$\overline{D}_h^i = \begin{cases} \overline{D}_{h(4.28)}\left(Q(x^*) = D^i\right) & \text{if } \overline{D}^i \cap \Gamma = \emptyset, \\ \overline{D}_{h(4.29)}\left(Q(x^*) = D^i\right) & \text{if } \overline{D}^i \cap \Gamma \neq \emptyset, \quad i = 1, \ldots, I, \end{cases} \tag{4.36}$$

that ensures the monotonicity of the operators $\Lambda_{(2)}^{*R} = \Lambda_{(2)}^*(L_{(2)}^{*R})$ and $\Lambda = \Lambda_{(4.20)}(L)$ on $\overline{D}_h^i$.

On the slab $\overline{D} = \overline{D}_{(2.4)}^{(0)}$ on the basis of the grids $\overline{D}_h^i = \overline{D}_{h(4.36)}^i$, for $i = 1, \ldots, I$, we construct the grid

$$\overline{D}_h = \overline{D}_{h(2.97)}^{\{0\}}\left(\overline{D}_{h(4.36)}^i\right), \ i = 1, \ldots, I). \tag{4.37}$$

With regard to the relations (2.98), where

$$\Lambda_i = \Lambda_{i(2.98)} \equiv \Lambda_{(4.20)}\left(L_{(4.25)}; \lambda_{(4.26)}^i\right), \quad i = 1, \ldots, I, \tag{4.38}$$

solving the discrete problems (2.98b), where $\Lambda_i = \Lambda_{i(4.38)}$, we find the functions

$$z^{[k]}(x) = z_{(2.98)}^{[k]}(x), \quad x \in \overline{D}_h, \quad k = 1, 2, 3, \ldots$$

and the interpolant

$$\breve{z}^{[k]}(x) = \breve{z}^{[k]}_{(2.98)}(x), \quad x \in \overline{D}, \quad k = 1, 2, 3, \dots. \tag{4.39}$$

The function $z^{[k]}(x)$, $x \in \overline{D}_h$, is the solution of the iterative difference scheme $\{(2.98), (2.97), (4.38), (4.36)\}$, i.e., the iterative scheme of the domain decomposition method on the overlapping subdomains.

The iterative difference scheme $\{(2.98), (2.97), (4.38), (4.36)\}$ is defined by canonical elements, i.e., the operator $\Lambda_{(4.20)}$ and the grids $\overline{D}_{h(4.28)}(\overline{\omega}_1^u)$ and $\overline{D}_{h(4.29)}(\overline{\omega}_{1(4.22)})$, on the basis of which the difference scheme is constructed. Set $N^i = \min_s N_s^i$ and $N = \min_i N^i$, where $N_s^i + 1$ is the number of nodes in the grid $\overline{D}_h^i$ along the x_s-axis, for $s = 1, \dots, n$. Assume that the following condition holds, which is similar to (4.33):

$$N \geq 2\,M\,\max_i(\lambda^i)^{-1} \max_{s, \overline{D}} \left[a_{ss}^{-1}(x) \, |b_s(x)| \right], \quad M = M_{(2.26)},$$

$$s = 1, \dots, n, \quad i = 1, \dots, I.$$

Theorem 4.4.1 *Let $a_{sk} \in C^1(\overline{D})$, b_s, c, c^0, $f \in C(\overline{D})$, for $s, k = 1, \dots, n$; the coefficients $a_{sk}(x)$ and $a_{sk(4.25)}^{\lambda^i}(x)$ satisfy, respectively, the conditions (2.57), (2.93), and (4.35). For the solution of the problem (2.2), (2.4), let the estimates of Theorem 2.2.1 hold for $K = 4$. Then the iterative difference scheme $\{(2.98), (2.97), (4.38), (4.36)\}$ on the grid $\overline{D}_{h(4.37)}$ converges ε-uniformly at the rate $\mathcal{O}\left(N^{-2} \ln^2 N + q^k \right)$ as $N, k \to \infty$:*

$$|u(x) - z^{[k]}(x)| \leq M\,[N^{-2} \ln^2 N + q^k], \quad x \in \overline{D}_h.$$

The interpolant $\breve{z}_{(4.39)}(x)$ of the solution of the iterative difference scheme converges ε-uniformly as $N, k \to \infty$ with the estimate

$$|u(x) - \breve{z}^{[k]}(x)| \leq M\,[N^{-2} \ln^2 N + q^k], \quad x \in \overline{D}.$$

The number of iterations k^f required for the solution of the iterative difference scheme with the estimate

$$|u(x) - \breve{z}^{[k^f]}(x)| \leq M\,N^{-2} \ln^2 N, \quad x \in \overline{D},$$

is ε-uniformly bounded. The value k^f satisfies the estimate

$$k^f \leq M \ln N.$$

We now construct a difference scheme for the boundary value problem (2.2) in the domain $\overline{D}_{(2.103)}$ with a curvilinear boundary. The fulfillment of the dominance condition of the diagonal elements of the elliptic operator, i.e., the condition (2.71), as well as restrictions for the coefficients of the mixed derivatives on the domain boundary, is not assumed. To solve the problem, we apply an iterative difference scheme similar to $\{(2.98), (2.97), (2.114)\}$.

Using an appropriate transformation of the coordinates on the set $D^i \bigcup \Gamma^{iL}$ in a neighborhood of the domain boundary, the boundary value problem (2.2), (2.103) can be brought to the form (2.105):

$$L_X U(X) = F(X), \quad X \in D^i_X, \tag{4.40a}$$

$$U(X) = \Phi(X), \quad X \in \Gamma^{iL}_X. \tag{4.40b}$$

Here $D^i_X \bigcup \Gamma^{iL}_X$ is the rectangular parallelepiped adjoining to the boundary

$$L_X \equiv \varepsilon^2 L_{2X} + L_{0X}, \quad L_{0X} = -C^0(X),$$

$$L_{2X} \equiv \sum_{s,k=1}^{n} A_{sk}(X) \frac{\partial^2}{\partial X_s \partial X_k} + \sum_{s=1}^{n} B_s(X) \frac{\partial}{\partial X_s} - C(X).$$

The coefficients A_{sk} of the operator L_X satisfy the strong ellipticity condition

$$A_0 \sum_{s=1}^{n} \xi_s^2 \le \sum_{s,k=1}^{n} A_{sk}(X)\, \xi_s\, \xi_k \le A^0 \sum_{s=1}^{n} \xi_s^2, \quad X \in \overline{D}^i_X, \quad A_0 > 0. \tag{4.40c}$$

On the domain boundary one has the condition

$$A_{1s}(X) = 0, \quad X \in \Gamma^{iL}_X, \quad s = 2, \ldots, n. \tag{4.40d}$$

We write the operator L_X as the decomposition similar to (4.25):

$$L_X = L^1_X + L^2_X. \tag{4.41a}$$

Here

$$L^1_X = L^1_X(\lambda^i) \equiv \varepsilon^2 \sum_{s,k=1}^{n} A^{\lambda^i}_{sk}(X) \frac{\partial^2}{\partial X_s \partial X_k}, \tag{4.41b}$$

$$L^2_X = L^2_X(\lambda^i) \equiv \tag{4.41c}$$

$$\equiv \varepsilon^2 \left\{ \lambda^i \sum_{s=1}^{n} A_{ss}(X) \frac{\partial^2}{\partial X_s^2} + \sum_{s=1}^{n} B_s(X) \frac{\partial}{\partial X_s} - C(X) \right\} - C^0(X),$$

where

$$A^{\lambda^i}_{ss}(X) = (1 - \lambda^i)\, A_{ss}(X), \quad A^{\lambda^i}_{sk}(X) = A_{sk}(X), \quad s \ne k.$$

By virtue of the conditions (4.40c), (4.40d), the regular operator $L^{*R}_{X(2)}$, i.e.,

$$L^{*R}_{X(2)} = L^*_{X(2)(2.50)}\left(\varepsilon^{-2} L^1_{X(4.41b)}(\lambda^i)\right) \equiv \varepsilon^{-2} L^1_{X(4.41b)}(\lambda^i) \tag{4.42a}$$

can be brought to almost canonical form on $\overline{D}_X^i$ (for sufficiently small diameters of the sets $\overline{D}_X^i$ and the values λ^i), and the operator $L_{X(2)}^{*R[1]}$, i.e.,

$$L_{X(2)}^{*R[1]} = L_{X(2)(2.62)}^{*[1]}\left(L_{X(2)(4.42a)}^{*R}\right), \tag{4.42b}$$

can be brought to strongly almost canonical form in $X_2, \ldots, X_n$ in a neighborhood of the boundary Γ_X^{iL}.

Assume that the operator $L_{X(2)(4.42a)}^{*R}$ has almost canonical form in the variables $X_1, \ldots, X_n$ on $\overline{D}_X^i$, and the operator $L_{X(2)(4.42b)}^{*R[1]}$ has strongly almost canonical form in $X_2, \ldots, X_n$ in a neighborhood of Γ_X^{iL}. Let the coefficients of the operator $L_{X(2)}^{*R}$ defined on $\overline{D}_X^i = \{X : d_{*s}^i \leq X_s \leq d_s^{i*}, \ s = 1, \ldots, n\}$ satisfy the condition (similar to (2.108)):

the coefficients $A_{sk}^{\lambda i}(X)$ of the operator $L_{X(2)(4.42a)}^{*R}$ on the set $\overline{D}_X^i$

satisfy the relations (2.78), {(2.55), (2.79)} if $J^{i*} \neq \emptyset$,

and the relations (2.55) if $J^{i*} = \emptyset$, $\tag{4.43}$

where $a_{sk}(x), \ldots, \overline{Q}(x^*)$ is $A_{sk}^{\lambda i}(X), \ldots, \{\overline{Q}(x^*)\}_X$,

under the condition $\{\overline{Q}(x^*)\}_X = \overline{D}_X^i$, $J^{i*} = J^{i*}(D_X^i) = \{k = 1\}$.

The condition $J^{i*}(D_X^i) = \{k = 1\}$ means that the piecewise-uniform grid $\overline{\omega}_{X1}$ in the X_1-axis condenses (for small ε) in a neighborhood of the left endpoint of the interval $[d_{*1}^i, d_1^{i*}]$.

By virtue of (4.40d), the coefficients of the operator $L_{X(2)(4.41a)}^*$ on the face Γ_X^{iL} satisfy the condition

$$A_{1s}^{\lambda i}(X) = 0, \quad X \in \Gamma_X^{iL}, \quad s = 2, \ldots, n,$$

i.e., the condition (2.78), where the coefficients $a_{1k}(x)$ on the set Γ_k, for $k \in J^*$, are the coefficients $A_{1k}^{\lambda i}(X)$ on the set Γ_X^{iL}. In the case of the operator $L_{X(2)}^*$, the following condition (similar to (2.55c) for $\overline{Q}(x^*) = \overline{D}^i$, where $\overline{D}^i \subseteq \overline{D}$) imposed on the parameters h_s takes the form:

$$h_s \in \left[\max_{\overline{D}_X^i} \rho_{*s}(X) \, h^*, \ \min_{\overline{D}_X^i} \rho_s^*(X) \, h^*\right], \quad s = 1, \ldots, n. \tag{4.44}$$

The fulfillment of the condition (4.43) allows us to construct on the set $\overline{D}_X^i$ a piecewise-uniform grid that ensures the monotonicity of the operators

$$\Lambda_{X(2)}^{*R} = \Lambda_{X(2)(2.51)}^*\left(L_{X(2)(4.42a)}^{*R}\right) \equiv \varepsilon^{-2} \Lambda_{X(4.27)}^1\left(L_{X(4.41b)}^1(\lambda^i)\right),$$

and also

$$\Lambda_X = \Lambda_X\left(L_{X(4.41)}(\lambda^i)\right) \equiv \Lambda_{X(4.20)}\left(L_{X(4.41a)}\right) =$$
$$= \Lambda_{X(4.27)}^1\left(L_{X(4.41b)}^1(\lambda^i)\right) + \Lambda_{X(4.32)}^2\left(L_{X(4.41c)}^2(\lambda^i)\right). \tag{4.45}$$

The operators $\Lambda^{*R}_{X(2)}$ and Λ_X are constructed using the operators $L^{*R}_{X(2)(4.42a)}$ and $L_{X(4.41a)}$, respectively.

In the case of the conditions (4.43) and (2.82), where $J^* = J^{i*}(D^i_X) = \{j = 1\}$, imposed on the parameters of the meshes $\overline{\omega}_{X1} = \overline{\omega}^l_{X1}(\overline{\omega}_{(4.22)})$ and $\overline{\omega}_{Xs} = \overline{\omega}^u_{Xs}$, for $s = 2,\ldots,n$, where $\overline{\omega}_{Xs} = \overline{\omega}^i_{Xs}(d^i_{*s}, d^{i*}_s)$, for $s = 1,\ldots,n$, the piecewise-uniform grid $\big($the grid (2.89), where $J^* = \{j = 1\}\big)$:

$$\overline{D}^i_{Xh} = \overline{D}^i_{Xh}\big(\{Q(x^*)\}_X = D^i_X, L^{*R}_{X(2)(4.42a)}; J^{i*}\big) = \qquad (4.46)$$

$$= \overline{D}^i_{Xh(2.89)}\big(\{Q(x^*)\}_X = D^i_X, L^{*R}_{X(2)(4.42a)};$$

$$\overline{\omega}^i_{X1} = \overline{\omega}^l_{X1}(\overline{\omega}_{(4.22)}), \quad \overline{\omega}^i_{Xs} = \overline{\omega}^u_{Xs}, \quad for \ \ s = 2,\ldots,n,$$

under the conditions (4.44), (2.82),

$$J^{i*} = \{j = 1\} \ \ in \ the \ case \ of \ the \ condition \ (4.43);$$

in the relations (2.78), $\{(2.55), (2.79)\}$ *from* (2.89)

$$a_{sk}(x) \ \ and \ \ \overline{Q}(x^*) \ \ are \ replaced \ by \ \ A^{\lambda^i}_{sk}(X) \ \ and \ \ \{\overline{Q}(x^*)\}_X\Big),$$

which is constructed directly on the set $\overline{D}^i_X$, is consistent $\big($in the variables $X_1, \ldots, X_n\big)$ on the set $\overline{D}^i_X$. The grid $\overline{D}^i_{Xh(4.46)}$ is generated by the meshes $\overline{\omega}^l_{X1} = \overline{\omega}^l_{X1}(\overline{\omega}_{(4.22)})$ and $\overline{\omega}^u_{Xs}$, for $s = 2,\ldots,n$, i.e.,

$$\overline{D}^i_{Xh(4.46)} = \overline{\omega}^l_{X1} \times \overline{\omega}^u_{X2} \times \ldots, \overline{\omega}^u_{Xn}.$$

For the subproblem (4.40) we construct the monotone difference approximation

$$\Lambda_X Z(X) = F(X), \quad X \in D^i_{Xh}; \quad Z(X) = \Phi(X), \quad X \in \Gamma^{iL}_{Xh}. \qquad (4.47)$$

Here $\Lambda_X = \Lambda_{X(4.20)}(L_X)$ and $\overline{D}^i_{Xh} = \overline{D}^i_{Xh(4.46)}$.

Returning to the variables $x = (x_1,\ldots,x_n)$, we obtain the monotone difference scheme

$$\Lambda^* z^*(x) = f(x), \quad x \in D^i_{Xh\,X^{-1}}; \quad z^*(x) = \varphi(x), \quad x \in \Gamma^{iL}_{Xh\,X^{-1}}. \qquad (4.48a)$$

Here

$$D^i_{Xh\,X^{-1}} = D^i \cap \overline{D}^i_{Xh\,X^{-1}}, \quad \Gamma^{iL}_{Xh\,X^{-1}} = \Gamma \cap \overline{D}^i_{Xh\,X^{-1}}, \qquad (4.48b)$$

$$\overline{D}^i_{Xh\,X^{-1}} = \{\overline{D}^i_{Xh}\}_{X^{-1}}; \quad z^*(x) = Z(X(x)), \quad x \in \overline{D}^i_{Xh\,X^{-1}}.$$

We assume that on subdomains $\overline{D}^i$, which have no common points with the boundary Γ, the operator L is almost canonical; let on these subdomains the following condition hold (similar to (2.87), where $J^* = \emptyset$):

> *the coefficients* $a_{sk}^{\lambda^i}(x)$ *of the operator* $L_{(2)(4.34a)}^{*R}$ *on the set* $\overline{D}^i$
> *satisfy the relations* (2.55), $\qquad\qquad\qquad\qquad\qquad\qquad\qquad$ (4.49)
> *where* $\overline{Q}(x^*) = \overline{D}^i$, $a_{sk}(x) = a_{sk(4.25)}^{\lambda^i}(x)$.

Such condition admits the construction of the uniform grids $\overline{D}_h^i$, i.e.,

$$\overline{D}_h^i = \overline{D}_{h(4.28)}\big(Q(x^*) = D^i,\ L_{(4.25,\,4.26)}^1(\lambda^i)\big), \qquad (4.50)$$

that are consistent with the monotonicity condition for the operators

$$\Lambda^i = \Lambda\big(L_{(4.25,\,4.26)}(\lambda^i)\big) \equiv \Lambda_{(4.20)}. \qquad (4.51)$$

Next, for the construction of iterative scheme we use the algorithm $\{(2.98),$ $(2.97), (2.114)\}$.

Let the subdomains $\overline{D}^i$, for $i = 1, \ldots, I$, form a covering of the set $\overline{D}$, moreover, the subdomains $\overline{D}^i$, for $i = 1, \ldots, I_1$, adjoin to the boundary Γ, and the subdomains $\overline{D}^i$, for $i = I_1 + 1, \ldots, I$, have no common points with the boundary Γ; and let the distance between these sets and Γ be independent of ε. We assume that the condition for the width of overlapping to subdomains holds, i.e., the minimal overlapping of subdomains that cover D is independent of the value of the parameter ε.

The sets $\overline{D}^i$ for $i = I_1 + 1, \ldots, I$, are rectangular parallelepipeds formed by coordinate planes $x_s =$const, for $s = 1, \ldots, n$. The sets $\overline{D}^i$, for $i = 1, \ldots, I_1$, in new variables $X^i = X^i(x)$ are the sets $\overline{D}_{X^i}^i$ which are rectangular parallelepipeds formed by coordinate planes $X_s^i = const$, $s = 1, \ldots, n$.

Note that the coefficients of the operator $L_{X^i(4.41)}$ on the set $\overline{D}_{X^i}^i$, for $i = 1, \ldots, I_1$, satisfy the condition (4.43), and the coefficients of the operator $L_{(4.25,\,4.26)}(\lambda^i)$ on the set $\overline{D}^i$, for $i = I_1 + 1, \ldots, I$, satisfy the condition (4.49).

Next, we construct the grid sets

$$\overline{D}_h^i \equiv \overline{D}_{h(2.114e)}^i = \begin{cases} \overline{D}_{h(2.114a)}^i & \text{for } i = 1, \ldots, I_1, \\[2mm] \overline{D}_{h(2.114c)}^i & \text{for } i = I_1 + 1, \ldots, I \end{cases} \qquad (4.52a)$$

and the difference operators

$$\Lambda_i \equiv \Lambda_{i(2.114f)} = \begin{cases} \Lambda_{i(2.114b)} & \text{for } i = 1, \ldots, I_1, \\[2mm] \Lambda_{i(2.114d)} & \text{for } i = I_1 + 1, \ldots, I. \end{cases} \qquad (4.52b)$$

Here

$$\overline{D}_{h(2.114a)}^i = \overline{D}_{X^i h\{X^i\}^{-1}(4.48b)}^i \quad \text{for} \quad i = 1, \ldots, I_1,$$

$$\overline{D}_{h(2.114c)}^i = \overline{D}_{h(4.50)}^i \qquad\qquad \text{for} \quad i = I_1 + 1, \ldots, I;$$

$$\Lambda_{i(2.114b)} = \Lambda^*_{(4.48a)} \quad \text{with} \quad x \in \overline{D}^i_{X^i h \{X^i\}^{-1}}, \quad \text{for} \quad i = 1, \dots, I_1,$$

$$\Lambda_{i(2.114d)} = \Lambda_{i(4.51)} \quad \text{with} \quad x \in \overline{D}^i_h, \qquad\qquad \text{for} \quad i = I_1 + 1, \dots, I.$$

In the iterative process (2.98), (2.97), (2.114), the interpolants $\breve{z}^{[k]}_i(x)$, $x \in \overline{D}$, for $k \geq 1$, are defined by the relations (2.114g).

The function $z^{[k]}(x) = z^{[k]}_{(2.98e)}(x)$, $x \in \overline{D}_h$, where $\overline{D}_h = \overline{D}_{h(2.97a)} \left(\overline{D}^i_{h(4.52a)}, i = 1, \dots, I \right)$, is the solution of the iterative difference scheme $\{(2.98), (2.97), (2.114), (4.52)\}$.

We denote by $N^i_s + 1$ the number of nodes in the grid $\overline{D}^i_{X^i h}$, for $i = 1, \dots, I_1$ (the grid $\overline{D}^i_h$, for $i = I_1 + 1, \dots, I$), along the X^i_s-axis (respectively, along the x_s-axis), for $s = 1, \dots, n$. Set $N^i = \min_s N^i_s$ and $N = \min_i N^i$. Assume that the following condition holds:

$$N \geq 2M \left\{ \max_i (\lambda^i)^{-1} \max_{s,i,\overline{D}^i} \left[A^{-1}_{ss}(X^i) \, |B_s(X^i)| \right]_{i=1,\dots,I_1}, \right.$$

$$\left. \max_i (\lambda^i)^{-1} \max_{s,\overline{D}} \left[a^{-1}_{ss}(x) \, |b_s(x)| \right]_{i=I_1+1,\dots,I} \right\}, \quad M = M_{(2.26)}.$$

The iterative difference scheme is defined by canonical elements, i.e., the operators $\Lambda_{(4.51)}(L(\lambda^i))$ and $\Lambda_{X^i(4.45)}(L_{X^i(4.41)})$ and the consistent uniform and piecewise-uniform grids $\overline{D}^i_{h(4.50)}$ and $\overline{D}^i_{X^i h(4.46)}$.

The following theorem on convergence holds.

Theorem 4.4.2 *Let $a_{sk} \in C^1(\overline{D})$, $b_s, c, c^0, f \in C(\overline{D})$, for $s, k = 1, \dots, n$, $\Gamma \in C^2$, the coefficients $a^{\lambda^i}_{sk}(x)$, $x \in \overline{D}^i$, and $A^{\lambda^i}_{sk}(X^i)$, $X^i \in \overline{D}_{X^i}$, satisfy, respectively, the conditions (4.49) for $i = I_1 + 1, \dots, I$, and (4.43), where $X = X^i$, for $i = 1, \dots, I_1$. For the solution of the problem (2.2), (2.103), let the estimates of Theorem 2.2.2 hold for $K = 4$. Then the difference scheme $\{(2.98), (2.97), (2.114), (4.52)\}$ on the grid $\overline{D}_{h(2.97a,2.114e)}$, in which the mesh $\overline{\omega}^l_{X^i 1}$ in $\overline{D}^i_{X^i h}$, for $i = 1, \dots, I_1$, is defined by the mesh $\overline{\omega}^i_{1(4.22)}$, converges ε-uniformly at the rate $\mathcal{O}\left(N^{-2} \ln^2 N + q^k\right)$ as $N, k \to \infty$:*

$$|u(x) - z^{[k]}(x)| \leq M \left(N^{-2} \ln^2 N + q^k \right), \quad x \in \overline{D}_h.$$

For the interpolant $\breve{z}^{[k]}(x)$, $\breve{z}^{[k]}(x) = \breve{z}^{[k]}_{(2.114g)}(x)$, $x \in \overline{D}$, one has the estimate

$$|u(x) - \breve{z}^{[k]}(x)| \leq M \left(N^{-2} \ln^2 N + q^k \right), \quad x \in \overline{D}.$$

The number of iterations required for the solution of the iterative difference scheme with the estimate

$$|u(x) - z^{[k^f]}(x)| \leq M \, N^{-2} \ln^2 N, \quad x \in \overline{D}_h,$$

satisfies the estimate

$$k^f \leq M \ln N.$$

Chapter 5

Parabolic reaction-diffusion equations

In this chapter, a method for the construction of ε-uniformly convergent difference schemes for parabolic reaction-diffusion equations is developed using classical finite difference approximations on special condensing grids. Conditions are given that are sufficient for the ε-uniform convergence of the difference schemes. For problems on a slab and a parallelepiped, monotone ε-uniformly convergent difference schemes are constructed.

5.1   Problem formulation

In an n-dimensional domain D, we consider the boundary value problem for the parabolic *reaction-diffusion equation*

$$L\,u(x, t) = f(x, t), \quad (x, t) \in G, \tag{5.1a}$$

$$u(x, t) = \varphi(x, t), \quad (x, t) \in S. \tag{5.1b}$$

Here

$$G = D \times (0, T], \quad S = S(G) = \overline{G} \setminus G;$$

$$L \equiv \varepsilon^2 L_2 + L_1, \quad L_1 \equiv -c^1(x, t) - p(x, t)\frac{\partial}{\partial t},$$

$$L_2 \equiv \sum_{s,k=1}^{n} a_{sk}(x, t)\frac{\partial^2}{\partial x_s \partial x_k} + \sum_{s=1}^{n} b_s(x, t)\frac{\partial}{\partial x_s} - c(x, t).$$

The coefficients of the equation satisfy the ellipticity condition

$$a_0 \sum_{s=1}^{n} \xi_s^2 \leq \sum_{s,k=1}^{n} a_{sk}(x, t)\,\xi_s\,\xi_k \leq a^0 \sum_{s=1}^{n} \xi_s^2, \quad (x, t) \in \overline{G}, \quad a_0 > 0, \tag{5.1c}$$

and also the conditions

$$p(x, t) \geq p_0 > 0, \quad c(x, t), \quad c^1(x, t) \geq 0, \quad (x, t) \in \overline{G}.$$

The coefficients and the right-hand side f on the set $\overline{G}$ are assumed to be sufficiently smooth, just as the boundary function φ on the closures of the smooth parts of the boundary S, i.e., the data of the problem (5.1) are assumed to be sufficiently smooth. We suppose that the problem data are bounded in the case of an unbounded domain D.

Consider the boundary value problem on the slab

$$D = \{x :\ d_{*1} < x_1 < d_1^*,\quad |x_s| < \infty,\quad s = 2, \ldots, n\} \tag{5.2}$$

and on the parallelepiped

$$D = \{x :\ d_{*s} < x_s < d_s^*,\quad s = 1, \ldots, n\}. \tag{5.3}$$

Let $\Gamma_j = \Gamma_{j(3.6)}$, where $j = 1, \ldots, J$, be the faces of the set D, and let

$$S = S^L \cup S_0,\quad S^L = \Gamma \times (0,\, T], \tag{5.4a}$$

where S^L and S_0 are the lateral and the lower parts of the boundary, here $S^L = \cup_j S_j$, $S_j = \Gamma_j \times (0,\, T]$, $j = 1, \ldots, J$ and $S_0 = \overline{S}_0$. Denote by S^c a set that is formed by mutual intersection of smooth parts of the boundary S, and

$$S^c = S^{Lc} \cup S_0^c, \tag{5.4b}$$

where $S^{Lc} = \Gamma^c \times (0,\, T]$ and $S_0^c = \Gamma \times \{t = 0\}$ are sets of "edges" on the lateral part of the boundary S^L and "edges" (boundaries) on the lower part of the boundary S_0.

In some cases, as well as the continuity condition for the function $\varphi(x, t)$ on the set S^c, additional compatibility conditions shall be imposed to guarantee the sufficient smoothness of the solution to the boundary value problem.

For the boundary value problem (5.1), it is required to construct a difference scheme that converges ε-uniformly.

5.2   Estimates of solutions and derivatives

Let us estimate the solution of the problem (5.1) on the slab (5.2) and on the parallelepiped (5.3). When deriving estimates, we use *a priori* estimates for regular problems [67, 37].

We denote by $C^{l_0, l_0/2}(\overline{G})$, where $l_0 = l + \alpha$, $l \geq 0$ is an integer number, and $\alpha \in (0, 1)$, the Banach space whose elements are continuous functions $u(x, t)$ in $\overline{G}$ that have in G continuous derivatives

$$(\partial^{k+k_0} / \partial x_1^{k_1} \ldots \partial x_n^{k_n} \partial t^{k_0})\, u(x, t),\quad k + 2k_0 \leq l,$$

which are in the sense of Hölder continuous with exponent α. For these elements a finite value is taken by the norm

$$|u|_{l+\alpha} = |u|_\alpha + \sum_{k+2k_0=1}^{l}\ \sum_{k_1+\ldots+k_n=k} \left| \frac{\partial^{k+k_0}}{\partial x_1^{k_1} \ldots \partial x_n^{k_n} \partial t^{k_0}} u \right|_\alpha.$$

Here

$$\overline{|u|}_\alpha = |u|_0 + \overline{H}_\alpha(u), \quad |u|_0 = \sup_G |u(x,t)|,$$

$$\overline{H}_\alpha(u) = \sup_{(x,t),(x',t')\in G} \frac{|u(x,t) - u(x',t')|}{d^\alpha((x,t),(x',t'))},$$

and $d((x,t),(x',t'))$ is the distance between the points (x,t) and (x',t')

$$d((x,t),(x',t')) = \left(|x-x'|^2 + \varepsilon^2\,|t-t'|\right)^{1/2}.$$

Let the function $u(x,t)$, $(x,t) \in \overline{G}$ on the boundary S take the values of $\varphi(x,t)$; moreover, on the set $S^g \subset S$ the function $\varphi(x,t)$ is "good", i.e., $\varphi \in C^{l_0,l_0/2}(S^g)$, and also, $S^g \in C^{l_0,l_0/2}$, $S^g = \overline{S}^g$, where $l_0 = l + \alpha$, $l \geq 0$ and $\alpha \in (0,1)$. The set S^g can coincide with either S_0 or $\overline{S}_j$, or can be the empty set. We denote by $C^{l_0,l_0/2}(G)$ the Banach space whose elements are continuous functions in $G \bigcup S^g$ that take "good" boundary values on the set S^g, and have continuous derivatives $(\partial^{k+k_0}/\partial x_1^{k_1} \dots \partial x_n^{k_n} \partial t^{k_0})\,u(x,t)$ for $k + 2k_0 \leq l$, which are in the sense of Hölder continuous with exponent α. For these elements a finite value is taken by the norm

$$|u|_{l+\alpha}^g = |u|_\alpha^g + \sum_{k+2\,k_0=1}^{l} \sum_{k_1+\dots+k_n=k} \left| d^{k+2\,k_0} \frac{\partial^{k+k_0}}{\partial x_1^{k_1}\dots\partial x_n^{k_n}\,\partial t^{k_0}}\,u \right|_\alpha^g.$$

Here

$$|u|_\alpha^g = |u|_0 + H_\alpha^g(u),$$

$$H_\alpha^g(u) = \sup_{(x,t),(x',t')\in G} d_{(x,t),(x',t')}^\alpha \frac{|u(x,t) - u(x',t')|}{d^\alpha((x,t),(x',t'))},$$

$$|d^p\,v|_\alpha^g = |d^p\,v|_0^g + H_\alpha^g(d^p\,v), \quad |d^p\,v|_0^g = \sup_{(x,t)\in G} d_{(x,t)}^p\,|v(x,t)|,$$

$$H_\alpha^g(d^p\,v) = \sup_{(x,t),(x',t')\in G} d_{(x,t),(x',t')}^{p+\alpha} \frac{|v(x,t) - v(x',t')|}{d^\alpha((x,t),(x',t'))},$$

$$d((x,t),(x',t')) = \left(|x-x'|^2 + \varepsilon^2\,|t-t'|\right)^{1/2},$$

$$d_{(x,t)} = d_{(x,t)}^g = \inf_{(x',t')\in S\setminus S^g} d((x,t),(x',t')),$$

$$d_{(x,t),(x',t')} = d_{(x,t),(x',t')}^g = \min\left(d_{(x,t)}, d_{(x',t')}\right).$$

For the function $\varphi(x,t)$, $(x,t) \in S$ satisfying the condition $\varphi \in C^{l_0,l_0/2}$ (S^g) for $S^g = S_0$, we also shall use equivalent notations $\varphi(\cdot,0) \in C^{l_0}(\overline{D})$ and $\varphi \in C^{l_0}(S_0)$.

We give a **definition**. Set $\varphi_0(x) = \varphi(x,t)$, $(x,t) \in S_0$. Let the function $\varphi(x,t)$, $(x,t) \in S$ satisfy the condition $\varphi(\cdot, 0) \in C^{l_0}(\overline{D})$ (i.e., $\varphi_0 \in C^{l_0}(\overline{D})$), and for the function $\varphi(x,t)$ considered on $\overline{S}^L$, the derivatives $(\partial^{k_0}/\partial t^{k_0})\varphi(x,t)$, where $k_0 \leq l/2$, are defined for $(x,t) \in S_0^c$. Using the function $\varphi_0(x)$ prescribed on the set S_0 and the equation (5.1a), we find the derivative in t of the function $u(x,t)$ on S_0. We denote it by $(\partial/\partial t)\,\varphi_{0,t=0}(x)$. Furthermore, differentiating the equation (5.1a) in x and t, we find the derivatives in t up to order $k_0 \leq [l/2]_i$, where $[a]_i$ is the integer part of the number $a \geq 0$; we denote these derivatives by $(\partial^{k_0}/\partial t^{k_0})\,\varphi_{0,t=0}(x)$, $x \in \overline{D}$. We say that the data of the boundary value problem satisfy a *compatibility condition on the set S_0^c* guaranteeing the continuity of the derivatives in t up to order K_0 of the function $u(x, t)$ on S_0^c, or, briefly, the problem data satisfy a *compatibility condition on S_0^c for the derivatives in t up to order K_0* [67, 37], if one has the condition

$$\frac{\partial^{k_0}}{\partial t^{k_0}} \varphi(x,t) = \frac{\partial^{k_0}}{\partial t^{k_0}} \varphi_{0,t=0}(x), \quad (x,t) \in S_0^c, \;\; 0 \leq k_0 \leq K_0. \tag{5.5}$$

Under the conditions given above, we have $K_0 \leq [l/2]_i$.

Using a maximum principle, one can verify that the problem solution is stable, i.e., that

$$|u(x,\, t)| \leq M \max \Big[\max_{\overline{G}} |f(x,\, t)|, \; \max_{S} |\varphi(x,\, t)| \Big], \quad (x,\, t) \in \overline{G},$$

and that it continuously approaches the boundary data on S^L (the lateral boundary of the set $\overline{G}$), i.e.,

$$|u(x,\, t) - \varphi(x^*, t)| \leq M\,\varepsilon^{-1}\, r\big(x,\, \Gamma\big), \quad (x,\, t) \in \overline{G}, \tag{5.6}$$

where (x^*, t) is a point on $\overline{S}^L$ nearest to the point $(x,\, t) \in \overline{G}$, thus $x^* = x^*\big(x, \Gamma\big)$.

Under the condition a_{sk}, b_s, c, c^1, p, $f \in C^{l_0, l_0/2}(\overline{G})$, $\varphi \in C(S)$, we have $u \in C^{l_1, l_1/2}(G)$, where $l_1 = l_0 + 2$, $l_0 = l + \alpha$, $l \geq 0$ is an integer number, and $\alpha \in (0,1)$. Under the additional condition $\varphi \in C^{\alpha_1}(S_0)$, $\varphi \in C^{\alpha_1, \alpha_1/2}(\overline{S}_j)$, $j = 1, \ldots, J$, where $J = 2$ in the case of the slab while $J = 2n$ in the case of the parallelepiped, we have $u \in C^{\alpha_2, \alpha_2/2}(\overline{G}) \cap C^{l_1, l_1/2}(G)$, $\alpha_2 \in (0, \alpha_1)$, $\alpha_1 \in (0, 1)$ [67, 37]. Let $\varphi \in C^{l_1}(S_0)$, $\varphi \in C^{l_1, l_1/2}(\overline{S}_j)$, $j = 1, \ldots, J$, i.e., $S_0 = S_0^g$, $\overline{S}_j = \overline{S}_j^g$ in the case of the function $u(x,t)$, $(x,t) \in \overline{G}$. Taking into account interior *a priori* estimates and the estimates up to the smooth parts of the boundary (derived using Schauder estimates in the variables ξ, t, where $\xi_s = \varepsilon^{-1} x_s$, $s = 1, \ldots, n$), we find

$$\left| \frac{\partial^{k+k_0}}{\partial x_1^{k_1} \ldots \partial x_n^{k_n} \partial t^{k_0}} u(x,\, t) \right| \leq M \left[\varepsilon^{-k} + r^{-k}(x, \Gamma) \right] \left[1 + \varepsilon^{2k_0}\, r^{-2k_0}(x, \Gamma) \right],$$

$$(x,\, t) \in \overline{G}, \quad x \notin \Gamma \;\; for \;\; k + k_0 > 0, \;\; 0 \leq k + 2k_0 \leq K, \tag{5.7}$$

where $K = l + 2$.

But if in the case of the problem on the slab (5.2), the boundary function $\varphi(x, t)$ is sufficiently smooth on each smooth part of the boundary S, namely, $\varphi \in C^{l_0+2}(S_0)$ and $\varphi \in C^{l_0+2, l_0/2+1}(\overline{S}_j)$, where $j = 1, 2$, and, moreover, for the data of the problem on the set S_0^c compatibility conditions up to order K_0 are fulfilled, where $K_0 = [l/2]_i + 1$, then $u \in C^{l_0+2, l_0/2+1}(\overline{G})$ [67, 37]. Using the *a priori* estimates up to the boundary (derived in the variables ξ, t), we obtain the estimate

$$\left| \frac{\partial^{k+k_0}}{\partial x_1^{k_1} \ldots \partial x_n^{k_n} \partial t^{k_0}} u(x, t) \right| \leq M \varepsilon^{-k}, \quad (x, t) \in \overline{G}, \quad 0 \leq k + 2k_0 \leq K, \quad (5.8)$$

where $K = l + 2$.

Let us find estimates in a neighborhood of the boundary layers. The derivation of the estimates of the solutions for problems on the slab and on the parallelepiped is analogous to those in Theorems 2.2.1 and 3.2.1.

Consider the problem on the slab (5.2). Assume that the problem data are sufficiently smooth and the boundary function $\varphi(x, t)$ is continuous, but the fulfillment of compatibility conditions on the set S_0^c is, in general, not assumed.

Write the problem solution as the sum of the functions

$$u(x, t) = U(x, t) + V(x, t), \quad (x, t) \in \overline{G}, \quad (5.9)$$

where $U(x, t)$ and $V(x, t)$ are the regular and singular parts of the solution of the problem. The function $U(x, t)$ is the restriction to $\overline{G}$ of the function $U^e(x, t)$, $(x, t) \in \overline{G}^e$, where $\overline{G}^e = G^e \cup S^e$, $G^e = D^e \times (0, T]$ and $D^e = \mathbb{R}^n$. The function $U^e(x, t)$, $(x, t) \in \overline{G}^e$ is the bounded solution of the problem

$$L^e U^e(x, t) = f^e(x, t), \quad (x, t) \in G^e, \quad U^e(x, t) = \varphi^e(x), \quad (x, t) \in S^e. \quad (5.10)$$

The domain G^e is an extension of the domain G beyond the boundary S^L. The coefficients and the right-hand side in (5.10) are smooth extensions of those in (5.1a), preserving their properties. Let $L^e \equiv \varepsilon^2 L_2^e + L_1^e$, and assume that $L_2^e = \triangle$ and $L_1^e = -c - \partial/\partial t$ outside an m-neighborhood of the set $\overline{G}$ and that there the function $f^e(x, t)$ vanishes. The function $\varphi^e(x)$ is sufficiently smooth on the boundary $S^e = S_0^e$ and coincides with the function $\varphi(x, t)$ on the set S_0; the function $\varphi^e(x)$ is assumed to be equal to zero outside an m-neighborhood of the set $\overline{D}$. The function $V(x, t)$, $(x, t) \in \overline{G}$, is the solution of the problem

$$LV(x, t) = 0, \quad (x, t) \in G, \quad V(x, t) = \varphi_V(x, t), \quad (x, t) \in S,$$

where $\varphi_V(x, t) = \varphi(x, t) - U(x, t)$, $(x, t) \in S$. The function $V(x, t)$ vanishes on the set S_0.

Write the function $U^e(x, t)$ as the sum of the functions

$$U^e(x, t) = U_0^e(x, t) + v_1^e(x, t), \quad (x, t) \in \overline{G}^e.$$

The functions $U_0^e(x,t)$ and $v_1^e(x,t)$ are solutions of the following problems:

$$L_1^e U_0^e(x,\,t) = f^e(x,\,t), \quad (x,\,t) \in G^e,$$

$$U_0^e(x,\,t) = \varphi^e(x), \quad (x,\,t) \in S^e;$$

$$L^e v_1^e(x,\,t) = -\varepsilon^2 L_2^e U_0^e(x,\,t), \quad (x,\,t) \in G^e,$$

$$v_1^e(x,\,t) = 0, \quad (x,\,t) \in S^e.$$

Let the data of the problem (5.1), (5.2) satisfy the condition a_{sk}, b_s, c, c^1, p, $f \in C^{l_0,l_0/2}(\overline{G})$, $\varphi \in C^{l_0}(S_0)$, $l_0 = l + \alpha$, $l \geq 2$ where $\alpha \in (0,\,1)$, and let the extensions of these data to the set $\overline{G}^e$ belong to the class $C^{l_0,l_0/2}(\overline{G}^e)$. Then for the functions $U_0^e(x,t)$ and $v_1^e(x,t)$, one has U_0^e, $v_1^e \in C^{l_0,l_0/2}(\overline{G}^e)$.

The functions $U_0^e(x,t)$ and $v_1^e(x,t)$ satisfy the estimates

$$\left| \frac{\partial^{k+k_0}}{\partial x_1^{k_1} \dots \partial x_n^{k_n} \partial t^{k_0}} U_0^e(x,\,t) \right| \leq M, \quad |v_1^e(x,\,t)| \leq M\,\varepsilon^2, \quad (x,\,t) \in \overline{G}^e,$$

$$0 \leq k + 2k_0 \leq K.$$

Taking into account interior *a priori* estimates, for derivatives of $v_1^e(x,t)$ we find the estimate

$$\left| \frac{\partial^{k+k_0}}{\partial x_1^{k_1} \dots \partial x_n^{k_n} \partial t^{k_0}} v_1^e(x,\,t) \right| \leq M\,\varepsilon^{2-k}, \quad (x,\,t) \in \overline{G}^e, \quad 0 \leq k + 2k_0 \leq K.$$

Thus $U \in C^{l_0,l_0/2}(\overline{G})$, and for the function $U(x,t)$ we have the estimate

$$\left| \frac{\partial^{k+k_0}}{\partial x_1^{k_1} \dots \partial x_n^{k_n} \partial t^{k_0}} U(x,\,t) \right| \leq M \left[1 + \varepsilon^{2-k} \right], \tag{5.11}$$

$$(x,\,t) \in \overline{G}, \quad 0 \leq k + 2k_0 \leq K,$$

where $K = l$.

For the function $V(x,t)$, one gets the estimate

$$|V(x,\,t)| \leq M \exp\left(-m\varepsilon^{-1} r(x,\,\Gamma) \right), \quad (x,\,t) \in \overline{G},$$

where m is an arbitrary constant.

For the data of the boundary value problem (5.1), (5.2), besides the conditions given above, let the condition $\varphi \in C^{l_0,l_0/2}(\overline{S}_j)$, $j = 1, 2$ also be fulfilled. Moreover, for the problem data on the set S_0^c, compatibility conditions up to order K_0, where $K_0 = [l/2]_i$, are assumed to be fulfilled. In that case $V \in C^{l_0,l_0/2}(\overline{G})$. With regard to the *a priori* estimates up to the boundary (derived in the variables ξ, t, where $\xi_s = \varepsilon^{-1} x_s$ and $s = 1, \dots, n$), we obtain

$$\left| \frac{\partial^{k+k_0}}{\partial x_1^{k_1} \dots \partial x_n^{k_n} \partial t^{k_0}} V(x,\,t) \right| \leq M\varepsilon^{-k} \exp\left(-m\varepsilon^{-1} r(x,\,\Gamma) \right),$$

$$(x,\,t) \in \overline{G}, \quad 0 \leq k + 2k_0 \leq K.$$

The derivation of φ_V and our previous estimates yield a bound on the derivatives of $\varphi_V(x,t)$, $(x,t) \in \overline{S}^L$ (not including differentiation in x_1)

$$\left| \frac{\partial^{k+k_0}}{\partial x_2^{k_2} \dots \partial x_n^{k_n} \partial t^{k_0}} \varphi_V(x,t) \right| \le M\left[1 + \varepsilon^{2-k}\right], \tag{5.12}$$

$$(x,t) \in \overline{S}^L, \quad 0 \le k + 2k_0 \le K, \quad k = k_2 + \dots + k_n.$$

Taking (5.12) into account, we find the refined estimate

$$\left| \frac{\partial^{k+k_0}}{\partial x_1^{k_1} \dots \partial x_n^{k_n} \partial t^{k_0}} V(x,t) \right| \le M\left[\varepsilon^{-k_1} + \varepsilon^{2-k}\right] \exp\left(-m\varepsilon^{-1} r(x,\Gamma)\right),$$

$$(x,t) \in \overline{G}, \quad 0 \le k + 2k_0 \le K, \tag{5.13}$$

where $K = l$ and m is an arbitrary constant.

In the case when the fulfillment of compatibility conditions on S_0^c is not assumed, we have $V \in C^{\alpha_1, \alpha_1/2}(\overline{G}) \cap C^{l_0, l_0/2}(G)$, moreover, $S_0 = S_0^g$, $\overline{S}^L = \overline{S}^{Lg}$, $l_0 = l + \alpha$, $l \ge 0$, and $\alpha, \alpha_1 \in (0,1)$. For the function $V(x,t)$, one obtains the estimate

$$\left| \frac{\partial^{k+k_0}}{\partial x_1^{k_1} \dots \partial x_n^{k_n} \partial t^{k_0}} V(x,t) \right| \le M\left[\varepsilon^{-k} + r^{-k}(x,\Gamma)\right] \times$$

$$\times \left[1 + \varepsilon^{2k_0} r^{-2k_0}(x,\Gamma)\right] \exp\left(-m\varepsilon^{-1} r(x,\Gamma)\right), \quad (x,t) \in \overline{G},$$

$$x \notin \Gamma \text{ for } k + k_0 > 0, \quad 0 \le k + 2k_0 \le K.$$

Note that the function $V(x,t)$ is equal to zero on S_0, and for the function $\varphi_V(x,t)$ the estimate (5.12) holds. Taking into account the *a priori* estimate up to the smooth parts of the boundary, we find the estimate

$$\left| \frac{\partial^{k+k_0}}{\partial x_1^{k_1} \dots \partial x_n^{k_n} \partial t^{k_0}} V(x,t) \right| \le M\left[\varepsilon^{-k_1} + \varepsilon^{1-k} + r^{-k}(x,\Gamma)\right] \times \tag{5.14}$$

$$\times \left[1 + \varepsilon^{2k_0} r^{-2k_0}(x,\Gamma)\right] \exp\left(-m\varepsilon^{-1} r(x,\Gamma)\right), \ (x,t) \in \overline{G},$$

$$x \notin \Gamma \text{ for } k + k_0 > 0, \quad 0 \le k + 2k_0 \le K,$$

where $K = l$, and m is an arbitrary constant.

The following theorem holds.

Theorem 5.2.1 *Let a_{sk}, b_s, c, c^1, p, $f \in C^{l_0, l_0/2}(\overline{G})$, $s, k = 1, \dots, n$, $\varphi \in C^{l_0, l_0/2}(\overline{S}^L) \cap C^{l_0}(S_0) \cap C(S)$, $l_0 = l + \alpha$, $l \ge K$, $K \ge 2$, $\alpha \in (0,1)$. Then for the function $u(x,t)$, i.e., the solution of the problem (5.1), (5.2) on the slab, and for the components in the representation (5.9), the estimates (5.6), (5.7),*

(5.11), and (5.14) *hold. In the case when on the set S_0^c the problem data satisfy the compatibility conditions* (5.5) *up to order* $[K/2]_i$, *the estimates* (5.8), (5.11), (5.13) *hold.*

In the case of the problem on the parallelepiped (5.3), we write its solution as the decomposition

$$u(x,t) = U(x,t) + V(x,t), \quad (x,t) \in \overline{G}. \tag{5.15}$$

The function $U(x,t)$, i.e., the regular component, is constructed similarly to $U(x,t)$ in the representation (5.9). The function $V(x,t)$, i.e., the singular component of the solution, is the solution of the problem

$$LV(x, t) = 0, \quad (x, t) \in G, \quad V(x, t) = \varphi_V(x, t), \quad (x, t) \in S,$$

where $\varphi_V(x, t) = \varphi(x, t) - U(x, t)$, $(x, t) \in S$.

Under the conditions a_{sk}, b_s, c, c^1, p, $f \in C^{l_0, l_0/2}(\overline{G})$, $\varphi \in C^{l_0}(S_0) \cap C(S)$, $l_0 = l + \alpha$, $l \geq 2$, $\alpha \in (0, 1)$, we have $U \in C^{l_0, l_0/2}(\overline{G})$ and $V \in C^{l_0, l_0/2}(G)$. For the functions $U(x,t)$ and $V(x,t)$, one obtains the estimates

$$\left| \frac{\partial^{k+k_0}}{\partial x_1^{k_1} \dots \partial x_n^{k_n} \partial t^{k_0}} U(x, t) \right| \leq M \left[1 + \varepsilon^{2-k} \right], \quad (x, t) \in \overline{G};$$

$$\left| \frac{\partial^{k+k_0}}{\partial x_1^{k_1} \dots \partial x_n^{k_n} \partial t^{k_0}} V(x, t) \right| \leq M \left[\varepsilon^{-k} + r^{-k}(x, \Gamma) \right] \times$$

$$\times \left[1 + \varepsilon^{2k_0} r^{-2k_0}(x, \Gamma) \right] \exp\left(- m \varepsilon^{-1} r(x, \Gamma) \right), \quad (x, t) \in \overline{G},$$

$$x \notin \Gamma \; \text{for} \; k + k_0 > 0, \quad 0 \leq k + 2k_0 \leq K.$$

where $K = l$, and m is an arbitrary constant. Further, we need more refined estimates for the singular component of the solution.

Let for the problem data as well as the condition given above, the additional condition $\varphi \in C^{l_0, l_0/2}(\overline{S}_j)$ holds, where $j = 1, \dots, 2n$. In that case, $V \in C^{\alpha_1, \alpha_1/2}(\overline{G})$, where $\alpha_1 \in (0, 1)$. Taking into account the condition $\varphi_V \in C^{l_0, l_0/2}(\overline{S}_j)$, we have $\overline{S}_j = \overline{S}_j^g$, $j = 1, \dots, 2n$ in the case of the function $V(x,t)$, $(x,t) \in \overline{G}$.

Set

$$G_{(j\dots r)} = D_{(j\dots r)} \times (0, T], \quad j, \dots, r = 1, \dots, 2n,$$

where $D_{(j\dots r)} = D_{(j\dots r)(3.6)}$ and $\overline{G}_{(j\dots r)} = G_{(j\dots r)} \cup S_{(j\dots r)}$.

Write the function $V(x,t)$, $(x,t) \in \overline{G}$, as the sum of the functions:

$$V(x, t) = V_0(x, t) + v_2(x, t), \quad (x, t) \in \overline{G}, \tag{5.16}$$

where $V_0(x, t)$ and $v_2(x, t)$ are the main term of the singular part of the solution and the remainder term. The functions $V_0(x, t)$ and $v_2(x, t)$, $(x, t) \in \overline{G}$ are solutions of homogeneous equations with homogeneous conditions.

Let us present the function $V_0(x, t)$, $(x, t) \in \overline{G}$, as the sum of the boundary layer functions of dimension from 1 to n:

$$V_0(x, t) = \sum_{j=1,\ldots,2n} V_{(j)}(x, t) + \sum_{\substack{j,\ldots,r=1,\ldots,2n \\ j<\ldots<r,\ 1<|j\ldots r|\le n \\ S_j \cap \ldots \cap S_r \ne \emptyset}} V_{(j\ldots r)}(x, t), \quad (x, t) \in \overline{G}, \quad (5.17)$$

where $V_{(j)}(x, t)$ and $V_{(j\ldots r)}(x, t)$ are the functions of the one-dimensional and the corner parabolic boundary layers. The function $V_{(j\ldots r)}(x, t)$ is the restriction to $\overline{G}$ of the function $V^e_{(j\ldots r)}(x, t)$, $(x, t) \in \overline{G}^{(1)}$, that is the solution of the problem

$$L^e V^e_{(j\ldots r)}(x, t) = 0, \qquad\qquad (x, t) \in G^{(1)},$$

$$V^e_{(j\ldots r)}(x, t) = \varphi^e_{(j\ldots r)}(x, t), \quad (x, t) \in S(G^{(1)}),$$

where $\overline{G}^{(1)} = \overline{G}_{(j\ldots r)}$, $L^e = L^e_{(5.10)}$, $j,\ldots,r = 1,\ldots,2n$ and $S_j \cap \ldots \cap S_r \ne \emptyset$; the functions $V^e_{(j\ldots r)}(x, t)$ exponentially decrease when moving away from the set $S(G^{(1)})$ for $|j\ldots r| = 1$ and from the set $\cap_q\{S_q(G^{(1)})\}$, where $q = j,\ldots,r$ for $|j\ldots r| > 1$, here $|j\ldots r| = |j\ldots r|_{(3.11)}$. The functions $V^e_{(j\ldots r)}(x, t)$ and $\varphi^e_{(j\ldots r)}(x, t)$ are smooth on the faces $G^{(1)}$ and exponentially decrease when moving away from the set $\cap_q\{S_q(G^{(1)})\}$, $q = j,\ldots,r$, $1 < |j\ldots r| \le n$; on the faces $\overline{S}_q \subset \overline{S}^L(G)$ they satisfy the condition

$$U(x, t) + \varphi^e_{(j)}(x, t) = \varphi(x, t), \quad (x, t) \in \overline{S}_q, \quad q = j, \ |j\ldots r| = 1;$$

$$U(x, t) + \sum_{q=j,\ldots,r} V_{(q)}(x, t) + \sum_{\substack{q,\ldots,k=j,\ldots,r \\ q<\ldots<k,\ 1<|q\ldots k|\le|j\ldots r|-1 \\ S_q \cap \ldots \cap S_k \ne \emptyset}} V_{(q\ldots k)}(x, t) +$$

$$+ \varphi^e_{(j\ldots r)}(x, t) = \varphi(x, t), \quad (x, t) \in \cup_q \overline{S}_q, \quad q = j,\ldots,r, \ |j\ldots r| \ge 2.$$

For the functions $V_{(j\ldots r)}(x, t)$ and $v_2(x, t)$ from (5.16), (5.17), the estimates are valid

$$|V_{(j\ldots r)}(x, t)| \le M \exp\left(-m\varepsilon^{-1} r\left(x, \cap_{q=j,\ldots,r}\Gamma_q\right)\right), \quad (x, t) \in \overline{G},$$

$$j,\ldots,r = 1,\ldots,2n, \quad 1 \le |j\ldots r| \le n;$$

$$|v_2(x, t)| \le M\varepsilon^2, \quad (x, t) \in \overline{G},$$

where m is an arbitrary constant.

Taking into account interior *a priori* estimates and the estimates up to the smooth parts of the boundary $S(G_{j\ldots r})$ (derived in the variables ξ, t), for the derivatives of the function $V^e_{(j...r)}(x, t)$, $(x, t) \in \overline{G}_{(j...r)}$, we find the estimate

$$\left| \frac{\partial^{k+k_0}}{\partial x_1^{k_1} \dots \partial x_n^{k_n} \partial t^{k_0}} V^e_{(j...r)}(x, t) \right| \leq M \left[\varepsilon^{-k} + r^{-k}\left(x, \Gamma(D_{(j...r)})\right) \right] \times$$

$$\times \left[1 + \varepsilon^{2k_0} r^{-2k_0}\left(x, \Gamma(D_{(j...r)})\right) \right] \exp\left(- m\varepsilon^{-1} r(x, \cap_{q=j,...,r} \Gamma_q(D_{(j...r)})) \right),$$

$$(x, t) \in \overline{G}_{(j...r)}, \quad x \notin \Gamma(D_{(j...r)}) \quad \text{for} \quad k + k_0 > 0, \quad 0 \leq k + 2k_0 \leq K.$$

Let us refine this estimate. Write the function $V^e_{(j...r)}(x, t)$ as the decomposition

$$V^e_{(j...r)}(x, t) = V^e_{(j...r)0}(x, t) + v_{V_{(j...r)}}(x, t), \quad (x, t) \in \overline{G}_{(j...r)},$$

where $V^e_{(j...r)0}(x, t)$ and $v_{V_{(j...r)}}(x, t)$ are the main and the remainder terms. The function $V^e_{(j...r)0}(x, t)$ is the solution of the problem

$$L^e_{(j...r)} V^e_{(j...r)0}(x, t) = 0, \qquad (x, t) \in G_{(j...r)},$$

$$V^e_{(j...r)0}(x, t) = \varphi^e_{(j...r)}(x, t), \quad (x, t) \in S(G_{(j...r)}),$$

where

$$L^e_{(j...r)} \equiv \varepsilon^2 \left[\sum_{s,k} a^e_{sk}(x, t) \frac{\partial^2}{\partial x_s \partial x_k} + \sum_s b^e_s(x, t) \frac{\partial}{\partial x_s} - c^e(x, t) \right] -$$

$$-c^{1e}(x,t) - p^e(x,t) \frac{\partial}{\partial t}, \quad s,k = \overline{j}, \dots, \overline{r}; \quad \overline{j} = \overline{j}_{(3.15)}(j),$$

the variables x_s, $s \neq j, \dots, r$ are regarded as parameters. For the function $V^e_{(j...r)0}(x, t)$, we obtain the estimate

$$\left| \frac{\partial^{k+k_0}}{\partial x_1^{k_1} \dots \partial x_n^{k_n} \partial t^{k_0}} V^e_{(j...r)0}(x,t) \right| \leq M \left[\varepsilon^{-(\overline{k}_j + \dots + \overline{k}_r)} + \right.$$

$$+ r^{-(\overline{k}_j + \dots + \overline{k}_r)}\left(x, \Gamma(D_{(j...r)})\right) \right] \left[1 + \varepsilon^{2k_0} r^{-2k_0}\left(x, \Gamma(D_{(j...r)})\right) \right] \times$$

$$\times \exp\left(- m\varepsilon^{-1} r\left(x, \cap_{q=j,...,r} \Gamma_q(D_{(j...r)})\right) \right),$$

$$(x,t) \in \overline{G}_{(j...r)}, \quad x \notin \Gamma(D_{(j...r)}) \quad \text{for} \quad k + k_0 > 0, \quad k \leq l,$$

where $\overline{k}_p = \overline{k}_{p(3.16)}(k_s)$ for $p = j, \dots, r$, and $k = \sum_{s=1}^n k_s$.
For the function $v_{V_{(j...r)}}(x,t)$, we have the estimate

$$\left| \frac{\partial^{k+k_0}}{\partial x_1^{k_1} \dots \partial x_n^{k_n} \partial t^{k_0}} v_{V_{(j...r)}}(x,t) \right| \leq M\varepsilon \left[\varepsilon^{-k} + r^{-k}\left(x, \Gamma(D_{(j...r)})\right) \right] \times$$

$$\times \left[1 + \varepsilon^{2k_0} r^{-2k_0}\left(x, \Gamma(D_{(j...r)})\right) \right] \exp\left(- m\varepsilon^{-1} r(x, \cap_{q=j,...,r} \Gamma_q(D_{(j...r)})) \right),$$

$$(x,t) \in \overline{G}_{(j...r)}, \quad x \notin \Gamma(D_{(j...r)}) \quad \text{for} \quad k + k_0 > 0, \quad 0 \leq k \leq K, \quad K = l.$$

Taking into account the estimates of the derivatives for the functions $V^e_{(j...r)0}(x, t)$ and $v_{V_{(j...r)}}(x, t)$, one obtains the estimate

$$\left| \frac{\partial^{k+k_0}}{\partial x_1^{k_1} \dots \partial x_n^{k_n} \partial t^{k_0}} V^e_{(j...r)}(x, t) \right| \leq M \left[\varepsilon^{-(\bar{k}_j + \dots + \bar{k}_r)} + \varepsilon^{1-k} + \right.$$

$$+ r^{-(\bar{k}_j + \dots + \bar{k}_r)} \left(x, \Gamma(D_{(j...r)}) \right) + \varepsilon\, r^{-k} \left(x, \Gamma(D_{(j...r)}) \right) \left] \times \right.$$

$$\times \left[1 + \varepsilon^{2k_0}\, r^{-2k_0} (x, \, \Gamma(D_{(j...r)})) \right] \exp \left(-m \varepsilon^{-1} r(x, \, \cap_{q=j,\dots,r} \Gamma_q(D_{(j...r)})) \right),$$

$$(x, t) \in \overline{G}_{(j...r)}, \quad x \notin \Gamma(D_{(j...r)}) \quad \text{for} \ \ k + k_0 > 0, \ \ 0 \leq k + 2k_0 \leq K,$$

where $\bar{k}_p = \bar{k}_{p(3.16)}(k_s)$ for $p = j, \dots, r$, and $k = \sum_{s=1}^n k_s$.

For $v_2(x, t)$ in (5.16), one gets the following estimate

$$\left| \frac{\partial^{k+k_0}}{\partial x_1^{k_1} \dots \partial x_n^{k_n} \partial t^{k_0}} v_2(x, t) \right| \leq M \varepsilon^2 \left[\varepsilon^{-k} + r^{-k}(x, \Gamma) \right] \left[1 + \varepsilon^{2k_0}\, r^{-2k_0}(x, \Gamma) \right],$$

$$(x, t) \in \overline{G}, \quad x \notin \Gamma \ \ \text{for} \ \ k + k_0 > 0, \quad 0 \leq k + 2k_0 \leq K.$$

Thus the solution of the boundary value problem can be presented as the sum of the functions

$$u(x, t) = U_{(0)}(x, t) + V_0(x, t), \quad (x, t) \in \overline{G}, \tag{5.18}$$

where

$$U_{(0)}(x, t) = U(x, t) + v_2(x, t), \quad (x, t) \in \overline{G}.$$

For the function $u(x, t)$ and its components $U_{(0)}(x, t)$ and $V_{(j...r)}(x, t)$ in the representations (5.18) and (5.17), the following estimates

$$\left| \frac{\partial^{k+k_0}}{\partial x_1^{k_1} \dots \partial x_n^{k_n} \partial t^{k_0}} u(x, t) \right| \leq M \left[\varepsilon^{-k} + r^{-k}(x, \Gamma) \right] \times \tag{5.19a}$$

$$\times \left[1 + \varepsilon^{2k_0}\, r^{-2k_0}(x, \Gamma) \right],$$

$$\left| \frac{\partial^{k+k_0}}{\partial x_1^{k_1} \dots \partial x_n^{k_n} \partial t^{k_0}} U_{(0)}(x, t) \right| \leq M \left[1 + \varepsilon^{2-k} + \varepsilon^2\, r^{-k}(x, \Gamma) \right] \times \tag{5.19b}$$

$$\times \left[1 + \varepsilon^{2k_0}\, r^{-2k_0}(x, \Gamma) \right],$$

$$\left| \frac{\partial^{k+k_0}}{\partial x_1^{k_1} \dots \partial x_n^{k_n} \partial t^{k_0}} V_{(j...r)}(x, t) \right| \leq M \left[\varepsilon^{-(\bar{k}_j + \dots + \bar{k}_r)} + \varepsilon^{1-k} + \right. \tag{5.19c}$$

$$+ r^{-(\bar{k}_j + \dots + \bar{k}_r)} (x, \Gamma) + \varepsilon\, r^{-k} (x, \Gamma) \left] \left[1 + \varepsilon^{2k_0}\, r^{-2k_0}(x, \Gamma) \right] \times \right.$$

$$\times \exp \left(-m \varepsilon^{-1} r(x, \cap_{q=j,\dots,r} \Gamma_q) \right), \quad (x, t) \in \overline{G}, \ \ x \notin \Gamma \ \ \text{for} \ \ k + k_0 > 0,$$

$$0 \leq k + 2k_0 \leq K, \quad j, \dots, r = 1, \dots, 2n, \ \ 1 \leq |j \dots r| \leq n;$$

$$|u(x, t) - \varphi(x^*, t)| \le M \varepsilon^{-1} r(x, \Gamma), \quad (x, t) \in \overline{G}, \qquad (5.19\text{d})$$

are valid, where $x^* = x^*_{(5.6)}(x, \Gamma)$ is a point on $\Gamma(D)$ nearest to the point $x \in \overline{D}$, $(x, t) \in \overline{G}$. In (5.19) $\overline{k}_p = \overline{k}_{p(3.16)}(k_s)$, $p = j, \ldots, r$, $k = \sum_{s=1}^{n} k_s$, $K = l$, and the constant m can be chosen arbitrarily.

The following theorem holds.

Theorem 5.2.2 *Let* a_{sk}, b_s, c, c^1, p, $f \in C^{l_0, l_0/2}(\overline{G})$, $s, k = 1, \ldots, n$, $\varphi \in C^{l_0, l_0/2}(\overline{S}_j) \cap C^{l_0}(S_0) \cap C(S)$, $j = 1, \ldots, 2n$, $l_0 = l + \alpha$, $l \ge K$, $K \ge 2$, $\alpha \in (0, 1)$. Then for the function $u(x, t)$, i.e., the solution of the problem (5.1) on the parallelepiped (5.3), and for the components $U_{(0)}(x, t)$ and $V_{(j \ldots r)}(x, t)$ in the representations (5.18) and (5.17) of the function $u(x, t)$, the estimates (5.19) hold.*

The smoothness of the data of the problem (5.1), (5.3) does not, in general, imply the smoothness of the solution on the set $\overline{G}$ (see, e.g., [67, 37]).

In some problems, for the function $u(x, t)$, one has the inclusion

$$u \in C^{l_0, l_0/2}(\overline{G}), \quad l_0 = l + \alpha, \quad l = K, \quad \alpha \in (0, 1), \qquad (5.20)$$

if on the set $S^c(\overline{G})$, the data of the problem (5.1), (5.3) satisfy special compatibility conditions. In some cases, the inclusion (5.20) allows us to obtain estimates of derivatives of the components in (5.18) and (5.17) on the set $\overline{G}$.

In the case of the problem (5.1), (5.3), assume that compatibility conditions in t up to order $K_0 = [K/2]_i$ are satisfied on the set S_0^c. In that case, for the solution of the boundary value problem and for the components in (5.18) and (5.17), the estimates (5.19) are satisfied, where $r(x, \Gamma)$ is $r(x, \Gamma^c)$. This estimate is used outside an $(m\varepsilon)$-neighborhood of the set S^{Lc}. In the cases $n = 2, 3$, compatibility conditions in x on the set S^{Lc}, which are necessary and sufficient for (5.20) for $K = 2$, can be written out based on results of [212, 213]. For example, if

$$L \equiv \varepsilon^2 \triangle - c^1(x, t) - p(x, t)\frac{\partial}{\partial t}, \quad (x, t) \in \overline{G}, \qquad (5.21)$$

$$r(x, \Gamma^c) \le m\varepsilon, \quad t \in [0, T],$$

then in a neighborhood of the set S^{Lc} one rewrites the parabolic problem corresponding to (5.21) as an elliptic problem where the term $p(x, t)\partial/\partial t$ has been moved to the right-hand side and t is regarded as a parameter. Applying compatibility conditions from [212, 213] to the elliptic problem and also to the problem obtained by differentiation in t of the problem (5.1), (5.21), it is not difficult to write down compatibility conditions on S^c for the problem (5.1), (5.21), which guarantee that (5.20) holds. Using the *a priori* estimates of the solution up to the smooth parts of the boundary (for $r(x, \Gamma^c) \ge m_1 \varepsilon$) and the estimates in a neighborhood of "edges" S^{Lc} (for $r(x, \Gamma^c) \le m_2 \varepsilon$, where

$m_2 > m_1$), then for the functions $V^e_{(j...r)}(x, t)$ and $v_2(x, t)$ for $n = 2, 3$, one can justify the estimates

$$\left| \frac{\partial^{k+k_0}}{\partial x_1^{k_1} \dots \partial x_n^{k_n} \partial t^{k_0}} V^e_{(j...r)}(x, t) \right| \leq M \left[\varepsilon^{-(\overline{k}_j + \dots + \overline{k}_r)} + \varepsilon^{2-k} \right] \times$$

$$\times \exp\left(-m\varepsilon^{-1} r\left(x, \cap_{q=j,\dots,r} \Gamma_q(D_{(j...r)})\right)\right), \quad (x, t) \in \overline{G}_{(j...r)},$$

$$\left| \frac{\partial^{k+k_0}}{\partial x_1^{k_1} \dots \partial x_n^{k_n} \partial t^{k_0}} v_2(x, t) \right| \leq M \varepsilon^{2-k}, \quad (x, t) \in \overline{G}, \quad 0 \leq k + 2k_0 \leq K,$$

where $K = l$.

In the case of the conditions (5.21), if the membership (5.20) holds, the functions $u(x, t)$, $U_{(0)}(x, t)$ and $V_{(j...r)}(x, t)$ then satisfy the estimates

$$\left| \frac{\partial^{k+k_0}}{\partial x_1^{k_1} \dots \partial x_n^{k_n} \partial t^{k_0}} u(x, t) \right| \leq M \varepsilon^{-k}, \tag{5.22a}$$

$$\left| \frac{\partial^{k+k_0}}{\partial x_1^{k_1} \dots \partial x_n^{k_n} \partial t^{k_0}} U_{(0)}(x, t) \right| \leq M \left[1 + \varepsilon^{2-k} \right], \tag{5.22b}$$

$$\left| \frac{\partial^{k+k_0}}{\partial x_1^{k_1} \dots \partial x_n^{k_n} \partial t^{k_0}} V_{(j...r)}(x, t) \right| \leq M \left[\varepsilon^{-(\overline{k}_j + \dots + \overline{k}_r)} + \varepsilon^{2-k} \right] \times \tag{5.22c}$$

$$\times \exp\left(-m\varepsilon^{-1} r\left(x, \cap_{q=j,\dots,r} \Gamma_q\right)\right),$$

$$(x, t) \in \overline{G}, \quad 0 \leq k + 2k_0 \leq K, \quad j, \dots, r = 1, \dots, 2n, \quad 1 \leq |j \dots r| \leq n.$$

In (5.22) we have $\overline{k}_p = \overline{k}_{p(3.16)}(k_s)$ for $p = j, \dots, r$, while $K = l$, and the constant m can be chosen arbitrarily.

The following theorem holds.

Theorem 5.2.3 *Let $n \leq 3$. Let the assumptions of Theorem 5.2.2, the condition (5.21), and the membership (5.20) be fulfilled. Then the function $u(x, t)$, i.e., the solution of the problem (5.1) on the parallelepiped (5.3), and the components $U_{(0)}(x, t)$ and $V_{(j...r)}(x, t)$ in the representations (5.18) and (5.17) satisfy the estimates (5.22).*

5.3   ε-uniformly convergent difference schemes

When constructing ε-uniformly convergent difference schemes for singularly perturbed parabolic reaction-diffusion equations, we use a technique that is similar to the one used for singularly perturbed elliptic reaction-diffusion equations. For the ε-uniform convergence of difference schemes whose construction is based on classical difference approximations of the boundary value problem (5.1), it is necessary to use meshes that condense in a neighborhood of the boundary layers.

In order that the difference schemes on such meshes be ε-uniformly monotone, it is required that the coefficients of the mixed derivatives in the equation at the domain boundary are subject to special conditions as are the relations between the step-sizes in the different space coordinates.

5.3.1 Grid approximations of the boundary value problem

In the case of the parallelepiped (5.3) on the set $\overline{G}$, we introduce the rectangular grid

$$\overline{G}_h = \overline{D}_h \times \overline{\omega}_0, \quad \overline{D}_h = \overline{\omega}_1 \times \ldots \times \overline{\omega}_n, \qquad (5.23a)$$

where $\overline{\omega}_0$ is a mesh on the interval $[0, T]$ on the t-axis with an arbitrary distribution of nodes, while $\overline{\omega}_s$, for $s = 1, \ldots, n$, is a mesh on the interval $[d_{*s}, d_s^*]$ on the x_s-axis with an arbitrary distribution of nodes. In the case of the slab (5.2), we use the grid

$$\overline{G}_h = \overline{D}_h \times \overline{\omega}_0, \quad \overline{D}_h = \overline{\omega}_1 \times \omega_2 \times \ldots \times \omega_n, \qquad (5.23b)$$

where $\overline{\omega}_0 = \overline{\omega}_{0(5.23a)}$, $\overline{\omega}_1 = \overline{\omega}_{1(5.23a)}$ and ω_s, for $s = 2, \ldots, n$, is a mesh on the x_s-axis. Set $h_t = \max_j h_t^j$, where $h_t^j = t^{j+1} - t^j$ with $t^j, t^{j+1} \in \overline{\omega}_0$. Let $N_0 + 1$ be the number of nodes in the mesh $\overline{\omega}_0$. Set $h = \max_s h_s$, $h_s = \max_i h_s^i$, $h_s^i = x_s^{i+1} - x_s^i$. Let $N_s + 1$ be the number of nodes in the mesh $\overline{\omega}_{s(5.23a)}$, $s = 1, \ldots, n$, in the case of the problem on the parallelepiped; in the case of the problem on the slab $N_1 + 1$ is the number of nodes in the mesh $\overline{\omega}_{1(5.23b)}$ and $N_s + 1$ is the minimal number of nodes in the mesh $\overline{\omega}_{s(5.23b)}$, $s = 2, \ldots, n$, per unit length; $N = \min_s N_s$, $s = 1, \ldots, n$. Assume that the conditions $h \leq MN^{-1}$ and $h_t \leq MN_0^{-1}$ are fulfilled. Set $G_h = G \cap \overline{G}_h$ and $S_h = S \cap \overline{G}_h$. On the grid $\overline{G}_{h(5.23)}$ for the problem (5.1), we consider the difference scheme

$$\Lambda z(x, t) = f(x, t), \ (x, t) \in G_h, \quad z(x, t) = \varphi(x, t), \ (x, t) \in S_h. \qquad (5.24)$$

Here

$$\Lambda = \Lambda_{(5.25)}(L) \equiv \varepsilon^2 \left\{ \sum_{s=1}^n a_{ss}(x, t) \delta_{\overline{x_s}\,\widehat{x_s}} + \right. \qquad (5.25)$$

$$+ 2^{-1} \sum_{\substack{s,k=1 \\ s \neq k}}^n \left[\left(a_{sk}^+(x, t) (\delta_{xs\,xk} + \delta_{\overline{xs}\,\overline{xk}}) + a_{sk}^-(x, t)(\delta_{xs\,\overline{xk}} + \delta_{\overline{xs}\,xk}) \right) \right] +$$

$$\left. + \sum_{s=1}^n \left[b_s^+(x, t) \delta_{xs} + b_s^-(x, t) \delta_{\overline{xs}} \right] - c(x, t) \right\} - c^1(x, t) - p(x, t) \delta_{\overline{t}},$$

where $\delta_{\bar{t}} z(x, t) = \left(h_t^{j-1}\right)^{-1} \left[z(x, t^j) - z(x, t^{j-1})\right], \quad h_t^{j-1} = t^j - t^{j-1},$
$t^j, t^{j-1} \in \bar{\omega}_0.$ The operator

$$L_{(2)}^* \equiv \sum_{s,k=1}^{n} a_{sk}(x, t) \partial^2 / \partial x_s \partial x_k \tag{5.26}$$

is the elliptic part of the operator $L_{(5.1)}$. We approximate it by the discrete operator

$$\Lambda_{(2)}^* = \Lambda_{(2)}^*(L_{(2)}^*) \equiv \sum_{s=1}^{n} a_{ss}(x, t)\delta_{\overline{x_s}\,\widehat{x_s}} + \tag{5.27}$$

$$+2^{-1} \sum_{\substack{s,k=1 \\ s \neq k}}^{n} \left[a_{sk}^+(x, t)(\delta_{x_s\,x_k} + \delta_{\overline{x_s}\,\overline{x_k}}) + a_{sk}^-(x, t)(\delta_{x_s\,\overline{x_k}} + \delta_{\overline{x_s}\,x_k})\right].$$

The difference scheme (5.24), (5.25), (5.23)} is, in general, not ε-uniformly monotone. The monotonicity of the operator $\Lambda_{(5.25)}$ on the grid $\overline{G}_h$ is violated, e.g., when one has

$$a_{sk}(x, t) \geq 0, \quad (x, t) \in \overline{G}, \quad s, k = 1, \ldots, s \neq k,$$

but the following condition is violated:

$$\min_{h_s^{(i_s)}, k, G_h} \left[a_{kk}(x, t) 2 \left(h_k^{i_k} + h_k^{i_k-1}\right)^{-1} - \sum_{\substack{s=1 \\ s \neq k}}^{n} a_{sk}(x, t) \left(h_s^{(i_s)}\right)^{-1}\right] \geq 0,$$

$$(x, t) \in \overline{G}, \quad x = (x_1^{i_1}, \ldots, x_n^{i_n}), \quad k = 1, \ldots, n,$$

where $h_s^{(i_s)} = h_s^{i_s}$, or $h_s^{(i_s)} = h_s^{i_s-1}$, $s = 1, \ldots, n$.

Imposing some conditions on the differential operator $L_{(2)}^*$, we construct grids $\overline{G}_h$ with a special distribution of nodes in the meshes x_s, for $s = 1, \ldots, n$, on which the difference operators $\Lambda_{(2)(5.27)}^*$ and $\Lambda_{(5.25)}$ and, hence, the difference scheme (5.24) are ε-uniformly monotone. These conditions are given below.

5.3.2 Consistent grids on a slab

Let us consider the problem (5.2) on the slab $\overline{D}_{(5.2)}$.

On the set $\overline{G}$ we introduce the uniform grid

$$\overline{G}_h = \overline{D}_h \times \bar{\omega}_0, \quad \overline{D}_h = \bar{\omega}_1 \times \omega_2 \times \ldots \times \omega_n, \tag{5.28}$$

where ω_s, for $s = 1, \ldots, n$, are uniform meshes with step-size h_s and $\bar{\omega}_0$ is a uniform mesh with step-size h_t. Let Q be a convex subdomain of G. On the set $\overline{Q}$ we introduce the uniform grid

$$\overline{G}_h(Q) = \{\overline{Q}\}_h = \overline{Q} \cap \overline{G}_{h(5.28)}. \tag{5.29}$$

The discrete operator $\Lambda^*_{(2)(5.27)}$ corresponds to the operator $L^*_{(2)(5.26)}$ on the grid $\overline{G}_{h(5.29)}(Q)$.

The elliptic operator $L^*_{(2)(5.26)}$, considered at the point (x_0, t_0), can be brought to canonical form by an appropriate linear transformation $X = X(x, t_0)$. In the transformed operator $L^*_{(2)X}$, the maxima of the coefficients of the mixed derivatives at the point (X, t) depend on the distance to the point (X_0, t_0), where $X_0 = X(x_0, t_0)$, and these maxima, in general, increase as this distance increases.

Let $L^*_{(2)(5.26)}$ be already the canonical operator at the point (x_0, t_0). Then in a small neighborhood of (X_0, t_0), the operator $L^*_{(2)}$ has "almost canonical" form, i.e., the coefficients of the mixed derivatives are small compared with the coefficients of the second-order derivatives in each variable.

We shall assume that for an arbitrary point $(x^*, t^*) \in \overline{G}$, one can find
(i) a sufficiently small neighborhood $Q(x^*, t^*)$, where $\overline{Q}(x^*, t^*) \subseteq \overline{G}$,
(ii) n pairs of functions $\rho_{*s}(x, t)$ and $\rho_s^*(x, t)$, $(x, t) \in \overline{Q}(x^*, t^*)$,
for $s = 1, \ldots, n$,
such that
(iii) for these functions one has

$$\max_{\overline{Q}(x^*, t^*)} \rho_{*s}(x, t) \leq \min_{\overline{Q}(x^*, t^*)} \rho_s^*(x, t), \quad s = 1, \ldots, n, \tag{5.30a}$$

(iv) the coefficients $a_{sk}(x, t)$, for $s, k = 1, \ldots, n$, satisfy the relation

$$\min_{\overline{Q}(x^*, t^*)} a_{kk}(x, t) h_k^{-1} - \sum_{\substack{s=1 \\ s \neq k}}^{n} \max_{\overline{Q}(x^*, t^*)} |a_{sk}(x, t)| h_s^{-1} > 0, \tag{5.30b}$$
$$k = 1, \ldots, n,$$

where
(v) the values h_s, for $s = 1, \ldots, n$, in (5.30b) satisfy the condition

$$h_s \in \left[\max_{\overline{Q}(x^*, t^*)} \rho_{*s}(x, t) h^*, \min_{\overline{Q}(x^*, t^*)} \rho_s^*(x, t) h^* \right], \quad s = 1, \ldots, n, \tag{5.30c}$$

and $h^* > 0$ is an arbitrary number.

The functions $\rho_{*s}(x, t)$ and $\rho_s^*(x, t)$ are defined by the coefficients of the operator $L^*_{(2)(5.26)}$:

$$\rho_{*s}(x, t) = \rho_{*s}\left(x, t; L^*_{(2)}\right), \quad \rho_s^*(x, t) = \rho_s^*\left(x, t; L^*_{(2)}\right), \quad s = 1, \ldots, n.$$

The values $\rho_{*s}(x, t)$ and $\rho_s^*(x, t)$ can be chosen to satisfy the relations

$$\max_{\overline{Q}(x^*, t^*)} \rho_{*s}(x, t) \to 0, \quad \min_{\overline{Q}(x^*, t^*)} \rho_s^*(x, t) \to \infty$$

$$for \quad \max_{s, k, s \neq k, \overline{Q}(x^*, t^*)} |a_{sk}(x, t)| \to 0.$$

In the general case, one can ensure the fulfillment of the relations (5.30) if one brings the operator $L^*_{(2)(5.26)}$ to the canonical form at the point (x^*, t^*) (or close to the canonical form) and chooses the neighborhood $Q(x^*, t^*)$ to be sufficiently small.

We say that the operator $L^*_{(2)(5.26)}$ on the set $\overline{G}$ has *local almost canonical form* (in the variables $x_1, \ldots, x_n$), if for an arbitrary point $(x^*, t^*) \in \overline{G}$ one can find a neighborhood $Q(x^*, t^*)$ and n pairs of the functions $\rho_{*s}(x, t)$ and $\rho^*_s(x, t)$, $x \in \overline{Q}(x^*, t^*)$, such that the relations (5.30) hold for the coefficients of the operator $L^*_{(2)}$.

Note that when defining the local almost canonical operator, in the relations (5.30) we use the values h_s, for $s = 1, \ldots, n$, which, in general, have no relations to grids on the set $\overline{Q}(x^*, t^*)$. Grids on the set $\overline{Q}(x^*, t^*)$ are considered below.

For the set $\overline{Q} \cap \overline{G}_h$ and the operator $\Lambda^*_{(2)(5.27)}$ (the operator $\Lambda_{(5.25)}$), we consider the node (x_0, t_0) as interior, if all nodes of the stencil, for which the point (x_0, t_0) is centre, of the operator $\Lambda^*_{(2)(5.27)}$ (the operator $\Lambda_{(5.25)}$) belong to the set $\overline{Q} \cap \overline{G}_h$.

Theorem 5.3.1 *Assume that the grid $\overline{G}_h$ is uniform on the set $\overline{Q}(x^*, t^*)$ in each of the variables, i.e.,*

$$\overline{G}_h = \overline{G}_h\big(Q(x^*, t^*),\ L^*_{(2)}\big) = \overline{G}_h\big(Q(x^*, t^*),\ L^*_{(2)},\ x_1, \ldots, x_n\big) = \qquad (5.31)$$

$$= \Big\{\{\overline{Q}(x^*, t^*)\}_{h(5.29)} \text{ under the condition } (5.30c), \text{ and when } (5.30) \text{ holds}\Big\}.$$

Then the operator $\Lambda_{(5.25)}(L)$ (the operator $\Lambda^*_{(2)(5.27)}(L^*_{(2)})$) is ε-uniformly monotone on the set of the interior nodes.

Thus, the uniform grid $\overline{G}_{h(5.31)}$ is consistent on $\overline{Q}(x^*, t^*)$ in $x_1, \ldots, x_n$ with the monotonicity condition for the operators $\Lambda^*_{(2)(5.27)}$ and $\Lambda_{(5.25)}$.

We are interested in monotone difference approximations of the boundary value problem (5.1) on the slab (5.2). The following statement is valid.

Lemma 5.3.1 *For the mesh $\overline{\omega}_1$ that defines the grid $\overline{G}_{h(5.23b)}$, the condition*

$$\left. \begin{array}{l} \varepsilon^{-1} \max_{d_{*1} < x^i_1 \le d_{*1} + M\varepsilon} h^{i-1}_1 \\[2mm] \varepsilon^{-1} \max_{d^*_1 - M\varepsilon \le x^i_1 < d^*_1} h^i_1, \end{array} \right\} \underset{\varepsilon}{\to} 0 \ \text{ for } N_1 \to \infty, \ x^i_1 \in \overline{\omega}_1, \qquad (5.32)$$

on the distribution of nodes is necessary for the ε-uniform approximation of the operators $L^*_{(2)(5.26)}$ and $L_{(5.1)}$ by the operators $\Lambda^*_{(2)(5.27)}$ and $\Lambda_{(5.25)}(L)$ on the problem solution.

Note that on the meshes, which are uniform in an $(M\varepsilon)$-neighborhood of the faces S_j, for $j = 1, 2$, *the condition* (5.32) *is* *necessary for the ε-uniform convergence of the difference scheme* (5.24), (5.25).

The condition (5.32) *is* *necessary for the ε-uniform resolvability of the boundary value problem* (5.1), (5.2) by the classical difference scheme (5.24), (5.25).

We shall consider the scheme (5.24), (5.25) on the grid

$$\overline{G}_h = \overline{D}_h \times \overline{\omega}_0, \quad \overline{D}_h = \overline{D}_h(\overline{\omega}_{1(2.48)}) = \overline{\omega}_1 \times \omega_2 \times \ldots \times \omega_n, \qquad (5.33)$$

which is condensing in a neighborhood of the faces S_j, $j = 1, 2$. Here $\overline{\omega}_1 = \overline{\omega}_{1(2.48)}$ is the piecewise-uniform mesh with $h_1 = \max\limits_i h_1^i$ and the ω_s, for $s = 2, \ldots, n$, are uniform meshes with step-size h_s.

Note that on the grid (5.33) the difference scheme (5.24), (5.25), when it is monotone, converges ε-uniformly. *A necessary condition for the monotonicity of the difference scheme* on the grid that satisfies the condition (5.32) is established in the following theorem.

Theorem 5.3.2 *If the coefficients $a_{1s}(x, t)$, for $s = 2, \ldots, n$ ($a_{1s}(x, t) = a_{s1}(x, t)$), do not satisfy the condition*

$$a_{1s}(x, t) = 0, \quad (x, t) \in \overline{S}^L, \quad s = 2, \ldots, n, \qquad (5.34)$$

then the difference scheme (5.24), (5.25) *on the grid $\overline{G}_{h(5.23b)}$ under the condition* (5.32) *(in particular, on the grid $\overline{G}_{h(5.33)}$) is not ε-uniformly monotone for any distribution of nodes in the meshes ω_s, for $s = 1, \ldots, n$.*

Let us consider the boundary value problem in the case when the coefficients $a_{1s}(x, t)$ satisfy the condition (5.34) (the canonicity condition of the truncated operator $L_{(2)}^{*[1]}$ on the faces $\overline{S}_j$, $j = 1, 2$), and also one has

$$a_{1s} \in C^1(\overline{G}), \quad s = 2, \ldots, n. \qquad (5.35)$$

We construct a grid that is consistent in $x_1, \ldots, x_n$, uniform in $x_2, \ldots, x_n$ and condenses in the boundary layer in x_1. When constructing such a mesh, one uses the piecewise-uniform grid (5.33).

As well as the condition (5.34) (i.e., the operator $L_{(2)}^*$ on the set $\overline{S}^L$ is canonical in x_1), assume that the coefficients of the operator $L_{(2)(5.26)}^*$ satisfy the condition that for an arbitrary point $(x^*, t^*) \in \overline{G}$, one can find
  (i) sufficiently small values m^*, $\nu > 0$,
  (ii) a neighborhood $Q(x^*, t^*)$,
and
  (iii) n pairs of functions $\rho_{*s}(x, t)$ and $\rho_s^*(x, t)$, for $s = 1, \ldots, n$, such that

(iv) the coefficients $a_{sk}(x, t)$, for $s, k = 1, \ldots, n$, of the operator $L^*_{(2)}$ satisfy the relations (5.30) (i.e., the operator $L^*_{(2)}$ is locally almost canonical in $x_1, \ldots, x_n$ on the set $\overline{G}$);

(v) the values m^*, $\nu > 0$ do not appear in (5.30).

We modify (iv) in an m^*-neighborhood of $\overline{S}^L$, where we assume that the coefficients $a_{sk}(x, t)$ satisfy (5.30) for $s, k = 2, \ldots, n$, and that one has the additional relation

$$\left[\min_{\substack{\overline{Q}(x^*, t^*) \\ r(x, \Gamma) \leq m^*}} a_{kk}(x, t) - \nu \right] h_k^{-1} - \sum_{\substack{s=2 \\ s \neq k}}^{n} \max_{\substack{\overline{Q}(x^*, t^*) \\ r(x, \Gamma) \leq m^*}} |a_{sk}(x, t)| \, h_s^{-1} > 0, \quad (5.36)$$

$$k = 2, \ldots, n.$$

Thus, in the above mentioned relations (5.30) and (5.36), considered in a neighborhood of $\overline{S}^L$, the coefficients $a_{sk}(x, t)$ are involved only for $s, k \geq 2$, i.e., only the coefficients of the truncated operator

$$L^{*[1]}_{(2)} = L^{*[1]}_{(2)}(L^*_{(2)})$$

appear. In the case when the coefficients of the operator $L^{*[1]}_{(2)}$ satisfy the conditions $\{(5.30), (5.36)\}$ for $s, k \geq 2$, the operator $L^{*[1]}_{(2)}$ is strongly local almost canonical in $x_2, \ldots, x_n$ in a neighborhood of the set $\overline{S}^L$.

The fulfillment of $\{(5.30), (5.36)\}$ in the case of the condition (5.34) can be ensured by a linear transformation of the variables $x_2, \ldots, x_n$ that brings the operator $L^*_{(2)}$ to the canonical form at the point $(x^*, t^*) \in \overline{S}^L$ (or to a form close to canonical), and by choosing the neighborhood $Q(x^*, t^*)$ and the values m^*, ν to be sufficiently small.

The condition (5.34) and the condition

the coefficients $a_{sk}(x, t)$ of the operator $L^*_{(2)}$ on the set

$$\overline{Q}(x^*, t^*) \subseteq \overline{G} \text{ satisfy the relations } \{(5.30), (5.36)\}$$

(5.37)

together allow us to construct grids on the set $\overline{Q}(x^*, t^*)$ that are consistent piecewise-uniform on the x_1-axis and thereby ensure the monotonicity of the operators $\Lambda^*_{(2)(5.27)}$ and $\Lambda_{(5.25)}$.

Let the conditions (5.34) and (5.37) be satisfied. Consider the grid

$$\overline{G}_h(Q) = \{\overline{Q}\}_h = \overline{Q} \cap \overline{G}_{h(5.33)}, \quad (5.38)$$

which is piecewise-uniform in x_1. We shall provide conditions on the parameters of this grid under which the operator $\Lambda^*_{(2)(5.27)}$ is monotone.

First, consider the operator $\Lambda^*_{(2)(5.27)}$ on the grid $\overline{G}_{h(5.33)}$. Let the step-sizes h_s in the meshes ω_s, for $s = 2, \ldots, n$, and the step-size h_1, i.e., the maximal step-size in the mesh $\overline{\omega}_{1(2.48)}$, satisfy the condition (5.30c). Note that $h_1 = h_{(2)}$ in the mesh (2.48). Let the value $\sigma = \sigma_{(2.48)}$ satisfy the condition

$$\sigma \leq m^*_{(5.36)}, \tag{5.39a}$$

and let the condition (5.35) hold. By virtue of conditions (5.34), (5.35) one has

$$|a_{1s}(x, t)| \leq M_1\sigma \text{ for } r\big((x, t), \overline{S}^L\big) \leq \sigma, \text{ where } M_1 = \max_{s,\, s\neq 1, \overline{G}} \left| \frac{\partial}{\partial x_1} a_{1s}(x, t) \right|,$$

and it follows that

$$\left|a_{1k}(x, t)\right| h_{(1)}^{-1} \leq M_1\, h_{(1)}^{-1}\sigma, \quad k \neq 1,$$

where $h_{(1)}$ is the step-size in the mesh $\overline{\omega}_{1(2.48)}$ on the intervals $\big[d_{*1},\, d_{*1} + \sigma\big]$ and $\big[d_1^* - \sigma,\, d_1^*\big]$. The inequality

$$\nu\, h_k^{-1} - \max_{\substack{\overline{Q}(x^*,\, t^*) \\ x_1 \notin (d_{*1}+\sigma,\, d_1^*-\sigma)}} \left|a_{1k}(x, t)\right| h_{(1)}^{-1} > 0, \quad k = 2,\ldots, n,$$

is fulfilled if $M_1\, h_{(1)}^{-1}\sigma < \nu \min\limits_{k=2,\ldots,n} [h_k^{-1}]$, i.e., when one has

$$N_{*1} \equiv \sigma\, h_{(1)}^{-1} < M_1^{-1}\, \nu \min_{k=2,\ldots,n} [h_k^{-1}], \tag{5.39b}$$

where $N_{*1} + 1$ is the number of nodes in the mesh $\overline{\omega}_{1(2.48)}$ on the interval $[d_{*1},\, d_{*1} + \sigma]$ or $[d_1^* - \sigma,\, d_1^*]$.

Under the above conditions on the values σ and N_{*1}, the operators $\Lambda^*_{(2)(5.27)}$ and $\Lambda_{(5.25)}$ are ε-uniformly monotone on the grid $\overline{G}_{h(5.38)}(Q)$, where Q is $\overline{Q}(x^*, t^*)$ from $\{(5.30), (5.36)\}$. Thus, in the case of $\{(5.30), (5.36)\}$, the condition (5.39) that is imposed on the parameters of the mesh $\overline{\omega}_{1(2.48)}$ is a *sufficient condition for the local ε-uniform monotonicity of the operators* $\Lambda^*_{(2)}$ *and* Λ *on the piecewise-uniform mesh* $\overline{G}_{h(5.38)}(Q)$.

In the case of the conditions (5.34), (5.35), (5.37), when constructing the consistent mesh (5.38) we use the meshes

$$\overline{\omega}_1 = \big\{\overline{\omega}_{1(2.48)} \text{ under the conditions } (5.30c),\, (5.39)\big\}. \tag{5.40}$$

Then the grid

$$\overline{G}_h = \overline{G}_h\big(Q(x^*, t^*),\, L^*_{(2)}\big) = \tag{5.41}$$

$$= \overline{G}_h\big(Q(x^*, t^*),\, L^*_{(2)},\, \overline{\omega}_{1(5.40)},\, x_1, \ldots, x_n\big) =$$

$$= \big\{\overline{G}_{h(5.33)} \cap \overline{Q}(x^*, t^*),\, \overline{\omega}_1 = \overline{\omega}_{1(2.48)},\, \omega_s = \omega_s^u,\, s = 2,\ldots, n,$$

under the conditions (5.30c), (5.39) *in the case of* $\{(5.30), (5.36)\}\}$

is consistent on the set $\overline{Q}(x^*, t^*)$ in the variables $x_1, \ldots, x_n$ with the monotonicity condition for the operators $\Lambda^*_{(2)(5.27)}$ and $\Lambda_{(5.25)}$.

Theorem 5.3.3 *Let the conditions* (5.34), (5.35), (5.37) *hold. Then the operator* $\Lambda_{(5.25)} = \Lambda(L)$ *on the grid* $\overline{G}_h = \overline{G}_{h(5.41)}(Q(x^*, t^*), L^*_{(2)})$ *is ε-uniformly monotone.*

Remark 5.3.1 Let the coefficients of the mixed derivatives of the operator $L^*_{(2)}$ satisfy the condition

$$\max_{\overline{G}} |a_{sk}(x, t)| < (n-1)^{-1} \min_{\overline{G}} \left[\alpha_s \, \alpha_k^{-1} \, a_{ss}(x, t), \; \alpha_k \, \alpha_s^{-1} \, a_{kk}(x, t) \right], \quad (5.42)$$

$$s, k = 1, \ldots, n, \quad s \neq k,$$

where $\alpha_1, \ldots, \alpha_n$ are some positive numbers, i.e., we have dominance of the diagonal terms in the matrix of the coefficients of the operator $L^*_{(2)(5.26)}$ on the whole set $\overline{G}$. Then, under the conditions (5.34), (5.35), one can choose the value ν and the functions $\rho_{*s}(x, t)$ and $\rho_s^*(x, t)$, for $s = 1, \ldots, n$, such that the condition (5.37) is valid for

$$\overline{Q}(x^*, t^*) = \overline{G}. \quad (5.43)$$

The following theorem holds.

Theorem 5.3.4 *Let* $a_{sk} \in C^1(\overline{G})$, b_s, c, c^1, p, $f \in C(\overline{G})$, *for* $s, k = 1, \ldots, n$. *Let the coefficients* $a_{sk}(x, t)$ *satisfy the conditions* (5.42), (5.34), *and* (5.37), *where* $\overline{Q} = \overline{G}$, *and let the solution of the problem* (5.1), (5.2) *satisfy the estimates* (5.6), (5.11), (5.14) *(or the estimates* (5.11), (5.13)*) of Theorem 5.2.1 for* $K = 4$. *Then the difference scheme* (5.24), (5.25) *on the grid* $\overline{G}_{h(5.41)}$, *where* $\overline{Q} = \overline{G}$, $\overline{\omega}_{1(2.48)} = \overline{\omega}_{1(2.44)}$, *and* $m_{1(2.44)} = m_{1(5.14)}$, *converges ε-uniformly. For the discrete solutions, the following estimate is valid:*

$$|u(x, t) - z(x, t)| \leq M \left[N^{-1} \ln N + N_0^{-1} \right]^\nu, \quad (x, t) \in \overline{G}_h, \quad (5.44)$$

where $\nu = 5^{-1}$ *in the case of the estimates* (5.6), (5.11), (5.14), *while* $\nu = 1$ *in the case of the estimates* (5.11), (5.13).

Under the hypotheses of Theorem 5.3.4, let

$$\overline{z}(x, t) = \overline{z}_{(2.42)}(x, t; z(\cdot), \overline{G}_h), \quad (x, t) \in \overline{G},$$

be the interpolant of the function $z(x, t)$, $(x, t) \in \overline{G}_h$, which is the solution of the difference scheme (5.24), (5.25) on the grid (5.41), (5.43). Then one has the estimate

$$|u(x, t) - \overline{z}(x, t)| \leq M \left[N^{-1} \ln N + N_0^{-1} \right]^\nu, \quad (x, t) \in \overline{G}, \quad \nu = \nu_{(5.44)}.$$

5.3.3 Consistent grids on a parallelepiped

In the case of the boundary value problem (5.1) on the parallelepiped (5.3), we are interested in monotone difference approximations on piecewise-uniform meshes. We shall consider the difference scheme (5.24), (5.25) on the grids

$$\overline{G}_h = \overline{G}_h\big(\overline{\omega}_s = \overline{\omega}_{1(2.48)}(d_{*s}, d_s^*), \quad s = 1, \ldots, n\big), \qquad (5.45)$$

that condense in a neighborhood of the faces S_j, for $j = 1, \ldots, 2n$.

The following statements are valid.

Lemma 5.3.2 *The condition*

$$\left.\begin{array}{c} \varepsilon^{-1} \displaystyle\max_{d_{*s} < x_s^i \le d_{*s} + M\varepsilon} h_s^{i-1} \\[2mm] \varepsilon^{-1} \displaystyle\max_{d_s^* - M\varepsilon \le x_s^i < d_s^*} h_s^i \end{array}\right\}_{\varepsilon} \to 0 \ as \ N \to \infty, \ x_s^i \in \overline{\omega}_s, \ s = 1, \ldots, n, \quad (5.46)$$

on the distribution of nodes in the meshes $\overline{\omega}_s$ is necessary for the ε-uniform approximation of the operators $L^*_{(2)(5.26)}$ and L by the operators $\Lambda^*_{(2)(5.27)}$ and $\Lambda_{(5.25)}$ applied to the problem solution.

The condition (5.46) holds for the grids $\overline{G}_{h(5.45)}$.

Theorem 5.3.5 *For the grid $\overline{G}_{h(5.23a)}$, let the condition (5.46) hold. Then the condition*

$$a_{sk}(x, t) = 0, \quad (x, t) \in \overline{S}^{(p)}, \quad p = s, \quad or \quad p = k, \qquad (5.47)$$

$$s, k, p = 1, \ldots, n, \quad s \ne k, \quad where$$

$$S^{(p)} = \Gamma^{(p)} \times (0, T], \quad for \ \Gamma^{(p)} = \Gamma^{(p)}_{(3.48)},$$

on the coefficients of the mixed derivatives of the operator L is necessary for the ε-uniform monotonicity of the operators $\Lambda^*_{(2)(5.27)}$ and $\Lambda_{(5.25)}$.

In the case of the boundary value problem on the parallelepiped (5.3) under the condition (5.47) (the canonicity condition of the truncated operator $L^{*[p]}_{(2)}$ on the faces $\overline{S}_j$ of the set $S^{(p)}$), for the system of difference equations (5.24), (5.25) it is required to construct a grid that is consistent in $x_1, \ldots, x_n$ and condenses in a neighborhood of the boundary S^L. When constructing the consistent grid, we use the piecewise-uniform grid (5.45).

Next, we impose conditions on both the operator $L^*_{(2)(5.26)}$ and the grids $\overline{G}_h$, which are piecewise uniform in all spatial variables, that are sufficient for the existence of consistent grids guaranteeing the monotonicity of the operators $\Lambda^*_{(2)(5.27)}$ and $\Lambda_{(5.25)}$.

Using an appropriate linear transformation of the variables $x_1, \ldots, x_{p-1}$, $x_{p+1}, \ldots, x_n$ which we write as $X = X(x, t_0)$ with $X_p = x_p$, the operator

$L^*_{(2)}$ at the point $(x_0, t_0) \in \overline{G}$ in a neighborhood of the set $\overline{S}^{(p)}$ (for $n > 2$) can be brought to canonical form in the variables $X_1, \ldots, X_{p-1}, X_{p+1}, \ldots, X_n$, where the coefficients of the mixed derivatives not involving differentiation in X_p vanish at the point (X_0, t_0) with $X_0 = X(x_0, t_0)$. Note that the coefficients of the mixed derivatives involving differentiation in X_p are equal to zero on $\overline{S}^{(p)}_X$, i.e., on the faces orthogonal to the X_p-axis.

Now that faces have been dealt with, we move on to edges of the parallelepiped.

At the point $(x_0, t_0) \in \overline{G}$ in a neighborhood of the set $\overline{S}^{(j \ldots r)}$, where $S^{(j \ldots r)} = \Gamma^{(j \ldots r)} \times (0\, T]$ and $\Gamma^{(j \ldots r)} = \Gamma^{(j \ldots r)}_{(3.49)}$ (when $|j \ldots r| \leq n - 2$), using a transformation of the variables x_s, for $s = 1, \ldots, n$, and $s \neq j, \ldots, r$ (written as $X = X(x, t_0)$, where $X_s = x_s$, with $s = j, \ldots, r$), one can bring the operator $L^*_{(2)}$ to canonical form in the variables X_s, for $s = 1, \ldots, n$, with $s \neq j, \ldots, r$, where the coefficients of the mixed derivatives not involving differentiation in X_s, for $s = j, \ldots, r$ vanish at the point (X_0, t_0). The coefficients of the mixed derivatives involving differentiation in X_s, for $s = j, \ldots, r$, are equal to zero on $\overline{S}^{(j \ldots r)}_X$.

Assume that the operator $L^*_{(2)}$ has the above canonical form at the point $(x_0, t_0) \in \overline{G}$, which depends on the relative position of the point (x_0, t_0) and the faces $S_j \subset S^L$. If a neighborhood of the point (x_0, t_0) is sufficiently small then in this neighborhood the operator $L^*_{(2)}$ has almost canonical form (in all spatial variables), i.e., the coefficients of the mixed derivatives $\partial^2 / \partial x_s \partial x_k$, for $s, k = 1, \ldots, n$, with $s \neq k$, are small compared with the coefficients of the second-order derivatives $\partial^2 / \partial x_s^2$, for $s = 1, \ldots, n$.

Under the condition (5.47), let the operator $L^*_{(2)(5.26)}$ have local almost canonical form in $x_1, \ldots, x_n$ in the set $\overline{G}$ and local almost canonical form in $x_1, \ldots, x_{p-1}, x_{p+1}, \ldots, x_n$ in a neighborhood of the faces $\overline{S}^{(p)}$, and in x_s, for $s = 1, \ldots, n$, with $s \neq j, \ldots, r$, in a neighborhood of the edges $\overline{S}^{(j \ldots r)}$.

We shall assume that, in a neighborhood of $\overline{S}^{(j \ldots r)}$, the condition of strongly local almost canonicity of the operator $L^*_{(2)}$ is fulfilled. That is, in the case of the condition (5.47), for an arbitrary point $(x^*, t^*) \in \overline{G}$, one can find sufficiently small values m^* and $\nu > 0$, a sufficiently small neighborhood $Q(x^*, t^*)$ and n pairs of the functions $\rho_{*s}(x, t)$ and $\rho_s^*(x, t)$, for $s = 1, \ldots, n$, for which the relation (5.30) holds, and one also has the following condition (similar to (5.36)):

$$\left[\min_{\substack{\overline{Q}(x^*, t^*) \\ r(x, \Gamma^{(j \ldots r)}) \leq m^*}} a_{kk}(x, t) - \nu \right] h_k^{-1} - \sum_{\substack{s=1,\, s \neq k \\ s \neq j, \ldots, r}}^{n} \max_{\substack{\overline{Q}(x^*, t^*) \\ r(x, \Gamma^{(j \ldots r)}) \leq m^*}} |a_{sk}(x, t)| \, h_s^{-1} > 0,$$

$$k, j, \ldots, r = 1, \ldots, n, \quad k \neq j, \ldots, r, \qquad (5.48)$$

$$j < \ldots < r, \quad 1 \leq |j \ldots r| \leq n - 2.$$

The condition (5.47) and the condition

> *the coefficients $a_{sk}(x, t)$ of the operator $L^*_{(2)}$ on the set*
>
> $\overline{Q}(x^*, t^*) \subseteq \overline{G}$ *satisfy the relations $\{(5.30), (5.48)\}$*   (5.49)

together allow us to construct on the set $\overline{Q}(x^*, t^*)$ piecewise-uniform (in all variables) consistent grids and thereby ensure the monotonicity of the operators $\Lambda^*_{(2)(5.27)}$ and $\Lambda_{(5.25)}$.

In the case of the conditions (5.47), (5.49), we now impose conditions on the parameters of the piecewise-uniform grid $\overline{G}_{h(5.45)}$ under which the operator $\Lambda^*_{(2)(5.27)}$ is monotone.

Let the step-sizes h_s, i.e., maximal step-sizes in the meshes

$$\overline{\omega}_s = \overline{\omega}_{1(2.48)}(d_{*s}, d^*_s; \sigma_s, N_s), \quad s = 1, \ldots, n,$$

satisfy the conditions (5.30), (5.48). Note that $h_s = h_{(2)s}$ in the mesh $\overline{\omega}_s$. Assume that the value σ_s that defines the transition points in the mesh $\overline{\omega}_s$ satisfies the condition

$$\sigma_s \leq m^*_{(5.48)}, \quad s = 1, \ldots, n, \tag{5.50a}$$

and that

$$a_{sk} \in C^1(\overline{G}), \quad s, k = 1, \ldots, n, \quad s \neq k.$$

By virtue of the condition (5.47), one has

$$\left| a_{sk}(x, t) \right| \leq M_1 \sigma_s \quad \text{for} \quad r\big((x, t), \overline{S}_s \cup \overline{S}_{s+n}\big) \leq \sigma_s, \quad s \neq k,$$

where

$$M_1 = \max_{s,k,r,\, s \neq k,\, \overline{G}} \left| \frac{\partial}{\partial x_r} a_{sk}(x, t) \right|.$$

Then

$$\left| a_{sk}(x, t) \right| h^{-1}_{(1)s} \leq M_1 h^{-1}_{(1)s} \sigma_s \quad \text{for} \quad r\big((x, t), \overline{S}_s \cup \overline{S}_{s+n}\big) \leq \sigma_s.$$

The inequality

$$\nu h^{-1}_k - \sum_{s=j,\ldots,r} \max_{\substack{\overline{Q}(x^*, t^*) \\ x_s \notin (d_{*s}+\sigma_s,\, d^*_s-\sigma_s)}} \left| a_{sk}(x, t) \right| h^{-1}_{(1)s} > 0,$$

$$k, j, \ldots, r = 1, \ldots, n, \quad k \neq j, \ldots, r \quad \text{for} \quad r\big((x, t)\,\overline{S}^{(j\ldots r)}\big) \leq m^*,$$

where $\overline{Q}(x^*, t^*)$ is taken from (5.30), (5.48), is satisfied if

$$M_1 \sum_{s=1}^n \left[\sigma_s h^{-1}_{(1)s} \right] < \nu \min_{s=1,\ldots,n} \left[h^{-1}_s \right],$$

i.e., when one has the condition

$$\sum_{s=1}^{n} N_{*s} < M_1^{-1} \nu \min_{k} \left[h_k^{-1} \right], \quad k = 1, \dots, n. \tag{5.50b}$$

Here $N_{*s} + 1$ is the number of nodes in the mesh $\overline{\omega}_s$ on each of the intervals $[d_{*s}, d_{*s} + \sigma_s]$ and $[d_s^* - \sigma_s, d_s^*]$; $h_{(1)s} = \sigma_s N_{*s}^{-1}$ and $h_s = h_{(2)s} \leq (d_s^* - d_{*s}) (N_s - 2 N_{*s})^{-1}$, with $N_s \leq M N_{*s}$.

Let the parameters of the meshes $\overline{\omega}_s$, for $s = 1, \dots, n$, that generate the piecewise-uniform mesh $\overline{G}_{h(5.45)}$ be chosen to satisfy the conditions (5.30c), (5.50), where $h_s = h_{(2)s}$. Write these meshes $\overline{\omega}_s$ in the following form:

$$\overline{\omega}_s = \{\overline{\omega}_s = \overline{\omega}_{1(2.48)}(d_s^*, d_s^*; \sigma_s, N_s) \ \text{ under the conditions}$$
$$\text{(5.30c), (5.50)}\}, \quad \text{for } s = 1, \dots, n. \tag{5.51}$$

In the case of
(i) the condition (5.47) on the coefficients $a_{sk}(x, t)$ on the boundary $\overline{S}^L$,
(ii) the condition (5.49) on the coefficients of the operator $L_{(2)}^*$ and the parameters h_s, m^*, ν on the set $\overline{G}$,
(iii) the conditions (5.30c), (5.50) on the meshes $\overline{\omega}_s$ that generate the piecewise-uniform grid $\overline{G}_{h(5.45)}$,
the operators $\Lambda_{(2)(5.27)}^*$ and $\Lambda_{(5.25)}$ are ε-uniformly monotone on the grid $\overline{G}_{h(5.45)} \cap \overline{Q}$, where $Q = Q(x^*, t^*)$ is taken from $\{(5.30), (5.48)\}$. Thus, in the case of the conditions (5.47), (5.49), the grid

$$\overline{G}_h = \overline{G}_h \left(Q(x^*, t^*), \ L_{(2)}^*, \ \overline{\omega}_{s(5.51)}, \ x_1, \dots, x_n \right) = \tag{5.52}$$

$$= \{\overline{Q}(x^*, t^*) \cap \overline{G}_{h(5.45)} \ \text{under the conditions (5.30c), (5.50)}$$

$$\text{in the case of the relations } \{(5.30), (5.48)\}\}$$

is consistent on the set $\overline{Q}(x^*, t^*)$ in the variables $x_1, \dots, x_n$ with the monotonicity condition on the operators $\Lambda_{(2)(5.27)}^*$ and $\Lambda_{(5.25)}$.

We consider the difference scheme (5.24), (5.25) on the grid

$$\overline{G}_h = \overline{G}_{h(5.52)} \left(Q(x^*, t^*), \ L_{(2)}^*, \ \overline{\omega}_{s(5.51)}, \ x_1, \dots, x_n \right), \tag{5.53}$$

where $\overline{\omega}_s = \overline{\omega}_{1(2.48)}$ is $\overline{\omega}_{1(2.44)}(d_{*s}, d_s^*; m)$, and m is an arbitrary constant. In the case of the condition (5.42), one can choose the set Q in (5.53) as $\overline{Q} = \overline{G}$.

The following theorem holds.

Theorem 5.3.6 *Let the following conditions be fulfilled: a) $a_{sk} \in C^1(\overline{G})$, b_s, c, c^1, p, $f \in C(\overline{G})$, for $s, k = 1, \dots, n$; b) the coefficients $a_{sk}(x, t)$ satisfy* *the conditions (5.42), (5.47) and {(5.30), (5.48)}, where $\overline{Q} = \overline{G}$; c) for the solution of the problem (5.1), (5.3), the estimates of Theorem 5.2.2 (Theorem 5.2.3) are valid for $K = 4$. Then the difference scheme (5.24), (5.25) converges ε-uniformly on the grid $\overline{G}_{(5.53)}$, where $\overline{Q} = \overline{G}$. The discrete solutions satisfy the estimate*

$$|u(x, t) - z(x, t)| \leq M \left[N^{-1} \ln N + N_0^{-1} \right]^{\nu}, \quad (x, t) \in \overline{G}_h, \tag{5.54}$$

where $\nu = 5^{-1}$ in the case of the estimates (5.19) of Theorem 5.2.2, while $\nu = 1$ in the case of the estimates (5.22) of Theorem 5.2.3.

Under the hypotheses of Theorem 5.3.6, let

$$\overline{z}(x, t) = \overline{z}_{(2.42)}\big((x, t); z(\cdot), \overline{G}_h\big), \quad (x, t) \in \overline{G},$$

be the interpolant of the function $z(x, t)$, $(x, t) \in \overline{G}_h$, which is the solution of the difference scheme (5.24), (5.25) on the grid $\overline{G}_{(5.53)}$, where $\overline{Q} = \overline{G}$. Then one has the estimate

$$|u(x, t) - \overline{z}(x, t)| \leq M \left[N^{-1} \ln N + N_0^{-1} \right]^{\nu}, \quad (x, t) \in \overline{G}, \quad \nu = \nu_{(5.54)}.$$

5.4 Consistent grids on subdomains

In Section 5.3, we have constructed special meshes that condense in a neighborhood of all faces of the lateral boundary $\overline{S}^L$. These meshes guarantee the monotonicity of the operator $\Lambda^*_{(2)(5.27)}$. For later use (see, e.g., Chapter 7), we shall need special meshes that condense in a neighborhood of only some faces of the lateral boundary $\overline{S}^L$ and also guarantee the monotonicity of the operator $\Lambda^*_{(2)(5.27)}$. Let us now discuss the construction of such consistent meshes.

5.4.1 The problem on a slab

For the boundary value problem (5.1) considered on subdomains of the slab (5.2), we impose conditions on both the coefficients $a_{sk}(x, t)$ and the meshes $\overline{\omega}_1, \omega_2, \ldots, \omega_n$ that generate consistent meshes guaranteeing the monotonicity of the operator $\Lambda^*_{(2)(5.27)}$. The mesh $\overline{\omega}_1$ on the subdomains is constructed based on the elementary meshes on the x_1-axis, i.e., $\overline{\omega}_1^u$, $\overline{\omega}_1^l$, $\overline{\omega}_1^r$, or $\overline{\omega}_1^b$.

Set

$$S^L = \cup_j S_j, \quad S_j = \Gamma_j \times (0, T], \quad j = 1, 2, \tag{5.55}$$

where Γ_1 and Γ_2 are the left and right parts of the boundary Γ. Consider a grid $\overline{G}_h$ that is piecewise-uniform in x_1 and condenses in a neighborhood of one or both of the sides $\overline{S}_1$ and $\overline{S}_2$, or is uniform. We denote by Γ^e those parts of the boundary Γ of the set $\overline{D}_{(5.2)}$ in whose neighborhoods it is required to condense the mesh $\overline{\omega}_1$. Let

$$J^* = J^*(D) = J^*(D; \Gamma^e) \tag{5.56}$$

be the set of indexes j related to Γ_j in whose neighborhood the mesh $\overline{\omega}_1$ condenses; here $\Gamma_j \subseteq \Gamma^e$.

Suppose that on the boundary $\overline{S}^L$ the coefficients $a_{sk}(x, t)$ of the operator $L^*_{(2)}$, where $L^*_{(2)} = L^*_{(2)(5.26)}(L_{2(5.1)})$, satisfy the condition

$a_{sk}(x, t) = 0,\ (x, t) \in \overline{S}_j,\ j \in J^*,\ s = 1\ or\ k = 1,\ s \neq k,\ if\ J^* \neq \emptyset;$

restrictions on $a_{sk}(x, t)$ *on* $\overline{S}^L$ *are not imposed if* $J^* = \emptyset,$ \qquad (5.57)

i.e., the operator $L^*_{(2)}$ is canonical on the side $\overline{S}_j$ for $j \in J^*$, where $J^* \neq \emptyset$.

Assume that on the set $\overline{Q}(x^*, t^*) \subseteq \overline{G}$ the coefficients $a_{sk}(x, t)$ satisfy the relations (5.30) (the condition of local almost canonicity of the operator $L^*_{(2)}$). In the case when $J^* \neq \emptyset$, we assume that the following additional condition holds, which is similar to (5.36):

$$\left[\min_{\substack{\overline{Q}(x^*, t^*) \\ r(x, \Gamma_j) \leq m^* \\ j \in J^*}} a_{kk}(x, t) - \nu \right] h_k^{-1} - \sum_{\substack{s=2 \\ s \neq k}}^n \max_{\substack{\overline{Q}(x^*, t^*) \\ r(x, \Gamma_j) \leq m^* \\ j \in J^*}} |a_{sk}(x, t)| h_s^{-1} > 0,$$

$$k = 2, \ldots, n, \quad J^* \neq \emptyset. \tag{5.58}$$

The condition (5.30) together with (5.58) is the condition of strongly local almost canonicity of the truncated operator $L^{*[1]}_{(2)}$ in a neighborhood of the set $\overline{S}_j$, where $j \in J^*$.

The condition

the coefficients $a_{sk}(x, t)$ *of the operator* $L^*_{(2)}$ *on the set* $\overline{G}$ \qquad (5.59)

satisfy the relations (5.57), $\{(5.30),\ (5.58)\}$ *if* $J^* \neq \emptyset$, *or*

the relations (5.30) *if* $J^* = \emptyset$, *where* $\overline{Q}(x^*, t^*) \subseteq \overline{G}$, $J^* = J^*_{(5.56)}(D),$

allows us on the set $\overline{Q}(x^*, t^*)$ to construct piecewise-uniform consistent grids that ensure the monotonicity of the operators $\Lambda^*_{(2)(5.27)}$ and $\Lambda_{(5.25)}$. We now construct such meshes.

Let the condition (5.59) hold. In the case of the condition $J^* \neq \emptyset$ (when the mesh $\overline{\omega}_1$ is piecewise uniform), we assume that the step-sizes $h_1 = h_{(2)1}$ and h_s in the meshes $\overline{\omega}_1$ and ω_s, for $s = 2, \ldots, n$, satisfy the condition (5.30c) on the set $\overline{Q}(x^*, t^*)$ and, moreover, for the mesh $\overline{\omega}_1$, one has the condition (similar to (5.39a))

$$\sigma = \sigma^l \le m^*_{(5.58)}, \quad for \ \ j = 1, \ \ j \in J^*,$$
$$\sigma = \sigma^r \le m^*_{(5.58)}, \quad for \ \ j = 2, \ \ j \in J^*, \ if \ J^* \ne \emptyset; \tag{5.60a}$$

and also the following condition (similar to (5.39b)):

$$N_{*1} \equiv \sigma \, h^{-1}_{(1)} < M^{-1}_1 \nu \min_{k=2,\ldots,n} [h^{-1}_k], \quad j \in J^*, \tag{5.60b}$$

where $N_{*1} + 1$ is the number of nodes on the interval $[d_{*1}, d_{*1} + \sigma]$ in the case $\{j = 1\} \subseteq J^*$, and on the interval $[d^*_1 - \sigma, d^*_1]$ for $\{j = 2\} \subseteq J^*$; here $M_1 = M_{1(5.39)}$. In the case when $J^* = \{j = 1, 2\}$, set $\sigma^l = \sigma^r$.

Furthermore, we assume that the condition (5.59) is fulfilled. In the mesh $\overline{G}_h$ along the x_1-axis, we use the piecewise-uniform or uniform meshes

$$\overline{\omega}_1 = \begin{cases} \overline{\omega}^l_1 \ \ if \ \ J^* = \{j = 1\} \\ \overline{\omega}^r_1 \ \ if \ \ J^* = \{j = 2\} \\ \overline{\omega}^b_1 \ \ if \ \ J^* = \{j = 1, 2\} \\ \overline{\omega}^u_1 \ \ under \ the \ condition \ (5.30c) \ if \ J^* = \emptyset; \end{cases} \quad \begin{matrix} under \ the \ conditions \\ (5.30c), \ (5.60) \\ if \ J^* \ne \emptyset, \end{matrix} \tag{5.61a}$$

the mesh $\overline{\omega}_{1(5.61)}$ condenses only in a neighborhood of the sets Γ_j for which $j \in J^*$, where $J^* \ne \emptyset$.

On the slab, when constructing consistent meshes that guarantee the monotonicity of the operator $\Lambda^*_{(2)} = \Lambda^*_{(2)(5.27)}(\Lambda_{(5.25)})$, we apply the piecewise-uniform mesh

$$\overline{\omega}_{1(5.61)}\big(d_{*1}, d^*_1\big) = \overline{\omega}_{1(2.83)}\big(d_{*1}, d^*_1; \overline{\omega}_{1(2.48)}\big). \tag{5.61b}$$

The parameters σ, $h_{(1)1}$, and $h_{(2)1}$ of the mesh $\overline{\omega}_{1(2.83)}\big(d_{*1}, d^{(*)}_1\big)$ are the same as those in the meshes $\overline{\omega}^l_1$ and $\overline{\omega}^r_1$.

Theorem 5.4.1 *In the case of the conditions (5.35), (5.59), the grid*

$$\overline{G}_h = \overline{G}_h\big(Q(x^*, t^*), L^*_{(2)}; J^*\big) = \tag{5.62}$$

$$= \overline{G}_h\Big(Q(x^*, t^*), L^*_{(2)}, \overline{\omega}_1 = \overline{\omega}_{1(5.61)}(\overline{\omega}_{1(2.48)})\Big),$$

$$\omega_s = \omega^u_s, \ s = 2, \ldots, n, \quad under \ the \ conditions \ (5.30c), \ (5.60)$$

in the case of the relations (5.57), $\{(5.30), (5.58)\}$ if $J^* \ne \emptyset$,

and under the conditions (5.30c) in the case

of the relations (5.30) if $J^* = \emptyset$; $\quad J^* = J^*_{(5.56)}(D; \Gamma^e)\big),$

is consistent on the set $\overline{Q}(x^*, t^*)$ with the monotonicity condition for the operator $\Lambda^*_{(2)(5.27)}$.

5.4.2 The problem on a parallelepiped

For the boundary value problem (5.1) considered on subdomains in $\overline{G}_{(5.3)}$, we impose conditions on both the coefficients $a_{sk}(x,\,t)$ and the meshes $\overline{\omega}_s$ that ensure, on the subdomains, the consistency of the meshes that are constructed based on the elementary meshes $\overline{\omega}_s^u$, $\overline{\omega}_s^l$, $\overline{\omega}_s^r$, and $\overline{\omega}_s^b$.

In the case of the grid $\overline{G}_h$, let the piecewise-uniform mesh on the x_s-axis condense in a neighborhood of the face $\overline{S}_s$ and/or $\overline{S}_{s+n}$, or this mesh is uniform, for $s = 1, \ldots, n$. We denote by Γ^e those faces of the boundary Γ of the set $\overline{D}_{h(5.3)}$ in whose neighborhoods the meshes $\overline{\omega}_s$, for $s = 1, \ldots, n$, condense; here $\overline{\omega}_s = \overline{\omega}_s(d_{*s},\,d_s^*)$. We denote by J^*, where

$$J^* = J^*(D) = J^*(D;\,\Gamma^e), \tag{5.63}$$

the set of indexes j, related to Γ_j in whose neighborhoods the grid $\overline{D}_h$ condenses in the direction orthogonal to the face Γ_j; here $\Gamma_j \subseteq \Gamma^e$.

We assume that the coefficients $a_{sk}(x,\,t)$ of the operator $L_{(2)}^*$ on the boundary $\overline{S}^L$ satisfy the following condition, which is similar to (5.47) (the condition of the canonicity of the operator $L_{(2)}^*$ on the faces $\overline{S}_j$, for $j \in J^*$)

$$a_{sk}(x,\,t) = 0, \quad (x,\,t) \in \overline{S}_j, \quad j \in J^*; \tag{5.64}$$

$$s = p_j, \quad \text{or} \quad k = p_j, \quad s \neq k, \quad j \in J^*, \quad \text{if } J^* \neq \emptyset;$$

$$\text{restrictions on } a_{sk}(x,\,t) \text{ on } \overline{S}^L \text{ are not imposed if } J^* = \emptyset,$$

where $p_j = p_{j(3.59)}(j, n)$.

Suppose also that the coefficients $a_{sk}(x,\,t)$, which define the operator $L_{(2)}^*$, satisfy the condition (5.30) on the set $\overline{Q}(x^*,\,t^*) \subseteq \overline{G}$, independent of the structure of the set J^*. Furthermore, if $J^* \neq \emptyset$, we assume that the following additional condition holds, which is similar to (5.48):

$$\left[\min_{\substack{\overline{Q}(x^*,\,t^*) \\ r(x,\,\cap_{j_i}\Gamma_{j_i}) \leq m^* \\ j_i \in J^*,\ i=1,\ldots,r}} a_{kk}(x,\,t) - \nu \right] h_k^{-1} - \sum_{\substack{s=1 \\ s \neq k, p_{j_i}}}^{n} \max_{\substack{\overline{Q}(x^*,\,t^*) \\ r(x,\,\cap_{j_i}\Gamma_{j_i}) \leq m^* \\ j_i \in J^*,\ i=1,\ldots,r}} |a_{sk}(x,\,t)|\, h_s^{-1} > 0,$$

$$k, p_{j_1}, \ldots, p_{j_r} = 1, \ldots, n, \quad k \neq p_{j_1}, \ldots, p_{j_r}, \quad 1 \leq r \leq n-2, \tag{5.65}$$

$$\cap_{j_i}\Gamma_{j_i} \neq \emptyset, \quad J^* \neq \emptyset,$$

where $p_j = p_{j(3.59)}(j, n)$, and m^*, ν are sufficiently small values. The condition (5.30) together with (5.65) is the condition of strongly local almost canonicity of the operator $L_{(2)}^{*[p_{j_1}, \ldots, p_{j_r}]}$ in a neighborhood of the set $\overline{S}_{j_1} \cap \ldots \cap \overline{S}_{j_r}$, where $j_1, \ldots, j_r \in J^*$.

The condition

> the coefficients $a_{sk}(x, t)$ of the operator $L^*_{(2)}$ on the set $\overline{G}$
>
> satisfy
>
> the relations (5.64), {(5.30), (5.65)} if $J^* \neq \emptyset$,   (5.66)
>
> and the relations (5.30) if $J^* = \emptyset$,
>
> where $\overline{Q}(x^*, t^*) \subseteq \overline{G}$, $J^* = J^*_{(5.63)}(D)$

allows us on the set $\overline{Q}(x^*, t^*)$ to construct piecewise-uniform consistent grids that guarantee the monotonicity of the operators $\Lambda^*_{(2)(5.27)}$ and $\Lambda_{(5.25)}$.

Next, on the set $\overline{Q}(x^*, t^*) \subseteq \overline{G}$, we construct consistent grids that are defined by the structure of the set J^*. Let the condition (5.66) hold.

In the case of the condition $J^* \neq \emptyset$ (i.e., when generating the grid $\overline{G}_h$, at least one of the meshes $\overline{\omega}_s$, for $s = 1, \ldots, n$, is piecewise uniform), we assume that the step-sizes $h_{(2)s}$ in the piecewise-uniform meshes $\overline{\omega}_s$ satisfy the condition (5.30c); furthermore, for the meshes $\overline{\omega}_s$ the following condition (similar to (5.50a)) holds:

$$\begin{aligned} \sigma_s = \sigma^l_s \leq m^*_{(5.65)} \quad &\text{for } s = p_j, \text{ if } p_j = j, \; j \in J^*; \\ \sigma_s = \sigma^r_s \leq m^*_{(5.65)} \quad &\text{for } s = p_j, \text{ if } p_j \neq j, \; j \in J^*, \end{aligned} \qquad (5.67a)$$

and also the following condition (similar to (5.50b)) is satisfied:

$$\sum_{s=p_j} N_{*s} < M_1^{-1} \nu \min_k [h_k^{-1}], \quad k = 1, \ldots, n, \; j \in J^*. \qquad (5.67b)$$

Here $M_1 = M_{1(5.50)}$, $p_j = p_{j(3.59)}$, for $j \in J^*$, and $N_{*s} + 1$ is the number of nodes in the mesh $\overline{\omega}_s$ on the interval $[d_{*s}, d_{*s} + \sigma_s]$ if the mesh $\overline{\omega}_s$ condenses in a neighborhood of the left endpoint of $[d_{*s}, d^*_s]$, or on the interval $[d^*_s - \sigma_s, d^*_s]$ if the mesh $\overline{\omega}_s$ condenses in a neighborhood of the right endpoint of the interval. In the case when the mesh $\overline{\omega}_s$ condenses in neighborhoods of both endpoints of the interval $[d_{*s}, d^*_s]$, we set

$$\sigma^l_s = \sigma^r_s = \sigma_s \quad \text{for } s = p_{j_1} = p_{j_2}, \text{ where } j_1, j_2 \in J^*, \text{ and } j_1 \neq j_2. \qquad (5.67c)$$

When constructing consistent grids, we apply piecewise-uniform or uniform grids along the x_s-axis (depending on the structure of the set J^*):

$$\overline{\omega}_s = \begin{cases} \left. \begin{array}{l} \overline{\omega}^l_s \;\; \text{if } J^*_s = \{j = s\} \\ \overline{\omega}^r_s \;\; \text{if } J^*_s = \{j = s + n\} \\ \overline{\omega}^b_s \;\; \text{if } J^*_s = \{j = s, s + n\} \end{array} \right\} \quad \begin{array}{l} \text{under the conditions} \\ (5.30\text{c}), \; (5.67) \\ \text{if } J^* \neq \emptyset, \end{array} \\ \overline{\omega}^u_s \;\; \text{under the condition } (5.30\text{c}) \;\; \text{if } J^*_s = \emptyset, \; J^* \neq \emptyset, \\ \hphantom{\overline{\omega}^u_s \;\;} \text{or if } J^* = \emptyset; \end{cases} \qquad (5.68)$$

$$J^* = J^*_{(5.63)}(D; \Gamma^e).$$

The mesh $\overline{\omega}_{s(5.68)}$ is piecewise uniform on at least one of the x_s-axes for $J^* \neq \emptyset$. The mesh $\overline{\omega}_{s(5.68)}$ is constructed based on the mesh $\overline{\omega}_{1(2.48)}$ similar to $\overline{\omega}_{1(2.83)}$

$$\overline{\omega}_s = \overline{\omega}_s\left(d_{*s}, d_s^*\right) \equiv \overline{\omega}_{1(2.83)}\left(d_{*s}, d_s^*; \overline{\omega}_{1(2.48)}\right).$$

Theorem 5.4.2 *In the case of the conditions* (5.35), (5.66), *the grid*

$$\overline{G}_h = \overline{G}_h\left(Q(x^*, t^*),\ L^*_{(2)};\ J^*\right) = \overline{G}_h\left(Q(x^*, t^*),\ L^*_{(2)},\ \overline{\omega}_s = \overline{\omega}_{s(5.68)},\right.$$

for $s = 1, \ldots, n$ *under the conditions* (5.30c), (5.67)

in the case of the relations (5.64), $\{(5.30),\ (5.65)\}$ *if* $J^* \neq \emptyset$, (5.69)

and under the condition (5.30c) *in the case*

of the relations (5.30) *if* $J^* = \emptyset$; $\qquad J^* = J^*_{(5.63)}\left(D;\ \Gamma^e\right)\right)$

is consistent on the set $\overline{Q}(x^*, t^*)$ *with the monotonicity condition for the operator* $\Lambda^*_{(2)(5.27)}$.

Chapter 6

Elliptic convection-diffusion equations

In this chapter, we develop a method for the construction of ε-uniformly convergent schemes for singularly perturbed convection-diffusion equations. In the present section, we consider boundary value problems for elliptic equations in domains that do not contain characteristic boundary parts.

6.1 Problem formulation

In an n-dimensional domain D with a boundary Γ we consider the Dirichlet problem for an elliptic *convection-diffusion equation*

$$L\,u(x) = f(x), \quad x \in D, \tag{6.1a}$$

$$u(x) = \varphi(x), \quad x \in \Gamma. \tag{6.1b}$$

Here

$$L \equiv \varepsilon L_2 + L_1, \quad L_1 \equiv \sum_{s=1}^{n} b_s^1(x)\frac{\partial}{\partial x_s} - c^1(x), \tag{6.1c}$$

$$L_2 \equiv \sum_{s,k=1}^{n} a_{sk}(x)\frac{\partial^2}{\partial x_s \partial x_k} + \sum_{s=1}^{n} b_s(x)\frac{\partial}{\partial x_s} - c(x).$$

The coefficients of the operator L satisfy the ellipticity condition

$$a_0 \sum_{s=1}^{n} \xi_s^2 \le \sum_{s,k=1}^{n} a_{sk}(x)\,\xi_s\,\xi_k \le a^0 \sum_{s=1}^{n} \xi_s^2, \quad x \in \overline{D}, \quad a_0 > 0, \tag{6.2a}$$

as well as the condition

$$\sum_{s=1}^{n} \left(b_s^1(x)\right)^2 \ge b_0^2, \quad b_0 > 0, \quad c(x), c^1(x) \ge 0, \quad x \in \overline{D}. \tag{6.2b}$$

The coefficients and the right-hand side in the equation are assumed to be sufficiently smooth; the parameter $\varepsilon \in (0, 1]$.

Assume that the solution of the problem is bounded ε-uniformly and the characteristics of the operator L_1 are of a limited length if the domain D is unbounded.

Let Γ_j, with $j = 1, \ldots, J$, be smooth faces that generate the boundary Γ, where $\Gamma = \bigcup_j \Gamma_j$ and $\Gamma_j = \overline{\Gamma}_j$. We denote by Γ^c the set of vertexes and edges of the domain D. The function $\varphi(x)$ is sufficiently smooth on each from the faces Γ_j, $j = 1, \ldots, J$, and is continuous on Γ. The fulfillment of other compatibility conditions on the set Γ^c is, in general, not assumed (under these assumptions, in general, $u \notin C^2(\overline{D})$).

Depending on the behaviour of the characteristics of the operator L_1 in a neighborhood of the boundary Γ, the boundary is divided into the subsets Γ^+, Γ^-, and Γ^0. We define positive direction of the characteristics of the operator L_1 by the vector $\mathbf{b}$, where $\mathbf{b}(x) = -(b_1^1(x), \ldots, b_n^1(x))$. The set Γ^- (or Γ^+) is the part of the boundary Γ through which the characteristics leave (enter) the domain D. The set Γ^0 is defined by the relation $\Gamma^0 = \overline{\Gamma} \backslash \{\Gamma^+ \cup \Gamma^-\}$. This set is generated by the characteristics of the reduced equation. We assume that each face Γ_j, with $j = 1, \ldots, J$, of the domain D belongs entirely to one of the sets Γ^+, Γ^-, or Γ^0. The characteristics of the reduced equation intersect the faces Γ_j in $\Gamma^+ \cup \Gamma^-$ with nonzero angles. Assume that $\Gamma^0 = \emptyset$.

For small values of the parameter ε, a boundary layer appears in a neighborhood of the set Γ^-.

We shall consider the boundary value problem (6.1) on a slab

$$\overline{D} = \left\{ x : \ d_{*1} \le x_1 \le d_1^*, \ |x_s| < \infty, \ s = 2, \ldots, n \right\} \qquad (6.3)$$

and on an n-dimensional rectangular parallelepiped

$$\overline{D} = \{ x : \ d_{*s} \le x_s \le d_s^*, \ s = 1, \ldots, n \}. \qquad (6.4)$$

For the boundary value problem (6.1) on the sets $\overline{D}_{(6.3)}$ and $\overline{D}_{(6.4)}$, it is required to construct a difference scheme that converges ε-uniformly.

6.2 Estimates of solutions and derivatives

In this chapter, we use spaces that were introduced in Chapter 2.

6.2.1 The problem solution on a slab

First, we consider the problem (6.1) on the slab $D_{(6.3)}$. The coefficients and the right-hand side f in (6.1a), as well as the boundary function φ are assumed to be sufficiently smooth and bounded on $\overline{D}$ and Γ, respectively. We assume that

$$b_1^1(x) \ge b_0 > 0, \quad x \in \Gamma. \qquad (6.5)$$

Thus,

$$\Gamma = \Gamma^- \cup \Gamma^+,$$

where $\Gamma^- = \Gamma_1$ and $\Gamma^+ = \Gamma_2$ are the left and right parts of the boundary Γ. Write the problem solution as the sum of the functions

$$u(x) = U(x) + V(x), \quad x \in \overline{D}, \tag{6.6}$$

where $U(x)$ and $V(x)$ are the regular and singular parts of the solution. The function $U(x)$, $x \in \overline{D}$, is the restriction of the function $U^e(x)$, $x \in \overline{D}^e$, to the set $\overline{D}$. Here $U^e(x)$ is the solution of the problem

$$L^e U^e(x) = f^e(x), \quad x \in D^e, \tag{6.7a}$$

$$U^e(x) = \varphi^e(x), \quad x \in \Gamma^e. \tag{6.7b}$$

The domain D^e is an extension of the domain D beyond the boundary Γ^-; the distance between the boundaries Γ^- and Γ^{e-} is strictly greater than zero, and is independent of ε, here $\Gamma^+ = \Gamma^{e+}$, $\Gamma^{e+} = \Gamma^+(D^e)$, and $\Gamma^{e-} = \Gamma^{e-}(D^e)$. The coefficients and the right-hand side in (6.7a) are smooth continuations of those in the equation (6.1a), preserving their properties. The boundary function $\varphi^e(x)$ is sufficiently smooth and coincides with the function $\varphi(x)$ on the set Γ^+. The function $V(x)$, $x \in \overline{D}$, is the solution of the boundary value problem

$$LV(x) = 0, \quad x \in D, \tag{6.8a}$$

$$V(x) = \varphi_V(x) \equiv \varphi(x) - U(x), \quad x \in \Gamma. \tag{6.8b}$$

One can represent the function $U(x)$ by the following expansion

$$U(x) = U_0(x) + \varepsilon U_1(x) + v_1(x) = U^{(1)}(x) + v_1(x), \quad x \in \overline{D}.$$

The functions $U_0(x)$, $U_1(x)$, and $v_1(x)$ are restrictions to the set $\overline{D}$ of the functions $U_0^e(x)$, $U_1^e(x)$, and $v_1^e(x)$ that are solutions of the problems

$$L_1^e U_0^e(x) = f^e(x), \qquad x \in \overline{D}^e \setminus \Gamma^{e+}, \quad U_0^e(x) = \varphi^e(x), \quad x \in \Gamma^{e+};$$

$$L_1^e U_1^e(x) = -L_2^e U_0^e(x), \quad x \in \overline{D}^e \setminus \Gamma^{e+}, \quad U_1^e(x) = 0, \qquad x \in \Gamma^{e+};$$

$$L^e v_1^e(x) = -\varepsilon^2 L_2^e U_1^e(x), \quad x \in D^e, \quad v_1^e(x) = \varphi^e(x) - U^{(1)e}(x), \quad x \in \Gamma^e.$$

Here $L_i^e = L_i$, $x \in \overline{D}$, $i = 1, 2$, $L_1 = L_{1(6.1)}$, $L_2 = L_{2(6.1)}$.

When the problem data are sufficiently smooth then the function $U^{(1)}(x)$ is sufficiently smooth, and its derivatives satisfy the estimates

$$\left| \frac{\partial^k}{\partial x_1^{k_1} \dots \partial x_n^{k_n}} U^{(1)}(x) \right| \le M, \quad x \in \overline{D}.$$

For the function $v_1(x)$ one has the estimate

$$|v_1(x)| \le M \varepsilon^2, \quad x \in \overline{D}.$$

Taking into account the *a priori* estimates (in the variables $\xi = \xi(x)$ and $\xi_s = \varepsilon^{-1} x_s$, for $s = 1, \ldots, n$), we obtain

$$\left| \frac{\partial^k}{\partial x_1^{k_1} \ldots \partial x_n^{k_n}} v_1(x) \right| \le M \varepsilon^{2-k}, \quad x \in \overline{D}. \tag{6.9}$$

Hence, for the function $U(x)$ the following estimate holds:

$$\left| \frac{\partial^k}{\partial x_1^{k_1} \ldots \partial x_n^{k_n}} U(x) \right| \le M \left[1 + \varepsilon^{2-k} \right], \quad x \in \overline{D}, \quad 0 \le k \le K. \tag{6.10}$$

For the function $V(x)$ the estimate

$$|V(x)| \le M \exp \left(- m_1 \varepsilon^{-1} r(x, \Gamma^-) \right), \quad x \in \overline{D}, \tag{6.11a}$$

is valid, where m_1 is an arbitrary number in the interval $(0, m_0)$, and the constant m_0 is given by the relation

$$m_0 = \min_{\overline{D}} [b_1^1(x) a_{11}^{-1}(x)]. \tag{6.11b}$$

With regard the *a priori* estimates (in the variables $\xi = \xi(x)$), we have

$$\left| \frac{\partial^k}{\partial x_1^{k_1} \ldots \partial x_n^{k_n}} V(x) \right| \le M \varepsilon^{-k} \exp \left(- m_1 \varepsilon^{-1} r(x, \Gamma^-) \right), \quad x \in \overline{D}. \tag{6.12}$$

Taking into account the estimate

$$\left| \frac{\partial^k}{\partial x_2^{k_2} \ldots \partial x_n^{k_n}} \varphi_V(x) \right| \le M \left[1 + \varepsilon^{2-k} \right], \quad x \in \Gamma^-, \quad k = k_2 + \ldots + k_n,$$

we have

$$\left| \frac{\partial^k}{\partial x_1^{k_1} \ldots \partial x_n^{k_n}} V(x) \right| \le M \left[\varepsilon^{-k_1} + \varepsilon^{2-k} \right] \exp \left(- m_1 \varepsilon^{-1} r(x, \Gamma^-) \right), \tag{6.13}$$

$$x \in \overline{D} \quad 0 \le k \le K,$$

where $m_1 = m_{1(6.11)}$.

The following theorem holds.

Theorem 6.2.1 *Let the condition* (6.5) *hold, and let* a_{sk}, b_s, b_s^1, c, c^1, $f \in C^{l_0}(\overline{D})$, *for* s, $k = 1, \ldots, n$, *and* $\varphi \in C^{l_0}(\Gamma)$, *where* $l_0 = l + \alpha$ *with* $l \ge K + 2$, $K \ge 2$ *and* $\alpha \in (0, 1)$. *Then for the components of the solution to the problem* (6.1), (6.3) *in the representation* (6.6), *the estimates* (6.10) *and* (6.13) *are fulfilled.*

6.2.2 The problem on a parallelepiped

Let us estimate the solution of the problem (6.1) on the parallelepiped (6.4). For simplicity, we assume that the following condition is fulfilled:

$$b_s^1(x) \geq b_0 > 0, \quad x \in \overline{D} \quad s = 1, \ldots, n. \tag{6.14}$$

In this case

$$\Gamma = \Gamma^- \cup \Gamma^+, \quad \Gamma^- = \bigcup_{q=1,\ldots,n} \Gamma_q, \quad \Gamma^+ = \bigcup_{q=n+1,\ldots,2n} \Gamma_q,$$

where $\Gamma_q = \Gamma_{q(3.6)}$, for $q = 1, \ldots, 2n$. The derivation of *a priori* estimates is analogous to one in the case of the boundary value problem (3.1), (3.2).

Let the data of the problem (6.1), (6.4) satisfy the condition

$$a_{sk}, \; b_s, \; b_s^1, \; c, \; c^1, \; f \in C^{l+\alpha}(\overline{D}), \quad s, \; k = 1, \ldots, n, \quad \varphi \in C^2(\Gamma_j), \quad j = 1, \ldots, 2n.$$

Then

$$u \in C^{\alpha_1}(\overline{D}) \cap C^{l+2+\alpha}(D), \quad l > 0, \quad \alpha, \alpha_1 \in (0,1).$$

The solution of the problem (6.1), (6.4) satisfies the estimates similar to (3.3)–(3.5):

$$|u(x)| \leq M \max \left[\max_{\overline{D}} |f(x)|, \; \max_{\Gamma} |\varphi(x)| \right], \quad x \in \overline{D}; \tag{6.15}$$

$$\left| \frac{\partial^k}{\partial x_1^{k_1} \ldots \partial x_n^{k_n}} u(x) \right| \leq M \left[\varepsilon^{-k} + r^{-k}(x, \Gamma) \right], \quad x \in \overline{D},$$

$$x \notin \Gamma \; for \; k > 0, \quad 0 \leq k \leq K;$$

$$\left| u(x) - \varphi(x^*(x)) \right| \leq M \varepsilon^{-1} r(x, \Gamma), \quad x \in \overline{D},$$

where $x^*(x) = x^*_{(3.5)}(x)$ is a point on the set Γ nearest to the point $x \in \overline{D}$, and $K = l + 2$.

Let us investigate the behaviour of the solution in a neighborhood of the boundary layer and outside of it. Assume that

$$D_{(j\ldots r)} = \bigcap_{q=j,\ldots,r} D_{(q)}, \qquad j, \ldots, r \leq 2n;$$
$$D_{[j\ldots r]} = D_{(j\ldots r)} \cap D_{(n+1,\ldots,2n)}, \; j, \ldots, r \leq n, \tag{6.16}$$

where

$$D_{(j)} = \{x : \; d_{*j} < x_j < \infty\} \qquad for \; j \leq n,$$
$$D_{(j)} = \{x : \; -\infty < x_{j-n} < d^*_{j-n}\} \; for \; j > n.$$

On the set $\overline{D}$, we write the problem solution as the sum

$$u(x) = U(x) + V(x), \quad x \in \overline{D},$$

where $U(x)$ and $V(x)$ are the regular and singular parts of the solution. The function $U(x)$, $x \in \overline{D}$, is the restriction of the function $U^e(x)$, $x \in \overline{D}^e$, to the set $\overline{D}$, where $\overline{D}^e = D_{n+1,\ldots,2n}$. Here $U^e(x)$ is the solution of the problem

$$L^e\, U^e(x) = f^e(x), \quad x \in D^e, \tag{6.17a}$$

$$U^e(x) = \varphi^e(x), \quad x \in \Gamma^e. \tag{6.17b}$$

The domain D^e is an extension of the domain D beyond the boundary Γ^-; the faces $\Gamma^e_{j_0}$ contain Γ_{j_0}, for $j_0 = n+1, \ldots, 2n$. The coefficients and the right-hand side of the equation (6.17a) are smooth continuations of those in the coefficients and the right-hand side of the equation (6.1a), preserving their properties. We assume that the functions $f^e(x)$ and $\varphi^e(x)$ equal to zero in a m-neighborhood of the set D. The faces from the set $\Gamma^e(x)$ belong to the set $\Gamma^{e+}(x)$. The boundary function $\varphi^e(x)$ is sufficiently smooth on each of the faces $\Gamma^e_j(x)$ and coincides with the function $\varphi(x)$ on the set Γ^+. The function $V(x)$, $x \in \overline{D}$, is the solution of the problem

$$L\, V(x) = 0, \quad x \in D, \quad V(x) = \varphi(x) - U(x), \quad x \in \Gamma.$$

We now estimate the functions $U(x)$, $V(x)$ and their derivatives. Let the data of the problem (6.1), (6.4) satisfy the condition

$$a_{sk},\, b_s,\, b^1_s,\, c,\, c^1,\, f \in C^{l+\alpha}(\overline{D}), \quad \varphi \in C^{l+\alpha}(\Gamma_j),$$

$$j = 1, \ldots, 2n, \quad l \geq 4, \quad \alpha \in (0,1).$$

The function $U(x)$ is bounded ε-uniformly:

$$|U(x)| \leq M, \quad x \in \overline{D}.$$

We represent the function $U(x)$ by the formal expansion

$$U(x) = U^{(1)}(x) + v_1(x), \quad x \in \overline{D},$$

where $U^{(1)}(x) = U_0(x) + \varepsilon U_1(x)$ is the main term of the regular component of the solution to the boundary value problem. The functions U_0, U_1, and v_1 are obtained as solutions of the following problems:

$$L_1\, U_0(x) = f(x), \quad x \in \overline{D} \setminus \Gamma^+, \quad U_0(x) = \varphi(x), \quad x \in \Gamma^+; \tag{6.18}$$

$$L_1\, U_1(x) = -L_2\, U_0(x), \quad x \in \overline{D} \setminus \Gamma^+, \quad U_1(x) = 0, \quad x \in \Gamma^+; \tag{6.19}$$

$$L\, v_1(x) = -\varepsilon^2\, L_2\, U_1(x), \quad x \in D, \quad v_1(x) = U(x) - U^{(1)}(x), \quad x \in \Gamma.$$

The equations from (6.18), (6.19) are hyperbolic. Solutions of the problems (6.18), (6.19) and, hence, the function $U^{(1)}(x)$ are sufficiently smooth in the case when on the intersection of the faces Γ_j from Γ^+, the data of the problems

(6.18), (6.19) satisfy special conditions, namely, compatibility conditions for a hyperbolic system.

Let us give the definition of compatibility conditions for the functions $f(x)$ and $\varphi(x)$ in the case of the hyperbolic system (6.18), (6.19).

Let the functions $u^1(x)$, $u^2(x)$ be solutions of the following problems

$$L_1 u^1(x) = f(x), \quad x \in \overline{D} \setminus \Gamma^+, \quad u^1(x) = \varphi(x), \quad x \in \Gamma^+; \quad (6.20)$$

$$L_1 u^2(x) = f^2(x), \quad x \in \overline{D} \setminus \Gamma^+, \quad u^2(x) = \varphi^2(x), \quad x \in \Gamma^+, \quad (6.21)$$

where $f^2(x) = -L_2 u^1(x)$, $x \in \overline{D}$, and $\varphi^2(x) = 0$, $x \in \Gamma^+$. On each of the faces $\Gamma_{j+n} \subset \Gamma^+$, the derivatives are defined:

$$\frac{\partial^k}{\partial x_1^{k_1} \dots \partial x_{j-1}^{k_{j-1}} \partial x_{j+1}^{k_{j+1}} \dots \partial x_n^{k_n}} u^i(x) = \frac{\partial^k}{\partial x_1^{k_1} \dots \partial x_{j-1}^{k_{j-1}} \partial x_{j+1}^{k_{j+1}} \dots \partial x_n^{k_n}} \varphi^i(x),$$

$$x \in \Gamma_{j+n}, \quad i = 1, 2.$$

By virtue of the equations (6.20), (6.21), we obtain the derivatives

$$\frac{\partial^k}{\partial x_1^{k_1} \dots \partial x_n^{k_n}} u^i(x), \quad x \in \Gamma_{j+n}.$$

We say that the data of the problems (6.20), (6.21), i.e., coefficients of the operators L_1 and L_2 and the functions $f(x)$, $x \in \overline{D}$, and $\varphi(x)$, $x \in \Gamma^+$, satisfy on the set Γ^c compatibility conditions guaranteeing the continuity of the derivatives up to order K of the function $U^{(1)}(x)$ on Γ^{+c} if the derivatives

$$\frac{\partial^k}{\partial x_1^{k_1} \dots \partial x_n^{k_n}} u^i(x), \quad 0 \leq k \leq K + 2(1 - i), \quad i = 1, 2,$$

are continuous on Γ^{+c}, where Γ^{+c} is a set that is formed by mutual intersections of all faces Γ_j from Γ^+, i.e., $\Gamma^{+c} = \Gamma^c(D_{(n+1,\dots,2n)}) \cap \Gamma^+$.

Assume that the following condition is fulfilled:

The data of the problems (6.18), (6.19) *on the set* Γ^{+c}

satisfy compatibility conditions (6.22)

of the derivatives up to order $l - 2$ *of the function* $U^{(1)}(x)$.

Under the above assumptions, the functions $U_0(x)$ and $U_1(x)$ satisfy the condition

$$U_0, U_1 \in C^{K+\alpha}(\overline{D}), \quad \alpha \in (0, 1), \quad K \geq 0;$$

furthermore, for the function $U^{(1)}(x)$ the following estimate is fulfilled:

$$\left| \frac{\partial^k}{\partial x_1^{k_1} \dots \partial x_n^{k_n}} U^{(1)}(x) \right| \leq M, \quad x \in \overline{D}, \quad 0 \leq k \leq K, \quad (6.23a)$$

where $K = l - 2$. For $l \geq 4$, for the function $v_1(x)$ one has the estimate

$$|v_1(x)| \leq M\varepsilon^2, \quad x \in \overline{D}. \tag{6.23b}$$

By virtue of interior *a priori* estimates, for derivatives of $v_1^e(x)$ we find the estimates

$$\left| \frac{\partial^k}{\partial x_1^{k_1} \dots \partial x_n^{k_n}} v_1(x) \right| \leq M\varepsilon^2 \left[\varepsilon^{-k} + r^{-k}(x, \Gamma) \right], \quad x \in \overline{D}, \tag{6.23c}$$

$$x \notin \Gamma \text{ for } k > 0, \quad 0 \leq k \leq K.$$

The estimate for derivatives of the function $U(x)$ follows from the estimates (6.23) for the derivatives of the functions $U^{(1)}(x)$ and $v_1(x)$.

Write the function $V(x)$, $x \in \overline{D}$, as a sum of the functions

$$V(x) = V_0(x) + v_2(x), \quad x \in \overline{D}.$$

Here the functions $V_0(x)$ and $v_2(x)$ are solutions of the problems

$$L V_0(x) = 0, \quad x \in D, \quad V_0(x) = \varphi(x) - U^{(1)}(x), \quad x \in \Gamma;$$

$$L v_2(x) = 0, \quad x \in D, \quad v_2(x) = -U(x) + U^{(1)}(x), \quad x \in \Gamma.$$

For the function $v_2(x)$ one has the estimate

$$\left| \frac{\partial^k}{\partial x_1^{k_1} \dots \partial x_n^{k_n}} v_2(x) \right| \leq M\varepsilon^2 \left[\varepsilon^{-k} + r^{-k}(x, \Gamma) \right], \quad x \in \overline{D},$$

$$x \notin \Gamma \text{ for } k > 0, \quad 0 \leq k \leq K.$$

We represent the function $V_0(x)$ as the sum of regular and corner boundary layers

$$V_0(x) = \sum_{j=1,\dots,n} V_{(j)}(x) + \sum_{\substack{j,\dots,r=1,\dots,n \\ j<\dots<r,\ 1<|j\dots r|\leq n \\ \Gamma_j,\dots,\Gamma_r \subset \Gamma^- \\ \Gamma_j \cap \dots \cap \Gamma_r \neq \emptyset}} V_{(j\dots r)}(x), \quad x \in \overline{D}, \tag{6.24}$$

where $V_{(j)}(x)$ and $V_{(j\dots r)}(x)$ are the regular and corner boundary layers; here $|j\dots r| = |j\dots r|_{(3.11)}$ is a dimension of the layer $V_{(j\dots r)}(x)$, and $\Gamma_j = \Gamma_{j(3.6)}(D_{(6.4)})$, for $j = 1, \dots, n$. The function $V_{(j\dots r)}(x)$ is the restriction to $\overline{D}$ of the function $V_{(j\dots r)}^e(x)$, $x \in \overline{D}^{(1)}$, that is the solution of the problem

$$L^e V_{(j\dots r)}^e(x) = 0, \quad x \in D^{(1)},$$

$$V_{(j\dots r)}^e(x) = \varphi_{(j\dots r)}^e(x), \quad x \in \Gamma(D^{(1)}),$$

where $D^{(1)} = D_{[j...r]}$. The functions $V^e_{(j...r)}(x)$ equal to zero at $\Gamma^+(D^{(1)})$ and decrease exponentially as $\varepsilon^{-1}\rho$ increases, where $\rho = r(x, \cap_{q=j,...,r} \Gamma_q(D^{(1)}))$, for $1 \le |j...r| \le n$.

The functions $\varphi^e_{(j...r)}(x)$ are sufficiently smooth on faces of the set $D^{(1)}$, and on the faces $\Gamma_q = \Gamma_q(D_{(6.4)})$ from $\Gamma^-(D^{(1)})$ they satisfy the condition

$$U^{(1)}(x) + \varphi^e_{(j)}(x) = \varphi(x), \quad x \in \Gamma_q, \quad q = j, \quad |j...r| = 1;$$

$$U^{(1)}(x) + \sum_{q=j,...,r} V_{(q)}(x) + \sum_{\substack{q,...,k=j,...,r \\ q<...<k,\ 1<|q...k| \le \\ \le |j...r|-1}} V_{(q...k)}(x) + \varphi^e_{(j...r)}(x) = \varphi(x),$$

$$x \in \Gamma_q, \quad q = j,...,r, \quad |j...r| \ge 2.$$

The functions $V_{(j...r)}(x)$ satisfy the estimate

$$\left| V_{(j...r)}(x) \right| \le M \exp\left(-m\varepsilon^{-1} r(x, \cap_{q=j,...,r}\Gamma_q) \right), \quad x \in \overline{D}, \quad (6.25)$$

$$j,...,r = 1,...,n, \quad 1 \le |j...r| \le n;$$

the constant m in (6.25) can be chosen arbitrary in the interval $(0, m_0)$

$$m_0 = (a^0)^{-1} \min_{s,\overline{D}} \left[b^1_s(x) \right], \quad a^0 = a^0_{(6.2)}.$$

Taking into account interior *a priori* estimates and the estimates up to smooth parts of the boundary $\Gamma(D_{|j...r|})$, we find the following estimate for the derivatives:

$$\left| \frac{\partial^k}{\partial x_1^{k_1} ... \partial x_n^{k_n}} V^e_{(j...r)}(x) \right| \le M \left[\varepsilon^{-k} + r^{-k}(x, \Gamma(D^{(1)})) \right] \times$$

$$\times \exp\left(-m\varepsilon^{-1} r(x, \cap_{q=j,...,r}\Gamma_q(D^{(1)})) \right), \quad x \in \overline{D}^{(1)},$$

$$x \notin \Gamma(D^{(1)}) \quad \text{for } k > 0, \quad 0 \le k \le K; \quad D^{(1)} = D_{[j...r]}.$$

Let us refine this estimate. Write the function $V^e_{(j...r)}(x)$ as the decomposition

$$V^e_{(j...r)}(x) = V^e_{(j...r)0}(x) + v_{V_{(j...r)}}(x), \quad x \in \overline{D}_{[j...r]},$$

where $V^e_{(j...r)0}(x)$ and $v_{V_{(j...r)}}(x)$ are the main and the remainder terms. The function $V^e_{(j...r)0}(x)$ is the solution of the problem

$$L^e_{(j...r)} V^e_{(j...r)0}(x) = 0, \quad x \in G_{[j...r]},$$

$$V^e_{(j...r)0}(x) = \varphi^e_{(j...r)}(x), \quad x \in S(G_{[j...r]}),$$

where

$$L^e_{(j\ldots r)} \equiv \varepsilon \left[\sum_{s,k} a^e_{sk}(x) \frac{\partial^2}{\partial x_s \, \partial x_k} + \sum_s b^e_s(x) \frac{\partial}{\partial x_s} \right] + \sum_s b^{1e}_s(x) \frac{\partial}{\partial x_s},$$

$$s, k = j, \ldots, r;$$

the variables x_s, $s \neq j, \ldots, r$ are regarded as parameters. For the function $V^e_{(j\ldots r)0}(x)$, we obtain the estimate

$$\left| \frac{\partial^k}{\partial x_1^{k_1} \ldots \partial x_n^{k_n}} V^e_{(j\ldots r)0}(x) \right| \leq M \left[\varepsilon^{-(k_j + \ldots + k_r)} + \right.$$

$$\left. + r^{-(k_j + \ldots + k_r)}(x, \Gamma(D_{[j\ldots r]})) \right] \exp\left(- m\varepsilon^{-1} r\left(x, \cap_{q=j,\ldots,r} \Gamma_q(D_{[j\ldots r]})\right)\right),$$

$$x \in \overline{D}_{[j\ldots r]}, \quad x \notin \Gamma(D_{[j\ldots r]}) \quad for \quad k > 0, \quad 0 \leq k + 2k_0 \leq K.$$

For the function $v_{V_{(j\ldots r)}}(x)$, we have the estimate

$$\left| \frac{\partial^k}{\partial x_1^{k_1} \ldots \partial x_n^{k_n}} v_{V_{(j\ldots r)}}(x) \right| \leq M\varepsilon \left[\varepsilon^{-k} + r^{-k}(x, \Gamma(D_{[j\ldots r]})) \right] \left[\varepsilon^{-k_0} + \right.$$

$$\left. + \varepsilon^{k_0} r^{-2k_0}(x, \Gamma(D_{[j\ldots r]})) \right] \exp\left(- m\varepsilon^{-1} r\left(x, \cap_{q=j,\ldots,r} \Gamma_q(D_{[j\ldots r]})\right)\right),$$

$$x \in \overline{D}_{[j\ldots r]}, \quad x \notin \Gamma(D_{[j\ldots r]}) \quad for \quad k > 0, \quad 0 \leq k \leq K.$$

For the functions $V_{(j\ldots r)}(x)$, $x \in \overline{D}$, one obtains the estimate

$$\left| \frac{\partial^k}{\partial x_1^{k_1} \ldots \partial x_n^{k_n}} V_{(j\ldots r)}(x) \right| \leq M \left[\varepsilon^{-(k_j + \ldots + k_r)} + r^{-(k_j + \ldots + k_r)} + \right. \qquad (6.26a)$$

$$\left. + \varepsilon^{1-k} + \varepsilon \, r^{-k}(x, \Gamma) \right] \exp\left(- m\,\varepsilon^{-1} r\left(x, \cap_{q=j,\ldots,r}\Gamma_q\right)\right), \quad x \in \overline{D},$$

$$x \notin \Gamma \quad for \quad k > 0, \quad 0 \leq k \leq K, \quad j, \ldots, r = 1, \ldots, n, \quad 1 \leq |j \ldots r| \leq n,$$

where $K = l - 2$ and $m = m_{(6.25)}$.

Thus, the solution of the boundary value problem can be presented as the sum of the functions

$$u(x) = U_{(0)}(x) + V_0(x), \quad x \in \overline{D}, \qquad (6.27)$$

where $U_{(0)}(x) = U^{(1)}(x) + v_1(x) + v_2(x)$. For the function $u(x)$ and its components $U_{(0)}(x)$ and $V_{(j\ldots r)}(x)$ in the representations (6.27), (6.24), we have the estimates (6.26a) and also the estimates

$$\left| \frac{\partial^k}{\partial x_1^{k_1} \ldots \partial x_n^{k_n}} U_{(0)}(x) \right| \leq M \left[1 + \varepsilon^{2-k} + \varepsilon^2 \, r^{-k}(x, \Gamma) \right], \qquad (6.26b)$$

$$x \in \overline{D}, \quad x \notin \Gamma \ \ for \ k > 0;$$

$$\left| u(x) - \varphi\big(x^*(x)\big) \right| \le M\, \varepsilon^{-1} r(x, \Gamma), \quad x \in \overline{D}, \quad 0 \le k \le K,$$

where $x^*(x) = x^*_{(3.5)}(x)$ and $K = K_{(6.26a)}$.

In the case when the condition (6.22) is not fulfilled, the functions $f(x)$, $x \in \overline{D}$ and $\varphi(x)$, $x \in \Gamma$, are approximated by the smooth functions $f^\lambda(x)$, $x \in \overline{D}$, and $\varphi^\lambda(x)$, $x \in \Gamma_j$, where $\varphi^\lambda \in C(\Gamma)$, for which the condition (6.22) holds, as well as the relations

$$\left| \frac{\partial^k}{\partial x_1^{k_1} \dots \partial x_n^{k_n}} f^\lambda(x) \right| \le M\,(1 + \lambda^{1-k}),$$

$$\left| f^\lambda(x) - f(x) \right| \le M\,\lambda, \quad x \in \overline{D};$$

$$\left| \frac{\partial^k}{\partial x_1^{k_1} \dots \partial x_{j-1}^{k_{j-1}} \partial x_{j+1}^{k_{j+1}} \dots \partial x_n^{k_n}} \varphi^\lambda(x) \right| \le M\,(1 + \lambda^{1-k}), \quad x \in \Gamma_q,$$

$$q = j,\, j+n, \quad j = 1, \dots, n, \quad k = k_1 + \dots + k_{j-1} + k_{j+1} + \dots + k_n,$$

$$\left| \varphi^\lambda(x) - \varphi(x) \right| \le M\,\lambda, \quad x \in \Gamma,$$

where the parameter λ takes arbitrary values from (0, 1).

Let

$$u^\lambda(x), \quad x \in \overline{D}, \tag{6.28}$$

be the solution of the problem (6.1), (6.4), where

$$f(x) = f^\lambda(x), \quad x \in \overline{D}, \quad \varphi(x) = \varphi^\lambda(x), \quad x \in \Gamma.$$

The function $u^\lambda(x)$ satisfies the inequality

$$\left| u(x) - u^\lambda(x) \right| \le M\,\lambda, \quad x \in \overline{D}. \tag{6.29}$$

For the function $u^\lambda(x)$ and its components $U_{(0)}(x)$ and $V_{(j\dots r)}(x)$, the estimates (6.26) hold, where $u(x)$ is replaced by $u^\lambda(x)$, under the condition

$$M_{(6.26)} = M(\lambda) = M_0\, \lambda^{-1-k}, \quad 0 \le k \le K, \tag{6.30}$$

where the constant M_0 is independent of λ.

The following theorem holds.

Theorem 6.2.2 *Let the condition* (6.14) *hold, and let* a_{sk}, b_s, b_s^1, c, c^1, $f \in C^{l_0}(\overline{D})$, $s, k = 1, \dots, n$, $\varphi \in C(\Gamma)$, $\varphi \in C^{l_0}(\Gamma_j)$, $j = 1, \dots, 2n$, $l_0 = l + \alpha$, $l \ge K + 2$, $K \ge 2$, $\alpha \in (0,1)$. *Then the function* $u(x)$, *i.e., the solution of the problem* (6.1), (6.4), *can be approximated by the function* $u^\lambda(x)$; *moreover, for* $u^\lambda(x)$ *and its components* $U_{(0)}(x)$ *and* $V_{(j\dots r)}(x)$ *in the representations* (6.27), (6.24), *where* $u(x)$ *is replaced by* $u^\lambda(x)$, *the inequality* (6.29) *and the* estimates (6.26), (6.30) *hold; in* (6.26) $u(x)$ *is replaced by* $u^\lambda(x)$. *Under the additional condition* (6.22), *the function* $u(x)$ *and its components* $U_{(0)}(x)$ *and* $V_{(j...r)}(x)$ *in the representations* (6.27), (6.24) *satisfy the estimates* (6.26).

For $n = 2, 3$ for the problem (6.1), (6.4), where

$$L \equiv \varepsilon \triangle + L_1, \quad x \in \overline{D}, \quad r(x, \Gamma^c) \leq m\varepsilon, \quad L_1 = L_{1(6.1)}, \quad (6.31)$$

one can write out necessary and sufficient compatibility conditions on the set Γ^c based on results of [212, 213] under which the following membership holds:

$$u \in C^{l_0}(\overline{D}), \quad l_0 = l + \alpha, \quad l \geq 0, \quad \alpha \in (0, 1). \quad (6.32)$$

If the condition (6.22) and (6.32) are fulfilled then the functions $U_{(0)}(x)$, $V_{(j...r)}(x)$ satisfy the estimates

$$\left| \frac{\partial^k}{\partial x_1^{k_1} \ldots \partial x_n^{k_n}} U_{(0)}(x) \right| \leq M [1 + \varepsilon^{2-k}], \quad (6.33)$$

$$\left| \frac{\partial^k}{\partial x_1^{k_1} \ldots \partial x_n^{k_n}} V_{(j...r)}(x) \right| \leq M [\varepsilon^{-(k_j + \ldots + k_r)} + \varepsilon^{1-k}] \times$$

$$\times \exp\left(-m\varepsilon^{-1} r\left(x, \cap_{q=j,\ldots,r} \Gamma_q\right)\right), \quad x \in \overline{D},$$

$$0 \leq k \leq K, \quad j,\ldots,r = 1,\ldots,n,$$

where $m = m_{(6.25)}$ and $K = l - 2$.

Note that for the case $n = 2$ in [180], compatibility conditions on the sets Γ^c and Γ^{c+} are given that are sufficient for required smoothness of the solution $u(x)$ and its components $U_{(0)}(x)$ and $V_{(j...r)}(x)$ for $|j \ldots r| \leq 2$.

Theorem 6.2.3 *Let* $n \leq 3$, *and let the hypotheses of Theorem 6.2.2 and also the conditions* (6.22), (6.31) *and the membership* (6.32) *be satisfied. Then for the function* $u(x)$, *i.e., the solution of the problem* (6.1), (6.4), *and components* $U_{(0)}(x)$ *and* $V_{(j...r)}(x)$ *in the representations* (6.27), (6.24), *the estimates* (6.33) *hold.*

6.3 On construction of ε-uniformly convergent difference schemes under their monotonicity condition

Unlike the boundary value problems that were considered in Chapters 2 and 5, in the case of problems for the equation (6.1) the terms of the differential equation (having derivatives orthogonal to the boundary) in a neighborhood of the boundary Γ^- are not bounded ε-uniformly. Because of this, a technique is required for the construction and justification of special schemes that is somewhat different from the technique used in Chapters 2 and 5.

6.3.1   Analysis of necessary conditions for ε-uniform convergence of difference schemes

Let us consider grid approximations for the boundary value problem (6.1), (6.3).

On the set $\overline{D}_{(6.3)}$ we introduce the grid

$$\overline{D}_h = \overline{\omega}_1 \times \omega_2 \times \ldots \times \omega_n, \tag{6.34}$$

where $\overline{\omega}_1$ and ω_s, for $s = 2, \ldots, n$, are spatial meshes on the interval $[d_{*1}, d_1^*]$ and on the x_s-axis, respectively. Set $h = \max_s h_s$ and $h_s = \max_i h_s^i$, for $s = 1, \ldots, n$. Here $N_1 + 1$ is the number of nodes in the mesh $\overline{\omega}_1$ and $N_s + 1$, for $s = 2, \ldots, n$, is the maximal number of nodes in the mesh ω_s on per unit length; $N = \min_s N_s$, for $s = 1, \ldots, n$. Let $h \le MN^{-1}$. On the grid $\overline{D}_{h(6.34)}$ we consider a difference scheme approximating the problem (6.1), (6.3):

$$\Lambda z(x) = f(x), \quad x \in D_h, \quad z(x) = \varphi(x), \quad x \in \Gamma_h. \tag{6.35}$$

Here

$$\Lambda \equiv \varepsilon\left\{ \sum_{s=1}^n a_{ss}(x)\delta_{\overline{x_s}x_s} + 2^{-1} \sum_{\substack{s,k=1 \\ s \neq k}}^n \left[a_{sk}^+(x)\left(\delta_{x_s x_k} + \delta_{\overline{x_s}\,\overline{x_k}}\right) + \right. \tag{6.36}$$

$$\left. + a_{\overline{sk}}(x)\left(\delta_{x_s \overline{x_k}} + \delta_{\overline{x_s} x_k}\right)\right] + \sum_{s=1}^n \left[b_s^+(x)\,\delta_{x_s} + b_s^-(x)\,\delta_{\overline{x_s}}\right] - c(x) \right\} +$$

$$+ \sum_{s=1}^n \left[b_s^{1+}(x)\,\delta_{x_s} + b_s^{1-}(x)\,\delta_{\overline{x_s}}\right] - c^1(x).$$

The operator $\Lambda_{(6.36)}$ on the grid $\overline{D}_{h(6.34)}$, in general, is not ε-uniformly monotone.

Lemma 6.3.1 *For the problem (6.1), (6.3), let the condition (6.5) be satisfied. The following condition for the nodes in the mesh $\overline{\omega}_1$:*

$$\varepsilon^{-1} \max_{\substack{d_{*1} < x_1^i \le \\ \le d_{*1} + M\varepsilon}} h_1^{i-1} \to 0 \quad as \quad N_1 \to \infty, \quad x_1^i \in \overline{\omega}_1, \tag{6.37}$$

that defines the grid $\overline{D}_{h(6.34)}$, is necessary in order that the difference scheme (6.35), (6.34) resolved the boundary value problem ε-uniformly.

The condition (6.37) is necessary in order that the grid $\overline{D}_{h(6.34)}$ was ε-uniformly informative. From this, it follows the necessity of this condition for solvability of the grid problem.

According to the condition (6.37), for ε-uniform solvability of a difference scheme constructed by a classical approximation of the operator L, in the case of the mesh $\overline{\omega}_1$ uniform in an $M\varepsilon$-neighborhood of the left endpoint of $[d_{*1}, d_1^*]$ (for $b_1^1(x) \geq \alpha > 0$, $x \in \overline{D}$), it is necessary that the following condition was fulfilled:

$$\varepsilon^{-1} h_1^* \underset{\varepsilon}{\to} 0 \quad as \quad N_1 \to \infty,$$

where h_1^* is step-size in the mesh $\overline{\omega}_1$ on the interval $[d_{*1}, d_1^*]$.

In the case when a difference scheme constructed by a classical approximation of a boundary value problem resolves the boundary value problem ε-uniformly, terms of the grid equation are not bounded ε-uniformly.

The following Theorem holds.

Theorem 6.3.1 *In the case of the problem* (6.1), (6.3), (6.5), *for a class of difference schemes on the grids* $\overline{D}_{h(6.34)}$ *satisfying the condition* (6.37), *whose difference equations are constructed by a term-wise classical approximation of differential equations, there exist no grids on which the difference scheme, on the interior mesh nodes, approximates termwise the differential equation* ε-*uniformly.*

To prove Theorem 6.3.1 it is sufficient to consider the expression $(\partial/\partial x_1 - \delta_{x_1})u(x)$ for $x \in (x_1^i, x_2, \ldots, x_n) \in D_h$, where $d_{*1} < x_1^i < d_{*1} + M\varepsilon$ for the value i corresponding to the maximal h_1^i.

Classical difference schemes do not have the property of the ε-uniform approximation of the boundary value problem (6.1) that does not allow to use **Statement** A from Section 2.1 in order to justify the ε-uniform convergence of the schemes. Thus, for the boundary value problem (6.1) there arises a necessity to develop a technique to investigate convergence of constructing schemes (and deriving *a priori* estimates of errors of approximate solutions), that allows to establish the ε-uniform convergence of the schemes. The same difficulties appear in the case of boundary value problems for parabolic convection-diffusion equations.

We now discuss difference schemes on piecewise uniform meshes.

Let

$$\overline{\omega}_1 = \overline{\omega}_1(d_1, d_2) = \overline{\omega}_1(d_1, d_2; \sigma, N_1) \tag{6.38a}$$

be a *piecewise-uniform* mesh on the interval $[d_1, d_2]$ on the x_1-axis; the step-sizes in the mesh $\overline{\omega}_{1(6.38a)}$ on the intervals $[d_1, d_1 + \sigma]$ and $[d_1 + \sigma, d_2]$, where $0 \leq \sigma \leq d_2 - d_1$, are $h_{(1)}$ and $h_{(2)}$, respectively; here $N_1 + 1$ is the number of nodes in the mesh $\overline{\omega}_1$. Using the meshes $\overline{\omega}_{1(6.38a)}$, one can construct the simplest meshes that condense in the boundary layer.

Lemma 6.3.2 *Let the scheme* (6.35) *be constructed by a classical approxima-tion of the boundary value problem* (6.1), (6.3), (6.5) *on the grid* $\overline{D}_{h(6.34)}$, *where* $\overline{\omega}_1 = \overline{\omega}_{1(6.38a)}\,(d_{*1},\,d_1^*;\,\sigma,\,N_1)$. *For* $\sigma = \sigma(\varepsilon)$ *(σ is independent of N_1), there exist no grids on which the scheme converges ε-uniformly.*

The proof of this lemma is similar to that of Lemma 2.3.4.

Theorem 6.3.2 *Let the scheme* (6.35) *be constructed by a classical approxi-mation of the boundary value problem* (6.1), (6.3), (6.5) *on the grid*

$$\overline{D}_{h(6.34)}, \quad \text{where } \overline{\omega}_1 = \overline{\omega}_{1(6.38a)}(d_{*1},\,d_1^*;\,\sigma,\,N_1).$$

In order that the scheme be convergent ε-uniformly it is necessary for the value σ *to be dependent on both* ε *and* N_1, *i.e.,* $\sigma = \sigma_0(\varepsilon,\,N_1)$. *In the class of the functions*

$$\sigma = \sigma_0(\varepsilon,\,N_1) = \min[m_1,\,\sigma^*(\varepsilon,\,N_1)], \quad \sigma^*(\varepsilon,\,N_1) = \varepsilon\,\sigma_1(N_1), \qquad (6.38b)$$

where m_1 is an arbitrary constant from the interval $(0,\,d_1^* - d_{*1})$, *the fulfillment of the condition*

$$\sigma_1(N_1) \to \infty, \quad N_1^{-1}\sigma_1(N_1) \to 0 \text{ as } N_1 \to \infty \qquad (6.38c)$$

is necessary for the scheme to be convergent ε-uniformly.

On the interval $[d_{*1},\,d_1^*]$, we shall consider the meshes

$$\overline{\omega}_1 = \overline{\omega}_{1(6.38a-c)}(d_{*1},\,d_1^*;\,\sigma,\,N_1), \qquad (6.38d)$$

whose step-sizes $h_{(1)}$ and $h_{(2)}$ satisfy the condition $h_{(1)} \le h_{(2)}$. The mesh $\overline{\omega}_{1(6.38)}$ is similar to $\overline{\omega}_{1(2.40)}$; however it condenses only in a neighborhood of the left endpoint of $[d_{*1},\,d_1^*]$.

The grid

$$\overline{D}_h = \overline{D}_{h(6.34)}, \quad \text{where } \overline{\omega}_1 = \overline{\omega}_{1(6.38)}(d_{*1},\,d_1^*), \qquad (6.39)$$

is ε-uniformly informative for the solution of the boundary value problem (6.1), (6.3).

In the case of the grid operator $\Lambda_{(6.36)}$, for the solution of the boundary value problem on the grid $\overline{D}_{h(6.39)}$ under the condition $|V(x^*)| \ge m_1$, $x^* \in \Gamma^-$, one obtains the estimate

$$|(L - \Lambda)\,u(x)| \ge m_2\,\varepsilon^{-2}\,h_{(1)} - M\,N^{-1} \ge$$

$$\ge m\,\min\left[\varepsilon^{-2}\,N_1^{-1},\ \varepsilon^{-1}\,N_1^{-1}\,\sigma_1(N_1)\right] - M\,N^{-1},$$

$$x \in D_h, \quad x_1 \le d_{1*} + M_1\,\varepsilon, \quad x_s = x_s^*, \quad s = 2,\dots,n.$$

Thus, we have

$$\max_{G_h} |(L - \Lambda)\,u(x)| \ge m\,\varepsilon^{-1}\,N_1^{-1} \text{ for } \varepsilon = \mathcal{O}(N_1^{-1}), \quad \overline{G}_h = \overline{G}_{h(6.39)}.$$

Theorem 6.3.3 *In the case of the boundary value problem* (6.1), (6.3), *for a class of difference schemes, whose discrete equations are constructed by a classical approximation of differential equations on the grids* $D_{h(6.39)}$ *piecewise-uniform in* x_1, *there are no grids on which the difference scheme approximates the boundary value problem* ε*-uniformly.*

Difficulties in the construction of ε-uniformly convergent difference schemes that are discussed in this section are also observed in the case of the problem on a parallelepiped.

6.3.2   The problem on a slab

Let us estimate the solution error for the difference scheme (6.35), (6.36) on the grid (6.39).

In this subsection, we assume that the operator $\Lambda_{(6.36)}$ on the grid $\overline{D}_{h(6.39)}$ is ε-uniformly monotone.

Write the solution of the difference scheme (6.35), (6.36), (6.39) as the decomposition

$$z(x) = z_U(x) + z_V(x), \quad x \in \overline{D}_h, \tag{6.40a}$$

that corresponds to the decomposition (6.6) of the solution to problem (6.1), (6.3). Here $z_U(x)$ and $z_V(x)$ are the regular and singular components of the discrete solution, and they are solutions of the problems

$$\Lambda z_U(x) = f(x), \quad x \in D_h, \quad z_U(x) = U(x), \quad x \in \Gamma_h; \tag{6.40b}$$

$$\Lambda z_V(x) = f(x), \quad x \in D_h, \quad z_V(x) = V(x), \quad x \in \Gamma_h. \tag{6.40c}$$

For the components in the representation (6.6), let the estimates of Theorem 2.2.1 for $K = 3$ be satisfied. Taking into account the estimates of derivatives of the function $U(x)$, we find

$$|U(x) - z_U(x)| \le M N^{-1}, \quad x \in \overline{D}_h.$$

Let us consider the discrete boundary layer $z_V(x)$, $x \in \overline{D}_h$.
Let the following condition be fulfilled:

$$\sigma = \varepsilon \, \sigma_1(N_1), \quad \sigma = \sigma_{(6.38)}, \quad \sigma_1(N_1) = \sigma_{1(6.38)}(N_1),$$

moreover, $\sigma < m_{1(6.38)}$. When we estimate $V(x) - z_V(x)$ outside a neighborhood of the set Γ^-, it is suitable to use that the functions $V(x)$ and $z_V(x)$ decay towards zero as x moves away from the set Γ^-.

In the case of the grid $\overline{D}_{h(6.39)}$, as a majorant of the discrete boundary layer $z_V(x)$, we can take the solution of the one-dimensional problem

$$\Lambda v(x_1) \equiv \{\varepsilon \, \delta_{x_1\widehat{x_1}} + m_1 \, \delta_{x_1}\} \, v(x_1) = 0, \quad x_1 \in \overline{\omega}_1^e, \tag{6.41}$$

$$v(x_1) = 1, \quad x_1 = d_{*1},$$

where the function $v(x_1)$ tends to zero as $x_1 \to \infty$, and m_1 is an arbitrary number in the interval $(0, m_0)$ with $m_0 = m_{0(6.11)}$. The mesh $\overline{\omega}_1^e$ on the set $[d_{*1}, \infty)$ is the extension of the mesh $\overline{\omega}_{1(6.38)}(d_{*1}, d_1^*)$ for $x > d_1^*$. The step-size in the mesh $\overline{\omega}_1^e$ for $x_1 \geq d_{*1} + \sigma$ equals to $h_{(2)(6.38)}$. The solution of this problem considered on the set $\overline{\omega}_1$ is the function

$$v(x_1) = \begin{cases} 1 - \alpha\,(1 - \psi_1^{-n_1}), & x_1 \leq d_{*1} + \sigma, \\ [1 - \alpha\,(1 - \psi_1^{-\overline{n}_1})]\,\psi_2^{-n_2}, & x_1 > d_{*1} + \sigma, \quad x_1 \in \overline{\omega}_1. \end{cases}$$

Here

$$\overline{\omega}_1 = \overline{\omega}_{1(6.38)}(d_{*1}, d_1^*), \quad \psi_i = \psi_i(\varepsilon, h_{(i)}, m_1) = 1 + m_1\,\varepsilon^{-1}\,h_{(i)}, \quad i = 1, 2,$$

$$n_1 = n_1(x_1) = (h_{(1)})^{-1}\,(x_1 - d_{*1}), \quad \overline{n}_1 = n_1(\sigma),$$

$$n_2 = n_2(x_1) = (h_{(2)})^{-1}\,(x_1 - d_{*1} - \sigma), \quad m_1 = m_{1(6.41)}, \quad \sigma = \sigma_{1(6.38)},$$

the value $\alpha = \alpha(\varepsilon, h_{(1)}, h_{(2)}, \sigma)$ is found from the relation

$$(1 - \alpha)\big(\varepsilon + 2^{-1}\,m_1\,(h_{(1)} + h_{(2)})\big) - 2^{-1}\,m_1\,\alpha\,(h_{(2)} - h_{(1)})\,\psi_1^{-\overline{n}_1} = 0.$$

The function $z_V(x)$ satisfies the estimate

$$\big|z_V(x)\big| \leq M_1\,v(\sigma) \leq M_1\,\psi_1^{-\overline{n}_1} \leq M\,\exp\big(-m^1\,\sigma_1(N_1)\big), \qquad (6.42)$$

$$x \in \overline{D}_h, \quad x_1 \geq d_{*1} + \sigma,$$

where m^1 is an arbitrary number in the interval $(0, m_{0(6.11)})$ satisfying the condition $m^1 \leq m_{1(6.41)}$, and $\sigma = \sigma_{0(6.38)}(\varepsilon, N_1)$.

Taking into account the estimate (6.13), we find

$$|V(x) - z_V(x)| \leq M\,\exp\big(-m^1\,\sigma_1(N_1)\big), \qquad (6.43)$$

$$x \in \overline{D}_h, \quad x_1 \geq d_{*1} + \sigma,$$

where $m^1 \leq m^1_{(6.42)}$. In an σ-neighborhood of the set Γ^-, one has the estimate

$$\big|\Lambda\big(V(x) - z_V(x)\big)\big| \leq M\varepsilon^{-1}\,\big[N_1^{-1}\,\sigma_1(N_1) + N^{-1}\big] \times$$

$$\times \exp\big(-m^1\,\varepsilon^{-1}\,r(x, \Gamma^-)\big), \quad x \in D_h, \quad x_1 < d_{*1} + \sigma.$$

Using as a majorant (up to a constant multiplier) the function

$$w(x) = \exp\big(-m^1\,\varepsilon^{-1}\,r(x, \Gamma^-)\big), \quad x \in \overline{D}_h, \quad r(x, \Gamma^-) \leq \sigma, \qquad (6.44)$$

and taking account of the estimate (6.43), in an σ-neighborhood of the set Γ^-, we obtain the estimate

$$|V(x) - z_V(x)| \leq M\,\Big[N^{-1} + N_1^{-1}\,\sigma_1(N_1) + \exp\big(-m^1\,\sigma_1(N_1)\big)\Big], \qquad (6.45)$$

$$x \in \overline{D}_h, \quad x_1 \leq d_{*1} + \sigma.$$

Note that the operator Λ on the function $w(x)$ is not ε-uniformly bounded. By virtue of (6.43), (6.45), for the component $z_V(x)$ we have the estimate

$$|V(x) - z_V(x)| \leq M \left[N^{-1} + N_1^{-1} \sigma_1(N_1) + \exp\left(-m^1 \sigma_1(N_1)\right) \right], \quad (6.46)$$

$$x \in \overline{D}_h.$$

Using majorant functions that are already defined on the set $\overline{D}_h$, we obtain similar estimates in the case when the parameters of the mesh $\overline{\omega}_{1(6.38)}$ satisfy the condition

$$\sigma = m_{1(6.38)}.$$

Thus, the component $z_V(x)$ converges ε-uniformly with the estimate (6.46).

For the solution of the difference scheme (6.35), (6.36), (6.39), we have the estimate

$$|u(x) - z(x)| \leq M \left[N^{-1} + N_1^{-1} \sigma_1(N_1) + \exp\left(-m^1 \sigma_1(N_1)\right) \right], \quad (6.47)$$

$$x \in \overline{D}_h.$$

It is convenient to consider the difference scheme on the grids

$$\overline{D}_h = \overline{\omega}_1 \times \omega_2 \times \ldots \times \omega_n, \quad (6.48)$$

where $\overline{\omega}_s$, for $s = 2, \ldots, n$, are uniform meshes, and the mesh

$$\overline{\omega}_1 = \overline{\omega}_1\left(d_{*1}, d_1^*\right) = \overline{\omega}_1\left(d_{*1}, d_1^*; m_1, m_2\right) \quad (6.49a)$$

is the mesh $\overline{\omega}_{1(6.38)}\left(d_{*1}, d_1^*; \sigma, N_1\right)$, where

$$\sigma = \sigma_{(6.38b)} = \min\left[m_1, \varepsilon \sigma_1(N_1)\right], \quad \sigma_1(N_1) = m_2^{-1} \ln N_1; \quad (6.49b)$$

here m_2 is an arbitrary constant from the interval $(0, m_0)$, with $m_0 = m_{0(6.11)}$, and $m_1 = m_{1(6.38)}$. We denote by $N_1^\sigma + 1$ the number of nodes in the mesh $\overline{\omega}_{1(6.49)}$ on the interval $[d_{*1}, d_{*1} + \sigma]$, where $\sigma = \min[m_1, m_2^{-1} \varepsilon \ln N_1]$. For the values m_1 and N_1^σ, one has the condition

$$N_1^\sigma = m_1 \left(d_1^* - d_{*1}\right)^{-1} N_1. \quad (6.49c)$$

For the solution of the difference scheme (6.35), (6.36), (6.39) on the grid (6.48) the following estimate holds:

$$|u(x) - z(x)| \leq M N^{-1} \ln N, \quad x \in \overline{D}_h. \quad (6.50)$$

Theorem 6.3.4 *For the components of the solution of the boundary value problem* (6.1), (6.3) *in the representation* (6.6), *let the estimates of Theorem 6.2.1 be satisfied for $K = 3$. Then the difference scheme* (6.35), (6.36), (6.39), *provided that the operator $\Lambda_{(6.36)}$ is ε-uniformly monotone, converges ε-uniformly. For the discrete solution in the case of the grid (or the grid* (6.48)) *the estimate* (6.47) *(the estimate* (6.50)) *holds.*

When the difference scheme (6.35), (6.36), (6.39) is ε-uniformly monotone then the ε-uniform convergence of the grid singular component $z_V(x)$ is obtained using a special form of majorant functions for the components $V(x)$, $z_V(x)$ and for their difference $V(x) - z_V(x)$.

6.3.3 The problem on a parallelepiped

We investigate difference schemes on the meshes that condense in a neighborhood of boundary layers in the case of the boundary value problem on a parallelepiped. Assume that for the solution of the problem (6.1), (6.4) the estimates of Theorem 6.2.2 are fulfilled.

On the parallelepiped $\overline{D}_{(6.4)}$ we introduce the rectangular mesh

$$\overline{D}_h = \overline{\omega}_1 \times \ldots \times \overline{\omega}_n, \tag{6.51}$$

where $\overline{\omega}_s$ is a mesh on the interval $[d_{*s}, d_s^*]$, for $s = 1, \ldots, n$. On the grid $\overline{D}_{h(6.51)}$ for the problem (6.1), (6.4) we consider the difference scheme

$$\Lambda z(x) = f(x), \quad x \in D_h, \quad z(x) = \varphi(x), \quad x \in \Gamma_h, \tag{6.52}$$

where $\Lambda = \Lambda_{(6.36)}$, and find sufficient conditions for its ε-uniform convergence.

We assume that the operator Λ is ε-uniformly monotone.

We shall consider the problem (6.1), (6.4) in the case when the problem solution satisfies the estimates (6.33) for $K = 3$.

Let the function $v(x)$, $x \in \overline{D}$, where $v \in C^3(D) \bigcap C(\overline{D})$, be one of the components in the decompositions (6.27), (6.24). We denote by $z_v(x)$ the solution of the difference problem

$$\Lambda z(x) = L v(x), \quad x \in D_h, \quad z(x) = v(x), \quad x \in \Gamma_h. \tag{6.53}$$

Write the solution of the difference scheme (6.52), (6.51) as the decomposition

$$z(x) = z_{U_{(0)}}(x) + z_{V_0}(x),$$

$$z_{V_0}(x) = \sum_j z_{V_{(j)}}(x) + \sum_{j,\ldots,r} z_{V_{(j\ldots r)}}(x), \quad x \in \overline{D}_h,$$

that corresponds to the decomposition (6.27), (6.24) of the solution of the boundary value problem.

Let us consider the difference scheme on the grid

$$\overline{D}_{h(6.51)}, \tag{6.54}$$

where $\overline{\omega}_s = \overline{\omega}_s(d_{*s}, d_s^*, \sigma_s, N_s) = \overline{\omega}_{1(6.49)}(d_{*s}, d_s^*, \sigma_s, N_s; m_s^1, m_2),$

for $s = 1, \ldots, n$, where m_2 and m_s^1 are arbitrary numbers from the intervals $(0, m_0)$ and $(0, d_s^* - d_{*s})$, respectively, and $m_0 = m_{0(6.26)}$.

We find the estimates $U_{(0)}(x) - z_{U_{(0)}}(x)$, $V_{(j)}(x) - z_{V_{(j)}}(x)$, and $V_{(j...r)}(x) - z_{V_{(j...r)}}(x)$ similar to the derivation of the estimates $U(x) - z_U(x)$ and $V(x) - z_V(x)$ for the problem on the slab.

In the case of the estimates (6.33) for the solution of the difference scheme (6.52), (6.54), one obtains the estimate

$$|u(x) - z(x)| \le M \, N^{-1} \ln N, \quad x \in \overline{D}_h. \tag{6.55}$$

For the solution of the problem and its components let the estimates (6.26) be fulfilled for $K = 3$.

For the solution of the difference scheme (6.52), (6.54), we find the estimate

$$|z(x) - \varphi(x^*(x))| \le M \, \varepsilon^{-1} r(x, \Gamma), \quad x \in \overline{D}_h, \quad x^*(x) = x^*_{(3.5)}(x).$$

With regard to the above estimate, we find

$$|u(x) - z(x)| \le M \varepsilon^{-1} r(x, \Gamma), \quad x \in \overline{D}_h.$$

Estimating $u(x) - z(x)$ for $r(x, \Gamma) \ge r_0$, where $r_0 > 0$ is an arbitrary value, we obtain

$$|u(x) - z(x)| \le M \left[\varepsilon^3 r_0^{-3} N^{-1} \ln N + \varepsilon^{-1} r_0 \right], \quad x \in \overline{D}_h, \quad r(x, \Gamma) \ge r_0.$$

Thus, for the solution of the difference scheme in the case of the *a priori* estimates (6.15), (6.26), the estimate holds

$$|u(x) - z(x)| \le M \left(N^{-1} \ln N \right)^\nu, \quad x \in \overline{D}_h, \tag{6.56}$$

where $\nu = 4^{-1}$.

Let the function $u^\lambda(x) = u^\lambda_{(6.28)}(x)$ approximate the function $u(x)$, i.e., the solution of the boundary value problem (6.1), (6.4). Consider the difference scheme (6.52), (6.54) in the case when for $u^\lambda(x)$ and for its components in the representations (6.27), (6.24), the estimates (6.29) and (6.26), (6.30) are fulfilled; in (6.27) and (6.26), $u(x)$ is replaced by $u^\lambda(x)$.

Let $z^\lambda(x)$, $x \in \overline{D}_h$, be the solution of the difference scheme

$$\Lambda z^\lambda(x) = f^\lambda(x), \quad x \in D_h,$$
$$z^\lambda(x) = \varphi^\lambda(x), \quad x \in \Gamma_h,$$

where $\Lambda = \Lambda_{(6.52)}$; $f^\lambda(x)$ and $\varphi^\lambda(x)$ are approximations of the functions $f(x)$ and $\varphi(x)$, respectively. For the function $z^\lambda(x)$ the estimate

$$|z(x) - z^\lambda(x)| \le M \lambda, \quad x \in \overline{D}_h, \tag{6.57}$$

holds that is similar to (6.29) for the function $u^\lambda(x)$; here and below the constant M is independent of λ.

Taking into account the *a priori* estimates for the function $u^\lambda(x)$, we find the following estimate similar to the derivation of the estimate (6.56):

$$\left| u^\lambda(x) - z^\lambda(x) \right| \leq M\lambda^{-1} \left(N^{-1} \ln N \right)^\nu, \quad x \in \overline{D}_h,$$

where $\nu = \nu_{(6.56)}$. With regard to the estimates (6.57) and (6.29), we obtain

$$\left| u(x) - z(x) \right| \leq M\left[\lambda^{-1} \left(N^{-1} \ln N \right)^\nu + \lambda \right], \quad x \in \overline{D}_h, \quad \nu = \nu_{(6.56)}.$$

Thus, for the solution of the difference scheme (6.52), (6.54) the estimate

$$\left| u(x) - z(x) \right| \leq M \left(N^{-1} \ln N \right)^\nu, \quad x \in \overline{D}_h, \tag{6.58}$$

holds, where $\nu = 8^{-1}$.

Theorem 6.3.5 *Let the condition (6.14) hold, and for the components of the solution to the boundary value problem (6.1), (6.4) in the representations (6.27), (6.24) (or, in the representations (6.27), where $u(x)$ is replaced by $u^\lambda(x)$, and (6.24)), let the estimates of Theorem 6.2.2 (Theorem 6.2.3), where $K = 3$, be satisfied. Then the difference scheme (6.52), (6.54), under the condition that the operator $\Lambda_{(6.52)}$ is ε-uniformly monotone, converges ε-uniformly. For the discrete solutions,*
(i) in the case of the a priori estimates (6.26), one has the estimate (6.56),
(ii) in the case of the estimates (6.29) and the estimates (6.26), (6.30) for the components of the function $u^\lambda(x)$, replacing $u(x)$ by $u^\lambda(x)$ in (6.26), then the estimate (6.58) is valid,
(iii) when the estimate (6.33) is fulfilled then one has the estimate (6.55).

6.4   Monotone ε-uniformly convergent difference schemes

We now impose conditions on the data of the boundary value problems (6.1), (6.3) and (6.1), (6.4), and on the parameters of the meshes constructed, in order to guarantee the ε-uniform monotonicity of the difference schemes (6.35), (6.34) and (6.52), (6.51) and the ε-uniform convergence of their solutions. We use the grids (2.84) for the slab (6.3) and (3.64) for the parallelepiped (6.4) for appropriate values of their parameters.

First, we consider the problem on the slab (6.3). To ensure the ε-uniform monotonicity of the difference schemes, we construct special piecewise-uniform meshes based on results of Subsection 2.4.3, Section 2.4, Chapter 2.

In the case of the condition (6.5), the grid $\overline{D}_h$ condenses in a neighborhood of the set Γ_1, i.e., $\Gamma^e_{(2.77)} = \Gamma_1$. Assume that the coefficients $a_{sk}(x)$ of the operator $L^*_{(2)}$ satisfy the condition

$$a_{sk}(x) = 0, \quad x \in \Gamma_j, \quad j \in J^*, \quad s = 1 \text{ or } k = 1, \ s \neq k, \tag{6.59}$$

on the boundary Γ. Here

$$J^* = J^*(D) = J^*_{(2.77)}(D; \Gamma^e) = \{j = 1\}. \tag{6.60}$$

The condition (6.59) is necessary for the ε-uniform monotonicity of the difference scheme $\{(6.35), (6.36), (6.34)\}$ provided that the difference scheme resolves the boundary value problem $\{(6.1), (6.3), (6.5)\}$ ε-uniformly.

Suppose that the coefficients $a_{sk}(x)$ satisfy the condition (2.55) on the set $\overline{Q}(x^*) \subseteq \overline{D}$, and also the additional condition

$$\left[\min_{\substack{\overline{Q}(x^*) \\ r(x,\Gamma_j) \leq m^* \\ j \in J^*}} a_{kk}(x) - \nu \right] h_k^{-1} - \sum_{\substack{s=2 \\ s \neq k}}^{n} \max_{\substack{\overline{Q}(x^*) \\ r(x,\Gamma_j) \leq m^* \\ j \in J^*}} |a_{sk}(x)| h_s^{-1} > 0, \tag{6.61}$$

$$k = 2, \ldots, n,$$

where m^* and ν are sufficiently small values. The conditions (2.55), (6.61), and (6.59) together allow us to construct grids $\overline{D}_h$, piecewise-uniform in x_1, on which the operator $\Lambda_{(6.36)}$ is ε-uniformly monotone.

Assume that the step-sizes $h_1 = h_{(2)1}$ and h_s in the meshes $\overline{\omega}_1 = \overline{\omega}_{1(6.38)}$ and ω_s, where $s = 2, \ldots, n$, satisfy the condition (2.55c), and also that the parameters of the mesh $\overline{\omega}_1$ satisfy the condition

$$\sigma = \sigma^l \leq m^*_{(6.61)}, \tag{6.62}$$

$$N_{*1} \equiv \sigma h_{(1)}^{-1} < M_1^{-1} \nu \min_{k=2,\ldots,n} [h_k^{-1}], \quad \nu = \nu_{(6.61)}, \quad M_1 = M_{1(2.67)};$$

here $N_{*1}+1$ is the number of nodes in the mesh $\overline{\omega}_1$ on the interval $[d_{*1}, d_{*1}+\sigma]$.

Enforcing the conditions (6.59) and (2.55), (6.61), (6.62) in the case of the grid $\overline{D}_{h(6.48)}$ allows us to choose parameters of the piecewise-uniform mesh $\overline{\omega}_1$ and the uniform meshes ω_s, where $s = 2, \ldots, n$, guaranteeing the ε-uniform monotonicity of the operator $\Lambda_{(6.36)}$.

In the grid $\overline{D}_{h(6.48)}$ along the x_1-axis we use the piecewise-uniform mesh

$$\overline{\omega}_1 = \overline{\omega}^l_{1(6.63)}(d_{*1}, d_1^*) \equiv \overline{\omega}_{1(6.38)}(d_{*1}, d_1^*). \tag{6.63}$$

The following Theorem is a special case of Theorem 2.4.6.

Theorem 6.4.1 *For the boundary value problem* (6.1), (6.3), *let the condition* (6.5) *hold, and let the coefficients* $a_{sk}(x)$ *of the operator* $L^*_{(2)}$ *on the set* $\overline{D}$ *satisfy the relations* (2.64), (6.59) *and* $\{(2.55), (6.61)\}$, *where* $\overline{Q}(x^*) \subseteq \overline{D}$. *Then the grid*

$$\overline{D}_h = \overline{D}_h(Q(x^*), L^*_{(2)}; J^*) = \overline{D}_h\Big(Q(x^*), L^*_{(2)}, \overline{\omega}_1 = \overline{\omega}_{1(6.63)}, \tag{6.64}$$

$$\omega_s = \omega_s^u, \, s = 2, \ldots, n, \quad \text{under the conditions} \,\, (2.55c), \,\, (6.62)$$

in the case of the relations (6.59), $\{(2.55), (6.61)\}$; $J^* = J^*_{(6.60)}\Big)$

is consistent on the set $\overline{Q}(x^*)$ *with the monotonicity condition for the operator* $\Lambda^*_{(2)}$.

Under the condition (2.71), one can choose the set Q in (6.64) as $\overline{Q} = \overline{D}$ in order to ensure the monotonicity of the difference scheme (6.35), (6.36) on the grid (6.64), where $\overline{Q} = \overline{D}$.

A corollary of Theorems 6.3.4 and 6.4.1 is the following result on the convergence of the scheme {(6.35), (6.36), (6.48)}.

Theorem 6.4.2 *Let* $a_{sk} \in C^1(\overline{D})$, b_s, b_s^1, c, c^1, $f \in C(\overline{D})$, $s,k = 1,\ldots,n$, *and let the condition* (6.5) *be satisfied, the coefficients* $a_{sk}(x)$ *satisfy the conditions* (2.71), (6.59) *and* {(2.55), (6.61)}, *where* $\overline{Q} = \overline{D}$, *and let for the solution of the problem* (6.1), (6.3), *the estimates of Theorem 6.2.1 for* $K = 3$. *Then the difference scheme* (6.35), (6.36) *on the grid* $\overline{D}_{h(6.64)}$, *where* $\overline{Q} = \overline{D}$ *and* $\overline{\omega}_{1(6.63)} = \overline{\omega}_{1(6.49)}$, *converges ε-uniformly at the rate* $\mathcal{O}(N^{-1} \ln N)$:

$$|u(x) - z(x)| \leq M\, N^{-1} \ln N, \quad x \in \overline{D}_h.$$

Under the hypotheses of Theorem 6.4.2, for the interpolant

$$\overline{z}(x) = \overline{z}_{(2.42)}(x;\, z(\cdot), \overline{D}_h), \quad x \in \overline{D},$$

where $z(x)$, $x \in \overline{D}_h$ is the the the solution of the difference scheme (6.35), (6.36) on the grid (6.64), the estimate holds

$$|u(x) - \overline{z}(x)| \leq M\, N^{-1} \ln N, \quad x \in \overline{D}.$$

Next, we consider the problem on the parallelepiped (6.4). Here, when constructing piecewise-uniform meshes ensuring the monotonicity of the difference schemes, we use results of Subsection 2.4.3, Section 2.4, Chapter 2.

In the case of the condition (6.14), the grid $\overline{D}_h$ condenses in a neighborhood of the faces Γ_j, for $j = 1,\ldots,n$, i.e., $\Gamma^e_{(3.57)} = \cup_{j=1,\ldots,n}\Gamma_j$. Assume that the coefficients $a_{sk}(x)$ of the operator $L^*_{(2)}$ satisfy the condition

$$a_{sk}(x) = 0, \quad x \in \Gamma_j, \quad j \in J^*, \quad s = 1 \text{ or } k = 1, \ s \neq k, \tag{6.65}$$

on the boundary Γ. Here

$$J^* = J^*(D) = J^*_{(3.57)}(D;\, \Gamma^e) = \quad \{j = 1,\ldots,n\}. \tag{6.66}$$

Suppose also that the coefficients $a_{sk}(x)$ satisfy the condition (2.55), and, moreover, the additional condition

$$\left[\min_{\substack{\overline{Q}(x^*) \\ r(x,\cap_{j_i}\Gamma_{j_i}) \leq m^* \\ j_i \in J^*,\ i=1,\ldots,r}} a_{kk}(x) - \nu \right] h_k^{-1} - \sum_{\substack{s=1 \\ s \neq k, j_i}}^{n} \max_{\substack{\overline{Q}(x^*) \\ r(x,\cap_{j_i}\Gamma_{j_i}) \leq m^* \\ j_i \in J^*,\ i=1,\ldots,r}} |a_{sk}(x)|\, h_s^{-1} > 0,$$

$$k, j_1, \ldots, j_r = 1, \ldots, n, \quad k \neq j_1, \ldots, j_r, \quad 1 \leq r \leq n-2, \quad j_i \in J^*, \qquad (6.67)$$

where m^* and ν are sufficiently small values.

Assume that the step-sizes $h_{(2)s}$ of the piecewise-uniform meshes $\overline{\omega}_s$ satisfy the condition (2.55c), and also that the parameters of the meshes $\overline{\omega}_s$ satisfy the condition

$$\sigma_s = \sigma_s^l \leq m^*_{(6.67)}, \qquad (6.68)$$

$$\sum_{s=1}^{n} N_{*s} < M_1^{-1} \nu \min_k[h_k^{-1}], \quad k = 1, \ldots, n, \quad \nu = \nu_{(6.67)}, \quad M_1 = M_{1(3.52)},$$

where $N_{*s} + 1$ is the number of nodes in the mesh $\overline{\omega}_s$ on $[d_{*s}, d_{*s} + \sigma_s]$. In the grid $\overline{D}_{h(6.54)}$ we use the piecewise-uniform meshes

$$\overline{\omega}_s = \overline{\omega}_{s(6.69)}^l(d_{*s}, d_s^*) \equiv \overline{\omega}_{s(6.54)}(d_{*s}, d_s^*), \quad s = 1, \ldots, n. \qquad (6.69)$$

The following Theorem is a special case of Theorem 3.4.2.

Theorem 6.4.3 *Let $a_{sk} \in C^1(\overline{D})$, $s \neq k$, and let the coefficients $a_{sk}(x)$ of the operator $L^*_{(2)}$ on the set $\overline{D}$ satisfy the relations (6.65), $\{(2.55),(6.67)\}$, where $\overline{Q}(x^*) \subseteq \overline{D}$. Then the grid*

$$\overline{D}_h = \overline{D}_h\Big(Q(x^*), L^*_{(2)}; J^*\Big) = \overline{D}_h\Big(Q(x^*), L^*_{(2)}, \overline{\omega}_s = \overline{\omega}_{s(6.69)}\Big), \quad s = 1, \ldots, n$$

under the conditions (2.55c), (6.68) $\qquad (6.70)$

in the case of the relations $\{(2.55), (6.67)\}$; $J^* = J^*_{(6.66)}\Big)$

is consistent on the set $\overline{Q}(x^*)$ with the monotonicity condition for the operator $\Lambda^*_{(2)}$.

Under the condition (2.71), one can choose the set $\overline{Q} = \overline{D}$ in (6.70).

A corollary of Theorems 6.3.5 and 6.4.3 is the following result on the convergence of the scheme (6.52), (6.54).

Theorem 6.4.4 *Let $a_{sk} \in C^1(\overline{D})$, b_s, b_s^1, c, c^1, $f \in C(\overline{D})$, s, $k = 1, \ldots, n$, and let the condition (6.14) be satisfied, the coefficients $a_{sk}(x)$ satisfy the conditions (2.71), (6.65) and $\{(2.55),(6.67)\}$, where $\overline{Q} = \overline{D}$, and let for the solution of the problem (6.1), (6.4) the estimates of Theorem 6.2.3 (the estimates of Theorem 6.2.2) be satisfied for $K = 3$. Then the difference scheme (6.52) on the grid $\overline{D}_{h(6.70)}$, where $\overline{Q} = \overline{D}$, converges ε-uniformly. For the discrete solutions, the estimate holds*

$$|u(x) - z(x)| \leq M\,(N^{-1}\ln N)^\nu, \quad x \in \overline{D}_h, \qquad (6.71)$$

where $\nu = 1$ in the case of the estimates (6.33), while $\nu = 4^{-1}$ in the case of the estimates (6.26) and $\nu = 8^{-1}$ in the case of the estimate (6.29) and the estimates (6.26), (6.30) for the components of the function $u^\lambda(x)$; here in (6.26) the function $u(x)$ is replaced by $u^\lambda(x)$).

Under the hypotheses of Theorem 6.4.4, for the interpolant

$$\overline{z}(x) = \overline{z}_{(2.42)}\left(x; z(\cdot), \overline{D}_h\right), \quad x \in \overline{D},$$

where $z(x)$, $x \in \overline{D}$ is the solution of the difference scheme (6.52) on the grid (6.70) with $\overline{Q} = \overline{D}$, the estimate holds

$$|u(x) - \overline{z}(x)| \le M\left(N^{-1} \ln N\right)^\nu, \quad x \in \overline{D}, \quad \nu = \nu_{(6.71)}.$$

Chapter 7

Parabolic convection-diffusion equations

7.1 Problem formulation

In an n-dimensional domain D, we consider the Dirichlet problem for the parabolic *convection-diffusion equation*

$$L\,u(x,\,t) = f(x,\,t), \quad (x,\,t) \in G, \tag{7.1}$$

$$u(x,\,t) = \varphi(x,\,t), \quad (x,\,t) \in S.$$

Here

$$\overline{G} = G \cup S, \quad G = D \times (0,\,T], \tag{7.2}$$

the operator L is defined by the relations

$$L \equiv \varepsilon L_2 + L_1;$$

$$L_2 \equiv \sum_{s,k=1}^{n} a_{sk}(x,\,t) \frac{\partial^2}{\partial x_s \partial x_k} + \sum_{s=1}^{n} b_s(x,\,t) \frac{\partial}{\partial x_s} - c(x,\,t);$$

$$L_1 \equiv \sum_{s=1}^{n} b_s^1(x,\,t) \frac{\partial}{\partial x_s} - c^1(x,\,t) - p(x,\,t) \frac{\partial}{\partial t}.$$

The coefficients of the operators L_2 and L_1 satisfy the conditions

$$a_0 \sum_{s=1}^{n} \xi_s^2 \leq \sum_{s,k=1}^{n} a_{sk}(x,\,t)\xi_s\xi_k \leq a^0 \sum_{s=1}^{n} \xi_s^2, \quad \sum_{s=1}^{n}(b_s^1(x,\,t))^2 \geq b_0^2, \tag{7.3}$$

$$p(x,\,t) \geq p_0, \quad c(x,\,t),\ c^1(x,\,t) \geq 0, \quad (x,\,t) \in \overline{G}, \quad a_0,\ b_0,\ p_0 > 0.$$

The coefficients and the right-hand side f on the set $\overline{G}$ are assumed to be sufficiently smooth, just as is the boundary function φ on the closures of the smooth parts of the boundary S, i.e., the data of the problem (7.1), (7.2), are assumed to be sufficiently smooth. We suppose that the problem data are bounded in the case of an unbounded domain D. In that case, the solution of the problem is assumed to be bounded.

The positive direction of characteristics of the operator L_1 is determined by the $(n + 1)$-dimensional vector $\mathbf{B}(x, t)$

$$\mathbf{B}(x, t) = (-b_1^1(x, t), \dots, -b_n^1(x, t), p(x, t)), \quad (x, t) \in \overline{G}.$$

Depending on the behaviour of the characteristics of the operator L_1 in a neighborhood of the boundary $\overline{S}^L$, where $S = S^L \cup S_0$ (see (5.4a)), this lateral boundary is split into subsets. By S^- (or S^+) we denote that part of the set $\overline{S}^L$ through which the characteristics leave (enter) the domain G. Define a set S^0 by the relation $S^0 = S^L \setminus \{S^- \cup S^+\}$. The set S^0 is formed by the characteristics of the operator L_1. Assume that $S^0 = \emptyset$.

For small values of the parameter ε, a boundary layer appears in a neighborhood of the set S^{L-}.

We consider the boundary value problem (7.1), (7.2) on the slab

$$\overline{D} = \{x : d_{*1} \le x_1 \le d_1^*, \ |x_s| < \infty, \ s = 2, \dots, n\}, \tag{7.4}$$

or on a rectangular parallelepiped

$$\overline{D} = \{x : d_{*s} \le x_s \le d_s^*, \ s = 1, \dots, n\}. \tag{7.5}$$

For the boundary value problem (7.1), (7.2), it is required to construct a difference scheme that converges ε-uniformly.

7.2 Estimates of the problem solution on a slab

In this section we use the spaces that were introduced in Chapter 5, with the only difference that for the interior bounds, the value $d((x, t), (x', t'))$ is here defined by the relation $d((x, t), (x', t')) = (|x - x'|^2 + \varepsilon |t - t'|)^{1/2}$.

Assume that the data of the boundary value problem (7.1), (7.2) satisfy the condition

$$a_{sk}, \ b_s, \ b_s^1, \ c, \ c^1, \ p, \ f \in C^{l_0, l_0/2}(\overline{G}), \quad l_0 = l + \alpha, \quad l \ge 0, \quad \alpha \in (0, 1).$$

Then $u \in C^{l_1, l_1/2}(G)$, $l_1 = l_0 + 2$. Under the additional condition

$$\varphi \in C^{\alpha_1, \alpha_1/2}(S_0) \cap C^{\alpha_1, \alpha_1/2}(\overline{S}_j), \quad j = 1, \dots, J,$$

where $\alpha_1 \in (0, 1)$, and $J = 2$ in the case of the slab while $J = 2n$ in the case of the parallelepiped, we have

$$u \in C^{\alpha_2, \alpha_2/2}(\overline{G}) \cap C^{l_1, l_1/2}(G), \quad \alpha_2 \in (0, \alpha_1).$$

Here

$$S_j = \Gamma_j \times (0, T], \quad \Gamma_j = \Gamma_{j(3.6)}, \quad j = 1, \dots, J.$$

Let

$$\varphi \in C^{l_1,l_1/2}(S_0) \cap C^{l_1,l_1/2}(\overline{S}_j), \quad j = 1, \dots, J.$$

Taking into account interior *a priori* estimates and the estimates up to the smooth parts of the boundary (derived in the variables ξ, τ where $\xi_s = \varepsilon^{-1} x_s$, $\tau = \varepsilon^{-1} t$, and $s = 1, \dots, n$), we find

$$\left| \frac{\partial^{k+k_0}}{\partial x_1^{k_1} \dots \partial x_n^{k_n} \partial t^{k_0}} u(x, t) \right| \leq M \left[\varepsilon^{-k} + r^{-k}(x, \Gamma) \right] \left[\varepsilon^{-k_0} + \varepsilon^{k_0} r^{-2k_0}(x, \Gamma) \right],$$

$$(x, t) \in \overline{G}, \quad x \notin \Gamma \quad \text{for} \quad k + k_0 > 0, \quad 0 \leq k + 2k_0 \leq K,$$

where $K = l + 2$.

Let us discuss estimates of the solution of the problem (7.1) on the slab (7.4). Assume that the data of the problem satisfy the condition

$$\begin{aligned}
&a_{sk}, \, b_s, \, b_s^1, \, c, \, c^1, \, p, \, f \in C^{l_1,l_1/2}(\overline{G}), \\
&\varphi \in C^{l_1+2,l_1/2+1}(S_0) \cap C^{l_1+2,l_1/2+1}(\overline{S}_j), \\
&j = 1, 2, \quad l_1 = l + \alpha, \quad l \geq 0, \quad \alpha \in (0, 1).
\end{aligned} \tag{7.6}$$

Assume also that on the set S_0^c, where $S_0^c = S_{0(5.4)}^c$, one has the condition

> The data of the problem (7.1), (7.4) satisfy on S_0^c
>
> a compatibility condition for the derivatives in t up to order K_0,   (7.7)
>
> where $K_0 = [l_1/2]_i + 1$.

Under these conditions, we have $u \in C^{l_1+2,l_1/2+1}(\overline{G})$ [67, 37]. Taking into account the *a priori* estimates up to the boundary (derived in the variables ξ, τ), we obtain the estimate

$$\left| \frac{\partial^{k+k_0}}{\partial x_1^{k_1} \dots \partial x_n^{k_n} \partial t^{k_0}} u(x, t) \right| \leq M \, \varepsilon^{-(k+k_0)}, \quad (x, t) \in \overline{G}, \quad 0 \leq k + 2k_0 \leq K,$$

where $K = l + 2$.

Further, we need more accurate estimates. Assume that the condition

$$b_1^1(x, t) \geq b_0, \quad (x, t) \in \overline{G} \tag{7.8}$$

is fulfilled. Thus,

$$S^L = S^{L-} \bigcup S^{L+},$$

where $S^{L-} = S_1$ and $S^{L+} = S_2$ are the left and right parts of the boundary S^L.

Write the solution of the problem as the sum of the functions

$$u(x,t) = U(x,t) + V(x,t), \quad (x,t) \in \overline{G}, \tag{7.9}$$

where $U(x,t)$ and $V(x,t)$ are the regular and singular parts of the solution. The function $U(x,t)$ is the restriction to $\overline{G}$ of the function $U^e(x,t)$, $(x,t) \in \overline{G}^e$, where $\overline{G}^e = G^e \cup S^e$, $G^e = D^e \times (0,T]$ is the extension of the domain G beyond the boundary S^{L-},

$$D^e = \{x : -\infty < x_1 < d_1^*, \ |x_s| < \infty, \ s = 2, \ldots, n\}.$$

The function $U^e(x,t)$, $(x,t) \in \overline{G}^e$ is the bounded solution of the problem

$$L^e U^e(x,t) = f^e(x,t), \quad (x,t) \in G^e,$$
$$U^e(x,t) = \varphi^e(x,t), \quad (x,t) \in S^e. \tag{7.10}$$

The function $\varphi^e(x,t)$ is sufficiently smooth on the boundary S^e, and it coincides with the function $\varphi(x,t)$ on the set $S^{L+} \cup S_0$. The function $V(x,t)$, $(x,t) \in \overline{G}$ is the solution of the problem

$$LV(x,t) = 0, \quad (x,t) \in G,$$
$$V(x,t) = \varphi_V(x,t) \equiv \varphi(x,t) - U(x,t), \quad (x,t) \in S.$$

The function $V(x,t)$ vanishes on the set $S^{L+} \cup S_0$.

In the case of the conditions (7.6), (7.7), one has u, U, $V \in C^{l_1+2,l_1/2+1}(\overline{G})$, and for the solution of the problem and its components in the representation (7.9), the following estimates are valid:

$$\left| \frac{\partial^{k+k_0}}{\partial x_1^{k_1} \ldots \partial x_n^{k_n} \partial t^{k_0}} u(x,t) \right| \leq M \varepsilon^{-(k+k_0)}, \tag{7.11}$$

$$\left| \frac{\partial^{k+k_0}}{\partial x_1^{k_1} \ldots \partial x_n^{k_n} \partial t^{k_0}} U(x,t) \right| \leq M \varepsilon^{-(k+k_0)},$$

$$\left| \frac{\partial^{k+k_0}}{\partial x_1^{k_1} \ldots \partial x_n^{k_n} \partial t^{k_0}} V(x,t) \right| \leq M \varepsilon^{-(k+k_0)} \exp\left(-m \varepsilon^{-1} r(x, \Gamma) \right),$$

$$(x,t) \in \overline{G}, \quad 0 \leq k + 2k_0 \leq K,$$

where $K = l_{(7.6)} + 2$, and m is an arbitrary constant in the interval $(0, m_0)$

$$m_0 = \min_{\overline{G}} \left[a_{11}^{-1}(x,t) \, b_1^1(x,t) \right].$$

Let us refine the estimates (7.11). Let the problem data satisfy the condition

$$a_{sk}, \ b_s, \ b_s^1, \ c, \ c^1, \ p, \ f \in C^{l_2}(\overline{G}), \tag{7.12}$$

$$\varphi \in C^{l_2}(S_0) \cap C^{l_2}(\overline{S}_2), \quad l_2 = l + \alpha, \quad l \geq 2, \quad \alpha \in (0,1).$$

One can represent the function $U(x,t)$ by the formal expansion

$$U(x,t) = U_0(x,t) + \varepsilon U_1(x,t) + v_1(x,t) = U^{(1)}(x,t) + v_1(x,t), \quad (x,t) \in \overline{G},$$

where $U^{(1)}(x, t)$ and $v_1(x, t)$ are the main and remainder terms of the regular component of the solution. The functions $U_0(x, t)$, $U_1(x, t)$, $v_1(x, t)$ are restrictions to the set $\overline{G}$ of the functions $U_0^e(x, t)$, $U_1^e(x, t)$, $v_1^e(x, t)$ that are solutions of the problems

$$L_1^e U_0^e(x, t) = f^e(x, t), \quad (x, t) \in \overline{G}^e \setminus \{S^{eL+} \cup S_0^e\} \tag{7.13}$$

$$U_0^e(x, t) = \varphi^e(x, t), \quad (x, t) \in S^{eL+} \cup S_0^e;$$

$$L_1^e U_1^e(x, t) = -L_2^e U_0^e(x, t), \quad (x, t) \in \overline{G}^e \setminus \{S^{eL+} \cup S_0^e\}, \tag{7.14}$$

$$U_1^e(x, t) = 0, \quad (x, t) \in S^{eL+} \cup S_0^e;$$

$$L^e v_1^e(x, t) = -\varepsilon^2 L_2^e U_1^e(x, t), \quad (x, t) \in G^e, \tag{7.15}$$

$$v_1^e(x, t) = \varphi^e(x, t) - U^{(1)e}(x, t), \quad (x, t) \in S^e.$$

Given sufficient smoothness of the data of the problem (7.1), (7.4), the solutions of the hyperbolic problems (7.13), (7.14) and the component $U^{(1)}(x, t)$ are sufficiently smooth provided that the data of the problem (7.13), (7.14) satisfy suitable compatibility conditions on the set S^{+c}, i.e., on the intersection of the faces S_0 and S^{L+}. Similar compatibility conditions for the hyperbolic system that defines the reduced problem will ensure the smoothness of the component $U^{(1)}(x, t)$ also in the case of the problem (7.1) on the parallelepiped (7.5).

It is appropriate here to give compatibility conditions for the hyperbolic system (7.13), (7.14) in the case of the problem on the parallelepiped under the assumption that

$$b_s^1(x, t) \geq b_0 > 0, \quad (x, t) \in \overline{G}, \quad s = 1, \ldots, n. \tag{7.16}$$

Set $S^+ = S^{L+} \cup S_0$. Let S^{+c} be the edges of S^+, i.e., S^{+c} is formed by taking pairwise intersection of the faces in S^+. More precisely, taking (7.16) into account, S^+ comprises the faces $\overline{S}_j$, $j \in J_0^+$, where $J_0^+ = \{j = 0, 1+n, \ldots, 2n\}$; these are the faces S_0 and all faces $\overline{S}_j$ from the set S^{L+}. Set $S^+ = \bigcup_q \overline{S}_q$, $q \in J_0^+$.

Let the functions $u^1(x, t)$ and $u^2(x, t)$ be solutions of the problems

$$L_1 u^1(x, t) = f(x, t), \quad (x, t) \in \overline{G} \setminus S^+, \tag{7.17}$$

$$u^1(x, t) = \varphi(x, t), \quad (x, t) \in S^+;$$

$$L_1 u^2(x, t) = f^2(x, t), \quad (x, t) \in \overline{G} \setminus S^+, \tag{7.18}$$

$$u^2(x, t) = \varphi^2(x, t), \quad (x, t) \in S^+;$$

where $f^2(x, t) = -L_2 u^1(x, t)$, $x \in \overline{G}$ and $\varphi^2(x, t) = 0$, $(x, t) \in S^+$. Assuming sufficient smoothness of the data of the problem (7.1), (7.5), on each of the faces $\overline{S}_{j+n} \subset S^{L+}$, where $j = 1, \ldots, n$, the following derivatives are defined:

$$\frac{\partial^{k+k_0}}{\partial x_1^{k_1} \ldots \partial x_{j-1}^{k_{j-1}} \partial x_{j+1}^{k_{j+1}} \ldots \partial x_n^{k_n} \partial t^{k_0}} u^i(x, t) =$$

$$= \frac{\partial^{k+k_0}}{\partial x_1^{k_1} \ldots \partial x_{j-1}^{k_{j-1}} \partial x_{j+1}^{k_{j+1}} \ldots \partial x_n^{k_n} \partial t^{k_0}} \varphi^i(x, t), \quad (x, t) \in \overline{S}_{j+n},$$

$$k = k_1 + \ldots + k_{j-1} + k_{j+1} + \ldots + k_n,$$

and on the face S_0 the following derivatives are defined:

$$\frac{\partial^k}{\partial x_1^{k_1} \ldots \partial x_n^{k_n}} u^i(x, t) = \frac{\partial^k}{\partial x_1^{k_1} \ldots \partial x_n^{k_n}} \varphi^i(x, t), \quad (x, t) \in S_0, \quad i = 1, 2.$$

By virtue of the equations (7.17), (7.18), we can compute the derivatives

$$\frac{\partial^{k+k_0}}{\partial x_1^{k_1} \ldots \partial x_j^{k_j} \ldots \partial x_n^{k_n} \partial t^{k_0}} u^i(x, t), \quad (x, t) \in \overline{S}_{j+n}, \quad \overline{S}_{j+n} \subset S^{L+},$$

$$\frac{\partial^{k+k_0}}{\partial x_1^{k_1} \ldots \partial x_n^{k_n} \partial t^{k_0}} u^i(x, t), \quad (x, t) \in S_0.$$

In the case of the problem (7.1) on the parallelepiped (7.5), we say that the data of the problem (7.17), (7.18), i.e., the coefficients of the operators L_1, L_2 and the functions $f(x, t)$, $(x, t) \in \overline{G}$ and $\varphi(x, t)$, $(x, t) \in S^+$, satisfy on the set S^{+c} *compatibility conditions for the derivatives of the component* $U^{(1)}(x, t)$ up to order K if the derivatives

$$\frac{\partial^{k+k_0}}{\partial x_1^{k_1} \ldots \partial x_n^{k_n} \partial t^{k_0}} u^i(x, t), \quad 0 \leq k + k_0 \leq K + 2(1 - i), \quad i = 1, 2$$

are continuous on S^{+c}. This ensures the continuity on S^{+c} of the derivatives in x and t of the function $U^{(1)}(x, t)$ up to order K,

In the case of the problem (7.1) on the slab (7.4), $S^{+c} = S_0 \cap \overline{S}_2$; we assume that

For the data of the problem (7.13), (7.14) *on the set* S^{+c}

compatibility conditions are fulfilled $\qquad\qquad$ (7.19)

for the derivatives $U^{(1)}(x, t)$ *up to order* $l - 2$, *where* $l = l_{(7.12)}$.

Under the assumptions given above, the functions $U_0(x, t)$, $U_1(x, t)$, and $U^{(1)}(x, t)$ satisfy the condition

$$U_0 \in C^{K+2+\alpha}(\overline{G}), \quad U_1, U^{(1)} \in C^{K+\alpha}(\overline{G}), \quad K \geq 0, \quad \alpha \in (0, 1);$$

moreover, the function $U^{(1)}(x, t)$ satisfies the estimate

$$\left| \frac{\partial^{k+k_0}}{\partial x_1^{k_1} \ldots \partial x_n^{k_n} \partial t^{k_0}} U^{(1)}(x, t) \right| \leq M, \quad (x, t) \in \overline{G}, \quad 0 \leq k + k_0 \leq K, \quad (7.20)$$

where $K = l_{(7.12)} - 2$.

For $l \geq 4$, the function $v_1(x, t)$ satisfies the estimate

$$|v_1(x, t)| \leq M \varepsilon^2, \quad (x, t) \in \overline{G}. \tag{7.21}$$

In the case of the conditions (7.6), (7.7), (7.12), and (7.19), where

$$l_{(7.6)} \geq l_{(7.12)} - 2, \quad l_{(7.12)} \geq 4, \tag{7.22}$$

we have $v_1 \in C^{l_3, l_3/2}(\overline{G})$, where $l_3 = l_{(7.12)} - 2 + \alpha$. For the function $v_1(x, t)$, we obtain the estimate

$$\left| \frac{\partial^{k+k_0}}{\partial x_1^{k_1} \ldots \partial x_n^{k_n} \partial t^{k_0}} v_1(x, t) \right| \leq M \varepsilon^{2-k-k_0}, \quad (x, t) \in \overline{G}, \quad 0 \leq k + 2 k_0 \leq K.$$

Taking into account this estimate and (7.20), for the component $U(x, t)$, we obtain the estimate

$$\left| \frac{\partial^{k+k_0}}{\partial x_1^{k_1} \ldots \partial x_n^{k_n} \partial t^{k_0}} U(x, t) \right| \leq M \left[1 + \varepsilon^{2-k-k_0} \right], \quad (x, t) \in \overline{G}, \tag{7.23}$$

$$0 \leq k + 2 k_0 \leq K,$$

where $K = l_{(7.12)} - 2$.

By virtue of (7.23), the function $\varphi_V(x, t)$, $(x, t) \in S^{L-}$, satisfies the estimate

$$\left| \frac{\partial^{k+k_0}}{\partial x_2^{k_2} \ldots \partial x_n^{k_n} \partial t^{k_0}} \varphi_V(x, t) \right| \leq M \left[1 + \varepsilon^{2-k-k_0} \right], \quad (x, t) \in S^{L-},$$

$$0 \leq k + 2 k_0 \leq K, \quad k = k_2 + \ldots + k_n.$$

From this estimate we obtain

$$\left| \frac{\partial^{k+k_0}}{\partial x_1^{k_1} \ldots \partial x_n^{k_n} \partial t^{k_0}} V(x, t) \right| \leq M \left[\varepsilon^{-k_1} + \varepsilon^{1-k} \right] \left[1 + \varepsilon^{1-k_0} \right] \times \tag{7.24}$$

$$\times \exp\left(-m \varepsilon^{-1} r(x, \Gamma_1) \right), \quad (x, t) \in \overline{G}, \quad 0 \leq k + 2 k_0 \leq K,$$

where $K = l_{(7.12)} - 2$, $m = m_{(7.11)}$.

If the conditions (7.7), (7.19) are not satisfied, then the functions $f(x, t)$, $(x, t) \in \overline{G}$ and $\varphi(x, t)$, $(x, t) \in S$ are approximated by smooth functions $f^\lambda(x, t)$, $(x, t) \in \overline{G}$ and $\varphi^\lambda(x, t)$, $(x, t) \in S$, for which the conditions (7.7),

(7.19) are fulfilled and also one has the estimates

$$\left| \frac{\partial^{k+k_0}}{\partial x_1^{k_1} \dots \partial x_n^{k_n} \partial t^{k_0}} f^\lambda(x, t) \right| \le M \left(1 + \lambda^{1-k-k_0}\right),$$

$$\left| f^\lambda(x, t) - f(x, t) \right| \le M \lambda, \quad (x, t) \in \overline{G};$$

$$\left| \frac{\partial^{k+k_0}}{\partial x_2^{k_2} \dots \partial x_n^{k_n} \partial t^{k_0}} \varphi^\lambda(x, t) \right| \le M \left(1 + \lambda^{1-k-k_0}\right), \quad (x, t) \in \overline{S}^L,$$

$$k_2 + \dots + k_n = k,$$

$$\left| \frac{\partial^k}{\partial x_1^{k_1} \dots \partial x_n^{k_n}} \varphi^\lambda(x, t) \right| \le M \left(1 + \lambda^{1-k}\right), \quad (x, t) \in S_0,$$

$$\left| \varphi^\lambda(x, t) - \varphi(x, t) \right| \le M \lambda, \quad (x, t) \in S,$$

where the parameter λ takes an arbitrary value from the interval $(0, 1)$.

Let $u^\lambda(x, t)$ be the solution of the problem (7.1), (7.4), where

$$f(x, t) = f^\lambda(x, t), \quad (x, t) \in \overline{G}, \quad \varphi(x, t) = \varphi^\lambda(x, t), \quad (x, t) \in S.$$

The function $u^\lambda(x, t)$ satisfies the inequality

$$|u(x, t) - u^\lambda(x, t)| \le M \lambda, \quad (x, t) \in \overline{G}. \tag{7.25}$$

For the components $U(x, t)$ and $V(x, t)$ in the representation (7.9), where $u(x, t)$ is replaced by $u^\lambda(x, t)$, the estimates (7.23), (7.24) are fulfilled with a constant M that depends on λ:

$$M_{(7.23)}, M_{(7.24)} = M_0 \lambda^{-1-k-2k_0}, \quad 0 \le k + 2k_0 \le K, \tag{7.26}$$

where the constant M_0 is independent of λ.

The following theorem holds.

Theorem 7.2.1 *For the data of the boundary value problem (7.1), (7.4) let the conditions (7.8) and also (7.6), (7.12) and (7.22) for $l_{(7.12)} \ge K+2$, where $K \ge 2$, be fulfilled. Then the function $u(x, t)$, i.e., the solution of the problem (7.1), (7.4), can be approximated by the function $u^\lambda(x, t)$; moreover, for the function $u^\lambda(x, t)$ and its components $U(x, t)$ and $V(x, t)$ in the representation (7.9), where $u(x, t)$ is replaced by $u^\lambda(x, t)$, the inequality (7.25) and the estimates (7.23), (7.24), and (7.26) hold. Under the additional conditions (7.7) and (7.19), the estimates (7.23) and (7.24) hold for the components $U(x, t)$ and $V(x, t)$ in the representation (7.9) of the function $u(x, t)$.*

7.3 Estimates of the problem solution on a parallelepiped

Let us consider the problem (7.1), (7.5) under the condition (7.16), assuming that the problem data are sufficiently smooth. In this case, $S_j \subset S^{L-}$, for $j = 1, \ldots, n$, and $S_j \subset S^{L+}$, for $j = n+1, \ldots, 2n$.

Assume that for the data of the problem (7.1), (7.5), one has the conditions

$$a_{sk}, \; b_s, \; b_s^1, \; c, \; c^1, \; p, \; f \in C^{l_1, l_1/2}(\overline{G}),$$

$$\varphi \in C^{l_1+2, \, l_1/2+1}(S_0) \cap C^{l_1+2, \, l_1/2+1}(\overline{S}_j), \quad j = 1, \ldots, 2n, \qquad (7.27)$$

$$l_1 = l + \alpha, \quad l \geq 0, \quad \alpha \in (0, 1).$$

Under these condition, one has

$$u \in C^{\alpha_1, \, \alpha_1/2}(\overline{G}) \cap C^{l_1+2, \, l_1/2+1}(G), \quad \alpha_1 \in (0, 1).$$

Taking into account the *a priori* estimates up to the smooth parts of the boundary S, we find

$$\left| \frac{\partial^{k+k_0}}{\partial x_1^{k_1} \ldots \partial x_n^{k_n} \partial t^{k_0}} u(x, t) \right| \leq M \left[\varepsilon^{-k} + r^{-k}(x, \Gamma) \right] \left[\varepsilon^{-k_0} + \varepsilon^{k_0} r^{-2k_0}(x, \Gamma) \right],$$

$$(x, t) \in \overline{G}, \quad x \notin \Gamma \;\; \text{for} \;\; k + k_0 > 0, \quad 0 \leq k + 2k_0 \leq K, \qquad (7.28)$$

where $K = l_{(7.27)} + 2$. Furthermore, the function $u(x, t)$ satisfies the estimate

$$\left| u(x, t) - \varphi(x^*, t) \right| \leq M \, \varepsilon^{-1} r(x, \Gamma), \quad (x, t) \in \overline{G}, \qquad (7.29)$$

where $x^* = x^*_{(5.6)}(x, \Gamma)$ is a point on Γ nearest to the point $x \in \overline{D}$, $(x, t) \in \overline{G}$.

Write the problem solution as the sum

$$u(x, t) = U(x, t) + V(x, t), \quad (x, t) \in \overline{G}, \qquad (7.30)$$

where $U(x, t)$ and $V(x, t)$ are the regular and singular parts of the solution of the problem.

The function $U(x, t)$, $(x, t) \in \overline{G}$, is the restriction to the set $\overline{G}$ of the function $U^e(x, t)$, $(x, t) \in \overline{G}^e$, that is the solution of the problem

$$L^e U^e(x, t) = f^e(x, t), \quad (x, t) \in G^e, \qquad (7.31a)$$

$$U^e(x, t) = \varphi^e(x, t), \quad (x, t) \in S^e. \qquad (7.31b)$$

The domain G^e is an extension of the domain G beyond the boundary S^{L-}, $G^e = D^e \times (0, T]$, $D^e = D_{(n+1, \ldots, 2n)(3.6)}$. The coefficients and the right-hand side in (7.31a) are smooth extensions of those in (7.1), preserving their properties. The function $\varphi^e(x, t)$, $(x, t) \in S^e$, is smooth on each of the faces S_j^e, and it coincides with the function $\varphi(x, t)$ on the set $S_0 \cup S^{L+}$. The function $V(x, t)$, $(x, t) \in \overline{G}$, is the solution of the problem

$$L V(x, t) = 0, \qquad\qquad\qquad (x, t) \in G, \qquad (7.32)$$

$$V(x, t) = \varphi_V(x, t) \equiv \varphi(x, t) - U(x, t), \quad (x, t) \in S.$$

The function $V(x, t)$ vanishes on the set $S^{L+} \cup S_0$.

Let the problem data satisfy the condition

$$a_{sk}, b_s, b_s^1, c, c^1, p, f \in C^{l_2}(\overline{G}),$$
$$\varphi \in C^{l_2}(S_0) \cap C^{l_2}(\overline{S}_j), \quad j = n+1, \ldots, 2n, \qquad (7.33)$$
$$l_2 = l + \alpha, \quad l \geq 2, \quad \alpha \in (0, 1).$$

We write the function $U(x, t)$ as a formal expansion

$$U(x, t) = U^{(1)}(x, t) + v_1(x, t), \quad (x, t) \in \overline{G}, \qquad (7.34)$$

where $U^{(1)}(x, t) = U_0(x, t) + \varepsilon U_1(x, t)$. The functions $U_0(x, t)$, $U_1(x, t)$, and $v_1(x, t)$ are restrictions to the set $\overline{G}$ of the functions $U_0^e(x, t)$, $U_1^e(x, t)$, and $v_1^e(x, t)$ that are solutions of the problems similar to (7.13), (7.14), (7.15):

$$L_1^e U_0^e(x, t) = f^e(x, t), \quad (x, t) \in \overline{G}^e \setminus S^{e+}, \qquad (7.35)$$

$$U_0^e(x, t) = \varphi^e(x, t), \quad (x, t) \in S^{e+};$$

$$L_1^e U_1^e(x, t) = -L_2^e U_0^e(x, t), \quad (x, t) \in \overline{G}^e \setminus S^{e+}, \qquad (7.36)$$

$$U_1^e(x, t) = 0, \qquad\qquad\qquad (x, t) \in S^{e+};$$

$$L^e v_1^e(x, t) = -\varepsilon^2 L_2^e U_1^e(x, t), \qquad (x, t) \in G^e,$$

$$v_1^e(x, t) = \varphi^e(x, t) - U^{(1)e}(x, t), \quad (x, t) \in S^e,$$

where $S^{e+} = S^{eL+} \cup S_0^e$, $L_1^e = L_{1(7.1)}$, $L_2^e = L_{2(7.1)}$ for $(x, t) \in \overline{G}$.

Let the boundary value problems (7.35) and (7.36) satisfy the condition

For the data of the problem (7.35) and (7.36) on the set S^{+c},

compatibility conditions are fulfilled $\qquad\qquad\qquad\qquad (7.37)$

for the derivatives $U^{(1)}(x, t)$ up to order $l - 2$, where $l = l_{(7.33)}$.

Then $U_0 \in C^{K+2+\alpha}(\overline{G})$, $U_1, U^{(1)} \in C^{K+\alpha}(\overline{G})$, $\alpha \in (0, 1)$, and the estimate holds

$$\left| \frac{\partial^{k+k_0}}{\partial x_1^{k_1} \ldots \partial x_n^{k_n} \partial t^{k_0}} U^{(1)}(x, t) \right| \leq M, \quad 0 \leq k + k_0 \leq K,$$

where $K = l_{(7.33)} - 2$. For $l_{(7.33)} \geq 4$, we have

$$|v_1(x, t)| \leq M \varepsilon^2, \quad (x, t) \in \overline{G}.$$

In the case of the conditions (7.27), (7.33), and (7.37), where

$$l_{(7.27)} \geq l_{(7.33)} - 2, \quad l_{(7.33)} \geq 4, \tag{7.38}$$

one has $v_1 \in C^{\alpha_1, \alpha_1/2}(\overline{G}) \cap C^{l_3, l_3/2}(G)$, where $l_3 = l_{(7.33)} - 2 + \alpha$. For the function $v_1(x, t)$, we obtain the estimate

$$\left| \frac{\partial^{k+k_0}}{\partial x_1^{k_1} \dots \partial x_n^{k_n} \partial t^{k_0}} v_1(x, t) \right| \leq M \varepsilon^2 \left[\varepsilon^{-k} + r^{-k}(x, \Gamma) \right] \left[\varepsilon^{-k_0} + \varepsilon^{k_0} r^{-2k_0}(x, \Gamma) \right],$$

$$(x, t) \in \overline{G}, \quad x \notin \Gamma \quad \text{for} \quad k + k_0 > 0, \quad 0 \leq k + 2k_0 \leq K, \tag{7.39}$$

where $K = l_{(7.33)} - 2$. The estimates of the derivatives of the functions $U^{(1)}(x, t)$ and $v_1(x, t)$ allow us to write down estimates of the derivatives of the component $U(x, t)$ in (7.30).

When considering the function $V(x, t)$, we assume that the conditions (7.27), (7.33), (7.37), and (7.38) hold.

Write the function $V(x, t)$ as the sum of the functions

$$V(x, t) = V_0(x, t) + v_2(x, t), \quad (x, t) \in \overline{G}.$$

Here $V_0(x, t)$ and $v_2(x, t)$ are solutions of the problems

$$L V_0(x, t) = 0, \quad (x, t) \in G,$$

$$V_0(x, t) = \varphi_{V_0}(x, t) \equiv \varphi(x, t) - U^{(1)}(x, t), \quad (x, t) \in S;$$

$$L v_2(x, t) = 0, \quad (x, t) \in G,$$

$$v_2(x, t) = -U(x, t) + U^{(1)}(x, t), \quad (x, t) \in S.$$

For the function $v_2(x, t)$, we have the estimate

$$\left| \frac{\partial^{k+k_0}}{\partial x_1^{k_1} \dots \partial x_n^{k_n} \partial t^{k_0}} v_2(x, t) \right| \leq M \varepsilon^2 \left[\varepsilon^{-k} + r^{-k}(x, \Gamma) \right] \left[\varepsilon^{-k_0} + \varepsilon^{k_0} r^{-2k_0}(x, \Gamma) \right],$$

$$(x, t) \in \overline{G}, \quad x \notin \Gamma \quad \text{for} \quad k + k_0 > 0, \quad 0 \leq k + 2k_0 \leq K, \tag{7.40}$$

where $K = K_{(7.39)}$.

One can represent the boundary layer function $V_0(x, t)$ by the sum of more simple boundary layers

$$V_0(x, t) = \sum_{j=1,\dots,n} V_{(j)}(x, t) + \sum_{\substack{j,\dots, r=1,\dots,n \\ j<\dots<r, \, 1<|j\dots r|\leq n \\ S_j \cap \dots \cap S_r \neq \emptyset}} V_{(j\dots r)}(x, t), \quad (x, t) \in \overline{G}. \tag{7.41}$$

The function $V_{(j...r)}(x, t)$ is the restriction to $\overline{G}$ of the function $V^e_{(j...r)}(x, t)$, $(x, t) \in \overline{G}^{(1)}$, that is the solution of the problem

$$L^e V^e_{(j...r)}(x, t) = 0, \qquad (x, t) \in G^{(1)},$$

$$V^e_{(j...r)}(x, t) = \varphi^e_{(j...r)}(x, t), \qquad (x, t) \in S(G^{(1)}),$$

where $G^{(1)} = G_{[j...r]}$ with $G_{[j...r]} = D_{[j...r]} \times (0, T]$, and $D_{[j...r]} = D_{[j...r](6.16)}$.
    The functions $V^e_{(j...r)}(x, t)$ are equal to zero on the set $S^{e+}(G^1)$, where $S^{e+}(G^1) = S^{eL+}(G^{(1)}) \cup S^e_0(G^{(1)})$, and they decrease exponentially as $\varepsilon^{-1} \rho$ increases, where $\rho = r\left(x, \cap_{q=j,...,r}\Gamma_q(D^{(1)})\right)$ with $D^{(1)} = D_{[j...r]}$, and $1 \leq |j...r| \leq n$. The functions $\varphi^e_{(j...r)}(x, t)$, $(x, t) \in S(G^{(1)})$, are sufficiently smooth on the faces $S(G^{(1)})$, and on the faces $S_q(G^{(1)})$ in $\overline{S}^{L-}(G^{(1)})$, they satisfy the condition

$$U^{(1)}(x, t) + \varphi^e_{(j)}(x, t) = \varphi(x, t), \qquad (x, t) \in S_q, \quad q = j, \quad |j...r| = 1;$$

$$U^{(1)}(x, t) + \sum_{q=j,...,r} V_{(q)}(x, t) + \sum_{\substack{q,...,k=j,...,r \\ q<...<k, 1<|q...k|\leq|j...r|-1}} V_{(q...k)}(x, t) +$$

$$+\varphi^e_{(j...r)}(x, t) = \varphi(x, t), \qquad (x, t) \in S_q, \quad q = j,...,r, \quad |j...r| \geq 2.$$

For the function $V_{(j...r)}(x, t)$, the estimate holds

$$\left|V_{(j...r)}(x, t)\right| \leq M \exp\left(-m\varepsilon^{-1} r(x, \cap_{q=j,...,r}\Gamma_q)\right), \qquad (x, t) \in \overline{G}, \quad (7.42)$$

where m is an arbitrary constant in the interval $(0, m_0)$ with

$$m_0 = \min_{s,\overline{G}}\left[(a^0)^{-1} b^1_s(x, t)\right], \qquad a^0 = a^0_{(7.3)}, \qquad s = 1,...,n.$$

Taking into account interior estimates and the estimates up to the smooth parts of the boundary $S(G_{[j...r]})$, we obtain the estimate

$$\left|\frac{\partial^{k+k_0}}{\partial x_1^{k_1}...\partial x_n^{k_n}\partial t^{k_0}} V_{(j...r)}(x, t)\right| \leq M\left[\varepsilon^{-k} + r^{-k}(x, \Gamma)\right] \times$$

$$\times \left[\varepsilon^{-k_0} + \varepsilon^{k_0} r^{-2k_0}(x, \Gamma)\right] \exp\left(-m\varepsilon^{-1} r(x, \cap_{q=j,...,r}\Gamma_q)\right),$$

$$(x, t) \in \overline{G}, \quad x \notin \Gamma \quad \text{for} \quad k+k_0 > 0, \quad 0 \leq k + 2k_0 \leq K,$$

where $K = K_{(7.39)}$ and $m = m_{(7.42)}$. Refine this estimate.
    Write the function $V^e_{(j...r)}(x, t)$ as the decomposition

$$V^e_{(j...r)}(x, t) = V^e_{(j...r)0}(x, t) + v_{V_{(j...r)}}(x, t), \qquad (x, t) \in \overline{G}_{[j...r]},$$

where $V^e_{(j...r)0}(x,t)$ and $v_{V_{(j...r)}}(x,t)$ are the main and remainder terms. The function $V^e_{(j...r)0}(x,t)$ is the solution of the problem

$$L^e_{(j...r)} \, V^e_{(j...r)0}(x,t) = 0, \qquad\qquad (x,t) \in G_{[j...r]},$$

$$V^e_{(j...r)0}(x,t) = \varphi^e_{(j...r)}(x,t), \quad (x,t) \in S(G_{[j...r]}),$$

where

$$L^e_{(j...r)} \equiv \varepsilon \left[\sum_{s,k} a^e_{sk}(x,t) \frac{\partial^2}{\partial x_s \, \partial x_k} + \sum_s b^e_s(x,t) \frac{\partial}{\partial x_s} \right] + \sum_s b^{1e}_s(x,t) \frac{\partial}{\partial x_s},$$

$$s,k = j, \ldots, r;$$

here the variables x_s for $s \neq j, \ldots, r$, and t appear as parameters. For the function $V^e_{(j...r)0}(x,t)$, by virtue of (7.42), we obtain the estimate

$$\left| \frac{\partial^{k+k_0}}{\partial x_1^{k_1} \ldots \partial x_n^{k_n} \, \partial t^{k_0}} V^e_{(j...r)0}(x,t) \right| \leq M \left[\varepsilon^{-(k_j + \ldots + k_r)} + \right.$$

$$\left. + r^{-(k_j + \ldots + k_r)}(x, \Gamma(D_{[j...r]})) \right] \exp\left(- m\varepsilon^{-1} r\left(x, \cap_{q=j,\ldots,r} \Gamma_q(D_{[j...r]})\right)\right),$$

$$(x,t) \in \overline{G}_{[j...r]}, \quad x \notin \Gamma(D_{[j...r]}) \quad \text{for} \quad k > 0, \quad 0 \leq k + 2k_0 \leq K.$$

For the function $v_{V_{(j...r)}}(x,t)$, we have the estimate

$$\left| \frac{\partial^{k+k_0}}{\partial x_1^{k_1} \ldots \partial x_n^{k_n} \, \partial t^{k_0}} v_{V_{(j...r)}}(x,t) \right| \leq M\varepsilon \left[\varepsilon^{-k} + r^{-k}(x, \Gamma(D_{[j...r]})) \right] \times$$

$$\times \left[\varepsilon^{-k_0} + \varepsilon^{k_0} r^{-2k_0}(x, \Gamma(D_{[j...r]})) \right] \exp\left(- m\varepsilon^{-1} r\left(x, \cap_{q=j,\ldots,r} \Gamma_q(D_{[j...r]})\right)\right),$$

$$(x,t) \in \overline{G}_{[j...r]}, \quad x \notin \Gamma(D_{[j...r]}) \quad \text{for} \quad k + k_0 > 0, \quad 0 \leq k + 2k_0 \leq K.$$

For the functions $V_{(j...r)}(x,t)$, $(x,t) \in \overline{G}$, we obtain the estimate

$$\left| \frac{\partial^{k+k_0}}{\partial x_1^{k_1} \ldots \partial x_n^{k_n} \, \partial t^{k_0}} V_{(j...r)}(x,t) \right| \leq M \left\{ \varepsilon^{-(k_j + \ldots + k_r)} + r^{-(k_j + \ldots + k_r)}(x, \Gamma) + \right.$$

$$\left. + \varepsilon \left[\varepsilon^{-k} + r^{-k}(x, \Gamma) \right] \left[\varepsilon^{-k_0} + \varepsilon^{k_0} r^{-2k_0}(x, \Gamma) \right] \right\} \exp\left(- m\varepsilon^{-1} r(x, \cap_{q=j,\ldots,r} \Gamma_q)\right),$$

$$(x,t) \in \overline{G}, \quad x \notin \Gamma \quad \text{for} \quad k + k_0 > 0, \quad 0 \leq k + 2k_0 \leq K, \qquad (7.43)$$

where $K = K_{(7.39)}$, and $m = m_{(7.42)}$.

Write the solution of the boundary value problem as the sum

$$u(x,t) = U_{(0)}(x,t) + V_0(x,t), \quad (x,t) \in G, \qquad (7.44)$$

where
$$U_{(0)}(x, t) = U^{(1)}(x, t) + v_1(x, t) + v_2(x, t).$$

In the case of the condition (7.37), for the function $u(x, t)$ and its components $U_{(0)}(x, t)$ and $V_{(j...r)}(x, t)$ in the representations (7.44) and (7.41), the estimates (7.28), (7.29), and (7.43) are valid, and also one has the estimate

$$\left| \frac{\partial^{k+k_0}}{\partial x_1^{k_1} \ldots \partial x_n^{k_n} \partial t^{k_0}} U_{(0)}(x, t) \right| \leq M \left\{ 1 + \varepsilon^{2-k-k_0} [1 + \varepsilon^k r^{-k}(x, \Gamma)] \times \right. \quad (7.45)$$

$$\left. \times [1 + \varepsilon^{2k_0} r^{-2k_0}(x, \Gamma)] \right\}, \quad (x, t) \in \overline{G}, \quad x \notin \Gamma \text{ for } k + k_0 > 0, \quad 0 \leq k + 2k_0 \leq K,$$

where $K = K_{(7.39)}$.

In the case when the condition (7.37) is violated, the function $u(x, t)$, i.e., the solution of the problem (7.1), (7.5), can be approximated by the function $u^\lambda(x, t)$, i.e., the solution of the problem (7.1), (7.5), where $f(x, t)$ and $\varphi(x, t)$ are replaced by $f^\lambda(x, t)$ and $\varphi^\lambda(x, t)$, with the error estimate

$$|u(x, t) - u^\lambda(x, t)| \leq M \lambda, \quad (x, t) \in \overline{G}; \quad (7.46)$$

furthermore, for $u^\lambda(x,t)$ and its components $U_{(0)}(x,t)$ and $V_{(j...r)}(x,t)$ in the representations (7.44) and (7.41), one has the estimates (7.29), (7.43), and (7.45), where $u(x, t)$ is replaced by $u^\lambda(x, t)$ in (7.29) and (7.44), and the constant M in the estimates (7.43) and (7.45) is defined by the relation

$$M = M_0 \lambda^{-1-(k+2k_0)}, \quad 0 \leq k + 2k_0 \leq K, \quad (7.47)$$

where M_0 is independent of λ.

The following theorem holds.

Theorem 7.3.1 *For the data of the boundary value problem (7.1), (7.5), let the conditions (7.16) and also (7.27), (7.33), (7.38) for $l_{(7.33)} \geq K + 2$, where $K \geq 2$, be fulfilled. Then the function $u(x, t)$, i.e., the solution of the problem (7.1), (7.5), can be approximated by the function $u^\lambda(x, t)$; moreover, for the function $u^\lambda(x, t)$ and its components $U_{(0)}(x, t)$ and $V_{(j...r)}(x, t)$ in the representations (7.44) and (7.41), the estimates (7.29), (7.43), and (7.45) hold, where $u(x, t)$ is replaced by $u^\lambda(x, t)$ in (7.29) and (7.44), and the constant M in the estimates is defined by the relation (7.47). Under the additional condition (7.37), for the components $U_{(0)}(x, t)$ and $V_{(j...r)}(x, t)$ of the function $u(x, t)$ in the representations (7.44) and (7.41), the estimates (7.29), (7.43), and (7.45) hold.*

Let $n \leq 3$ in the case of the problem (7.1) on the parallelepiped (7.5). Assume that the conditions of Theorem 7.3.1 are fulfilled along with (7.37), and also the following condition holds:

For the data of the problem (7.1), (7.5) on the set S_0^c,

compatibility conditions are fulfilled   (7.48)

for the derivatives in t up to order $[l/2]_i$, $l = l_{(7.33)}$.

In the case of (7.48), the solution of the problem and its components in the representations (7.44) and (7.41) satisfy the estimates

$$
\left| \frac{\partial^{k+k_0}}{\partial x_1^{k_1} \ldots \partial x_n^{k_n} \partial t^{k_0}} u(x, t) \right| \leq M \left[\varepsilon^{-k} + r^{-k}(x, \Gamma^c) \right],
$$

$$
\left| \frac{\partial^{k+k_0}}{\partial x_1^{k_1} \ldots \partial x_n^{k_n} \partial t^{k_0}} U_{(0)}(x, t) \right| \leq M \left\{ 1 + \varepsilon^{2-k} \left[1 + \varepsilon^k r^{-k}(x, \Gamma^c) \right] \right\},
$$

$$
\left| \frac{\partial^{k+k_0}}{\partial x_1^{k_1} \ldots \partial x_n^{k_n} \partial t^{k_0}} V_{(j\ldots r)}(x, t) \right| \leq M \left\{ \varepsilon^{-(k_j+\ldots+k_r)} + r^{-(k_j+\ldots+k_r)}(x, \Gamma^c) + \right.
$$

$$
\left. + \varepsilon \left[\varepsilon^{-k} + r^{-k}(x, \Gamma^c) \right] \right\} \exp \left(- m \varepsilon^{-1} r(x, \cap_{q=j,\ldots,r} \Gamma_q) \right),
$$

$$
(x, t) \in \overline{G}, \quad x \notin \Gamma^c \quad \text{for } k > 0, \quad 0 \leq k + 2k_0 \leq K,
$$

where $K = K_{(7.39)}$, and $m = m_{(7.42)}$. We shall use these estimates outside an $(m\varepsilon)$-neighborhood of the set $\overline{S}^{Lc}$.

In certain situations, the parabolic equation (7.1) can be considered as an elliptic equation whose right-hand side involves the derivative $(\partial/\partial t) u(x, t)$. In this case, one can use results from [212, 213] in order to obtain estimates in an $(m\varepsilon)$-neighborhood of the set $\overline{S}^{Lc}$. For example, for $n = 2, 3$ for the problem (7.1), (7.5), where

$$
L \equiv \varepsilon \triangle + L_1, \quad (x, t) \in \overline{G}, \quad r(x, \Gamma^c) \leq m\varepsilon, \quad L_1 = L_{1(7.1)}, \tag{7.49}
$$

on the basis of results from [212, 213], for the "elliptic" equation with the modified right-hand side it is possible to write down compatibility conditions on the set S^{Lc} that are necessary and sufficient for continuity of the derivatives in x in a neighborhood of the set $\overline{S}^{Lc}$, under which one has

$$
u \in C^{l_0, l_0/2}(\overline{G}), \quad l_0 = l + \alpha, \quad l > 0, \quad \alpha \in (0, 1), \tag{7.50}
$$

for $l \geq l_{(7.27)} + 2$. The membership (7.50) allows us to obtain estimates of the solution of the boundary value problem and its components in (7.44) and (7.41) in an $(m\varepsilon)$-neighborhood of the set $\overline{S}^{Lc}$.

In the case of the condition $r(x, \Gamma^c) \leq m\varepsilon$ of (7.49), if (7.50) holds, then for the problem solution $u(x, t)$ and its components $U_{(0)}(x, t)$ and $V_{(j\ldots r)}(x, t)$

on the set $\overline{G}$, we obtain the estimates

$$\left| \frac{\partial^{k+k_0}}{\partial x_1^{k_1} \ldots \partial x_n^{k_n} \partial t^{k_0}} u(x,t) \right| \le M\,\varepsilon^{-k}, \tag{7.51}$$

$$\left| \frac{\partial^{k+k_0}}{\partial x_1^{k_1} \ldots \partial x_n^{k_n} \partial t^{k_0}} U_{(0)}(x,t) \right| \le M\left[1 + \varepsilon^{2-k}\right],$$

$$\left| \frac{\partial^{k+k_0}}{\partial x_1^{k_1} \ldots \partial x_n^{k_n} \partial t^{k_0}} V_{(j\ldots r)}(x,t) \right| \le M\left[\varepsilon^{-(k_j+\ldots+k_r)} + \varepsilon^{1-k}\right] \times$$

$$\times \exp\left(-m\varepsilon^{-1} r(x, \cap_{q=j,\ldots,r}\Gamma_q)\right),$$

$$(x,t) \in \overline{G}, \quad 0 \le k + 2k_0 \le K, \quad j,\ldots,r = 1,\ldots,n,$$

where $m = m_{(7.42)}$, $K = l_{(7.33)} - 2$.
The following theorem holds.

Theorem 7.3.2 *Let $n \le 3$, and the conditions of Theorem 7.3.1 be fulfilled. Assume also that the conditions (7.37), (7.48), (7.49) and the inclusion (7.50) hold. Then the components $U_{(0)}(x,t)$ and $V_{(j\ldots r)}(x,t)$ in the representations (7.44) and (7.41) satisfy the estimates (7.51).*

7.4 Necessary conditions for ε-uniform convergence of difference schemes

Let us consider grid approximations for the boundary value problem (7.1) on the slab (7.4).

On the set $\overline{G}$ we introduce the grid

$$\overline{G}_h = \overline{D}_h \times \overline{\omega}_0, \quad \overline{D}_h = \overline{\omega}_1 \times \omega_2 \times \ldots \times \omega_n, \tag{7.52}$$

where $\overline{\omega}_0$ is a uniform mesh on the interval $[0, T]$, while $\overline{\omega}_1$ and ω_s, for $s = 2,\ldots,n$, are spatial meshes on the interval $[d_{*1}, d_1^*]$ and on the x_s-axis, respectively. Assume $h = \max_s h_s$, where $h_s = \max_i h_s^i$, for $s = 1,\ldots,n$. Let $N_1 + 1$ be the number of nodes in the mesh $\overline{\omega}_1$; $h_t = \max_j h_t^j$; $N_1 + 1$ is the number of nodes in the mesh $\overline{\omega}_1$; $N_s + 1$, where $s = 2,\ldots,n$, is the minimal number of nodes in the mesh ω_s on per unit length; $N = \min_s N_s$, for $s = 1,\ldots,n$; $N_0 + 1$ is the number of nodes in the mesh $\overline{\omega}_0$. Assume $h \le M N^{-1}$ and $h_t \le M N_0^{-1}$. On the grid $\overline{G}_{h(7.52)}$, for the problem (7.1), (7.4) we consider the difference scheme

$$\Lambda z(x,t) = f(x,t), \quad (x,t) \in G_h, \quad z(x,t) = \varphi(x,t), \quad (x,t) \in S_h. \tag{7.53}$$

Here

$$\Lambda = \Lambda(L) \equiv \varepsilon \left\{ \sum_{s=1}^{n} a_{ss}(x,\, t)\delta_{\overline{x_s}\widehat{x_s}} + \right. \tag{7.54}$$

$$+2^{-1} \sum_{\substack{s,k=1 \\ s \neq k}}^{n} \left[a_{sk}^{+}(x,\, t)\left(\delta_{x_s x_k} + \delta_{\overline{x_s}\overline{x_k}}\right) + a_{sk}^{-}(x,\, t)\left(\delta_{x_s \overline{x_k}} + \delta_{\overline{x_s} x_k}\right) \right] +$$

$$+\sum_{s=1}^{n} [b_s^{+}(x,\, t)\delta_{x_s} + b_s^{-}(x,\, t)\delta_{\overline{x_s}}] - c(x,\, t) \bigg\} +$$

$$+\sum_{s=1}^{n} [b_s^{1+}(x,\, t)\delta_{x_s} + b_s^{1-}(x,\, t)\delta_{\overline{x_s}}] - c^{1}(x,\, t) - p(x,\, t)\delta_{\overline{t}}.$$

The operator $\Lambda_{(7.54)}$ on the grid $\overline{D}_{h(7.52)}$ is, in general, not ε-uniformly monotone.

We have the following results which are similar to Lemma 6.3.1 and Theorem 6.3.1.

Lemma 7.4.1 *For the boundary value problem (7.1), (7.4), let the condition (7.8) be satisfied. For the nodes in the mesh $\overline{\omega}_1$ that defines the grid $\overline{G}_{h(7.52)}$, the condition*

$$\varepsilon^{-1} \max_{\substack{d_{*1} < x_1^i \leq \\ \leq d_{*1} + M\varepsilon}} h_1^{i-1} \xrightarrow[\varepsilon]{} 0 \quad as \quad N_1 \to \infty, \quad x_1^i \in \overline{\omega}_1, \tag{7.55}$$

is necessary in order that the difference scheme {(7.53), (7.54), (7.52)} be resolved the boundary value problem ε-uniformly.

Theorem 7.4.1 *In the case of the problem (7.1), (7.4), for a class of difference schemes on the grids $\overline{G}_{h(7.52)}$ satisfying the condition (7.55), whose difference equations are constructed by a term-wise classical approximation of differential equations, there exist no meshes on which the difference scheme on the interior mesh nodes approximates termwise the differential equation ε-uniformly.*

Classical difference schemes do not have the property of the ε-uniform approximation of the boundary value problem (7.1), (7.2).

In the case of difference schemes on the piecewise-uniform meshes, we have the results which are similar to Lemma 6.3.2 and Theorems 6.3.2 and 6.3.3.

Lemma 7.4.2 *Let the difference scheme (7.53) be constructed by the classical approximation of the boundary value problem (7.1), (7.4) on a family of the grids $\overline{G}_{h(7.52)}$, where the mesh $\overline{\omega}_1 = \overline{\omega}_{1(6.38a)}$ is piecewise uniform. In the case when $\sigma = \sigma(\varepsilon)$ (σ is independent of N_1), there exist no meshes on which the scheme converges ε-uniformly.*

Theorem 7.4.2 *In order that the scheme* (7.53), *which is constructed by the classical approximation of the boundary value problem* (7.1), (7.4) *on the grid*

$$\overline{G}_h = \overline{D}_h \times \overline{\omega}_0, \quad \overline{D}_h = \overline{D}_{h(7.52)}, \quad where \ \overline{\omega}_1 = \overline{\omega}_{1(6.38a)}(d_{*1}, d_1^*; \ \sigma, N_1),$$

be convergent ε-uniformly, it is necessary for the parameters of the piecewise-uniform mesh $\overline{\omega}_{1(6.38a)}$ to satisfy the conditions (6.38b) *and* (6.38c).

Theorem 7.4.3 *In the case of the boundary value problem* (7.1), (7.4), *for a class of difference schemes, whose discrete equations are constructed by a classical approximation of the differential equation from* (7.1) *on a family of the grids $\overline{G}_{h(7.52)}$, piecewise-uniform in x_1, i.e.,*

$$\overline{G}_h = \overline{G}_{h(7.52)}, \quad where \ \overline{\omega}_1 = \overline{\omega}_{1(6.38)}(d_{*1}, d_1^*; \ \sigma, N_1), \qquad (7.56)$$

there are no meshes on which the difference scheme approximates the boundary value problem ε-uniformly.

We shall study convergence of the difference scheme (7.53) on the piecewise-uniform meshes in the case when the difference scheme is monotone.

Let us estimate the solution error for the difference scheme $\{(7.53), (7.54), (7.56)\}$ under the assumption that the operator $\Lambda_{(7.54)}$ on the grid $G_{h(7.56)}$ is ε-uniformly monotone.

Write the solution of the scheme (7.53), (7.56) as the decomposition

$$z(x, t) = z_U(x, t) + z_V(x, t), \quad (x, t) \in \overline{G}_h, \qquad (7.57a)$$

that corresponds to (7.9). Here $z_U(x, t)$ and $z_V(x, t)$ are the regular and singular components of the discrete solution, and they are solutions of the problems

$$\Lambda \, z_U(x, t) = f(x, t), \quad (x, t) \in G_h, \qquad (7.57b)$$

$$z_U(x, t) = U(x, t), \quad (x, t) \in S_h;$$

$$\Lambda \, z_V(x, t) = 0, \quad (x, t) \in G_h, \qquad (7.57c)$$

$$z_V(x, t) = V(x, t), \quad (x, t) \in S_h,$$

where $U(x, t)$ and $V(x, t)$ are the components in the representation (7.9).

Let for the components in the representation (7.9), the estimates (7.23), (7.24) from Theorem 7.2.1 for $K = 4$ be satisfied.

Taking into account the estimates of derivatives of the function $U(x, t)$, we find

$$|U(x, t) - z_U(x, t)| \le M \left[N^{-1} + N_0^{-1} \right], \quad (x, t) \in \overline{G}_h.$$

Estimating $V(x, t) - z_V(x, t)$ outside a neighborhood of the set $\overline{S}^{L-}$, we take into account that the functions $V(x, t)$ and $z_V(x, t)$ decay towards zero as (x, t) moves away from $\overline{S}^{L-}$. As a majorant, we use the function $v(x, t) = v_{(6.41)}(x_1)$, $(x, t) \in \overline{G}_h$, $x_1 \in \overline{\omega}^e_{1(6.41)}$, where $v_{(6.41)}(x_1)$ is the solution of the problem (6.41). Here $\overline{\omega}^e_{1(6.41)}$ is the extension of the mesh $\overline{\omega}_{1(7.56)}$, and m_1 in (6.41) is an arbitrary number in the interval $(0, m_0)$, where $m_0 = m_{0(7.11)}$.

With these estimates for the functions $V(x, t)$ and $z_V(x, t)$ outside an σ-neighborhood of the set $\overline{S}^{L-}$, where $\sigma = \sigma_{(6.38)}$, we estimate $V(x, t) - z_V(x, t)$ in an σ-neighborhood of the set $\overline{S}^{L-}$, taking into account the estimate (7.24) for the derivative. Here, as a majorant, we use a function similar to (6.44).

For the component $z_V(x, t)$, we obtain the estimate

$$|V(x, t) - z_V(x, t)| \leq M \left[N^{-1} + N_1^{-1} \sigma_1(N_1) + \right. \tag{7.58}$$
$$\left. + \exp\left(-m^1 \sigma_1(N_1) \right) + N_0^{-1} \right], \quad (x, t) \in \overline{G}_h,$$

where m^1 is an arbitrary number in the interval $(0, m_{0(7.11)})$ satisfying the condition $m^1 \leq m_{1(6.41)}$, and $\sigma_1(N_1) = \sigma_{1(6.38)}(N_1)$.

For the solution of the difference scheme (7.53), (7.56), we obtain the estimate

$$|u(x, t) - z(x, t)| \leq M \left[N^{-1} + N_1^{-1} \sigma_1(N_1) + \right. \tag{7.59}$$
$$\left. + \exp\left(-m^1 \sigma_1(N_1) \right) + N_0^{-1} \right], \quad (x, t) \in \overline{G}_h.$$

Let us consider the difference scheme (7.53) on the grids

$$\overline{G}_h = \overline{D}_h \times \overline{\omega}_0, \quad \overline{D}_h = \overline{\omega}_1 \times \omega_2 \times \ldots \times \omega_n, \tag{7.60}$$

where $\overline{\omega}_0$ and ω_s, $s = 2, \ldots, n$, are uniform meshes, and

$$\overline{\omega}_1 = \overline{\omega}_1\left(d_{*1}, d_1^*\right) = \overline{\omega}_{1(7.61)}\left(d_{*1}, d_1^*; m_1, m_2\right) \tag{7.61a}$$

is the mesh $\overline{\omega}_{1(6.38)}\left(d_{*1}, d_1^*; \sigma, N_1\right)$, where

$$\sigma_1(N_1) = m_2^{-1} \ln N_1, \tag{7.61b}$$

m_2 is an arbitrary constant from the interval $(0, m_0)$, where $m_0 = m_{0(7.11)}$, and $m_1 = m_{1(6.38)}$. For the values m_1 and N_1^σ, where $N_1^\sigma + 1$ is the number of nodes in the interval $[d_{*1}, d_{*1} + \sigma]$ with $\sigma = \min[m_1, m_2^{-1} \varepsilon \ln N_1]$, the following condition is fulfilled:

$$N_1^\sigma = m_1 \left(d_1^* - d_{*1}\right)^{-1} N_1. \tag{7.61c}$$

For the solution of the difference scheme (7.53) on the grid (7.60), the estimate holds

$$|u(x, t) - z(x, t)| \leq M \left[N^{-1} \ln N + N_0^{-1} \right], \quad (x, t) \in \overline{G}_h. \tag{7.62}$$

In the case when the estimate (7.25) is fulfilled for the solution of the boundary value problem (7.1), (7.4), and for the components of the function $u^\lambda(x, t)$, the estimates (7.23), (7.24), (7.26) with $K = 4$ hold, then for the solution of the difference scheme (7.53), (7.60), we obtain the estimate

$$|u(x, t) - z(x, t)| \leq M \left[N^{-1} \ln N + N_0^{-1} \right]^\nu, \quad (x, t) \in \overline{G}_h, \qquad (7.63)$$

where $\nu = 6^{-1}$. The derivation of (7.63) in the case of the problem on the slab is a simplified variant of the derivation of the similar estimate in the case of the problem on the parallelepiped that is considered in the next subsection (see (7.71)).

Theorem 7.4.4 *For the components in the representation (7.9) of the solution of the boundary value problem (7.1), (7.4), let the estimates of Theorem 7.2.1 be satisfied for $K = 4$. Then the difference scheme (7.53), (7.54) on the grids (7.56) and (7.60), under the condition that the operator $\Lambda_{(7.54)}$ is ε-uniformly monotone, converges ε-uniformly. In the case of the a priori estimates (7.23), (7.24), for the solution of the difference scheme on the grid (7.56), the estimate (7.59) holds, while on the grid (7.60), the estimate (7.62) holds. In the case of the estimate (7.25) for $|u(x, t) - u^\lambda(x, t)|$, when one also has the a priori estimates (7.23), (7.24), (7.26) for the components of the function $u^\lambda(x, t)$, replacing $u(x, t)$ by $u^\lambda(x, t)$ in (7.23), (7.24), then the estimate (7.63) is valid for the solution of the difference scheme on the grid (7.60).*

7.5   Sufficient conditions for ε-uniform convergence of monotone difference schemes

We shall now study convergence of the difference scheme (7.53) on piecewise-uniform meshes in the case of the boundary value problem on the parallelepiped (7.5). Assume that the scheme is monotone.

On the set $\overline{G}$ we introduce the grid

$$\overline{G}_h = \overline{D}_h \times \overline{\omega}_0, \quad \overline{D}_h = \overline{\omega}_1 \times \ldots \times \overline{\omega}_n, \qquad (7.64)$$

where $\overline{\omega}_0 = \overline{\omega}_{0(7.52)}$ and $\overline{\omega}_s$ is a mesh on the interval $[d_{*s}, d_s^*]$, for $s = 1, \ldots, n$. For the problem (7.1), (7.5), on the grid $\overline{G}_{h(7.64)}$ we consider the difference scheme

$$\Lambda z(x, t) = f(x, t), \quad (x, t) \in G_h, \quad z(x, t) = \varphi(x, t), \quad (x, t) \in \Gamma_h, \qquad (7.65)$$

where $\Lambda = \Lambda_{(7.54)}$. We shall derive sufficient conditions for the ε-uniform convergence of the difference scheme. Assume that the operator Λ is ε-uniformly monotone.

We consider the problem (7.1), (7.5) in the case when the problem solution satisfies the estimates (7.51) for $K = 3$.

Let the function $v(x, t)$, $(x, t) \in \overline{G}$, where $v \in C^{4,2}(G) \cap C(\overline{G})$, be either of the components $U_{(0)}(x, t)$ and $V_{(j...r)}(x, t)$ in the decompositions (7.44), (7.41). We denote by $z_v(x, t)$, $(x, t) \in \overline{G}$, the solution of the difference problem

$$\Lambda z(x, t) = L v(x, t), \quad x \in D_h, \quad z(x, t) = v(x, t), \quad (x, t) \in S_h. \quad (7.66)$$

Similarly to the decomposition (7.44), (7.41) of the solution of the boundary value problem, we make the following decomposition of the solution of the difference scheme (7.65), (7.64):

$$z(x, t) = z_{U_{(0)}}(x, t) + z_{V_0}(x, t),$$

$$z_{V_0}(x, t) = \sum_j z_{V_{(j)}}(x, t) + \sum_{j,...,r} z_{V_{(j...r)}}(x, t), \quad x \in \overline{G}_h.$$

Consider the difference scheme on the grid

$$\overline{G}_h = \overline{D}_h \times \overline{\omega}_0, \quad (7.67)$$

where $\overline{\omega}_0 = \overline{\omega}_{0(7.60)}$, and the piecewise-uniform meshes $\overline{\omega}_s$ that generate the grid $\overline{D}_h$ are defined by the relation

$$\overline{\omega}_s = \overline{\omega}_s\big(d_{*s}, d_s^*, \sigma_s, N_s\big) = \overline{\omega}_{1(7.61)}\big(d_{*s}, d_s^*, \sigma_s, N_s; m_s^1, m_2\big),$$

where $s = 1, \ldots, n$, the constants m_2 and m_s^1 are arbitrary numbers in the intervals $(0, m_0)$ and $(0, d_s^* - d_{*s})$, respectively, for $m_0 = m_{0(7.42)}$.

We find the estimates $U_{(0)}(x, t) - z_{U_{(0)}}(x, t)$, $V_{(j)}(x, t) - z_{V_{(j)}}(x, t)$, and $V_{(j...r)}(x, t) - z_{V_{(j...r)}}(x, t)$ by a way similar to the derivation of the estimates $U(x, t) - z_U(x, t)$ and $V(x, t) - z_V(x, t)$ in the case of the problem on the slab. For the solution of the difference scheme (7.65), (7.67), the estimate holds

$$|u(x, t) - z(x, t)| \leq M \left[N^{-1} \ln N + N_0^{-1}\right], \quad (x, t) \in \overline{G}_h. \quad (7.68)$$

Let for the solution of the problem (7.1), (7.5) and its components in (7.44), (7.41) the estimates (7.45), (7.43) be fulfilled for $K = 4$.

For the solution of the difference scheme (7.65), (7.67), we find the estimate

$$\left|z(x, t) - \varphi\big(x^*(x), t\big)\right| \leq M \, \varepsilon^{-1} r(x, \Gamma), \quad (x, t) \in \overline{G}_h,$$

$$x^*(x) = x_{(5.6)}^*(x, \Gamma).$$

Taking into account this estimate, we obtain

$$|u(x, t) - z(x, t)| \leq M \varepsilon^{-1} r(x, \Gamma), \quad (x, t) \in \overline{G}_h.$$

Estimating $u(x, t) - z(x, t)$ for $r(x, \Gamma) \geq r_0$, where $r_0 > 0$ is an arbitrary value, we obtain

$$|u(x, t) - z(x, t)| \leq M \left[\varepsilon^3 r_0^{-3} N^{-1} \ln N + \varepsilon^4 r_0^{-4} N_0^{-1} + \varepsilon^{-1} r_0 \right],$$

$$(x, t) \in \overline{G}_h, \qquad r(x, \Gamma) \geq r_0.$$

Thus, for the solution of the difference scheme in the case of the *a priori* estimates (7.29), (7.43), (7.45), the estimate holds

$$|u(x, t) - z(x, t)| \leq M \left[N^{-1} \ln N + N_0^{-1} \right]^{\nu}, \qquad (x, t) \in \overline{G}_h, \qquad (7.69)$$

where $\nu = 5^{-1}$.

Let us consider the difference scheme (7.65), (7.67) in the case when for the function $u^{\lambda}(x, t)$ that approximates the function $u(x, t)$, i.e., the solution of the boundary value problem (7.1), (7.5), and for the components of $u^{\lambda}(x, t)$ in the representations (7.41) and (7.44), the estimates (7.46) and (7.29), (7.45), (7.43), (7.47) are fulfilled, respectively; in (7.29) and (7.44), $u(x, t)$ is $u^{\lambda}(x, t)$.

Let $z^{\lambda}(x, t)$, $(x, t) \in \overline{G}_h$, be the solution of the difference scheme

$$\Lambda z^{\lambda}(x, t) = f^{\lambda}(x, t), \quad (x, t) \in G_h,$$

$$z^{\lambda}(x, t) = \varphi^{\lambda}(x, t), \quad (x, t) \in S_h,$$

where $\Lambda = \Lambda_{(7.65)}$; the functions $f^{\lambda}(x, t)$ and $\varphi^{\lambda}(x, t)$ are approximations of $f(x, t)$ and $\varphi(x, t)$, respectively. The function $z^{\lambda}(x, t)$ satisfies the estimate

$$|z(x, t) - z^{\lambda}(x, t)| \leq M \lambda, \quad (x, t) \in \overline{G}_h, \qquad (7.70)$$

which is similar to (7.46) for the function $u^{\lambda}(x, t)$. In (7.70) and below the constant M is independent of the parameter λ.

Taking into account the *a priori* estimates for the function $u^{\lambda}(x, t)$, by a way similar to the derivation of (7.69), we find the estimate

$$|u^{\lambda}(x, t) - z^{\lambda}(x, t)| \leq M \lambda^{-1} \left[N^{-1} \ln N + N_0^{-1} \right]^{\nu}, \quad (x, t) \in \overline{G}_h,$$

where $\nu = \nu_{(7.69)}$. Taking into account the estimates (7.70) and (7.46), we obtain

$$|u(x, t) - z(x, t)| \leq M \left\{ \lambda^{-1} \left[N^{-1} \ln N + N_0^{-1} \right]^{\nu} + \lambda \right\}, \quad (x, t) \in \overline{G}_h,$$

where $\nu = \nu_{(7.69)}$. Thus, for the solution of the difference scheme (7.65), (7.67), the estimate holds

$$|u(x, t) - z(x, t)| \leq M \left[N^{-1} \ln N + N_0^{-1} \right]^{\nu}, \quad (x, t) \in \overline{G}_h, \qquad (7.71)$$

where $\nu = 10^{-1}$.

The following theorem holds.

Theorem 7.5.1 *Let for the components in the representations (7.44) and (7.41) (in the representations (7.41), and also (7.44), where $u(x, t)$ is replaced by $u^\lambda(x, t)$), for the solution of the boundary value problem (7.1), (7.5), the estimates of Theorem 7.3.2 (of Theorem 7.3.1) be satisfied for $K = 4$. Then the difference scheme $\{(7.65), (7.54), (7.67)\}$, provided that the operator $\Lambda_{(7.65)}$ is ε-uniformly monotone, converges ε-uniformly. For the discrete solutions in the case of the a priori estimates (7.29), (7.45), (7.43), the estimate (7.69) holds, while in the case of estimates (7.46) and also estimates (7.29), (7.45), (7.43), (7.47), the estimate (7.71) holds, where $u(x, t)$ is replaced by $u^\lambda(x, t)$ in (7.29) and (7.44); and when the a priori estimates (7.51) are fulfilled, the estimate (7.68) holds.*

7.6 Monotone ε-uniformly convergent difference schemes

We now impose conditions on the data of the boundary value problem (7.1), (7.2) on the slab (7.4) and on the parallelepiped (7.5), and on the parameters of the meshes constructed, in order to guarantee the ε-uniform monotonicity of the difference schemes $\{(7.53), (7.54), (7.52)\}$ and $\{(7.65), (7.54), (7.64)\}$ and the ε-uniform convergence of their solutions. When constructing piecewise-uniform meshes ensuring the monotonicity of the difference schemes, we use the grids (5.62) (for the slab) and (5.69) (for the parallelepiped) that are consistent with the monotonicity condition for the operator $\Lambda^*_{(2)(5.27)}$ for appropriate values of their parameters. Note that $L^*_{(2)}(L_{(7.1)}) = L^*_{(2)}(L_{(5.1)}) = L^*_{(2)(5.26)}$ and $\Lambda^*_{(2)}(\Lambda_{(7.54)}) = \Lambda^*_{(2)}(\Lambda_{(5.25)}) = \Lambda^*_{(2)(5.27)}$.

First, we consider the problem (7.1), (7.2) on the slab (7.4).

In the case of the condition (7.8), the grid $\overline{G}_h$ condenses in a neighborhood of the set S_1, i.e., $\Gamma^e_{(5.56)} = \Gamma_1$. Assume that the coefficients $a_{sk}(x, t)$ of the operator $L^*_{(2)}$, where $L^*_{(2)} = L^*_{(2)}(L_{2(7.1)})$, satisfy the condition

$$a_{sk}(x, t) = 0, \quad (x, t) \in \overline{S}_j, \quad j \in J^*, \quad s = 1 \ \ or \ \ k = 1, \ s \neq k, \qquad (7.72)$$

on the boundary $\overline{S}^L$. Here

$$J^* = J^*(D) = J^*_{(5.56)}(D; \Gamma^e) = \{j = 1\}. \qquad (7.73)$$

Note that the condition (7.72) is necessary for the ε-uniform monotonicity of the difference scheme $\{(7.53), (7.54), (7.52)\}$ provided that the difference scheme resolves the boundary value problem $\{(6.1), (6.3), (6.5)\}$ ε-uniformly.

Suppose that the coefficients $a_{sk}(x, t)$ satisfy the condition (5.30) on the set $\overline{G}$, and also the additional condition

$$
\left[\min_{\substack{\overline{Q}(x^*,t^*) \\ r(x,\Gamma_j)\leq m^* \\ j\in J^*}} a_{kk}(x,t)-\nu\right] h_k^{-1} - \sum_{\substack{s=2 \\ s\neq k}}^{n} \max_{\substack{\overline{Q}(x^*,t^*) \\ r(x,\Gamma_j)\leq m^* \\ j\in J^*}} |a_{sk}(x,t)| h_s^{-1} > 0, \quad (7.74)
$$

$$
k = 2,\ldots,n,
$$

where m^* and ν are sufficiently small values. The conditions (7.72), (5.30), and (7.74) together allow us to construct grids $\overline{G}_h$, piecewise-uniform in x_1, on which the operator $\Lambda_{(7.54)}$ is ε-uniformly monotone.

Assume that the step-sizes $h_1 = h_{(2)1}$ and h_s, which define the piecewise-uniform mesh $\overline{\omega}_1 = \overline{\omega}_{1(6.38)}$ and uniform meshes $\overline{\omega}_s = \overline{\omega}_s^u$ for $s = 2,\ldots,n$, satisfy the condition (5.30c), and also that the parameters of the mesh $\overline{\omega}_1$ satisfy the condition

$$
\sigma = \sigma^l \leq m^*_{(7.74)}, \tag{7.75}
$$

$$
N_{*1} \equiv \sigma\, h_{(1)}^{-1} < M_1^{-1}\nu \min_{k=2,\ldots,n} [h_k^{-1}], \quad M_1 = M_{1(5.39)}, \quad \nu = \nu_{(7.74)}.
$$

Enforcing the conditions (7.72) and (5.30), (7.74), (7.75) in the case of the grid $\overline{G}_{h(7.60)}$ allows us to choose parameters of the piecewise-uniform mesh $\overline{\omega}_1$ and the uniform meshes ω_s, where $s = 2,\ldots,n$, guaranteeing the ε-uniform monotonicity of the operator $\Lambda_{(7.54)}$.

When constructing the consistent grid on the slab that ensures the monotonicity of the operator $\Lambda^*_{(2)} = \Lambda^*_{(2)}(\Lambda_{(7.54)})$, along the x_1-axis we use the piecewise-uniform mesh

$$
\overline{\omega}_1 = \overline{\omega}^l_{1(7.76)}(d_{*1}, d_1^*) \equiv \overline{\omega}_{1(6.38)}(d_{*1}, d_1^*). \tag{7.76}
$$

The following Theorem is a special case of Theorem 5.4.1.

Theorem 7.6.1 *Let the condition (7.8) be fulfilled, and let the coefficients $a_{sk}(x,t)$ of the operator $L^*_{(2)}$ satisfy the condition $a_{sk} \in C^1(\overline{G})$ and the relations (7.72) and $\{(5.30),(7.74)\}$, where $\overline{Q}(x^*,t^*) \subseteq \overline{G}$. Then the grid*

$$
\overline{G}_h = \overline{G}_h\big(Q(x^*,t^*), L^*_{(2)}; J^*\big) = \overline{G}_h\big(Q(x^*,t^*), L^*_{(2)}, \overline{\omega}_1 = \overline{\omega}_{1(7.76)}, \tag{7.77}
$$

$$
\omega_s = \omega_s^u, \; s = 2,\ldots,n, \; \text{under the conditions (5.30c), (7.75)}
$$

$$
\text{in the case of the relations } \{(5.30),(7.74)\})
$$

is consistent on the set $\overline{Q}(x^*,t^*)$ with the monotonicity condition for the operator $\Lambda^*_{(2)}$.

Under the condition (5.42), one can choose the set Q in (7.77) as $\overline{Q} = \overline{G}$.

A corollary of Theorems 7.4.4 and 7.6.1 is the following result on the convergence of the scheme $\{(7.53),(7.54),(7.60)\}$.

Theorem 7.6.2 *Let*

$$a_{sk} \in C^1(\overline{G}), \quad b_s, \, b_s^1, \, c, \, c^1, \, f \in C(\overline{G}), \quad s, k = 1, \ldots, n,$$

and let the condition (7.8) be satisfied, the coefficients $a_{sk}(x, t)$ satisfy the conditions (5.42), (7.72), and $\{(5.30), (7.74)\}$, where $\overline{Q} = \overline{G}$, and let for the solution of the problem (7.1), (7.4), the estimates of Theorem 7.2.1 hold for $K = 4$. Then the difference scheme (7.53), (7.54) on the grid $\overline{G}_{h(7.77)}$, where $\overline{Q} = \overline{G}$ and $\overline{\omega}_{1(7.76)} = \overline{\omega}_{1(7.61)}$, converges ε-uniformly. For the discrete solutions, the estimate holds

$$|u(x, t) - z(x, t)| \le M \left[N^{-1} \ln N + N_0^{-1} \right]^{\nu}, \quad (x, t) \in \overline{G}_h, \tag{7.78}$$

where $\nu = 1$ in the case of the a priori estimates (7.23), (7.24) while $\nu = 6^{-1}$ in the case of the estimate (7.25) and the estimates (7.23), (7.24), and (7.26) for the components of the function $u^{\lambda}(x, t)$).

Under the hypotheses of Theorem 7.6.2, for the interpolant

$$\overline{z}(x, t) = \overline{z}_{(2.42)}(x; z(\cdot), \overline{G}_h), \quad (x, t) \in \overline{G},$$

where $z(x, t)$, $(x, t) \in \overline{G}_h$ is the the solution of the difference scheme (7.53), (7.54) on the grid (7.77) with $\overline{Q} = \overline{G}$, the following estimate holds:

$$|u(x, t) - \overline{z}(x, t)| \le M \left[N^{-1} \ln N + N_0^{-1} \right]^{\nu}, \quad (x, t) \in \overline{G}, \quad \nu = \nu_{(7.78)}.$$

Next, we consider the problem (7.1), (7.2) on the parallelepiped (7.5).

By virtue of (7.16), the grid $\overline{G}_h$ condenses in a neighborhood of the faces $\overline{S}_j$, for $j = 1, \ldots, n$, i.e., $\Gamma^e_{(5.63)} = \bigcup_{j=1,\ldots,n} \Gamma_j$. Assume that the coefficients $a_{sk}(x, t)$ of the operator $L^*_{(2)}$ satisfy the condition

$$a_{sk}(x, t) = 0, \quad (x, t) \in \overline{S}_j, \quad s = j, \quad \text{or} \quad k = j, \; s \ne k, \; j \in J^*, \tag{7.79}$$

on the boundary $\overline{S}^L$. Here

$$J^* = J^*(D) = J^*_{(5.63)}(D; \Gamma^e) = \quad \{j = 1, \ldots, n\}. \tag{7.80}$$

Suppose also that the coefficients $a_{sk}(x, t)$ satisfy the condition (5.30), and, moreover, the additional condition

$$\left[\min_{\substack{\overline{Q}(x^*, t^*) \\ r(x, \cap_{j_i} \Gamma_{j_i}) \le m^* \\ j_i \in J^*, \; i=1,\ldots,r}} a_{kk}(x, t) - \nu \right] h_k^{-1} - \sum_{\substack{s=1 \\ s \ne k, j_i}}^{n} \max_{\substack{\overline{Q}(x^*, t^*) \\ r(x, \cap_{j_i} \Gamma_{j_i}) \le m^* \\ j_i \in J^*, \; i=1,\ldots,r}} |a_{sk}(x, t)| \, h_s^{-1} > 0,$$

$$k, j_1, \ldots, j_r = 1, \ldots, n, \quad k \ne j_1, \ldots, j_r, \quad 1 \le r \le n - 2, \tag{7.81}$$

$$\cap_{j_i} \Gamma_{j_i} \ne \emptyset, \quad j_i \in J^*,$$

where m^* and ν are sufficiently small values.

Assume that the step-sizes $h_{(2)s}$ of the piecewise-uniform meshes $\overline{\omega}_s$ satisfy the condition (5.30c), and also that the parameters of the meshes $\overline{\omega}_s$ satisfy the condition

$$\sigma_s = \sigma_s^l \leq m^*_{(7.81)}, \tag{7.82}$$

$$\sum_{s=1}^{n} N_{*s} < M_1^{-1} \nu \min_{k}[h_k^{-1}], \quad k = 1, \ldots, n, \quad M_1 = M_{1(5.50)},$$

where $N_{*s} + 1$ is the number of nodes in the mesh $\overline{\omega}_s$ on $[d_{*s}, d_{*s} + \sigma_s]$.

When constructing the consistent mesh on the parallelepiped that ensure the monotonicity of the operator $\Lambda^*_{(2)} = \Lambda^*_{(2)}(\Lambda_{(7.54)})$, we use the piecewise-uniform meshes

$$\overline{\omega}_s = \overline{\omega}^l_{s(7.83)}(d_{*s}, d^*_s) \equiv \overline{\omega}_{1(6.38)}(d_{*s}, d^*_s), \quad s = 1, \ldots, n. \tag{7.83}$$

The following Theorem is a special case of Theorem 5.4.2.

Theorem 7.6.3 *Let the condition* (7.16) *be fulfilled, and let the coefficients* $a_{sk}(x, t)$ *of the operator* $L^*_{(2)}$ *satisfy the condition* $a_{sk} \in C^1(\overline{G})$ *and the relations* (7.79) *and* $\{(5.30), (7.81)\}$, *where* $\overline{Q}(x^*, t^*) \subseteq \overline{G}$. *Then the grid*

$$\overline{G}_h = \overline{G}_h(Q(x^*, t^*), L^*_{(2)}; J^*) = \overline{G}_h(Q(x^*, t^*), L^*_{(2)}, \tag{7.84}$$

$$\overline{\omega}_s = \overline{\omega}_{s(7.83)}, \quad s = 1, \ldots, n, \quad \textit{under the conditions } (5.30c), \ (7.82)$$

in the case of the relations $\{(5.30), \ (7.81)\}$)

is consistent on the set $\overline{Q}(x^*, t^*)$ *with the monotonicity condition for the operator* $\Lambda^*_{(2)}$.

Under the condition (5.42), one can choose the set $\overline{Q} = \overline{G}$ in (7.84).

A corollary of Theorems 7.5.1 and 7.6.3 is the following result on the convergence of the scheme (7.65), (7.67).

Theorem 7.6.4 *Let the conditions*

$$a_{sk} \in C^1(\overline{G}), \quad b_s, b_s^1, c, c^1, f \in C(\overline{G}), \quad s, k = 1, \ldots, n,$$

just as (7.16) *be satisfied, the coefficients* $a_{sk}(x, t)$ *satisfy the conditions* (5.42), (7.79) *and* $\{(5.30), (7.81)\}$, *where* $\overline{Q} = \overline{G}$, *and let for the solution of the problem* (7.1), (7.5), *the estimates of Theorem* 7.3.1 *(the estimates of Theorem* 7.3.2*) be satisfied for* $K = 4$. *Then the difference scheme* (7.65) *on the grid* $\overline{G}_{h(7.84)}$, *where* $\overline{Q} = \overline{G}$, $\overline{\omega}_{s(7.83)} = \overline{\omega}_{s(7.67)}$, *converges* ε-*uniformly. For the discrete solutions, the estimate holds*

$$|u(x, t) - z(x, t)| \leq M \left[N^{-1} \ln N + N_0^{-1}\right]^\nu, \quad (x, t) \in \overline{G}_h, \tag{7.85}$$

where $\nu = 5^{-1}$ in the case of the estimates (7.29), (7.45), and (7.43), while $\nu = 10^{-1}$ in the case of the estimate (7.46) and the estimates (7.29), (7.45), (7.43), and (7.47); here in (7.29), (7.44) the function $u(x, t)$ is replaced by $u^\lambda(x, t)$); in the case when the estimates (7.51) hold, we have $\nu = 1$.

Under the hypotheses of Theorem 7.6.4, for the interpolant

$$\overline{z}(x,t) = \overline{z}_{(2.42)}\left(x,t; z(\cdot), \overline{G}_h\right), \quad (x, t) \in \overline{G},$$

where $z(x, t)$, $(x, t) \in \overline{G}_h$ is the solution of the difference scheme (7.65) on the grid (7.84) with $\overline{Q} = \overline{G}$, the estimate holds

$$|u(x, t) - \overline{z}(x, t)| \le M \left[N^{-1} \ln N + N_0^{-1} \right]^\nu, \quad (x, t) \in \overline{G}, \quad \nu = \nu_{(7.85)}.$$

Part II

Advanced trends in ε-uniformly convergent difference methods

Chapter 8

Grid approximations of parabolic reaction-diffusion equations with three perturbation parameters

The grid approximations of an initial-boundary value problem are considered for a singularly perturbed parabolic reaction-diffusion equation with three parameters on a rectangular domain in x and t. Using the condensing mesh technique—condensing in both x and t—a difference scheme is constructed that converges uniformly with respect to the perturbation parameters.

8.1   Introduction

An initial-boundary value problem is considered for a singularly perturbed parabolic reaction-diffusion equation on a rectangular domain in x and t. The second-order spatial derivative, the temporal derivative, and the reaction term in the differential equation are multiplied by parameters ε_1^2, ε_2^2, and ε_3^2, respectively; these parameters are components of the perturbation vector-parameter $\bar\varepsilon = (\varepsilon_1, \varepsilon_2, \varepsilon_3)$, where $\varepsilon_1, \varepsilon_2 \in (0, 1]$, and $\varepsilon_3 \in [0, 1]$. Solutions of such problems are $\bar\varepsilon$-uniformly bounded only under the condition that the right-hand side in the equation is of order $\mathcal{O}\left(\varepsilon_1^2 + \varepsilon_2^2 + \varepsilon_3^2\right)$. Here we consider problems whose solutions are $\bar\varepsilon$-uniformly bounded that leads to the additional condition imposed on the perturbation parameters: $\varepsilon_1 + \varepsilon_2 + \varepsilon_3 = \mathcal{O}(1)$. The solutions of such parabolic problems typically have both boundary and initial layers and also initial-boundary layers depending on the relations between the scalar parameters ε_1, ε_2 and ε_3. Such layers were studied in [148].

It is known that in the case of singularly perturbed problems in the presence of parabolic boundary and/or initial layers, the fitted operator method is inapplicable in order to construct schemes convergent uniformly with respect to the perturbation parameters (see, e.g., [130, 135, 138, 41]). Because of this, here the condensing mesh technique is used. A description of a method using special meshes that condense in the boundary and interior layers can be found, e.g., in [15, 138, 87, 102, 33]. *A priori* estimates are constructed for the problem solution and its derivatives. Using such estimates and the *condensing mesh method* for a rectangular grid condensing in both x and t, a difference scheme is constructed that converges $\bar\varepsilon$-uniformly at the rate $\mathcal{O}\left(N^{-2}\ln^2 N + N_0^{-1}\ln N_0\right)$, where $N + 1$ and $N_0 + 1$ are the numbers of mesh points in x and t, respectively. For fixed values of the parameter $\bar{\varepsilon}$ the difference scheme converges at the rate $\mathcal{O}\left(N^{-2} + N_0^{-1}\right)$.

Like the boundary value problem that is considered here, grid approximations of boundary value problems for a parabolic reaction-diffusion equation on a strip, that is two-dimensional in space, and on a rectangle were studied in [148]; a difference scheme was constructed that converges $\bar{\varepsilon}$-uniformly but only at the rate $\mathcal{O}\left((N^{-1}\ln N)^{2/3} + (N_0^{-1}\ln N_0)^{1/2}\right)$.

8.2   Problem formulation. The aim of the research

On the domain $\overline{G}$, where

$$G = D \times (0,T], \quad \overline{G} = G\cup S, \quad D = (0, d), \tag{8.1}$$

we consider the initial-boundary value problem for the singularly perturbed parabolic equation

$$L_{(8.2)}\, u(x,t) = F(x,t), \quad (x,t) \in G, \tag{8.2a}$$

$$u(x,t) = \varphi(x,t), \quad (x,t) \in S. \tag{8.2b}$$

Here

$$L_{(8.2)} \equiv \varepsilon_1^2\, a(x,t)\,\frac{\partial^2}{\partial x^2} - \varepsilon_2^2\, p(x,t)\,\frac{\partial}{\partial t} - \varepsilon_3^2\, c(x,t), \quad F(x,t) = F(x,t;\bar{\varepsilon}).$$

The parameters ε_1, ε_2 and the parameter ε_3 are components of a vector-parameter $\bar{\varepsilon} = (\varepsilon_1, \varepsilon_2, \varepsilon_3)$ (or, briefly, the parameter $\bar{\varepsilon}$) and take arbitrary values in the open-closed interval $(0,1]$ and in the closed interval $[0,1]$, respectively. Assume that the coefficients $a(x,t)$, $c(x,t)$, $p(x,t)$ and also the right-hand side $F(x,t)$ for a fixed value of the parameter $\bar{\varepsilon}$ are sufficiently smooth on the set $\overline{G}$, moreover,

$$0 < a_0 \le a(x,t) \le a^0, \quad c(x,t) \ge 0, \quad 0 < p_0 \le p(x,t) \le p^0, \quad (x,t) \in \overline{G}, \tag{8.3a}$$

the boundary function $\varphi(x,t)$ is assumed to be sufficiently smooth on the sets $\overline{S}^L$ and S_0 and to be continuous on S, where $S^L = \Gamma\times(0,T]$, $S_0 = \overline{D}\times\{t = 0\}$, $S = S^L\cup S_0$; $\Gamma = \overline{D}\setminus D$.

By a solution of the initial-boundary value problem, we mean a function $u \in C(\overline{G})\cap C^{2,1}(G)$ that satisfies the differential equation on G and the initial-boundary condition on S. In the case when a discrete solution converges uniformly with respect to the parameters ε_i, for $i = 1, 2, 3$, we say that the solution converges $\bar{\varepsilon}$-uniformly. We are interested in numerical methods that converge $\bar{\varepsilon}$-uniformly in the maximum norm.

We now consider conditions imposed on the functions $F(x, t; \bar{\varepsilon})$ and $c(x, t)$. For simplicity we assume that

$$c(x, t) \geq c_0 > 0, \quad (x, t) \in \overline{G}, \quad for \;\; \varepsilon_3 \geq M (\varepsilon_1 + \varepsilon_2). \qquad (8.3b)$$

From the estimate of the solution to the boundary value problem (see the estimate (8.7) in Section 8.3), it follows that the solution is $\bar{\varepsilon}$-uniformly bounded if the function $F(x, t; \bar{\varepsilon})$ satisfies the condition

$$|F(x, t; \bar{\varepsilon})| \leq M (\varepsilon_1 + \varepsilon_2 + \varepsilon_3)^2, \quad (x, t) \in \overline{G}.$$

This *condition is unimprovable* with respect to the values ε_i, for $i = 1, 2, 3$.

Assume that the function $F(x, t; \bar{\varepsilon})$ takes a form

$$F(x, t; \bar{\varepsilon}) = (\varepsilon_1 + \varepsilon_2 + \varepsilon_3)^2 f(x, t), \quad (x, t) \in \overline{G}, \qquad (8.3c)$$

where $f(x, t)$ is a sufficiently smooth bounded function. Then the differential equation (8.2a) can be written as

$$L_{(8.4)} u(x, t) \equiv \left\{ \tilde{\varepsilon}_1^2 \, a(x, t) \frac{\partial^2}{\partial x^2} - \tilde{\varepsilon}_2^2 \, p(x, t) \frac{\partial}{\partial t} - \tilde{\varepsilon}_3^2 \, c(x, t) \right\} u(x, t) =$$

$$= f(x, t), \quad (x, t) \in G, \qquad (8.4)$$

where $\tilde{\varepsilon}_i = \varepsilon_i (\varepsilon_1 + \varepsilon_2 + \varepsilon_3)^{-1}$, for $i = 1, 2, 3$. Dimensionless components $\tilde{\varepsilon}_1$, $\tilde{\varepsilon}_2$, and $\tilde{\varepsilon}_3$ of the vector-parameter $\tilde{\varepsilon}$ satisfy the condition

$$\tilde{\varepsilon}_1, \tilde{\varepsilon}_2 > 0, \quad \tilde{\varepsilon}_3 \geq 0, \quad \tilde{\varepsilon}_1 + \tilde{\varepsilon}_2 + \tilde{\varepsilon}_3 = 1,$$

which is equivalent to the condition for the vector-parameter $\bar{\varepsilon}$

$$\varepsilon_1, \varepsilon_2 > 0, \quad \varepsilon_3 \geq 0, \quad \varepsilon_1 + \varepsilon_2 + \varepsilon_3 \approx O(1). \qquad (8.3d)$$

Our aim for the boundary value problem (8.2), (8.1) is to construct a difference scheme that converges $\bar{\varepsilon}$-uniformly with convergence rate close to that of a classical difference scheme for regular problems.

Since the character of the layers arising in the solution depends on the relations between the values $\varepsilon_1 + \varepsilon_2$ and ε_3, it is convenient to consider estimates of the solutions to problem (8.2), (8.1) separately in the case of the conditions

$$\varepsilon_3 \leq M (\varepsilon_1 + \varepsilon_2) \qquad (8.5)$$

and

$$\varepsilon_1 + \varepsilon_2 \leq M \varepsilon_3. \qquad (8.6)$$

When one or both of the parameters ε_1 and ε_2 tend to zero, the solution of the boundary value problem exhibits layers which are initial, boundary, and initial-boundary depending on the parameter ε_3. The initial-boundary layers are parabolic while the initial and boundary layers can be both parabolic and regular. See Remarks 8.3.1 and 8.3.2 in Section 8.3.

8.3   *A priori* estimates

We give *a priori estimates* for the problem solution and its derivatives that will be used for the construction and justification of $\bar{\varepsilon}$-uniformly convergent difference schemes. They are derived using techniques from the works [58, 67, 37, 148].

8.3.1. Using comparison theorems, we find that

$$|u(x,t)| \leq M\left[(\varepsilon_1 + \varepsilon_2 + \varepsilon_3)^{-2} \max_{\overline{G}} |F(x,t;\bar{\varepsilon})| + \max_S |\varphi(x,t)|\right], \qquad (8.7)$$

$$(x,t) \in \overline{G}.$$

The estimate (8.7) is unimprovable with respect to the parameters ε_i, for $i = 1, 2, 3$.

Assume that the problem data satisfy the condition

$$a, c, p, f \in C^{l+\alpha,(l+\alpha)/2}(\overline{G}), \qquad (8.8a)$$

$$\varphi \in C^{l+2+\alpha}(S_0) \cap C^{l+2+\alpha,(l+2+\alpha)/2}(\overline{S}^L) \cap C(S), \quad l \geq 0, \quad \alpha > 0,$$

and that on the set $S^c = \overline{S}^L \cap S_0$, the data satisfy compatibility conditions [67] that ensure

$$u \in C^{l+2+\alpha,(l+2+\alpha)/2}(\overline{G}) \qquad (8.8b)$$

for each fixed value of the parameters ε_i. Additional conditions will be given later.

Using *a priori* estimates up to the boundary [67, 37], we obtain the following estimate for the solution of the problem (8.2), (8.1)

$$\left|\frac{\partial^{k+k_0}}{\partial x^k \partial t^{k_0}} u(x,t)\right| \leq M \tilde{\varepsilon}_1^{-k} \tilde{\varepsilon}_2^{-2k_0}, \quad (x,t) \in \overline{G}, \ k + 2k_0 \leq K, \qquad (8.9)$$

where $K = l + 2, l = l_{(8.8)}$.

Theorem 8.3.1 *The solution of the initial-boundary value problem* (8.2), (8.1) *satisfies the estimate* (8.7). *If the hypothesis* (8.8) *is satisfied, then the estimate* (8.9) *is valid.*

8.3.2. When deriving the estimates based on asymptotic constructions, we assume that the following condition holds:

$$a, c, p, f \in C^{l+2+\alpha,(l+2+\alpha)/2}(\overline{G}), \qquad (8.10a)$$

$$\varphi \in C^{l+2+\alpha}(S_0) \cap C^{l+2+\alpha,(l+2+\alpha)/2}(\overline{S}^L) \cap C(S), \quad l \geq 0, \quad \alpha > 0,$$

moreover, on the set S^c the data satisfy compatibility conditions that ensure the smoothness

$$u \in C^{l+2+\alpha,(l+2+\alpha)/2}(\overline{G}). \tag{8.10b}$$

Here we find estimates for the smooth and singular components of the solution, assuming that the condition (8.5) holds.

Let

$$\varepsilon_2 \leq M \varepsilon_1. \tag{8.11}$$

In the case of conditions (8.5), (8.11) we have $\widetilde{\varepsilon}_1 \approx 1$.

Write the problem solution as the sum of the functions

$$u(x,t) = U(x,t) + W(x,t), \quad (x,t) \in \overline{G}, \tag{8.12}$$

where $U(x,t)$ and $W(x,t)$ are the regular and singular parts of the solution. The function $U(x,t)$ is the restriction to $\overline{G}$ of the function $U^e(x,t)$, $(x,t) \in \overline{G}^{e1}$, where $U^e(x,t)$ is the solution of the problem

$$L^e_{(8.4)} U^e(x,t) = f^e(x,t), \quad (x,t) \in G^{e1}, \quad U^e(x,t) = \varphi^e(x,t), \quad (x,t) \in S^{e1}.$$

Here $S^{e1} = S(G^{e1})$; the domain G^{e1} is an extension of the domain G beyond the set S_0, G^{e1} includes G together with its m_1-neighborhood; the coefficients of the operator $L^e_{(8.4)}$ and the function $f^e(x,t)$ are smooth extensions of those in (8.2), (8.1); $\varphi^e(x,t)$ is a smooth function, moreover, $\varphi^e(x,t) = \varphi(x,t)$, $(x,t) \in S^L$. The functions $f^e(x,t)$ and $\varphi^e(x,t)$ are assumed to be equal to zero outside an $(2^{-1}m_1)$-neighborhood of the set $\overline{G}$. The function $W(x,t)$ is the solution of the problem

$$L_{(8.4)}W(x,t) = 0, \quad (x,t) \in G, \quad W(x,t) = \varphi(x,t) - U(x,t), \quad (x,t) \in S.$$

Write the function $U^e(x,t)$ as the decomposition

$$U^e(x,t) = U^e_0(x,t) + v^e(x,t), \quad (x,t) \in \overline{G}^{e1},$$

where the functions $U^e_0(x,t)$ and $v^e(x,t)$ are solutions of the problems

$$L^e_{(8.13)} U^e_0(x,t) \equiv \left\{ \widetilde{\varepsilon}_1^2 \, a^e(x,t) \frac{\partial^2}{\partial x^2} - \widetilde{\varepsilon}_3^2 \, c^e(x,t) \right\} U^e_0(x,t) = \tag{8.13}$$

$$= f^e(x,t), \quad (x,t) \in \overline{G}^{e1} \setminus \overline{S}^{e1L},$$

$$U^e_0(x,t) = \varphi^e(x,t), \quad (x,t) \in \overline{S}^{e1L};$$

$$L^e_{(8.4)} v^e(x,t) = \widetilde{\varepsilon}_2^2 \, p^e(x,t) \frac{\partial}{\partial t} U^e_0(x,t), \quad (x,t) \in G^{e1},$$

$$v^e(x,t) = \varphi^e(x,t) - U^e_0(x,t), \quad (x,t) \in S^{e1}.$$

For the components $U_0^e(x,t)$ and $v^e(x,t)$ we obtain the estimates

$$\left| \frac{\partial^{k+k_0}}{\partial x^k \partial t^{k_0}} U_0^e(x,t) \right| \le M,$$

$$\left| \frac{\partial^{k+k_0}}{\partial x^k \partial t^{k_0}} v^e(x,t) \right| \le M \tilde{\varepsilon}_2^{2-2k_0}, \quad (x,t) \in \overline{G}^{e1}, \quad k + 2k_0 \le K.$$

From these estimates one has the estimate

$$\left| \frac{\partial^{k+k_0}}{\partial x^k \partial t^{k_0}} U(x,t) \right| \le M [1 + \tilde{\varepsilon}_2^{2-2k_0}], \quad (x,t) \in \overline{G}, \quad k + 2k_0 \le K. \quad (8.14a)$$

For the function $W(x,t)$ we obtain the estimate

$$\left| \frac{\partial^{k+k_0}}{\partial x^k \partial t^{k_0}} W(x,t) \right| \le M \tilde{\varepsilon}_2^{-2k_0} \exp(-m \tilde{\varepsilon}_2^{-2} t), \quad (x,t) \in \overline{G}, \quad (8.14b)$$

$$k + 2k_0 \le K,$$

where m is an arbitrary number satisfying the condition

$$m < m_{(8.14)}^0, \quad m_{(8.14)}^0 = (1 + M_{(8.5)})^{-2} (1 + M_{(8.11)})^{-2} \pi^2 d^{-2} a_0(p^0)^{-1}.$$

Let

$$\varepsilon_1 \le M \varepsilon_2. \quad (8.15)$$

In this case $\tilde{\varepsilon}_2 \approx 1$. Write the solution of the initial-boundary value problem as the sum of the functions

$$u(x,t) = U(x,t) + V(x,t), \quad (x,t) \in \overline{G}, \quad (8.16)$$

where $U(x,t)$ and $V(x,t)$ are the regular and singular parts of the solution. The function $U(x,t)$ is the restriction to $\overline{G}$ of the function $U^e(x,t)$, $(x,t) \in \overline{G}^{e2}$, where $U^e(x,t)$ is the solution of the problem

$$L_{(8.4)}^e U^e(x,t) = f^e(x,t), \quad (x,t) \in G^{e2}, \quad U^e(x,t) = \varphi^e(x,t), \quad (x,t) \in S^{e2}.$$

Here $S^{e2} = S(G^{e2})$; the domain G^{e2} is an extension of the domain G beyond the set S^L, the set G^{e2} includes the domain G together with its m_2-neighborhood; the coefficients of the operator $L_{(8.4)}^e$ and the right-hand side $f^e(x,t)$ are smooth extensions of the corresponding data in the problem (8.2), (8.1); $\varphi^e(x,t)$ is a smooth function, moreover, $\varphi^e(x,t) = \varphi(x,t)$, $(x,t) \in S_0$. The functions $f^e(x,t)$ and $\varphi^e(x,t)$ are assumed to be equal to zero outside an $(2^{-1}m_2)$-neighborhood of the set $\overline{G}$. The function $V(x,t)$ is the solution of the problem

$$L_{(8.4)} V(x,t) = 0, \quad (x,t) \in G, \quad V(x,t) = \varphi(x,t) - U(x,t), \quad (x,t) \in S.$$

For the functions $U(x,t)$ and $V(x,t)$ we obtain the estimates

$$\left| \frac{\partial^{k+k_0}}{\partial x^k \partial t^{k_0}} U(x,t) \right| \leq M \left[1 + \tilde{\varepsilon}_1^{2-k} \right], \tag{8.17}$$

$$\left| \frac{\partial^{k+k_0}}{\partial x^k \partial t^{k_0}} V(x,t) \right| \leq M \tilde{\varepsilon}_1^{-k} \exp \left(- m \tilde{\varepsilon}_1^{-1} r(x,\Gamma) \right), \quad (x,t) \in \overline{G},$$

$$k + 2k_0 \leq K,$$

where m is an arbitrary number and $r(x,\Gamma)$ is a distance from the point x to the boundary Γ. In the estimates (8.14), (8.17) we have $K = l + 2$, where $l = l_{(8.10)}$.

Theorem 8.3.2 *Let the data of the initial-boundary value problem* (8.2), (8.1) *and its solutions satisfy the condition* (8.10). *Then for the functions* $U(x,t)$, $V(x,t)$, $W(x,t)$, *i.e., the components in the representations* (8.12), (8.16), *under the conditions* (8.5), (8.11) *and* (8.5), (8.15), *the estimates* (8.14) *and* (8.17) *are satisfied, respectively.*

8.3.3. Next, we give estimates for the problem solution and its derivatives under the condition (8.6); in that case $\tilde{\varepsilon}_3 \approx 1$.

We assume that the following condition holds:

$$a, c, p, f \in C^{l+4,(l+4)/2}(\overline{G}), \tag{8.18a}$$

$$\varphi \in C^{l+2+\alpha}(S_0) \cap C^{l+2+\alpha,(l+2+\alpha)/2}(\overline{S}^L) \cap C(S), \quad l \geq 0, \quad \alpha > 0,$$

moreover, on the set S^c the data satisfy compatibility conditions that ensure the smoothness

$$u \in C^{l+2+\alpha,(l+2+\alpha)/2}(\overline{G}). \tag{8.18b}$$

The problem solution will now be decomposed in the following way:

$$u(x,t) = U(x,t) + W(x,t) + V(x,t) + Q(x,t), \quad (x,t) \in \overline{G}, \tag{8.19}$$

where $U(x,t)$ is the regular part of the solution while $W(x,t)$, $V(x,t)$, and $Q(x,t)$ are its singular components, i.e., the initial, boundary and initial-boundary ("corner") layers.

Extend the data of problem (8.2) beyond the boundary S (beyond the sets S^L and S_0) to a larger domain $\overline{G}^{e0}$: the function $\varphi^e(x,t)$ is smooth on $\overline{G}^{e0}$ with $\varphi^e(x,t) = \varphi(x,t)$, $(x,t) \in S$, and the functions $f^e(x,t)$ (which extends f) and $\varphi^e(x,t)$ are assumed to be equal to zero outside a sufficiently small neighborhood of the set $\overline{G}$. Define the function $U^e(x,t)$, $(x,t) \in \overline{G}^{e0}$, to be the solution of the problem

$$L^e_{(8.4)} U^e(x,t) = f^e(x,t), \quad (x,t) \in G^{e0}, \tag{8.20}$$

$$U^e(x,t) = \varphi^e(x,t), \quad (x,t) \in S^{e0}.$$

Now take $U(x,t)$ to be the restriction of $U^e(x,t)$ to $\overline{G}$.

Next, choose the domains G^{e1} and G^{e2} as extensions of G beyond the sets S^L and S_0, respectively, where $\overline{G}^{e1}$, $\overline{G}^{e2} \subset \overline{G}^{e0}$. The sets S_0 and S^L are parts of the boundaries of the extended domains G^{e1} and G^{e2}, respectively. The function $W(x,t)$ is the restriction to $\overline{G}$ of the function $W^e(x,t)$, $(x,t) \in \overline{G}^{e1}$, where $W^e(x,t)$ is the solution of the problem

$$L^e_{(8.4)} W^e(x,t) = 0, \qquad\qquad (x,t) \in G^{e1}, \qquad (8.21)$$

$$W^e(x,t) = \varphi^e(x,t) - U^e(x,t), \quad (x,t) \in S^{e1}.$$

The function $V(x,t)$ is the restriction to $\overline{G}$ of the function $V^e(x,t)$, $(x,t) \in \overline{G}^{e2}$, where $V^e(x,t)$ is the solution of the problem

$$L^e_{(8.4)} V^e(x,t) = 0, \qquad\qquad (x,t) \in G^{e2}, \qquad (8.22)$$

$$V^e(x,t) = \varphi^e(x,t) - U^e(x,t), \quad (x,t) \in S^{e2}.$$

The function $Q(x,t)$ is the solution of the problem

$$L_{(8.4)} Q(x,t) = 0, \qquad\qquad (x,t) \in G,$$

$$Q(x,t) = \varphi(x,t) - [U^e(x,t) + W^e(x,t) + V^e(x,t)], \quad (x,t) \in S.$$

We represent the function $U^e(x,t)$ (the solution of problem (8.20)) as an asymptotic expansion with respect to the parameter $\widetilde{\varepsilon}_1^{\,2}$:

$$U^e(x,t) = U^e_0(x,t) + \widetilde{\varepsilon}_1^2\, U^e_1(x,t) + v^e_U(x,t), \quad (x,t) \in \overline{G}^{e0}.$$

The functions $U^e_0(x,t)$ and $U^e_1(x,t)$ are solutions of the problems

$$L^e_{(8.23)} U^e_0(x,t) = f^e(x,t), \quad (x,t) \in \overline{G}^{e0} \setminus S^{e0}_0, \qquad (8.23)$$

$$U^e_0(x,t) = \varphi^e(x,t), \quad (x,t) \in S^{e0}_0;$$

$$L^e_{(8.23)} U^e_1(x,t) = -a^e(x,t)\frac{\partial^2}{\partial x^2} U^e_0(x,t), \quad (x,t) \in \overline{G}^{e0} \setminus S^{e0}_0,$$

$$U^e_1(x,t) = 0, \quad (x,t) \in S^{e0}_0.$$

Here $L^e_{(8.23)}$ is an extension of the operator $L \equiv -\widetilde{\varepsilon}_2^2\, p(x,t)\dfrac{\partial}{\partial t} - \widetilde{\varepsilon}_3^2\, c(x,t)$.

The problems (8.20) and (8.23) can be differentiated with respect to t. Estimating the functions $U^e_0(x,t)$, $U^e_1(x,t)$, and $v^e_U(x,t)$ in turn on $\overline{G}^{e0}$, one obtains the following estimate for the function $U(x,t)$

$$\left| \frac{\partial^{k+k_0}}{\partial x^k \partial t^{k_0}} U(x,t) \right| \le M\,[1 + \widetilde{\varepsilon}_1^{4-k}], \quad (x,t) \in \overline{G}, \quad k + 2k_0 \le K, \qquad (8.24a)$$

where $K = l + 2$ with $l = l_{(8.18)}$.

Decompose the solutions of (8.21) and (8.22) as the sums

$$W^e(x,t) = W_0^e(x,t) + v_W^e(x,t), \quad (x,t) \in \overline{G}^{e1},$$
$$V^e(x,t) = V_0^e(x,t) + v_V^e(x,t), \quad (x,t) \in \overline{G}^{e2}, \tag{8.25}$$

where the functions $W_0^e(x,t)$, $(x,t) \in \overline{G}^{e1}$ and $V_0^e(x,t)$, $(x,t) \in \overline{G}^{e2}$ are solutions of the problems

$$L_{(8.23)}^e W_0^e(x,t) = 0, \qquad\qquad (x,t) \in \overline{G}^{e1} \setminus S^{e1L},$$
$$W_0^e(x,t) = \varphi^e(x,t) - U^e(x,t), \quad (x,t) \in S_0^{e1L};$$
$$L_{(8.13)}^e V_0^e(x,t) = 0, \qquad\qquad (x,t) \in \overline{G}^{e2} \setminus S_0^{e2},$$
$$V_0^e(x,t) = \varphi^e(x,t) - U^e(x,t), \quad (x,t) \in S_0^{e2}.$$

Estimating the components in (8.25), we find the estimates

$$\left| \frac{\partial^{k+k_0}}{\partial x^k \partial t^{k_0}} W(x,t) \right| \leq M \, \widetilde{\varepsilon}_2^{-2k_0} \, [1 + \widetilde{\varepsilon}_2^{4-2k_0}] \exp\left(-m_2 \, \widetilde{\varepsilon}_2^{-2} \, t\right),$$
$$\left| \frac{\partial^{k+k_0}}{\partial x^k \partial t^{k_0}} V(x,t) \right| \leq M \, \widetilde{\varepsilon}_1^{-k_1} \, [1 + \widetilde{\varepsilon}_2^{4-2k_0}] \exp\left(-m_1 \, \widetilde{\varepsilon}_1^{-1} \, r(x,\Gamma)\right), \tag{8.24b}$$
$$(x,t) \in \overline{G}, \quad k + 2k_0 \leq K.$$

For the component $Q(x,t)$ we obtain the estimate

$$\left| \frac{\partial^{k+k_0}}{\partial x^k \partial t^{k_0}} Q(x,t) \right| \leq M \, \widetilde{\varepsilon}_1^{-k_1} \, \widetilde{\varepsilon}_2^{-2k_0} \, [1 + \widetilde{\varepsilon}_2^{4-2k_0}] \times \tag{8.24c}$$

$$\times \min\left[\exp\left(-m_1 \, \widetilde{\varepsilon}_1^{-1} \, r(x,\Gamma)\right), \exp(-m_2 \, \widetilde{\varepsilon}_2^{-2} \, t) \right], \quad (x,t) \in \overline{G}, \quad k + 2k_0 \leq K.$$

In (8.24b and 8.24c) we have $K = l + 2$, where $l = l_{(8.18)}$, while m_1 and m_2 are arbitrary numbers satisfying the condition $m_i = m_{i(8.24)} < m_{i(8.24)}^0$, for $i = 1, 2$, where

$$m_{1(8.24)}^0 = \left(1 + M_{(8.6)}\right)^{-1} c_0^{1/2} \, (a^0)^{-1/2}, \quad m_{2(8.24)}^0 = \left(1 + M_{(8.6)}\right)^{-2} c_0 \, (p^0)^{-1}.$$

Theorem 8.3.3 *Let conditions (8.18) be fulfilled for the data of the boundary value problem (8.2), (8.1) and its solution. Then, under the condition (8.6), the functions $U(x,t)$, $W(x,t)$, $V(x,t)$, and $Q(x,t)$, i.e., the components in the representation (8.19), satisfy the estimates (8.24).*

Remark 8.3.1 An examination of the main terms in the asymptotic representations of the singular components of the problem solution (see also [148]) reveals that boundary layers appear when $\widetilde{\varepsilon}_1 = o(1)$ while initial layers appear when $\widetilde{\varepsilon}_2 = o(1)$. If both conditions $\widetilde{\varepsilon}_1 = o(1)$ and $\widetilde{\varepsilon}_2 = o(1)$ are fulfilled, then the solution also contains initial-boundary layers. The initial-boundary layers are parabolic. The initial and boundary layers are parabolic if either $\tilde{\varepsilon}_1 \approx 1$ or $\tilde{\varepsilon}_2 \approx 1$; otherwise these layers are regular.   ∎

Remark 8.3.2 Unlike the parameters ε_1 and ε_2, the parameter ε_3 does not belong to singularly perturbed. But the parameter ε_3 determines types of the arising layers just as their appearance. Thus, under the condition $\varepsilon_1, \varepsilon_2 \approx \varepsilon$, $\varepsilon = o(1)$, the initial-boundary value problem is regular if $\varepsilon_3 = O(\varepsilon)$ and is singularly perturbed if $\varepsilon = o(\varepsilon_3)$. Under the condition either $\varepsilon_1 = o(\varepsilon_2)$ or $\varepsilon_2 = o(\varepsilon_1)$, the initial and boundary layers are parabolic if $\varepsilon_3 = O(\varepsilon_1 + \varepsilon_2)$. But if $\varepsilon_1 + \varepsilon_2 = o(\varepsilon_3)$ these layers are regular and, moreover, the initial-boundary parabolic layers arise.   ∎

8.4   Grid approximations of the initial-boundary value problem (8.2), (8.1)

8.4.1. We now construct a finite difference scheme that uses a classical approximation of the boundary value problem (8.2), (8.1) on rectangular grids. On the set $\overline{G}$ we introduce the grid

$$\overline{G}_h = \overline{D}_h \times \overline{\omega}_0 = \overline{\omega} \times \overline{\omega}_0, \tag{8.26}$$

where $\overline{\omega}$ and $\overline{\omega}_0$ are, in general, nonuniform meshes on the intervals $[0, d]$ and $[0, T]$, respectively. Set $h^i = x^{i+1} - x^i$, $h = \max_i h^i$ for $x^i, x^{i+1} \in \overline{\omega}$ and $h_t^j = t^{j+1} - t^j$, $h_t = \max_j h_t^j$ for $t^j, t^{j+1} \in \overline{\omega}_0$. Let $N + 1$ and $N_0 + 1$ be the number of nodes in the meshes $\overline{\omega}$ and $\overline{\omega}_0$. Assume that $h \leq MN^{-1}$, $h_t \leq MN_0^{-1}$. For the problem (8.2), (8.1) on the grid $\overline{G}_h$, we consider the difference scheme

$$\Lambda_{(8.27)} z(x, t) = F(x, t; \varepsilon), \ (x, t) \in G_h, \quad z(x, t) = \varphi(x, t), \ (x, t) \in S_h. \tag{8.27}$$

Here $G_h = G \cap \overline{G}_h$, $S_h = S \cap \overline{G}_h$,

$$\Lambda_{(8.27)} \equiv \varepsilon_1^2 \, a(x, t) \, \delta_{\overline{x}\widehat{x}} - \varepsilon_2^2 \, p(x, t) \, \delta_{\overline{t}} - \varepsilon_3^2 \, c(x, t), \quad (x, t) \in G_h;$$

$\delta_{\overline{x}\widehat{x}} z(x, t)$ and $\delta_{\overline{t}} z(x, t)$ are the second-order and the first-order (backward) difference derivatives, e.g.,

$$\delta_{\overline{x}\widehat{x}} z(x, t) = 2(h^i + h^{i-1})^{-1}[\delta_x z(x, t) - \delta_{\overline{x}} z(x, t)],$$

$$\delta_x z(x, t) = \left(h^i\right)^{-1} \left(z(x^{i+1}, t) - z(x, t)\right),$$

$$\delta_{\overline{x}} z(x, t) = \left(h^{i-1}\right)^{-1} \left(z(x, t) - z(x^{i-1}, t)\right), \quad x = x^i,$$

$$\delta_{\overline{x}} z(x, t) = \left(h_t^{j-1}\right)^{-1} \left(z(x, t) - z(x, t^{j-1})\right), \quad t = t^j.$$

The difference operator $\Lambda_{(8.27)}$ is $\bar{\varepsilon}$-uniformly monotone [108].

By using comparison theorems, one can verify that the solution of the problem (8.27), (8.26) is $\bar{\varepsilon}$-uniformly bounded:

$$|z(x,t)| \leq M, \quad (x,t) \in \overline{G}_h.$$

Taking into account the estimates of Theorem 8.3.1 for $K = 4$, one can show that

$$|u(x,t) - z(x,t)| \leq M \left[(\varepsilon_1 (\varepsilon_1 + \varepsilon_2 + \varepsilon_3)^{-1} + N^{-1})^{-1} N^{-1} + \right. \qquad (8.28a)$$

$$\left. + (\varepsilon_2^2 (\varepsilon_1 + \varepsilon_2 + \varepsilon_3)^{-2} + N_0^{-1})^{-1} N_0^{-1} \right], \quad (x,t) \in \overline{G}_h.$$

The same estimate, written down using the dimensionless parameters $\widetilde{\varepsilon}_i$, takes the compact form

$$|u(x,t) - z(x,t)| \leq M \left[(\widetilde{\varepsilon}_1 + N^{-1})^{-1} N^{-1} + (\widetilde{\varepsilon}_2^2 + N_0^{-1})^{-1} N_0^{-1} \right], \quad (8.28b)$$

$$(x,t) \in \overline{G}_h.$$

Thus the scheme (8.27), (8.26) converges under the condition $N^{-1} \ll \widetilde{\varepsilon}_1$, $N_0^{-1} \ll \widetilde{\varepsilon}_2^2$, or more precisely,

$$N^{-1} = o(\varepsilon_1 (\varepsilon_1 + \varepsilon_2 + \varepsilon_3)^{-1}), \quad N_0^{-1} = o(\varepsilon_2^2 (\varepsilon_1 + \varepsilon_2 + \varepsilon_3)^{-2}). \qquad (8.29)$$

On the uniform grid

$$\overline{G}_h, \qquad (8.30)$$

we have the estimate that is written down using the parameters ε_i:

$$|u(x,t) - z(x,t)| \leq M \left[(\varepsilon_1 (\varepsilon_1 + \varepsilon_2 + \varepsilon_3)^{-1} + N^{-1})^{-2} N^{-2} + \right. \qquad (8.31a)$$

$$\left. + (\varepsilon_2^2 (\varepsilon_1 + \varepsilon_2 + \varepsilon_3)^{-2} + N_0^{-1})^{-1} N_0^{-1} \right], \quad (x,t) \in \overline{G}_h,$$

and the estimate that is written down using the dimensionless parameters $\widetilde{\varepsilon}_i$:

$$|u(x,t) - z(x,t)| \leq M \left[(\widetilde{\varepsilon}_1 + N^{-1})^{-2} N^{-2} + (\widetilde{\varepsilon}_2^2 + N_0^{-1})^{-1} N_0^{-1} \right], \qquad (8.31b)$$

$$(x,t) \in \overline{G}_h.$$

It follows that the finite difference scheme (8.27), (8.26) converges under the condition (8.29). Note that from the estimates (8.28) and (8.31), it follows that the finite difference schemes (8.27), (8.26) and (8.27), (8.30) under the condition (8.29) converge ε_3-uniformly.

Theorem 8.4.1 *Let for the data of the boundary value problem (8.2), (8.1), the conditions (8.3) be satisfied and assume for the problem solution that the estimates of Theorem 8.3.1 are fulfilled for $K = 4$. Then, under the condition (8.29), the solutions of the finite difference scheme (8.27) on the grids (8.26) and (8.30) converge to the solution of the boundary value problem with the estimates (8.28) and (8.31), respectively.*

8.4.2. We now construct a grid that condenses in the boundary and initial layer regions and on which the solution of the finite difference scheme converges $\bar{\varepsilon}$-uniformly under the condition (8.5) (i.e., $\varepsilon_3 \leq M (\varepsilon_1 + \varepsilon_2)$). On the set $\overline{G}$ we introduce the grid

$$\overline{G}_h = \overline{D}^s_h \times \overline{w}^s_0 = \overline{w}^s \times \overline{w}^s_0, \tag{8.32a}$$

where $\overline{w}^s = \overline{w}^s(\sigma_1)$ and $\overline{w}^s_0 = \overline{w}^s_0(\sigma_2)$ are piecewise-uniform meshes on $[0, d]$ and $[0, T]$, respectively; here σ_1 and σ_2 are parameters depending on N, N_0, and $\bar{\varepsilon}$. The mesh sizes in $\overline{w}^s$ (see, e.g., [138, 135, 137]) are $h^{(1)} = 4\sigma_1 N^{-1}$ on the intervals $[0, \sigma_1]$ and $[d - \sigma_1, d]$, and $h^{(2)} = 2(d - 2\sigma_1)N^{-1}$ on $[\sigma_1, d - \sigma_1]$. The mesh sizes in $\overline{w}^s_0$ are $h^{(1)}_0 = 2\sigma_2 N_0^{-1}$ on the interval $[0, \sigma_2]$ and $h^{(2)}_0 = 2(T - \sigma_2)N_0^{-1}$ on $[\sigma_2, T]$. The values σ_1 and σ_2 are specified by

$$\begin{aligned}
\sigma_1 &= \sigma_{1(8.32)}(\bar{\varepsilon}, N) = \min \left[4^{-1} d,\; M_1\, \varepsilon_1 \, (\varepsilon_1 + \varepsilon_2 + \varepsilon_3)^{-1} \ln N \right], \\
\sigma_2 &= \sigma_{2(8.32)}(\bar{\varepsilon}, N_0) = \min \left[2^{-1} T,\; M_2\, \varepsilon_2^2 \, (\varepsilon_1 + \varepsilon_2 + \varepsilon_3)^{-2} \ln N_0 \right],
\end{aligned} \tag{8.32b}$$

where

$$M_1,\ M_2 \quad \text{are arbitrary constants.} \tag{8.32c}$$

The grid $\overline{G}_{h(8.32)}$ is constructed. Thus, the grid $\overline{G}_{h(8.32)}$ is defined by the parameters N, N_0, $\bar{\varepsilon}$ and by the constants M_1 and M_2, i.e.,

$$\overline{G}_{h(8.32)} = \overline{G}_{h(8.32)}(N, N_0, \bar{\varepsilon};\, M_1, M_2) = \overline{G}_{h(8.32)}(M_1, M_2).$$

Under the condition (8.5), from the estimates of Theorem 8.3.2 (for $K = 4$) one can deduce $\bar{\varepsilon}$-uniform convergence of the solution of the finite difference scheme (8.27), (8.32) to the solution of the initial-boundary value problem (8.2), (8.1). The convergence rate of the solution of the finite difference scheme is estimated using a technique from [138, 135, 137]. In the case when the value M_2 satisfies the condition

$$M_2 > (m^0_{(8.14)})^{-1}, \tag{8.33}$$

we obtain the $\bar{\varepsilon}$-uniform estimate

$$|u(x,t) - z(x,t)| \leq M \left[N^{-2} \ln^2 N + N_0^{-1} \ln N_0 \right], \quad (x,t) \in \overline{G}_h. \tag{8.34}$$

Theorem 8.4.2 *Let the components of the vector-parameter $\bar{\varepsilon}$ satisfy the condition (8.5). For the components of the solution of the boundary value problem (8.2), (8.1) in the representation (8.12), (8.16), assume that the estimates of Theorem 8.3.2 hold for $K = 4$. Then the solution of the finite difference scheme (8.27), (8.32) converges $\bar{\varepsilon}$-uniformly. If in addition the condition (8.33) is satisfied then the estimate (8.34) is valid.*

8.4.3. Under the condition (8.6), from the *a priori* estimates of Theorem 8.3.3 (for $K = 4$), it follows that the finite difference scheme (8.27), (8.32) converges $\bar{\varepsilon}$-uniformly. In the case of the grid $\overline{G}_{h(8.32)}$ under the condition

$$M_1 > 2\,(m^0_{1(8.24)})^{-1}, \quad M_2 > (m^0_{2(8.24)}) \tag{8.35}$$

the solution of the finite difference scheme satisfies the estimate (8.34).
    The following theorem holds.

Theorem 8.4.3 *Let the components of the vector-parameter $\bar{\varepsilon}$ satisfy the condition (8.6). For the components of the solution of the boundary value problem (8.2), (8.1) in the representation (8.19), assume that the estimates of Theorem 8.3.3 hold for $K = 4$. Then the solution of the finite difference scheme (8.27), (8.32) is $\bar{\varepsilon}$-uniformly convergent to the solution of (8.2), (8.1). The solution of the finite difference scheme {(8.27), (8.32), (8.35)} satisfies the estimate (8.34).*

8.4.4. From the results of 8.4.2, 8.4.3 it follows that in the case of the estimates of Theorems 8.3.2 and 8.3.3 for $K = 4$, the solution of the finite difference scheme (8.27), (8.32) converges $\bar{\varepsilon}$-uniformly.
    On the grid $\overline{G}_{h(8.32)}$ under the condition

$$M_1 > 2\,(m^0_{1(8.24)})^{-1}, \quad M_2 > \max\left[(m^0_{(8.14)})^{-1},\ (m^0_{2(8.24)})^{-1}\right], \tag{8.36}$$

the solution of the finite difference scheme satisfies the estimate (8.34).
    The following theorem is valid.

Theorem 8.4.4 *Let for the components of the solution of the boundary value problem (8.2), (8.1) in the representations (8.12), (8.16), and (8.19) the estimates of Theorems 8.3.2 and 8.3.3 be, respectively, satisfied for $K = 4$. Then the solution of the finite difference scheme (8.27), (8.32) converges to the solution of the boundary value problem $\bar{\varepsilon}$-uniformly. The solution of the finite difference scheme {(8.27), (8.32), (8.36)} satisfies the estimate (8.34).*

    For the finite difference scheme {(8.27), (8.32), (8.33)}, {(8.27), (8.32), (8.35)} and {(8.27), (8.32), (8.36)}, except the estimate (8.34), the classical estimate (8.31) also holds. Hence, for these schemes one has also the $\bar{\bar{\varepsilon}}$-dependent estimate

$$|u(x,t) - z(x,t)| \le M\left[(\tilde{\varepsilon}_1 + \ln^{-1} N)^{-2}\, N^{-2} + (\tilde{\varepsilon}_2^2 + \ln^{-1} N_0)^{-1}\, N_0^{-1}\right],$$

$$(x,t) \in \overline{G}_h.$$

For fixed values of the parameter $\bar{\bar{\varepsilon}}$ these schemes converge at the rate

$$\mathcal{O}\left(N^{-2} + N_0^{-1}\right).$$

Chapter 9

Application of widths for construction of difference
schemes for problems with moving boundary layers

A grid approximation of a boundary value problem is considered for a singularly perturbed parabolic reaction-diffusion equation in domains with *moving boundaries*. Using *widths* similar to Kolmogorov's widths, necessary and sufficient conditions for the ε-uniform convergence of approximations to the solution of the boundary value problem are established. Taking into account these conditions, a difference scheme is constructed on a mesh which is piecewise uniform in a coordinate system adapted to the moving boundary. This scheme *converges ε-uniformly in the maximum norm*. Ability to apply *the technique based on the widths* for an investigation of ε-uniformly convergent difference schemes is discussed.

9.1 Introduction

Special numerical methods for singularly perturbed parabolic equations have been studied intensively only for problems with stationary boundary and interior layers. There are some specific features in constructing special difference schemes in the case of moving boundary and interior layers. For the initial singularly perturbed problem with a moving concentrated source in a neighborhood of which a *moving interior layer* appears, special ε-uniformly convergent difference schemes were constructed in [159, 172, 203]. To construct such schemes in a neighborhood of the trajectory of the moving source, nonrectangular grids were used that condense along the x-axis. These schemes are fairly complicated, which draws our attention to methods for constructing simpler schemes and alternative numerical methods (based on *a posteriori* adapted grids) that converge ε-uniformly. Problems with the moving layers (see, e.g., [159, 172, 203, 176] in the case of interior layers and [188, 194] in the case of boundary layers) have solutions with more complicated behaviour compared to problems with stationary layers. Such behaviour of solutions impose more stronger requirements to grid constructs when constructing ε-uniformly convergent finite difference schemes.

In the monograph [107] (see also [108]), A.A. Samarskii have pointed out that, when constructing numerical methods for regular boundary value problems with sufficiently complicated solutions, in order to prevent appearance of nonphysical effects in the computed solutions, it is necessary to use discrete approximations that inherit monotonicity of the boundary value problem. For singularly perturbed boundary value problems, such natural requirements bring to quite complicated finite difference schemes (in the presence of mixed derivatives in equations, see, e.g., schemes in [112] for regular problems, and in [138] for singularly perturbed problems). Therefore, it is important to impose conditions on the schemes under construction, which are sufficient and close to necessary ones, that guarantee monotonicity and ε-uniform convergency of these schemes.

The use of widths that are similar to Kolmogorov's widths allowed to find and study necessary and sufficient conditions for grid constructs and, as a result, to construct schemes that converge ε-uniformly in the maximum norm. This was applied in the case of parabolic equations in [167, 172] (for problems with interior layers) and in [188, 194] (for problems with boundary layers) and also in the case of elliptic equations in [183] (for problems in domains with curvilinear boundaries).

When studying ε-uniformly convergent numerical methods, the potential usefulness of widths was already mentioned by N.S. Bakhvalov in the 1970s. An application of widths to the study of optimal L_2-convergence rates for approximations of solutions to singularly perturbed elliptic problems is given in [60, 61]; see also [82] and the bibliographies of these papers. But the widths in spaces with L_2-norm do not allow to investigate ε-uniform convergence of numerical methods in the maximum norm.

In this chapter, the boundary value problem (studied in [188, 194]) is considered for a singularly perturbed parabolic reaction-diffusion equation in a domain whose boundaries move in the positive direction of the x-axis. For small values of the parameter ε (i.e., the coefficient of the highest-order derivative in the equation, $\varepsilon \in (0, 1]$), a *moving boundary layer* appears in a neighborhood of the left lateral boundary S_1^L. The derivatives of the solution in x and t grow unboundedly in a neighborhood of the boundary layer as $\varepsilon \to 0$. For problems of this type, classical finite difference schemes based on uniform grids converge only when $N^{-1} + N_0^{-1} \ll \varepsilon$, where N and N_0 define the number of mesh points in x and t (see the statement of Theorem 9.4.1 in Section 9.4). It turns out that, in the class of difference schemes based on rectangular grids that condense in a neighborhood of S_1^L with respect to x and t, there are *no convergent schemes even under the condition* $P_0^{-1} \approx \varepsilon^{1/2}$, where P_0 is the number of nodes in the meshes used and $P_0 \approx N N_0$ (see Remark 9.5.4 to Theorem 9.5.2 in Section 9.5).

Here, applying the technique based on the widths, conditions are found which are necessary and sufficient conditions for the ε-uniform convergence of a difference scheme on grids that are not the tensor product of meshes in x and t. These conditions are used to construct (on a piecewise uniform mesh in a coordinate system adapted to the moving boundary) a scheme that converges ε-uniformly in the maximum norm. In Section 9.7, some remarks and generalizations are given related to applications of *the technique based on the widths* for an investigation of ε-uniformly convergent difference schemes.

Some aspects of the construction and investigation of ε-uniformly convergent difference schemes in the maximum norm for the problem under consideration by means of widths are discussed in [188, 194]. A difference scheme and numerical experiments for a problem with moving boundaries are given in [27]. For parabolic convection-diffusion problems with stationary boundaries, widths were applied in [195] to construct on piecewise-uniform meshes finite difference schemes such that are optimal with respect to the ε-uniform convergency rate.

9.2 A boundary value problem for a singularly perturbed parabolic reaction-diffusion equation

9.2.1 Problem (9.2), (9.1)

In the domain $\overline{G}$ with the boundary $S = \overline{G} \setminus G$, where

$$G = \{(x,t) : \beta_1(t) < x < \beta_2(t), \quad t \in (0,T]\}, \tag{9.1}$$

we consider the boundary value problem for the singularly perturbed parabolic equation

$$L\,u(x,t) \equiv \left\{ \varepsilon \frac{\partial^2}{\partial x^2} - \frac{\partial}{\partial t} \right\} u(x,t) = f(x,t), \quad (x,t) \in G, \tag{9.2}$$

$$u(x,t) = \varphi(x,t), \quad (x,t) \in S.$$

Here $f(x,t)$, $(x,t) \in \overline{G}$, $\varphi(x,t)$, $(x,t) \in S$, and $\beta_i(t)$, $t \in [0,T]$, $i = 1,2$, are sufficiently smooth functions that satisfy the conditions

$$|f(x,t)| \le M, \quad (x,t) \in \overline{G}, \quad |\varphi(x,t)| \le M, \quad (x,t) \in S; \tag{9.3}$$

$$0 < v_0 \le (d/dt)\,\beta_i(t) \equiv v_i(t) \le v^0, \quad m_1 \le \beta_2(t) - \beta_1(t) \le M_1,$$

$$t \in [0,T], \quad i = 1,2,$$

moreover, $\beta_1(0) = 0$, $\beta_2(0) = d$; the parameter ε takes arbitrary values in the open-closed interval $(0,1]$, and the derivatives $\beta_i'(t)$ specify the velocity of the moving lateral boundaries. Assume that the boundary S consists of the sets S^L and S_0, i.e., $S = S_0 \cup S^L$, where S_0 is the lower basis of G with $S_0 = \overline{S}_0$ and S^L is the lateral boundary; $S^L = S_1^L \cup S_2^L$, where S_1^L and S_2^L are, respectively, the left and the right boundaries of the set G.

We assume that compatibility conditions guarantee the sufficient smoothness of the solution for fixed ε (see [67]) on the set $S^c = S_0 \cap \overline{S}^L$, i.e., at the corner points $(0,0)$ and $(d,0)$.

As $\varepsilon \to 0$, a moving boundary layer appears in a neighborhood of the set S_1^L. This layer decreases exponentially when moving away from S_1^L as x increases and/or t decreases (see estimate (9.12) in Section 9.3).

9.2.2   Some definitions

Unlike regular problems, in the case of singularly perturbed problems the ε-uniform convergence of the grid solution $z(x,t)$ at the nodes of the grid $\overline{G}_h$ in the maximum discrete norm is, in general, inadequate to describe the ε-uniform convergence of the approximation constructed on the set $\overline{G}$. The convergence of the grid solution on the grid $\overline{G}_h$ does not imply the convergence of its interpolants on the set $\overline{G}$.

We give some *definitions*.

In the case when the interpolant $\overline{z}_{(9.22)}(x,t)$, $(x,t) \in \overline{G}$, converges on $\overline{G}$, we say that the difference *scheme resolves the boundary value problem* (for some values of the parameter ε); otherwise, we say that the difference scheme does not resolve the boundary value problem. When the interpolant $\overline{z}_{(9.22)}(x,t)$, $(x,t) \in \overline{G}$, converges on $\overline{G}$ ε-uniformly, we say that *the difference scheme resolves the boundary value problem ε-uniformly.*

Let $\overline{G}_h$ be some grid, and let $\overline{u}^h(x,t)$, $(x,t) \in \overline{G}$, be a linear interpolant constructed using the solution $u(x,t)$ of the boundary value problem at the nodes of the grid $\overline{G}_h$. If some such interpolant $\overline{u}^h(x,t)$ converges on $\overline{G}$ as N and $N_0 \to \infty$ (for some values of the parameter ε), we say that the *grid $\overline{G}_h$ is informative* (and if $\overline{u}^h(x,t)$ converges ε-uniformly then the grid $\overline{G}_h$ is ε-uniformly informative); otherwise, we say that the grid $\overline{G}_h$ is noninformative. Here N and N_0 define the number of mesh points in x and t, respectively.

The *informativity of the grid $\overline{G}_h$* is a *necessary condition* for the boundary value problem (9.2), (9.1) *to be resolved* by the difference scheme on $\overline{G}_h$.

We say that the solution of a difference scheme (or, briefly, the scheme itself) converges if the grid solution converges on $\overline{G}_h$ and the difference scheme resolves the boundary value problem. But if the solution converges only on the grid $\overline{G}_h$, however, the solvability of the boundary value problem is, in general, not assumed; we say that the *solution (or, the scheme) converges on the grid $\overline{G}_h$*.

Let $E_\varepsilon = \{\varepsilon : \varepsilon \in (0,1]\}$. Let $E_{\overline{N}}$ be a subset of the set of pairs of positive integers (N, N_0) satisfying the condition

$$N, N_0 \geq M_0.$$

Let functions $\psi_i(N^{-1}, N_0^{-1}, \varepsilon)$, for $i = 1, 2$, be defined on the set $E_{\overline{N}, \varepsilon} = E_{\overline{N}} \times E_\varepsilon$ and satisfy $\psi_i(N^{-1}, N_0^{-1}, \varepsilon) > 0$. The notation

$$\psi_1(N^{-1}, N_0^{-1}, \varepsilon) = \widehat{O}\big(\psi_2(N^{-1}, N_0^{-1}, \varepsilon)\big) \quad on \ E_{\overline{N}, \varepsilon}$$

means that one can find a point $\left(\tilde{N}_1^{-1}, \tilde{N}_2^{-1}, \tilde{\varepsilon}\right)$ such that the relation

$$\psi_1\left(N^{-1}, N_0^{-1}, \varepsilon\right) \left[\psi_2(N^{-1}, N_0^{-1}, \varepsilon)\right]^{-1} \to 0$$

$$as \ \left(N^{-1}, N_0^{-1}, \varepsilon\right) \to \left(\tilde{N}^{-1}, \tilde{N}_0^{-1}, \tilde{\varepsilon}\right),$$

$$with \ \left(N^{-1}, N_0^{-1}, \varepsilon\right), \ \left(\tilde{N}^{-1}, \tilde{N}_0^{-1}, \tilde{\varepsilon}\right) \in E_{\overline{N}, \varepsilon},$$

is fulfilled. For a grid function $z(x, t)$, $(x, t) \in \overline{G}_h$, i.e., a solution of a difference scheme, assume that the estimate

$$|u(x, t) - z(x, t)| \leq M \mu\left(N^{-1}, N_0^{-1}, \varepsilon\right), \quad (x, t) \in \overline{G}_h,$$

holds. Here the grid $\overline{G}_h$ is a tensor product of meshes in x and t, where $N+1$ and N_0+1 are the number of nodes in x and t, respectively. We say that this *estimate is unimprovable* with respect to the values N, N_0, ε if the estimate

$$|u(x, t) - z(x, t)| \leq M \mu_0\left(N^{-1}, N_0^{-1}, \varepsilon\right), \quad (x, t) \in \overline{G}_h,$$

is, in general, not valid in the case when

$$\mu_0\left(N^{-1}, N_0^{-1}, \varepsilon\right) = \widehat{o}\left(\mu(N^{-1}, N_0^{-1}, \varepsilon)\right) \ on \ E_{\overline{N}, \varepsilon}.$$

Assume that a solution of a difference scheme converges to the solution of the boundary value problem as N, $N_0 \to \infty$ and $\varepsilon \in E_\varepsilon$ provided that

$$N^{-1}, N_0^{-1} = o\left(\varepsilon^\nu\right), \quad \varepsilon \in E_\varepsilon,$$

but convergence does not, in general, occur under the condition

$$N^{-1}, N_0^{-1} = \mathcal{O}(\varepsilon^\nu).$$

In this case, we say that the difference scheme *converges with defect ν* with respect to the parameter ε as N, $N_0 \to \infty$ (or, briefly, the scheme converges with defect ν). When $\nu = 0$, the convergence is ε-uniform.

Let $\overline{G}_h$ be a grid (in general, not a rectangular one) on the set $\overline{G}$, and let P_0 be the number of mesh points in $\overline{G}_h$. Let the grid function $z(x, t)$, $(x, t) \in \overline{G}_h$, be a solution of some difference scheme that converges with the estimate

$$|u(x, t) - z(x, t)| \leq M \mu\left(P_0^{-1}, \varepsilon\right), \quad (x, t) \in \overline{G}_h. \tag{9.4}$$

Unimprovability of an estimate with respect to the values P_0, ε is defined similarly to the definition with respect to the values N, N_0, ε.

If the solution of a difference scheme converges to the solution of the boundary value problem, as $P_0 \to \infty$, $\varepsilon \in E_\varepsilon$, under the condition

$$P_0^{-1/2} = o(\varepsilon^\nu), \quad \varepsilon \in E_\varepsilon,$$

but does not, in general, converge under the condition

$$P_0^{-1/2} = \mathcal{O}(\varepsilon^{\nu}),$$

then the scheme is said to be convergent *with defect ν* (as $P_0 \to \infty$). Note that, as $N \approx N_0$, we have $N^{-1}, N_0^{-1} \approx P_0^{-1/2}$, which motivates the given definition of convergence with defect ν for the scheme on a mesh with P_0 points.

If the value ν can be chosen arbitrarily small, we say that the difference scheme, which is controlled by the value ν, converges *almost ε-uniformly with defect ν* (or, briefly, *almost ε-uniformly*). For example, the defect of the ε-uniform convergence to the classical scheme (9.17), (9.19) (with respect to N, N_0, and ε) equals 1.

9.2.3   The aim of the research

Errors in the solutions of finite difference schemes based on classical approximations of the problem (9.2), (9.1) depend on the parameter ε and become small only when ε is essentially greater than, e.g., the "effective" step-sizes of the meshes in x and t. So, by virtue of estimates (9.20) and (9.24), the classical difference scheme (9.17), (9.19) (see Section 9.4) converges under the condition $N^{-1} + N_0^{-1} \ll \varepsilon$, or more precisely,

$$\varepsilon^{-1} = o(\min[N, N_0]), \quad N, N_0 \to \infty. \tag{9.5}$$

If this condition is violated, the solution of the difference scheme does not converge to the solution of the problem (9.2), (9.1). Condition (9.5) is more restrictive than the condition $N^{-1} \ll \varepsilon$, or more precisely,

$$\varepsilon^{-1} = o(N), \quad N, N_0 \to \infty, \tag{9.6}$$

i.e., the convergence condition of the classical scheme for problems in domains with fixed boundaries. Note that there are no constraints on the t-mesh-size in the condition (9.6).

Thus, because of the above behavior of the grid solutions that approximate the solution of the differential problem with a moving boundary layer, it is necessary to construct special difference schemes in which the solution error is independent of ε. In particular, it is of interest to have schemes that converge under a weaker condition than (9.5), which is a condition for the convergence of solutions of the scheme (9.17), (9.19).

The conditions imposed on the grid approximations of the problem (9.2), (9.1) that are necessary and sufficient for ε-uniform (or close to it) convergence of grid solutions are of great importance.

Our aim for the boundary value problem (9.2), (9.1) is to find necessary conditions for the ε-uniform and almost ε-uniform convergence of solutions of difference schemes constructed using classical approximations of the differential equation and, in addition, to construct a difference scheme that converges ε-uniformly.

9.3 *A priori* estimates

Let us give *a priori* estimates on the solution of the problem (9.2), (9.1) that are used in the following constructions. On the set $\overline{G}$, we represent the solution as the sum of its regular and singular components:

$$u(x,t) = U(x,t) + W(x,t), \quad (x,t) \in \overline{G}. \tag{9.7}$$

In the problem (9.2), (9.1), we pass to the variables ξ, t in which the lateral boundaries are fixed. It is convenient to transform the variable x into the variable $\xi = \xi(x,t)$ defined by

$$\xi = \xi(x,t) = d \left(x - \beta_1(t) \right) \left(\beta_2(t) - \beta_1(t) \right)^{-1}, \quad (x,t) \in \overline{G}. \tag{9.8a}$$

We denote the inverse mapping of $\xi(x,t)$ by $\xi^{-1}(\xi,t) \equiv x(\xi,t)$. For the functions $v(x,t)$ and $Z(\xi,t)$ and the subdomains $G^0 \subseteq G$, we will use the notation

$$v\big(x(\xi,t),t\big) = v_\xi(\xi,t) = \{v(x,t)\}_\xi = \tilde{v}(\xi,t), \tag{9.8b}$$

$$Z\big(\xi(x,t),t\big) = Z_{\xi^{-1}}(x,t) = \{Z(\xi,t)\}_{\xi^{-1}},$$

$$G^0_\xi = \{G^0\}_\xi = \xi(G^0) = \{(\xi,t) : \ (x(\xi,t),t) \in G^0\}. \tag{9.8c}$$

We define

$$\tilde{G}^0_{\xi^{-1}} = \{\tilde{G}^0\}_{\xi^{-1}} = \xi^{-1}(\tilde{G}^0) = \{(x,t) : \ (\xi(x,t),t) \in \tilde{G}^0\},$$

where $\tilde{G}^0$ is some subset of a set $\overline{\overline{G}}$, and $\overline{\overline{G}} = \overline{G}_\xi = \{\overline{G}\}_\xi$.

In the variables ξ, t, the problem (9.2), (9.1) is transformed into the boundary value problem

$$\tilde{L}\tilde{u}(\xi,t) \equiv \left\{ \varepsilon A(\xi,t)\frac{\partial^2}{\partial \xi^2} + B(\xi,t)\frac{\partial}{\partial \xi} - \frac{\partial}{\partial t} \right\} \tilde{u}(\xi,t) = \tilde{f}(\xi,t), \quad (\xi,t) \in \tilde{G},$$

$$\tilde{u}(\xi,t) = \tilde{\varphi}(\xi,t), \quad (\xi,t) \in \tilde{S}. \tag{9.9}$$

Here

$$A(\xi, t) = \left\{ \left[\frac{\partial}{\partial x}\xi(x,t) \right]^2 \right\}_\xi, \quad B(\xi, t) = - \left\{ \left[\frac{\partial}{\partial t}\xi(x,t) \right] \right\}_\xi, \quad (\xi, t) \in \overline{\overline{G}}.$$

Owing to condition (9.3), we have

$$B(\xi, t) \geq B_0 > 0, \quad (\xi, t) \in \overline{\overline{G}}.$$

On the set

$$\overline{\widetilde{G}} = \widetilde{G} \cup \widetilde{S}, \quad \widetilde{G} = \widetilde{D} \times (0, T], \quad \widetilde{D} = \{\xi : 0 < \xi < d\}, \qquad (9.10)$$

the problem (9.9) is a boundary value problem for a singularly perturbed parabolic convection-diffusion equation in a domain with a fixed lateral boundary. The boundary layer appears in a neighborhood of the left side $\widetilde{S}_1^L$ of the lateral boundary $\widetilde{S}^L$, towards which the convective flow is directed. Let us estimate the regular and singular component in the variables ξ, t (see, e.g., [50, 52]).

Returning to the variables x, t, we obtain the estimates

$$\left| \frac{\partial^{k_1 + k_0}}{\partial x^{k_1} \partial t^{k_0}} U(x, t) \right| \leq M, \qquad (9.11)$$

$$\left| \frac{\partial^{k_1 + k_0}}{\partial x^{k_1} \partial t^{k_0}} W(x, t) \right| \leq M \varepsilon^{-k_1 - k_0} \exp\left(-m_1 \varepsilon^{-1} (x - \beta_1(t))\right),$$

$$(x, t) \in \overline{G}; \quad k_1 + 2k_0 \leq 4,$$

where m_1 is an arbitrary number in the interval $(0, m_0)$, and

$$m_0 = m_{1(9.3)} M_{1(9.3)}^{-1} \min_{[0, T]} \left[(d/dt)\beta_1(t) \right].$$

For the function $W(x, t)$, we also have the estimate

$$\left| \frac{\partial^{k_1 + k_0}}{\partial x^{k_1} \partial t^{k_0}} W(x, t) \right| \leq M \varepsilon^{-k_1 - k_0} \exp\left(-m \varepsilon^{-1} r((x, t), S_1^L)\right), \quad (9.12a)$$

$$(x, t) \in \overline{G}; \quad k_1 + 2k_0 \leq 4,$$

where $r((x, t), S_1^L)$ is the distance from the point (x, t) to the set S_1^L, and m is an arbitrary number in the interval $(0, m^0)$, where

$$m^0 = v_0 (1 + (v^0)^2)^{-1/2}, \quad v_0 = v_{0(9.3)}, \quad v^0 = v_{(9.3)}^0. \qquad (9.12b)$$

Thus, unlike problems in domains with fixed boundaries, the derivatives of the singular component $W(x, t)$ with respect to both variables x and t grow unboundedly in a neighborhood of the moving boundary layer as $\varepsilon \to 0$.

When deriving the estimates, for simplicity we assume that the following compatibility condition on the set S^c (see [67]) is fulfilled:

$$\frac{\partial^{k_1 + k_0}}{\partial x^{k_1} \partial t^{k_0}} \varphi(x, t) = 0, \quad \frac{\partial^{k_1 + k_0}}{\partial x^{k_1} \partial t^{k_0}} f(x, t) = 0, \quad (x, t) \in S^c, \quad k_1 + k_0 \leq l, \quad (9.13)$$

where $l > 6$,

Theorem 9.3.1 *Let for the data of the boundary value problem* (9.2), (9.1), *the conditions* $f \in C^{l+\alpha}(\overline{G})$, $\varphi \in C^{l+\alpha}(\overline{G})$, $\beta_i \in C^{l+\alpha}([0, T])$, *where* $l = K + 4$, $K \geq 2$ *and* $\alpha \in (0, 1)$, *hold, and let also the conditions* (9.3) *and* (9.13) *be fulfilled. Then the components in the representation* (9.7) *for the solution of the boundary value problem* (9.2), (9.1) *satisfy the estimates* (9.11) *and* (9.12).

9.4 Classical finite difference schemes

For the problem (9.2), (9.1), we consider classical difference schemes on different type grids and discuss some difficulties that arise in the numerical solution when ε is small.

9.4.1. On the strip

$$\overline{G}^{\infty} = I\!R \times [0, T]$$

we define rectangular *basis* grids $\overline{G}_h^b$ that will be used to construct the required grids. Let

$$\overline{G}_h = \overline{G}_h^b = \omega_1 \times \overline{\omega}_0, \tag{9.14}$$

where ω_1 and $\overline{\omega}_0$ are meshes with an arbitrary distribution of nodes on the x-axis and on the interval $[0, T]$, respectively. Set $h^i = x^{i+1} - x^i$, $h = \max_i h^i$ for x^i, $x^{i+1} \in \omega_1$, and $h_t^j = t^{j+1} - t^j$, $h_t = \max_j h_t^j$ for t^j, $t^{j+1} \in \overline{\omega}_0$. Let $N + 1$ and $N_0 + 1$ be the maximal number of nodes per unit length on the x-axis and the number of nodes in the mesh $\overline{\omega}_0$. Assume that $h \leq MN^{-1}$ and $h_t \leq MN_0^{-1}$. Of particular interest for us are the grids $\overline{G}_h^{bu}$ that are uniform with respect to x and t:

$$\overline{G}_h = \overline{G}_h^b = \overline{G}_h^{bu}; \tag{9.15}$$

the grid $\overline{G}_h^{bu}$ is $\overline{G}_{h(9.14)}^b$, where ω_1 and $\overline{\omega}_0$ are uniform meshes with step-sizes $h = N^{-1}$ and $h_t = TN_0^{-1}$.

On the set $\overline{G}$, we construct the grid (generated by the grid $\overline{G}_{h(9.14)}$)

$$\overline{G}_h = \overline{G}_h(\overline{G}_{h(9.14)}^b) = G_h \bigcup S_h. \tag{9.16}$$

The set G_h is the set of nodes (x^i, t^j) in $G \cap \overline{G}_h^b$ for which the segment $x^i \times (t^{j-1}, t^j]$ entirely belongs to G. The intersection of the lines $t = t^j$, $t^j \in \omega_0$ with the sides S^L is denoted by S_h^L. The set S_h is formed by S_h^L and the nodes (x^i, t^0) belonging to S_0 for which the segment $x^i \times [t^0, t^1)$ entirely belongs to $G \bigcup S_0$ (this set is denoted by S_{0h}); the nodes $(0, 0)$ and $(d, 0)$ are assumed to lie in S_{0h}. We set $S_h = S_{0h} \bigcup S_h^L$.

Problem (9.2), (9.1) is approximated by the implicit difference scheme

$$\Lambda z(x,t) \equiv \left\{ \varepsilon\, \delta_{\overline{x}\widehat{x}} - \delta_{\overline{t}} \right\} z(x,t) = f(x,t), \quad (x,t) \in G_h, \qquad (9.17)$$

$$z(x,t) = \varphi(x,t), \quad (x,t) \in S_h.$$

Here $\delta_{\overline{x}\widehat{x}}\, z(x,t)$ and $\delta_{\overline{t}}\, z(x,t)$ are the first-order and second-order difference derivatives,

$$\delta_{\overline{x}\widehat{x}}\, z(x,t) = 2\left(h^i + h^{i-1}\right)^{-1} \left\{ \delta_x - \delta_{\overline{x}} \right\} z(x,t),$$

$x = x^i$, and h^{i-1} and h^i are the left and right "arms" of the three-point stencil on G_h (of the operator $\delta_{\overline{x}\widehat{x}}$) centered at the nodes $(x^i, t^j) \in G_h$.

The difference scheme (9.17), (9.16) satisfies the maximum principle [108]. For the solution of this scheme, we have the estimate

$$|u(x,t) - z(x,t)| \leq M \left(\varepsilon + N^{-1} + N_0^{-1} \right)^{-1} \left[N^{-1} + N_0^{-1} \right], \qquad (9.18)$$

$$(x,t) \in \overline{G}_h,$$

where $\overline{G}_h = \overline{G}_{h(9.16)}$.

On the grid (generated by the grid $\overline{G}_{h(9.15)}$)

$$\overline{G}_h = \overline{G}_h^u = \overline{G}_h\big(\overline{G}_{h(9.15)}^{bu}\big), \qquad (9.19)$$

i.e., the grid $\overline{G}_{h(9.16)}\big(\overline{G}_{h(9.14)}^b\big)$, where $\overline{G}_{h(9.14)}^b$ is the grid $\overline{G}_{h(9.15)}^{bu}$, we obtain the estimate

$$|u(x,t) - z(x,t)| \leq M \left[\left(\varepsilon + N^{-1}\right)^{-2} N^{-2} + \left(\varepsilon + N_0^{-1}\right)^{-1} N_0^{-1} \right], \qquad (9.20)$$

$$(x,t) \in \overline{G}_h,$$

which is unimprovable with respect to N, N_0, and ε.

The schemes (9.17), (9.16) and (9.17), (9.19) converge under the condition N^{-1}, $N_0^{-1} \ll \varepsilon$, which is unimprovable, or more precisely,

$$\varepsilon^{-1} = o\big(\min[N, N_0] \big), \quad N, N_0 \to \infty. \qquad (9.21)$$

The convergence defect of the scheme (9.17) on the grids (9.16) and (9.19) is equal to unity.

Note that the operator Λ on the solution of the boundary value problem (9.2), (9.1) is not ε-uniformly bounded (unlike the problem (9.2) on the set $\overline{G}$ with a fixed lateral boundary).

9.4.2. Along with the solutions of the scheme (9.17), (9.16), we will consider their interpolants that can be constructed in the following way.

On the basis of the grid $\overline{G}_h$, we construct a *triangulation* of the domain $\overline{G}$ and use $r(x,t)$, $(x,t) \in \overline{G}_h$, to construct the interpolant $\overline{z}(x,t)$, $(x,t) \in \overline{G}$.

We cover the domain $\overline{G}$ by elementary rectangles, irregular quadrangles (with the sides not parallel to the coordinate axes), and triangles. Some vertices of the irregular quadrangles and triangles are the nodes belonging to the set S_h, and one of their sides belongs to the set S^L. We divide the irregular quadrangles into rectangles and irregular triangles. On the lines $t = t^j$, $t^j \in \overline{\omega}_0$, using the values $z(x, t)$, $(x, t) \in \overline{G}_h$, we construct linear (with respect to x) interpolants $\widetilde{z}(x, t)$, $(x, t) \in \overline{G}$, $t \in \overline{\omega}_0$. All the rectangles are partitioned into triangular elements by their diagonals. These regular and irregular triangular elements form a triangulation of the domain $\overline{G}$. On the triangular elements, we construct *linear interpolants* on the basis of the values $\widetilde{z}(x, t)$ at the vertices of the triangular elements on the sets $t = t^j$, $t^j \in \overline{\omega}_0$. The interpolant

$$\overline{z}(x, t) = \overline{z}_{(9.22)}(x, t; z(\cdot), \overline{G}_h), \quad (x, t) \in \overline{G} \tag{9.22}$$

is then constructed; see, e.g., [79].

In the case of the difference scheme (9.17), (9.16), the function $\overline{z}(x, t)$, $(x, t) \in \overline{G}$, satisfies the estimate

$$|u(x, t) - \overline{z}(x, t)| \leq M \left(\varepsilon + N^{-1} + N_0^{-1} \right)^{-1} \left[N^{-1} + N_0^{-1} \right], \tag{9.23}$$

$$(x, t) \in \overline{G}.$$

In the case of the scheme (9.17), (9.19), we have the error estimate

$$|u(x, t) - \overline{z}(x, t)| \leq M \left[\left(\varepsilon + N^{-1} \right)^{-2} N^{-2} + \left(\varepsilon + N_0^{-1} \right)^{-1} N_0^{-1} \right], \tag{9.24}$$

$$(x, t) \in \overline{G},$$

which is unimprovable with respect to N, N_0, and ε.

Thus, the rate of convergence of the solutions to the difference scheme (9.17), (9.16) (scheme (9.17), (9.19)) and their interpolants are of the same order.

The *optimal order* (with respect to P_0) *of convergence rate of the scheme* on the grid (9.19) is obtained under the condition $N^2 \approx \varepsilon^{-1} N_0$. In this case, we have the unimprovable (with respect to P_0 and ε) estimate

$$|u(x, t) - \overline{z}(x, t)| \leq M \left(\varepsilon^{4/3} + P_0^{-2/3} \right)^{-1} P_0^{-2/3}, \quad (x, t) \in \overline{G}, \tag{9.25}$$

where P_0 is the number of mesh points in $\overline{G}_h$ in the case of the scheme (9.17), (9.19), and $P_0 \approx N N_0$.

In the case of the scheme (9.17), (9.16), the least right-hand side in the estimate (9.23) is obtained under the condition $N \approx N_0$. Thus, the solution of the scheme (9.17), (9.16) satisfies the estimate

$$|u(x, t) - \overline{z}(x, t)| \leq M \left(\varepsilon + P_0^{-1/2} \right)^{-1} P_0^{-1/2}, \quad (x, t) \in \overline{G}, \tag{9.26}$$

which is weaker in comparison with the estimate (9.25).

The condition $P_0^{-1/2} \ll \varepsilon$, or more precisely,

$$\varepsilon^{-1} = o(P_0^{1/2}), \quad P_0 \to \infty, \tag{9.27}$$

is necessary and sufficient for the convergence of the scheme (9.17) on the grids (9.16) and (9.19) when the order of convergence rate is optimal with respect to P_0. The convergence defect of these schemes in P_0 and ε is equal to unity.

Theorem 9.4.1 *Let for the solution of the boundary value problem (9.2), (9.1), the a priori estimates (9.11) and (9.12) be satisfied for $K = 4$. Then the condition (9.21) (the condition (9.27)) is necessary and sufficient for the convergence of the scheme (9.17) on the grids (9.16), (9.19) (on the grids (9.16), (9.19) with the optimal order of convergence rate with respect to P_0). The grid solutions satisfy the estimates (9.18), (9.20), (9.23)–(9.26).*

9.5   Construction of ε-uniform and almost ε-uniform approximations to solutions of Problem (9.2), (9.1)

In this section we consider some specific features related to a triangulation of the domain $\overline{G}$ that arise when constructing ε-uniform approximations of solutions to the singularly perturbed problem (9.2), (9.1). In our considerations of the approximations, we will use an analog of Kolmogorov's widths (see [15, 10]).

9.5.1. Let $\mathcal{U}$ be a set of solutions in the class of boundary value problems (9.2), (9.1) (defined by the conditions (9.3)). We are interested in the approximation of $\mathcal{U}$ in the space X, that is, the set of continuous functions with the maximum norm. The solutions are assumed to be sufficiently smooth on $\overline{G}$ for fixed values of the parameter ε; the solutions and their components in the representation (9.7) satisfy the estimates (9.11) and (9.12).

Let us describe approximations to the solutions. Let $\overline{G}^h$ be a finite set of points (we say, *the grid*) on $\overline{G}$. The grids $\overline{G}^h$ may be both structured (generated by some regular family of lines) and unstructured. The number of *nodes* in the grid $\overline{G}^h$ on $\overline{G}$ is denoted by P; $\overline{G}^h = \overline{G}^h(P)$. Let T_P be a *triangulation* (partition) of the set $\overline{G}$ *generated by the grid* $\overline{G}^h$ (see, e.g., [79]); we assume that the mesh points in $\overline{G}^h$ are the vertices of the triangular elements, where the triangle sides are line segments that pass through the nodes of $\overline{G}^h$ if at least one of the endpoints of a triangle side belongs to G^h, or are segments of curves if both of the endpoints belong to the boundary

$\overline{S}^L$; here $G^h = \overline{G}^h \cap G$. Let some grid function $u^h(x,t)$, $(x,t) \in \overline{G}^h$, be defined on the set $\overline{G}^h$. By $\overline{u}^h(x,t)$, $(x,t) \in \overline{G}$, we denote the piecewise-linear interpolant that is linear on each triangle and is constructed using the values of $u^h(x,t)$ at the vertices of the triangular elements. The set of such piecewise-linear interpolants for the fixed triangulation T_P is denoted by U_P^h. The set of all feasible grid sets $\overline{G}^h$ (however, with the number of nodes equal to P on $\overline{G}$) and of triangulations T_P based on them will be denoted by $\mathcal{T}_P$; we say that $\mathcal{T}_P$ is the *set of triangulations* of the domain $\overline{G}$. The classes of feasible grid sets and triangulations on them are specified in detail below. This set of triangulations $\mathcal{T}_P$ and the set of interpolants U_P^h (for each triangulation in $\mathcal{T}_P$) approximate the *space* X. Define the *width* $d_P(\mathcal{U}, X)$ by

$$d_P(\mathcal{U}, X) = \inf_{\mathcal{T}_P} \sup_{u \in \mathcal{U}} \inf_{\overline{u}^h \in U_P^h} \| u - \overline{u}^h \|, \qquad (9.28)$$

where $\|\cdot\|$ is the maximum norm in $C(\overline{G})$. A definition of Kolmogorov's width can be found, for example, in [10, Chapter 3]. The quantity $d_P(\mathcal{U}, X)$ is the *error of the optimal approximation of the set $\mathcal{U}$ in the space X* using a grid with P nodes, or, briefly, the *error of the optimal approximation*.

Definitions. Let

$$d_P^i(\mathcal{U}, X) = d_P(\mathcal{U}, X; \overline{G}_i^h), \quad i = 1, 2,$$

be the widths induced by two families of grids $\overline{G}_i^h = \overline{G}_i^h(P)$, $i = 1, 2$. When the widths $d_P^i(\mathcal{U}, X)$ satisfy the estimate

$$m\, d_P^1(\mathcal{U}, X) \le d_P^2(\mathcal{U}, X) \le M\, d_P^1(\mathcal{U}, X), \quad P \ge M_1,$$

uniformly with respect to P for some sufficiently large M_1, we say that the widths $d_P^1(\mathcal{U}, X)$ and $d_P^2(\mathcal{U}, X)$ are *equivalent*.

By $\rho_1(T_P^j)$ and $\rho_2(T_P^j)$, we denote the radii of the inscribed and circumscribed circles for the triangular element T_P^j in the triangulation T_P, $j = 1, \ldots, J$, where $J = J(P)$ is the number of triangular elements in T_P (we take $J \approx P$). The triangulation T_P is said to be *isotropic* if the condition

$$\rho_1^{-1}(T_P^j)\, \rho_2(T_P^j) \le M, \quad j = 1, \ldots, J,$$

holds; however, the quantities $\rho_1^{-1}(T_P^j)$, $\rho_2(T_P^j)$ and *the anisotropy coefficient* $\eta(T_P^j) = \rho_1^{-1}(T_P^j)\rho_2(T_P^j)$ for each element T_P^j can differ significantly from element to element. The triangulation T_P is called *anisotropic* (with *the anisotropy coefficient* $\eta \ge M_0$, where the lower threshold M_0 may be sufficiently large) if

$$\eta \equiv \sup_{j=1,\ldots,J} \eta(T_P^j) \ge M_0,$$

and the constant M_0 does not depend on the parameters ε, P. We assume that the set $\mathcal{T}_P$, as well as the width $d_P(\mathcal{U}, X)$, is determined by η; thus,

$$\mathcal{T}_P = \mathcal{T}_P(\eta), \quad T_P = T_P(\eta), \quad d_P(\mathcal{U}, X) = d_P(\mathcal{U}, X; \eta).$$

Isotropic and anisotropic *triangulations on subsets* of $\overline{G}$ can be defined in a similar way. When the widths are considered on the subset $\overline{G}^0 \subset \overline{G}$ (in this case, we denote the width by $d_P(\mathcal{U}, X; \overline{G}^0)$), the quantity $\| u - \overline{u}^h \|$ in (9.28) is computed using only the triangular elements that belong entirely to $\overline{G}^0$.

A triangulation element satisfying the condition

$$\eta(T_P^j) \to \infty \quad \text{as} \quad P \to \infty \quad \text{and/or} \quad \varepsilon \to 0$$

is called *essentially anisotropic*; a triangulation T_P containing such elements is said to be *essentially anisotropic*.

The width $d_P(\mathcal{U}, X)$ tends to zero as $P \to \infty$; however, this convergence to zero is not ε-uniform. In general, the approximations converge as $P \to \infty$ only for certain relationships between P and ε.

Let the width $d_P(\mathcal{U}, X)$ satisfy the upper estimate

$$d_P(\mathcal{U}, X) \leq M \lambda \left(\varepsilon^{-\nu} P^{-1/2} \right), \quad P \to \infty, \quad \varepsilon \in (0, 1], \tag{9.29}$$

that is similar to the estimate (9.4) for the difference scheme. As in the convergence of solutions to difference schemes, one can define *the convergence of the width* (the error of the optimal approximation) *with defect ν* (of the ε-uniform convergence) with respect to the values of P and ε and also its ε-*uniform and almost ε-uniform convergence*. In the case of almost ε-uniform convergence, the value ν controls the triangulations T_P.

The most interesting approximations of the set $\mathcal{U}$ are those that converge —if possible—with minimal defect and, in particular, those that converge ε-uniformly.

9.5.2. Consider an estimate on the width when the moving boundary S_1^L is a line segment

$$\beta_1(t) = v_1 t, \quad t \in [0, T], \quad v_1 \in [v_0, v^0], \tag{9.30}$$

where $v_0 = v_{0(9.3)}$, $v^0 = v_{(9.3)}^0$.

In the case of isotropic triangulations of the domain $\overline{G}$, the width satisfies the lower estimate

$$d_P(\mathcal{U}, X) \geq m(1 + \varepsilon P)^{-1}. \tag{9.31}$$

On the set

$$\overline{G}^0 = \overline{G}_1(\rho_1), \quad \rho_1 = M\varepsilon \tag{9.32}$$

where

$$G_1(\rho_1) = \left\{ (x, t) : x \in \left(\beta_1(t), \beta_1(t) + \rho_1 \right), t \in (0, T] \right\}$$

is the right ρ_1-neighborhood of the set S_1^L, we obtain the estimate

$$d_P\left(\mathcal{U}, X; \overline{G}_{(9.32)}^0 \right) \geq m(1 + \varepsilon P)^{-1}, \tag{9.33}$$

which is unimprovable with respect to P and ε.

In the case of anisotropic triangulations, taking into account the explicit form of the main term of the asymptotic expansion (in powers of ε) of the singular component in the representation (9.7), under the condition (9.30) we find the estimate

$$d_P(\mathcal{U}, X) \geq m \min \left\{ \left(\varepsilon \eta + (\varepsilon \eta)^{-1} \right) P^{-1}, 1 \right\}. \tag{9.34}$$

On the set $\overline{G}^0_{(9.32)}$, we have the estimate

$$d_P(\mathcal{U}, X; \overline{G}^0) \geq m \min \left\{ \left(\varepsilon \eta + (\varepsilon \eta)^{-1} \right) P^{-1}, 1 \right\}, \tag{9.35}$$

which is unimprovable with respect to P, ε, and η.

9.5.3. Assume that the condition (9.30) is fulfilled. The estimates (9.33) and (9.35), under the condition

$$M_0 \leq \eta \leq M_1, \quad \text{where the constant } M_1 \text{ is independent of } P \text{ and } \varepsilon, \tag{9.36}$$

show that the *convergence defect* of the error of the optimal approximation on the set $\overline{G}^0_{(9.32)}$ is equal to 2^{-1}. Therefore, the convergence defect of the widths $d_P(\mathcal{U}, X)$ on an isotropic triangulation and on an anisotropic triangulation under the condition (9.36) is not less than 2^{-1}.

By virtue of the estimate (9.35), the condition (of the essential anisotropy of the triangulations)

$$\eta = \eta(\varepsilon, P), \quad \eta(\varepsilon, P) \to \infty \quad \text{for } P \to \infty \text{ and/or } \varepsilon \to 0; \tag{9.37}$$

$$P \to \infty, \quad \varepsilon \in (0, 1],$$

is necessary for the defect (of convergence of the width on the anisotropic triangulation T_P) to be less than 2^{-1}, and also for ε-uniform or almost ε-uniform convergence.

According to the condition (9.37) (necessary for convergence of the width with defect less than 2^{-1}) and from the estimates (9.34) and (9.35), which are unimprovable with respect to P, ε, and η, we define η by

$$\eta = \eta(\varepsilon, P) \equiv \varepsilon^{-1} \eta_0(P), \quad \text{where } \eta_0(P) \to \infty \quad \text{for } P \to \infty. \tag{9.38}$$

Then the additional condition $\eta_0(P) P$, $\eta_0^{-1}(P) P \gg 1$; more precisely,

$$\eta_0(P), \, \eta_0^{-1}(P) = o(P), \quad P \to \infty, \tag{9.39}$$

is necessary for the ε-uniform convergence of the width $d_P(\mathcal{U}, X; \overline{G}^0)$ on the set $\overline{G}^0 = \overline{G}^0_{(9.32)}$. Under the condition (9.38), taking into account the *a priori* estimates of Theorem 9.3.1, we obtain the ε-uniform upper estimate

$$d_P(\mathcal{U}, X; \overline{G}^0) \leq M P^{-1} \left[\eta_0(P) + \eta_0^{-1}(P) \right], \tag{9.40}$$

which is unimprovable with respect to P.

For the width $d_P(\mathcal{U}, X)$, under the condition (9.38) we obtain the ε-uniform estimate

$$d_P(\mathcal{U}, X) \leq M \min\left[P^{-1} \ln P\left[\eta_0(P) + \eta_0^{-1}(P)\right]; 1\right]. \qquad (9.41)$$

Under the additional condition $P^{-1} \ln P\left[\eta_0(P) + \eta_0^{-1}(P)\right] \ll 1$; more precisely,

$$\eta_0(P), \eta_0^{-1}(P) = o(P \ln^{-1} P), \quad P \to \infty, \qquad (9.42)$$

which is slightly stronger than the condition (9.39), the width $d_P(\mathcal{U}, X)$ converges ε-uniformly.

Theorem 9.5.1 *Let the components of the solution of the boundary value problem* (9.2), (9.1) *in the representation* (9.7) *satisfy the a priori estimates* (9.11) *and* (9.12), *where $K = 2$. In the case of the condition* (9.30), *the condition $\{(9.38), (9.39)\}$ is necessary and the condition $\{(9.38), (9.42)\}$ is sufficient for the ε-uniform convergence (as $P \to \infty$) of the width $d_P(\mathcal{U}, X)$ for the anisotropic triangulation* (9.38). *Under the condition* (9.38), *the widths $d_P(\mathcal{U}, X; \overline{G}^0)$ and $d_P(\mathcal{U}, X)$ satisfy the estimates* (9.40) *(unimprovable with respect to P) and* (9.41), *respectively.*

Remark 9.5.1 In the case of the condition (9.30), the necessary condition $\{(9.38), (9.39)\}$ and the sufficient condition $\{(9.38), (9.42)\}$, for the ε-uniform convergence of the width, are almost the same. From the unimprovable estimate (9.40) and the representation (9.38) for the value η, it follows that the triangulation elements on the set $\overline{G}^0$ are essentially anisotropic in the case when the width on $\overline{G}^0$ converges with defect $\nu < 2^{-1}$. ∎

9.5.4. Assume that the following condition holds:

$$\textit{the set } S_1^L \textit{ is a segment of a smooth curve} \qquad (9.43)$$

with a bounded nonzero curvature, and let the triangulation T_P be anisotropic and satisfy the condition

all the sides of the triangular elements in the triangulation
are line segments. $\qquad (9.44)$

On the triangulations $T_P = T_P(\eta)$ having anisotropy η, we define the width $d_P^*(\mathcal{U}, X)$ by

$$d_P^*(\mathcal{U}, X) = \inf_\eta \, d_P(\mathcal{U}, X; \eta). \qquad (9.45)$$

We say that the width $d_P^*(\mathcal{U}, X)$ is *optimal with respect to the anisotropy* on the triangulations $T_P(\eta)$, or, briefly, we say the *optimal width* $d_P^*(\mathcal{U}, X)$.

In the case of the conditions (9.43) and (9.44), taking into account the explicit form of the main term of the component $W_{(9.7)}(x, t)$, we obtain the following unimprovable estimate for the width $d_P^*(\mathcal{U}, X)$ considered on $\overline{G}_{(9.32)}^0$:

$$d_P^*(\mathcal{U}, X; \overline{G}^0) \geq m(1 + \varepsilon^{1/2}P)^{-1}. \tag{9.46}$$

Theorem 9.5.2 *Let the hypotheses of Theorem 9.5.1 be fulfilled. Then, under the conditions (9.43) and (9.44), the width $d_{P(9.45)}^*(\mathcal{U}, X; \overline{G}^0)$ satisfies the estimate (9.46).*

Remark 9.5.2 By virtue of the estimate (9.46), the convergence defect of the width $d_P^*(\mathcal{U}, X; \overline{G}^0)$ in the case of a curvilinear boundary S_1^L is 4^{-1}. Thus, for the width $d_P^*(\mathcal{U}, X)$ in the case of the triangulations generated by the triangular elements satisfying condition (9.44) and the interpolants $\overline{u}^h(x, t)$ that are linear on the elements T_P^j, a convergence defect less than 4^{-1} cannot be achieved in the case of condition (9.43). ∎

Remark 9.5.3 It follows from the considerations given above that, in order to construct optimal approximations whose widths have defect of convergence less than 4^{-1}, it is necessary to use triangulations T_P with curvilinear triangular elements T_P^j and/or nonlinear interpolants constructed on T_P^j from the values of the function $u^h(x, t)$. ∎

Remark 9.5.4 Theorem 9.5.2 implies that, subject to the condition (9.43), a convergence defect less than 4^{-1} cannot be achieved also for the difference schemes constructed by the classical approximation of the boundary value problem (9.2), (9.1) on (reasonable) meshes generating a triangulation that satisfies the condition (9.44). ∎

9.6 Difference scheme on a grid adapted in the moving boundary layer

We are interested in the difference schemes for which the interpolants of the grid solutions constructed on the triangulation elements generated by the grid nodes are convergent ε-uniformly. The results given in Section 9.5 imply that the optimal approximations of the boundary value problem (9.2), (9.1) constructed on the basis of regular anisotropic triangulations and linear interpolants converge with a defect not lower than 4^{-1} in the case of a curvilinear boundary S_1^L. Hence, it follows that, for the solutions of the scheme (9.17) on the grids generated by rectangular (basis) grids, a convergence defect of the interpolants of these grid solutions less than 4^{-1} is unachievable. Therefore, in the case of a curvilinear boundary S_1^L, for the convergence of the interpolants of grid solutions with defect lower than 4^{-1} (or ε-uniformly), it is necessary to use grids fitted to the boundary S_1^L that generate essentially anisotropic triangulation elements. These conditions, which are necessary for the ε-uniform approximations of the widths, are used in the construction of an ε-uniformly convergent scheme.

9.6.1. To construct a special scheme that converges ε-uniformly, we use the following approach. After passing to the variables ξ, t, $\xi = \xi(x, t)$, the problem (9.2), (9.1) is transformed into the problem (9.9), (9.10) in a domain with fixed lateral boundaries. For this problem, the optimal approximations of its solution converge ε-uniformly in the case of the triangulations based on the grids that are piecewise-uniform with respect to ξ and uniform with respect to t and on the interpolants that are linear on the triangulation elements. Having constructed an ε-uniformly convergent scheme for the problem (9.9), (9.10) (in such a "standard" difference scheme, the linear interpolant based on its grid solutions converges ε-uniformly) and then having returned to the initial variables, we obtain an ε-uniformly convergent scheme for the problem (9.2), (9.1); for this scheme, the interpolant is written in terms of the grid solutions but is no longer linear in x and t on the triangulation elements.

In terms of the initial variables, the resulting grids are no longer rectangular (the distribution of the grid nodes is adapted to the moving boundary S_1^L). Generally, this implies some inconvenience in the construction of grid domains and in the numerical solution of the problem. However, such a scheme may be constructed only in a small neighborhood of the boundary S_1^L; outside this neighborhood, rectangular (in the initial variables) grids and classical grid approximations of the problem may be used.

According to this approach, we pass from the problem (9.2), (9.1) to the problem (9.9), (9.10) for which we construct a scheme on *a priori* condensing grids. On the set $\overline{\overline{G}}$, we introduce the rectangular grid

$$\overline{\overline{G}}_h = \overline{\overline{\omega}}_1 \times \overline{\omega}_0, \tag{9.47}$$

where $\overline{\overline{\omega}}_1$ and $\overline{\omega}_0$ are meshes on the intervals $\overline{\overline{D}} = [0, d]$ and $[0, T]$, respectively; $\overline{\omega}_0 = \overline{\omega}_{0(9.15)}$, $\overline{\overline{\omega}}_1$ is a mesh with an arbitrary distribution of nodes satisfying the condition $h_\xi \leq MN^{-1}$, where $h_\xi = \max_i h_\xi^i$, $h_\xi^i = \xi^{i+1} - \xi^i$, ξ^i, $\xi^{i+1} \in \overline{\overline{\omega}}_1$; and $N + 1$ is the number of mesh points in $\overline{\overline{\omega}}_1$.

To solve the problem (9.9), (9.10), we use the difference scheme

$$\tilde{\Lambda} Z(\xi, t) \equiv \left\{ \varepsilon A(\xi, t)\, \delta_{\overline{\xi}\hat{\xi}} + B(\xi, t)\, \delta_\xi - \delta_{\overline{t}} \right\} Z(\xi, t) = \tilde{f}(\xi, t), \quad (\xi, t) \in \tilde{G}_h, \tag{9.48}$$

$$Z(\xi, t) = \tilde{\varphi}(\xi, t), \quad (\xi, t) \in \tilde{S}_h,$$

where $\tilde{G}_h = \tilde{G} \cap \overline{\overline{G}}_h$, $\tilde{S}_h = \tilde{S} \cap \overline{\overline{G}}_h$, and $\delta_{\overline{\xi}\hat{\xi}} Z(\xi, t)$ and $\delta_\xi Z(\xi, t)$, $\delta_{\overline{t}} Z(\xi, t)$ are the second and the first difference derivatives.

Furthermore, we consider the well-known standard piecewise-uniform grid (see, for example, [50, 52]).

9.6.2. On the set $\overline{\overline{G}}$, we introduce the standard grid $\overline{\overline{G}}_h^S$ condensing in a neighborhood of the boundary layer:

$$\overline{\overline{G}}_h^S = \overline{\overline{\omega}}_1^S \times \overline{\omega}_0, \tag{9.49}$$

where $\overline{\omega}_0 = \overline{\omega}_{0(9.47)}$ and $\overline{\overline{\omega}}_1^S = \overline{\overline{\omega}}_1^S(\sigma)$ is a piecewise-uniform mesh. The mesh-sizes of $\overline{\overline{\omega}}_1^S$ are constant on the intervals $[0, \sigma]$ and $[\sigma, d]$ and equal to $h_{(1)} = 2\,\sigma\,N^{-1}$ and $h_{(2)} = 2\,[d - \sigma]\,N^{-1}$, respectively; the value σ is chosen so as to satisfy the condition $\sigma = \sigma(\varepsilon, N) = \min\left[2^{-1}d,\ m^{-1}\varepsilon \ln N\right]$, where m is an arbitrary number in the interval $(0, m^0)$, and $m^0 = d^{-1}\,m_{1(9.3)}\,v_{0(9.3)}$.

The scheme (9.48), (9.49) converges ε-uniformly with the error estimate

$$|\widetilde{u}(\xi, t) - Z(\xi, t)| \le M\,\left[N^{-1} \ln N + N_0^{-1}\right], \quad (\xi, t) \in \overline{\overline{G}}_h.$$

For the interpolant $\overline{Z}(\xi, t)$, $(\xi, t) \in \overline{\overline{G}}$, which is linear in ξ and t on the triangular elements of the triangulation generated by the grid $\overline{\overline{G}}_h$, we have the estimate

$$|\widetilde{u}(\xi, t) - \overline{Z}(\xi, t)| \le M\,\left[N^{-1} \ln N + N_0^{-1}\right], \quad (\xi, t) \in \overline{\overline{G}}.$$

In the variables x and t, the grid

$$\overline{\overline{G}}_{h\,\xi^{-1}} = \left\{\overline{\overline{G}}_{h(9.49)}\right\}_{\xi^{-1}} \tag{9.50}$$

is not a tensor product of meshes in x and t. This grid is uniform in t and is piecewise-uniform in x for $t = t^j$, $t^j \in \overline{\omega}_0$. Passing to the variables x and t in the scheme (9.48), we come to the scheme

$$\widetilde{\Lambda}_{\xi^{-1}}\,Z_{\xi^{-1}}(x, t) \equiv \varepsilon\,A_{\xi^{-1}}(x, t)\left\{\delta_{\overline{\xi}\widehat{\xi}}\,Z(\xi, t)\right\}_{\xi^{-1}} + B_{\xi^{-1}}(x, t)\left\{\delta_\xi\,Z(\xi, t)\right\}_{\xi^{-1}} -$$

$$-\left\{\delta_{\overline{t}}\,Z(\xi, t)\right\}_{\xi^{-1}} = f(x, t), \quad (x, t) \in \widetilde{G}_{h\,\xi^{-1}}, \tag{9.51}$$

$$Z_{\xi^{-1}}(x, t) = \varphi(x, t), \quad (x, t) \in \widetilde{S}_{h\,\xi^{-1}}.$$

The function $Z^*(x, t) = Z_{\xi^{-1}}(x, t)$, $(x, t) \in \overline{G}_h^*$, where $\overline{G}_h^* = \overline{\overline{G}}_{h\,\xi^{-1}}$, i.e., the solution of the difference scheme (9.51), (9.50), satisfies the estimate

$$|u(x, t) - Z^*(x, t)| \le M\,\left[N^{-1} \ln N + N_0^{-1}\right], \quad (x, t) \in \overline{G}_h^*. \tag{9.52a}$$

For the interpolant $\overline{Z}^*(x, t) = \left\{\overline{Z}(\xi, t)\right\}_{\xi^{-1}}$, $(x, t) \in \overline{G}$, which is obtained from the function $\overline{Z}(\xi, t)$ by passing to the variable x and t, we have a similar estimate:

$$\left|u(x, t) - \overline{Z}^*(x, t)\right| \le M\,\left[N^{-1} \ln N + N_0^{-1}\right], \quad (x, t) \in \overline{G}. \tag{9.52b}$$

Theorem 9.6.1 *Let the hypothesis of Theorem 9.4.1 be fulfilled. Then the difference scheme* (9.51), (9.50) *approximating the boundary value problem* (9.2), (9.1) *converges ε-uniformly. The grid solution $Z^*(x,t)$, $(x,t) \in \overline{G}_h^*$, and the interpolant $\overline{Z}^*(x,t)$, $(x,t) \in \overline{G}$, satisfy the estimates* (9.52).

Remark 9.6.1 For the solution of the difference scheme (9.51), (9.50), we have the estimate (see (9.52))

$$|u(x,t) - Z^*(x,t)| \leq M \left[N^{-1} \ln N + N\, P^{-1} \right], \quad (x,t) \in \overline{G}_h^*,$$

where $P = N\, N_0$. Under the condition

$$N \approx N_0 \tag{9.53}$$

which is natural for regular problems, we obtain the ε-uniform estimate

$$|u(x,t) - Z^*(x,t)| \leq M\, P^{-1/2} \ln P, \quad (x,t) \in \overline{G}_h^*. \tag{9.54}$$

∎

9.7   Remarks and generalizations

In this section we give some remarks and generalizations related to application of the technique based on the widths for investigation of ε-uniformly convergent difference schemes.

9.7.1. From discussion of Section 9.3–9.6, one obtains the following variant of our approach to the construction of ε-uniformly convergent finite difference schemes. The problem (9.2), (9.1), by a change of variables, is transformed to a problem with stationary boundaries. For the new problem, a scheme is constructed that converges ε-uniformly. Next, we find the solution of this scheme and return to the original variables. Such an approach was used in [27] for a parabolic convection-diffusion equation in a domain with moving boundaries.

9.7.2. To construct ε-uniformly convergent schemes for problem (9.2), (9.1), one can use *the domain decomposition method* on overlapping subdomains. In the subdomain that includes the boundary layer, a finite difference scheme is constructed applying the grid constructs given in Section 9.5. In the subdomain not including boundary layer, a classical finite difference scheme based on uniform meshes is constructed similar to those in Section 9.4. One can show that in the case when the minimal width of the subdomain overlaps is not less than value of the parameter ε and the maximum of the step-sizes in x and t, the domain decomposition scheme converges ε-uniformly.

9.7.3. The approach based on widths is applied to construct ε-uniformly convergent finite difference scheme for *elliptic and parabolic equations* in domains with *curvilinear boundaries*. Analysis of widths allows us to determine conditions that are necessary and (for additional assumptions) sufficient for ε-uniform convergence of the finite difference schemes.

In [183], *widths* were studied for *a class of solutions to boundary value problems* of elliptic reaction-diffusion equations in two dimensional domains with curvilinear boundaries. In these problems, for ε-uniform convergence (in the maximum norm) of widths defined on grid sets, it is *necessary* that these grid sets be condensed in a neighborhood of the boundary layer and adapted to the domain boundary. It is not difficult to realize such adaptation of the grid sets if we pass to a local coordinate system in which the piece of the boundary becomes a line part. In the new coordinate system, on a simplest grid, which are uniform along the boundary and piecewise-uniform along the normal to the boundary, one can derive *sufficient conditions close to necessary* for ε-uniform convergence of widths. The transition point in the piecewise-uniform meshes depends both on the parameter ε and the value N_1, i.e., the number of nodes in the mesh along the normal to the boundary. It is necessary for these piecewise-uniform meshes to be consistent with the boundary in its σ_0^*-neighborhood, where

$$\sigma_0^* = \sigma_{(1)} + \sigma_{(2)}, \quad \sigma_{(1)} \approx \varepsilon\lambda(N_1), \quad \sigma_{(2)} \approx N_2^{-2};$$

$$\lambda(N_1) \to \infty, \quad N_1^{-1}\lambda(N_1) \to 0, \quad for \ N_1, N_2 \to \infty, \quad N_1 N_2 \approx P.$$

Here P is the number of nodes in the grid set. Outside the σ_0^*-neighborhood, the widths converge ε-uniformly already on grid sets with isotropic triangulation, e.g., generated by uniform grids. In [183], for a differential problem local difference schemes were constructed on such type grids, adapted in a neighborhood of the boundary. These schemes possess the required properties such as approximation and stability. Furthermore, using the domain decomposition method based on the local approximations of the problem, ε-uniformly convergent finite difference schemes were constructed for elliptic reaction-diffusion equations in a domain with a curvilinear boundary.

Similar approach is realized in [138] and also in Chapters 2–7 to construct ε-uniformly convergent finite difference schemes for elliptic and parabolic problems for reaction-diffusion and convection-diffusion equations in *n-dimensional domains with* smooth and *piecewise smooth boundaries*.

9.7.4. *The use of widths* allows us to justify *applicability of the fitted operator method* in the construction of ε-uniformly convergent finite difference schemes for parabolic convection-diffusion and reaction-diffusion problems. For an initial-boundary value problem in a domain with stationary boundaries, the singular component of the solution to the convection-diffusion problem is the regular boundary layer (described by an ordinary differential equation) and to the reaction-diffusion problem it is the parabolic boundary layer (described by a parabolic equation). The main term in an expansion of the singular component, i.e., *the regular layer*, is defined by only *one coefficient-parameter*, namely, the ratio of coefficients to the second- and first-order derivatives in x. The main term in an expansion of the singular component, i.e., *the parabolic layer*, is an infinite sum of singular components (we say, elementary singular components) that are defined by both *the coefficient-parameter* which is the ratio of coefficients to the second-order derivative in x and the first-order derivative in t, and by *the coefficients of the Taylor expansion* to the boundary function.

Under the condition that the problem solution does not involve the main term of the singular component (i.e., the first term of the expansion to the boundary layer function), the solutions of the initial-boundary value problems for convection-diffusion and reaction-diffusion equations generate the set $\mathcal{U}^0$, i.e., the set of regular solutions, whose widths $d_P(\mathcal{U}^0, X)$ converge ε-uniformly for finite values of anisotropy η to the domain triangulation.

For *the fitted operator method* applied to *the convection-diffusion problem*, the main term of the singular component is the solution of a homogenous difference equation. At the same time, the regular component of the solution to the finite difference scheme converges ε-uniformly to the regular component of the solution to the differential problem that implies the ε-uniform convergence of the fitted operator scheme.

When *the fitted operator method* is applied to *the reaction-diffusion problem*, by choosing a finite number of the fitted coefficients in a scheme, it is possible to attain that only some elementary singular components of the main term of the singular component (i.e., the parabolic-layer function) could be the solution of the homogenous difference equation.

The remaining elementary singular components, as well as the regular components of the solutions, generate the set $\mathcal{U}^1$, i.e., the set of the singular solutions, whose widths $d_P(\mathcal{U}^1, X)$ do not already converge ε-uniformly for finite values of the anisotropy η to the domain triangulations. Hence, for the reaction-diffusion problems with the boundary parabolic layers, there exist no schemes of the fitted operator method that converge ε-uniformly.

The proof of similar statement in [130, 138, 87] is carried out *in other ways* unlike given here. In [130, 138, 87], the direct check justifies that, for a finite number of the fitted coefficients, there exist elementary singular components in the singular term of the problem solution such that are not approximated ε-uniformly by discrete solutions.

9.7.5. When constructing ε-uniformly convergent fitted operator schemes for parabolic reaction-diffusion problems with parabolic layers, difficulties are observed for both *the boundary layer* and *the initial layer*. In [148], an initial-boundary value problem was considered for a parabolic reaction-diffusion equation with *perturbation parameters multiplied by the spatial and temporal derivatives*. In this problem, initial, boundary, and parabolic layers appear, depending on the relations between the parameters. In [148], using a technique similar to given in [130, 138, 87], it was shown that, in the presence of the initial parabolic layers, there exist no schemes of the fitted operator method that converge ε-uniformly. But the construction of ε-uniformly convergent schemes of the fitted operator method resolving the initial-boundary parabolic layers has no difficulties. The same results can be obtained using *a technique based on the widths*.

9.7.6. In [180], a boundary value problem is considered for an elliptic convection-diffusion equation with a perturbation vector-parameter $\bar{\varepsilon} = \varepsilon$, where $\bar{\varepsilon} = (\varepsilon_1, \varepsilon_2)$. The second-order and the first-order derivatives in the differential equation are multiplied by the component-parameters ε_1^2 and ε_2^2, respectively. Depending on the relations between the component-parameters, regular, parabolic, and *hyperbolic* boundary layers appear. In [180], on the basis of a technique for analyzing convergence of the fitted operator schemes for problems with parabolic layers (see, for example, [130, 138, 87]), the conclusion is derived that *in the presence of hyperbolic layers, the fitted operator method is inapplicable* for the construction of ε-uniformly convergent schemes. The same result is obtained naturally when *the width technique* is applied.

Chapter 10

High-order accurate numerical methods for singularly perturbed problems

In this chapter we consider various approaches to construct finite difference schemes with improved convergence order for a singularly perturbed parabolic convection-diffusion equation with sufficiently smooth and piecewise-smooth initial data. There are well elaborated methods:

(a) the defect correction,

(b) the Richardson extrapolation,

and newly created and developing methods:

(c) based on asymptotic constructs, and

(d) the additive splitting of singularities.

10.1   Introduction

For singularly perturbed boundary value problems, well studied ε-uniformly convergent numerical methods have the low accuracy. So, for a parabolic convection-diffusion problem with sufficiently smooth data, order of ε-uniform convergence for finite difference schemes on special meshes (basic schemes) does not exceed 1 (see, e.g., [138, 87, 102, 33]). But already for piecewise-smooth initial data this order is not higher than 2^{-1}. The low rate of ε-uniform convergence is the main restriction to use such methods in practice. Therefore, it is actually to develop ε-uniformly convergent high-order accurate numerical methods.

At present, for singularly perturbed problems a technique of constructing ε-uniformly convergent schemes with improved accuracy is well elaborated such as: (a) the defect correction method and (b) the Richardson extrapolation method.

The *defect correction method* was successfully applied to nonstationary reaction-diffusion problems in [46, 47] and to convection-diffusion problems in [50, 51, 52]. This method can be considered as *stabilization of an unstable difference scheme* constructed using an approximation of the first-order derivatives in x and t by *central difference derivatives* (see Remark 10.4.1 to Theorem 10.4.1 in Section 10.4). The advantage of the defect correction schemes is so that both the basic and corrected schemes are solved on the same mesh for one and the same equation but with different right-hand sides.

The Richardson method was studied for singularly perturbed boundary value problems:

for parabolic reaction-diffusion equations in [120],

for elliptic convection-diffusion equations in [151],

for elliptic reaction-diffusion equations in [53, 197];

moreover, in [197] this method was considered for a quasilinear singularly perturbed equation.

The *Richardson extrapolation method* is based on a proper representation of the main component of the error in the solution of the basic scheme in the form of an expansion with respect to effective step-sizes (in x and in t) of the meshes used. This representation allows us to construct a suitable linear combination of discrete solutions obtained on embedded meshes, and, as a result, to obtain a solution with improved accuracy (on the intersection of the meshes) as compared to the solution of the basic scheme. The advantage of this method is so that one and the same discrete problem is solved but on one-type embedded meshes.

In the case of singularly perturbed problems with sufficiently complicated singularities of solutions, methods based on asymptotic expansions bring more simple approximations of the solutions (see, e.g., [57]). Due to this, a new approach arose for constructing ε-uniformly convergent high-order accurate schemes, namely, (c) *the method based on the asymptotic expansion technique* (or briefly, *the method of asymptotic constructs*), which is now in an initial stage of development. This method was applied in [168, 175] to improve order of ε-uniform convergence rate for discrete solutions to singularly perturbed elliptic convection-diffusion equations in domains with characteristic [168] and noncharacteristic [175] parts of the boundary. Such approach seems attractive because, for small values of the parameter, we use approximations of "auxiliary" problems on subdomains (simpler those as compared with the solved problem) that describe main terms of the asymptotics. These problems on the subdomains (in a neighborhood of the boundary layers and out of them) are solved step by step, moreover, on uniform meshes. In this case, the computation of solutions is simplified as compared with the basic ε-uniformly convergent scheme.

The approaches (a)–(c) mentioned above are applied to construct ε-uniformly convergent schemes of improved accuracy for boundary value problems with sufficiently smooth data. But if the problem data are not sufficiently smooth and/or compatibility conditions are not satisfied then smoothness of the problem solution decreases that implies decrease of convergence order to difference schemes. For example, when the x-derivative of the initial function has a discontinuity of the first kind, then the ε-uniform convergence rate for the basic scheme on a piecewise-uniform mesh for the parabolic convection-diffusion equation becomes of order 2^{-1} (see, e.g., the estimate (10.72) for the scheme (10.68), (10.71) in the case of the problem (10.56), (10.55) in Section

10.9). Therefore, for singularly perturbed problems when the initial function or its derivatives have a discontinuity, an interest arises to develop special numerical methods that allow us (1) *to restore the convergence rate of the difference schemes* (that is lost due to decreasing smoothness of the problem solution) and, furthermore, (2) *to increase accuracy of the discrete solutions.*

The approach (d), which uses a *method of the additive splitting of singularities* (for a description of this approach see, e.g., [184]), turns out to be applicable to construct schemes of improved accuracy in the case of problems with a restricted smoothness of the data. In this approach the singular part of the problem solution generated by typical nonsmooth data is split off separately. For the remainder part of the solution, it is possible to apply schemes (including schemes of improved accuracy) that are developed for problems with sufficiently smooth data. The additive splitting method was used in [64, 71] to solve singularly perturbed problems for parabolic equations with discontinuous initial conditions.

These approaches (a)–(d), which use the technique based on piecewise-uniform meshes condensing in the layer region, allow us to construct effective numerical methods for a sufficiently wide class of boundary value problems for elliptic and parabolic equations whose solutions exhibit thin layer phenomena.

In this chapter, in order to demonstrate the use of the mentioned approaches to the construction of higher-order accurate schemes, we focus our attention on a singularly perturbed parabolic convection-diffusion equation with smooth and piecewise-smooth initial data. We give principles used in the construction of improved schemes. More detailed constructions and also achieved results can be found in the cited bibliography.

10.2 Boundary value problems for singularly perturbed parabolic convection-diffusion equations with sufficiently smooth data

10.2.1 Problem with sufficiently smooth data

Here we define "Problem with sufficiently smooth data" that will be also considered in the next chapters.

On the domain $\overline{G}$, where

$$G = D \times (0, T], \quad \overline{G} = G \cup S, \quad D = (0, d), \tag{10.1}$$

we consider the initial-boundary value problem for the singularly perturbed parabolic convection-diffusion equation

$$\begin{aligned} L\, u(x,t) &= f(x,t), \quad (x,t) \in G, \\ u(x,t) &= \varphi(x,t), \quad (x,t) \in S. \end{aligned} \tag{10.2}$$

Here

$$L_{(10.2)} \equiv \varepsilon\, a(x,t)\frac{\partial^2}{\partial x^2} + b(x,t)\frac{\partial}{\partial x} - c(x,t) - p(x,t)\frac{\partial}{\partial t}, \quad (x,t) \in G,$$

the functions $a(x,t)$, $b(x,t)$, $c(x,t)$, $p(x,t)$, $f(x,t)$, and $\varphi(x,t)$ are assumed to be sufficiently smooth on the sets $\overline{G}$ and S, respectively, moreover,

$$a_0 \le a(x,t) \le a^0, \quad b_0 \le b(x,t) \le b^0, \quad c_0 \le c(x,t) \le c^0, \qquad (10.3)$$

$$p_0 \le p(x,t) \le p^0, \quad (x,t) \in \overline{G}; \qquad a_0, b_0, c_0, p_0 > 0;$$

$$|f(x,t| \le M, \quad (x,t) \in \overline{G}; \qquad |\varphi(x,t)| \le M, \quad (x,t) \in S;$$

the parameter ε takes arbitrary values in the open-closed interval $(0,1]$.

Assume that the data of the problem (10.2), (10.1) on the set of corner points $S_* = S_0 \cap \overline{S}^L$ satisfy the compatibility conditions that ensure the required smoothness of the solution on $\overline{G}$ (see, e.g., [67]). Here $S = S_0 \cup S^L$, S_0 and S^L are the lower and lateral parts of the boundary; $S_0 = \overline{S}_0$, $S^L = S^l \cup S^r$, where S^l and S^r are the left and right parts of the lateral boundary.

For small values of the parameter ε, a regular boundary layer appears in a neighborhood of the set

$$S^l = \{(x,t): \ x = 0, \ 0 < t \le T\}.$$

By a solution of the boundary value problem, we mean a function $u \in C(\overline{G}) \cap C^{2,1}(G)$ that satisfies the differential equation on G and the initial-boundary condition on S. In the case when a discrete solution converges uniformly with respect to the parameter ε, we say that the solution converges $\overline{\varepsilon}$-uniformly. We are interested in numerical methods that converge $\overline{\varepsilon}$-uniformly in the maximum norm.

10.2.2 A finite difference scheme on an arbitrary grid

Here we give a finite difference scheme that is constructed based on a classical approximation of the problem (10.2), (10.1). On the set $\overline{G}$ we introduce the rectangular grid

$$\overline{G}_h = \overline{\omega} \times \overline{\omega}_0, \qquad (10.4)$$

where $\overline{\omega}$ and $\overline{\omega}_0$ are arbitrary, in general, nonuniform meshes on the intervals $[0,d]$ and $[0,T]$, respectively. Set $h^i = x^{i+1} - x^i$, $h = \max_i h^i$ for $x^i, x^{i+1} \in \overline{\omega}$, and $h_t^k = t^{k+1} - t^k$, $h_t = \max_k h_t^k$ for $t^k, t^{k+1} \in \overline{\omega}_0$. Let $N+1$ and N_0+1 be the number of nodes in the meshes $\overline{\omega}$ and $\overline{\omega}_0$, respectively. Assume that $h \le M\,N^{-1}$ and $h_t \le M\,N_0^{-1}$.

Problem (10.2), (10.1) is approximated by the finite difference scheme [108]

$$\Lambda\, z(x,t) = f(x,t), \quad (x,t) \in G_h, \qquad (10.5)$$

$$z(x,t) = \varphi(x,t), \quad (x,t) \in S_h.$$

Here $G_h = G \cap \overline{G}_h$, $S_h = S \cap \overline{G}$,

$$\Lambda \equiv \varepsilon\, a(x,t)\, \delta_{\overline{x}\widehat{x}} + b(x,t)\, \delta_x - c(x,t) - p(x,t)\, \delta_{\overline{t}}, \quad (x,t) \in G_h,$$

and $\delta_{\overline{x}\widehat{x}}\, z(x,t)$ is the central difference derivative on a nonuniform mesh, where $\delta_x\, z(x,t)$ and $\delta_{\overline{x}}\, z(x,t)$, $\delta_{\overline{t}}\, z(x,t)$ are the first (forward and backward) difference derivatives (see Section 8.4 in Chapter 8).

The finite difference scheme (10.5), (10.4) is monotone ε-uniformly. A definition of monotonicity for regular problems can be found in [108].

Here we introduce a simplest *interpolant*

$$\overline{z}(x,t) = \overline{z}(x,t;\; z(\cdot),\, \overline{G}_h), \quad (x,t) \in \overline{G}, \tag{10.6}$$

i.e., is a linear interpolant on triangular elements (triangulation of elementary rectangles from $\overline{G}$ generated by the points of the mesh $\overline{G}_h$) that is constructed using the discrete function $z(x,t)$, $(x,t) \in \overline{G}_h$.

10.2.3 Estimates of solutions on uniform grids

In the case of uniform grids in both variables:

$$\overline{G}_h = \overline{\omega}^u \times \overline{\omega}_0^u, \tag{10.7}$$

using the maximum principle, for the solution of the finite difference scheme (10.5), (10.7) we obtain the estimate

$$|u(x,t) - z(x,t)| \le M\left[\left(\varepsilon + N^{-1}\right)^{-1} N^{-1} + N_0^{-1}\right], \quad (x,t) \in \overline{G}_h, \tag{10.8}$$

which is unimprovable with respect to N, N_0, ε.

The interpolant $\overline{z}(x,t) = \overline{z}_{(10.6)}\left(x,t;\; z_{(10.5,\,10.7)}(\cdot),\, \overline{G}_h^u\right)$ satisfies the estimate

$$|u(x,t) - \overline{z}(x,t)| \le M\left[\left(\varepsilon + N^{-1}\right)^{-1} N^{-1} + N_0^{-1}\right], \quad (x,t) \in \overline{G}. \tag{10.9}$$

The difference scheme (10.5), (10.7) converges under the condition $N^{-1} \ll \varepsilon$, or more precisely:

$$\varepsilon^{-1} = o(N), \quad N \to \infty, \quad \varepsilon \in (0,1]. \tag{10.10}$$

10.2.4 Special ε-uniform convergent finite difference scheme

Let us give a finite difference scheme that converges ε-uniformly (see, e.g., [128, 138, 50, 51, 52]). On the set $\overline{G}$ we introduce the grid

$$\overline{G}_h = \overline{G}_h^s = \overline{\omega}^* \times \overline{\omega}_0, \tag{10.11a}$$

where $\overline{\omega}_0 = \overline{\omega}_{0(10.7)}^u$ and $\overline{\omega}^*$ is a piecewise-uniform mesh (Shishkin's mesh [83]) that is constructed as follows. The interval $[0, d]$ is divided into two parts $[0, \sigma]$

and $[\sigma, d]$. The mesh sizes in $\overline{\omega}^*$ are $h^{(1)} = 2\,\sigma\,N^{-1}$ on the interval $[0, \sigma]$ and $h^{(2)} = 2(d - \sigma)N^{-1}$ on $[\sigma, d]$. The value σ is specified by

$$\sigma = \sigma(\varepsilon,\, N,\, l) = \min\left[2^{-1}\,d,\, l\,m^{-1}\,\varepsilon\,\ln N\right], \qquad (10.11\text{b})$$

where m is an arbitrary number in $(0, m_0)$, with $m_0 = m_{(10.19)}$. Here

$$l = 1; \qquad (10.11\text{c})$$

in other piecewise-uniform grids, this parameter will be chosen. The mesh $\overline{\omega}^*$ and the grid $\overline{G}_h = \overline{G}_h(l = 1)$ are constructed.

For the solution of the scheme (10.5), (10.11) one has the ε-uniform estimate

$$|u(x, t) - z(x, t)| \leq M\left[N^{-1}\ln N + N_0^{-1}\right], \qquad (x, t) \in \overline{G}_h, \qquad (10.12\text{a})$$

and the ε-dependent estimate

$$|u(x, t) - z(x, t)| \leq M\left[N^{-1}\min[\varepsilon^{-1}, \ln N] + N_0^{-1}\right], \qquad (x, t) \in \overline{G}_h. \quad (10.12\text{b})$$

The estimates (10.12a) and (10.12b) are unimprovable with respect to N, N_0 and N, N_0, ε, respectively. The finite difference scheme (10.5), (10.11) converges with the convergence order close to one.

For the interpolant $\overline{z}(x, t) = \overline{z}_{(10.6)}\big(x, t; z_{(10.5,\ 10.11)}(\cdot),\ \overline{G}_h^u\big)$, we have the ε-uniform estimate

$$|u(x, t) - \overline{z}(x, t)| \leq M\left[N^{-1}\ln N + N_0^{-1}\right], \qquad (x, t) \in \overline{G}, \qquad (10.13\text{a})$$

and the ε-dependent estimate

$$|u(x, t) - \overline{z}(x, t)| \leq M\left[N^{-1}\min[\varepsilon^{-1}, \ln N] + N_0^{-1}\right], \qquad (x, t) \in \overline{G}. \quad (10.13\text{b})$$

From the estimates (10.8), (10.9) and (10.12), (10.13) it follows that the convergence rates of solutions to the difference schemes (10.5), (10.4) and (10.5), (10.11) and their interpolants are of one and the same order.

The following convergence theorem holds.

Theorem 10.2.1 *Let the solution $u(x, t)$ of the problem (10.2), (10.1) satisfy the estimates in Theorem 10.3.1, where $K = 4$. Then the difference scheme (10.5), (10.11) converges ε-uniformly, while the scheme (10.5), (10.7) converges for fixed values of the parameter ε. The discrete solutions satisfy the estimates (10.12) and (10.8).*

10.2.5   The aim of the research

Our aim for the boundary value problem (10.2), (10.1) is to construct a finite difference scheme that converges ε-uniformly with the convergence order higher than one.

10.3 *A priori* estimates for problem with sufficiently smooth data

Here we give *a priori* estimates of the solution and its derivatives for the problem (10.2), (10.1); the derivation of these estimates is similar to that in [138, 50, 51, 52].
Write the problem solution as the sum of the functions

$$u(x,t) = U(x,t) + V(x,t), \quad (x,t) \in \overline{G}, \tag{10.14}$$

where $U(x,t)$ and $V(x,t)$ are the regular and singular parts of the solution. The function $U(x,t)$, $(x,t) \in \overline{G}$, is the restriction to $\overline{G}$ of the function $U^e(x,t)$, $(x,t) \in \overline{G}^e$, where $\overline{G} \subset \overline{G}^e$. The function U^e is the solution of an "extended" problem that is obtained by extension of the problem (10.2), (10.1) beyond the left boundary S^l, preserving properties similar to (10.3):

$$L^e_{(10.2)} U^e(x,t) = f^e(x,t), \quad (x,t) \in G^e, \quad U^e(x,t) = \varphi^e(x,t), \quad (x,t) \in S^e.$$

The domain G^e is an extension of the domain G beyond the boundary S^l. The right-hand side $f^e(x,t)$ in the equation is a smooth extension of the function $f(x,t)$. The function $\varphi^e(x,t)$, $(x,t) \in S^e$, is smooth on each piecewise-smooth part of the boundary S^e, and coincides with the function $\varphi(x,t)$ on the set $S^e \cap S$.
The function $V(x,t)$, $(x,t) \in \overline{G}$, is the solution of the problem

$$L_{(10.2)} V(x,t) = 0, \quad (x,t) \in G,$$

$$V(x,t) = \varphi(x,t) - U(x,t), \quad (x,t) \in S.$$

For simplicity, we assume that the coefficients $a(x,t)$ and $b(x,t)$ satisfy the condition

$$a(x,t) = g(x)\, b(x,t), \quad (x,t) \in \overline{G}, \tag{10.15}$$

e.g., $a(x,t)$ and $b(x,t)$ are independent of t. Suppose also that the data of the problem (10.2), (10.1) are sufficiently smooth, i.e.,

$$a,\, b,\, c,\, p,\, f \in C^{l,l}(\overline{G}), \quad \varphi \in C^{l,l}(\overline{G}), \quad l = 3n+1, \tag{10.16}$$

where $n > 0$ is an integer. The value n defines the number of terms in asymptotic expansions of the components $U(x,t)$ and $V(x,t)$ in the decomposition (10.14) that are used for the construction of required estimates for derivatives of these components.
Let the condition

$$\frac{\partial^k}{\partial x^k}\, \varphi(x,t), \quad \frac{\partial^{k_0}}{\partial t^{k_0}}\, \varphi(x,t) = 0, \quad k + 2k_0 \leq n+1, \tag{10.17}$$

$$\frac{\partial^{k+k_0}}{\partial x^k\, \partial t^{k_0}}\, f(x,t) = 0, \quad k + 2k_0 \leq n-1, \quad (x,t) \in S_*,$$

be fulfilled, and also the following condition hold

$$\frac{\partial^k}{\partial x^k}\,\varphi(x,t), \ \ \frac{\partial^{k_0}}{\partial t^{k_0}}\,\varphi(x,t)=0, \ \ n+1<k+2k_0, k+k_0 \ \leq 3n+1, \quad (10.18)$$

$$\frac{\partial^{k+k_0}}{\partial x^k\,\partial t^{k_0}}\,f(x,t)=0, \ \ n-1<k+2k_0, \ \ k+k_0\leq 3n, \ \ (x,t)\in S_*^{(r)},$$

where $S_*^{(r)}=(0,d)$ is the right "corner" point at S_*. The condition (10.17) guarantees the required smoothness of the solution of the boundary value problem (10.2), (10.1), and the condition (10.17) and (10.18) ensure the smoothness of the components in (10.14).

In the case when

$$u,\, U,\, V \in C^{l^1,\,l^1/2}(\overline{G}), \ \ l^1=n+1+\alpha, \ \ \alpha\in(0,1);$$

for the components $U(x,t)$ and $V(x,t)$ in (10.14) one has

$$\left|\frac{\partial^{k+k_0}}{\partial x^k\partial t^{k_0}}\,U(x,t)\right| \leq M\,, \tag{10.19}$$

$$\left|\frac{\partial^{k+k_0}}{\partial x^k\partial t^{k_0}}\,V(x,t)\right| \leq M\,\varepsilon^{-k}\,\exp(-m\,\varepsilon^{-1}\,x), \ \ (x,t)\in\overline{G}, \ \ k+2k_0\leq K,$$

where $K=n+1$ and m is an arbitrary number in $(0,m_0)$ with $m_0=\min_{\overline{G}}[a^{-1}(x,t)\,b(x,t)]$.

Theorem 10.3.1 *Let the conditions (10.15)–(10.18) be satisfied for the data of the boundary value problem (10.2), (10.1). Then the solution of the problem and its components in the representation (10.14) satisfy the estimates (10.19).*

10.4 The defect correction method

10.4.1. For the difference scheme (10.2), (10.1), errors in the approximation of the partial derivatives $(\partial/\partial t)\,u(x,t)$ and $(\partial/\partial x)\,u(x,t)$ by the difference derivatives $\delta_{\overline{t}}\,z(x,t)$ and $\delta_x\,z(x,t)$ are associated with the truncation errors given by (an expansion of the approximation error in h):

$$\frac{\partial u}{\partial t}(x,t)-\delta_{\overline{t}}u(x,t)=2^{-1}h_t^{k-1}\frac{\partial^2 u}{\partial t^2}(x,t-\theta_0), \ \ \theta_0\in[0,h_t^{k-1}], \ \ t=t^k\in\omega^0,$$

$$\frac{\partial u}{\partial x}(x,t)-\delta_x u(x,t)=-2^{-1}h^i\frac{\partial^2}{\partial x^2}u(x+\theta_1,t), \ \ \theta_1\in[0,h^i], \ \ x=x^i\in\overline{\omega},$$
$$x^i\neq d.$$

The defect correction method (a) is based on expansion of the approximation error in powers of h that allows us to increase the approximation order.

Thus, we can evaluate a better approximation than (10.5) by the defect correction as follows

$$\Lambda_{(10.5)} z^c(x,t) = f(x,t) + 2^{-1} h^i b(x,t) \frac{\partial^2 u}{\partial x^2}(x,t) + 2^{-1}\tau p(x,t) \frac{\partial^2 u}{\partial t^2}(x,t),$$

with $x = x^i$. Here $x \in \overline{\omega}^*$ and $t \in \overline{\omega}_0^u$, where $\overline{\omega}^*$ and $\overline{\omega}_0^u$ are the meshes as in (10.11); $h^i = x^{i+1} - x^i$ is a local step-size in the mesh $\overline{\omega}^*$; $z^c(x,t)$ is the "corrected" solution. Instead of $(\partial^2/\partial t^2) u(x,t)$ and $(\partial^2/\partial x^2) u(x,t)$, we use the difference derivatives $\delta_{\overline{t}t} z(x,t)$ and $\delta_{\overline{x}\widehat{x}} z(x,t)$, respectively, where $z(x,t)$, $(x,t) \in G_{h(10.11)}$, is the solution of the difference scheme. We may expect that the new solution $z^c(x,t)$ has an ε-uniform consistency error of order $\mathcal{O}\left(N_0^{-2}\right)$ with respect to the variable t. Concerning the variable x, the consistency error of the corrected solution on uniform meshes is supposed to be of order $\mathcal{O}\left(N^{-2}\right)$ for fixed ε. But in the case of special piecewise-uniform grids the order of ε-uniform convergence with respect to x is expected to be of order $\mathcal{O}\left(N^{-2} \ln^2 N\right)$, i.e., the second order up to a logarithmic factor.

10.4.2. We present a difference scheme of the defect correction method.

For notational convenience, on the grid $\overline{G}_h$ we write the finite difference scheme (10.5), (10.4) in the form

$$\Lambda_{(10.5)}\, z^{(1)}(x,t) = f(x,t), \quad (x,t) \in G_h, \tag{10.20}$$

$$z^{(1)}(x,t) = \varphi(x,t), \quad (x,t) \in S_h,$$

where $z^{(1)}(x,t)$ is the uncorrected solution. This scheme will be used as the basic one when constructing a defect correction scheme of improved accuracy.

On the grid (10.4), we approximate the boundary value problem (10.2), (10.1) by the difference scheme

$$\Lambda_{(10.5)}\, z^{(2)}(x,t) = f(x,t) + \psi^1(x,t) + \psi^0(x,t), \quad (x,t) \in G_h, \tag{10.21a}$$

$$z^{(2)}(x,t) = \varphi(x,t), \quad (x,t) \in S_h.$$

Here

$$\psi^1(x,t) \equiv 2^{-1} h^{i-1} b(x,t)\, \delta_{\overline{x}\widehat{x}}\, z^{(1)}(x,t), \quad x = x^i, \quad (x,t) \in G_h; \tag{10.21b}$$

$z^{(1)}(x,t)$, $(x,t) \in G_h$ is the solution of the basic scheme (10.20), (10.4); $h^{i-1} = x^i - x^{i-1}$, $x^{i-1}, x^i \in \omega$. Note that

$$\delta_x z(x^i,t) - 2^{-1} h^{i-1} \delta_{\overline{x}\widehat{x}}\, z(x^i,t) = \delta_{\widehat{x}}\, z(x^i,t),$$

where $\delta_{\widehat{x}}\, z(x^i,t)$ is the first central difference derivative:

$$\delta_{\widehat{x}}\, z(x^i,t) = \left(h^i + h^{i-1}\right)^{-1} \left(z(x^{i+1},t) - z(x^{i-1},t)\right).$$

The function $\psi^0(x,t)$ has the form

$$\psi^0(x,t) \equiv 2^{-1} p(x,t) \left\{ \begin{array}{ll} h_t^k \dfrac{\partial^2}{\partial t^2} u(x,0), & t = t^k, \; k = 1 \\ h_t^{k-1} \delta_{2\bar{t}} z^{(1)}(x,t), & t = t^k, \; k > 1 \end{array} \right\}, \quad (x,t) \in G_h,$$

$$(10.21c)$$

where $\delta_{2\bar{t}} z(x,t)$ is the second-order backward difference derivative:

$$\delta_{2\bar{t}} z(x,t) = \delta_{\bar{t}}(\delta_{\bar{t}} z(x,t)) = \delta_{\widehat{tt}} z(x,t^{k-1}), \quad t = t^k \in \omega_0, \; k > 1.$$

On the mesh $\overline{\omega}_0 = \overline{\omega}_0^u$ with the step-size $\tau = T N_0^{-1}$, we have

$$\psi^0(x,t) \equiv 2^{-1} \tau \, p(x,t) \left\{ \begin{array}{ll} \dfrac{\partial^2}{\partial t^2} u(x,0), & t = \tau \\ \delta_{2\bar{t}} z^{(1)}(x,t), & t \geq 2\tau \end{array} \right\}, \quad (x,t) \in G_h, \quad (10.22)$$

where

$$\delta_{2\bar{t}} z(x,t) = \left(\delta_{\bar{t}} z(x,t) - \delta_{\bar{t}} z(x,t-\tau) \right)/\tau, \quad (x,t) \in \overline{G}_h, \; t \geq 2\tau.$$

The derivative $(\partial^2/\partial t^2)u(x,0)$ can be obtained from the equation (10.2). We call the function $z^{(2)}(x,t)$, $(x,t) \in G_{h(10.4)}$, the solution of the difference scheme (10.21), (10.20), (10.4) (or briefly, (10.21), (10.4)).

10.4.3. Using the defect correction method, we now construct a *difference scheme* of *improved accuracy* that converges ε-uniformly. For this we use the special piecewise-uniform grid

$$\overline{G}_h = \overline{G}_{h(10.11)}, \quad (10.23a)$$

where $\overline{\omega}^* = \overline{\omega}_{h(10.11)}^*(\sigma)$ provided that

$$\sigma = \sigma_{(10.23)}(\varepsilon, \, N) = \min\left[2^{-1} d, \; l \, m^{-1} \varepsilon \ln N \right], \quad (10.23b)$$

$m = m_{(10.11)}$ and $l \geq 3$ is an arbitrary number. Choosing the grid by such a way, we have thus constructed the defect correction difference scheme (10.21), (10.23).

For simplicity, we assume that the coefficients $a(x,t)$, $b(x,t)$ satisfy the condition (10.15), and the initial condition is homogeneous

$$\varphi(x,t) = 0, \quad (x,t) \in S_0. \quad (10.24)$$

Note that, under the condition (10.24), the derivative $\dfrac{\partial^2}{\partial t^2} u(x,0)$, which is required to compute the function $\psi^0(x,t)$ for $t = \tau$, is defined by the relation

$$\frac{\partial^2}{\partial t^2} u(x,0) = -p^{-1}(x,t) \left\{ \varepsilon \, a(x,t) \frac{\partial^2}{\partial x^2} + b(x,t) \frac{\partial}{\partial x} - c(x,t) + p(x,t) \frac{\partial}{\partial t} \right\} \times$$

$$\times \left\{ p^{-1}(x,t)\, f(x,t) \right\}, \quad (x,t) \in S_0.$$

For the solution of the difference scheme (10.21), (10.23), we have the ε-uniform estimate

$$|u(x,t) - z^{(2)}(x,t)| \le M \left[N^{-2} \ln^2 N + N_0^2 \right], \quad (x,t) \in \overline{G}_h. \qquad (10.25)$$

Theorem 10.4.1 *Let for the data of the boundary value problem* (10.2), (10.1), *the following conditions be satisfied:*
(i) $a, b, c, p, f \in C^{l,l/2}(\overline{G})$, $\varphi \in C^{l,l/2}(\overline{G})$, $l = 6 + \alpha$, $\alpha > 0$,
(ii) (10.3), (10.15), (10.24),
(iii) (10.17), *where* $n = 5$.
And let the components of the solution of the problem (10.2), (10.1) *in the representation* (10.14) *satisfy a priori estimates* (10.19), *where* $K = 6$. *Then the solution of the difference scheme* (10.21), (10.23) *converges to the solution of the boundary value problem ε-uniformly at the rate* $\mathcal{O}\left(N^{-2} \ln^2 N + N_0^{-2}\right)$. *The discrete solutions satisfy the estimate* (10.25).

Remark 10.4.1 When the derivatives $(\partial/\partial x)\, u(x,t)$ and $(\partial/\partial t)\, u(x,t)$ in the differential equation from (10.2) are approximated by the central difference derivatives, then we come to the discrete equation

$$\Lambda^c z(x,t) \equiv \left\{ \varepsilon\, a(x,t)\, \delta_{\overline{x}\widehat{x}} + b(x,t)\, \delta_{\mathring{x}} - c(x,t) - p(x,t)\, \delta_{\mathring{t}} \right\} z(x,t) = f(x,t),$$
$$(x,t) \in G_h, \quad t < T; \qquad (10.26)$$

$\delta_{\mathring{x}}\, z(x,t)$, $\delta_{\mathring{t}}\, z(x,t)$ are the central difference derivatives. The equation (10.26) approximates the equation in (10.2), on uniform meshes, with the second order of accuracy. But schemes based on approximations (10.26) are unstable. Note that

$$\left(\Lambda_{(10.5)} - \Lambda^c_{(10.26)}\right) z(x,t) = \left\{ b(x,t)2^{-1}h^{i-1}\delta_{\overline{x}\widehat{x}} + p(x,t)2^{-1}h_t^k\delta_{\overline{t}\widehat{t}} \right\} z(x,t),$$
$$(x,t) \in G_h, \quad t < T.$$

Thus, the defect correction scheme (10.21), (10.23) is a stabilized variant of the approximation to problem (10.2), (10.1) based on the central difference derivatives in x and t. ∎

Remark 10.4.2 Let the function $z(x,t)$, $(x,t) \in \overline{G}_h$ be the solution of the difference scheme (10.21), (10.23). We construct a function $\overline{z}(x,t)$ which is an extension of $z(x,t)$ on the set $\overline{G}$:

$$\overline{z}(x,t) = \overline{z}(x,t; z(\cdot)), \quad (x,t) \in \overline{G}. \qquad (10.27)$$

In the case when the difference scheme converges ε-uniformly at the rate $\mathcal{O}\left(N^{-2} \ln^2 N + N_0^{-2}\right)$, the function $\overline{z}(x,t)$ is a bilinear interpolant on elementary rectangles generated by straight lines passing through the nodes in the mesh $\overline{G}_h$ in parallel to the coordinate axes. Let the difference scheme converge on $\overline{G}_h$ at the rate $\mathcal{O}\left(N^{-2}\ln^2 N + N_0^{-k_0}\right)$, where $k_0 > 2$; in this case the function $\widetilde{z}(x,t)$ is constructed in the following way. On straight lines passing through the nodes in the the mesh $\overline{\omega}$, we construct interpolants $\widetilde{z}(x,t)$, $x \in \overline{\omega}$, $t \in [0,T]$. The function $\widetilde{z}(x,t)$ on the interval $[t^j, t^{j+1}]$, where $t^j, t^{j+1} \in \overline{\omega}_0$ and $x \in \overline{\omega}$, is the Lagrange interpolation polynomial that is constructed using the values of $z(x,t)$ in sequential k nodes $t^{j_1}, t^{j_1+1}, \ldots, t^{j_1+k-1} \in \overline{\omega}_0$ including the nodes t^j, t^{j+1} [16]. The function $\widetilde{z}(x,t)$ on $\overline{G}$ is the linear interpolant in x of the function $\widetilde{z}(x,t)$.

In the case when the hypotheses of Theorem 10.4.1 hold, for the interpolant

$$\overline{z}(x,t) = \overline{z}_{(10.27)}(x,t; z^{(2)}(\cdot)), \quad (x,t) \in \overline{G},$$

one has the estimate (10.25), where $z^{(2)}(x,t)$ and $\overline{G}_h$ are $\overline{z}(x,t)$ and $\overline{G}$, respectively. ∎

Remark 10.4.3 The defect correction method allows us to construct schemes whose solution interpolants converge to the solution of the boundary value problem ε-uniformly at the rate $\mathcal{O}\left(N^{-2}\ln^2 N + N_0^{-k_0}\right)$, where $k_0 > 2$ (the technique to construct schemes that converge ε-uniformly for $k_0 = 3$ can be found, e.g., in [46, 47, 50, 51, 52]). ∎

10.5   The Richardson extrapolation scheme

Here we give a Richardson extrapolation method (b) that is used to improve accuracy of the solution of the special basic scheme.

10.5.1. On the set $\overline{G}$ we construct grids

$$\overline{G}_h^i = \overline{G}_h^{si} = \overline{\omega}^{*i} \times \overline{\omega}_0^i, \quad i = 1,2, \tag{10.28a}$$

$$\overline{\omega}^{*i} = \overline{\omega}_{h(10.28)}^{*i}(\sigma), \quad \overline{\omega}_0^i = \overline{\omega}_0^{iu} \text{ is uniform mesh.}$$

Here $\overline{G}_h^2$ is $\overline{G}_{h(10.11a)}$, and $\overline{\omega}^{*2} = \overline{\omega}_{h(10.11)}^*(\sigma)$, where

$$\sigma = \sigma_{(10.11b)}(\varepsilon, N, l) \text{ for } l = 2; \tag{10.28b}$$

$\overline{G}_h^1$ is a "coarsened" mesh. The piecewise-uniform meshes $\overline{\omega}^{*1}$ and $\overline{\omega}^{*2}$ that are defined by the parameter σ have one and the same transition point. The step-sizes in the mesh $\overline{\omega}^{*1}$ on the intervals $[0, \sigma]$ and $[\sigma, d]$ are k times larger than the step-sizes in the mesh $\overline{\omega}^{*2}$, and the step-size of the mesh $\overline{\omega}_0^1$ (on the interval $[0, T]$) is k times larger than the step-size in the mesh $\overline{\omega}_0^2$. Here $k^{-1} N + 1$ and $k^{-1} N_0 + 1$ are the numbers of nodes in the meshes $\overline{\omega}^{*1}$ and $\overline{\omega}_0^1$, respectively, while $N + 1$ and $N_0 + 1$ are those in $\overline{\omega}^{*2}$ and $\overline{\omega}_0^{*2}$, respectively. Let

$$\overline{G}_h^0 = \overline{G}_h^1 \cap \overline{G}_h^2. \qquad (10.28c)$$

$\overline{G}_h^0 = \overline{G}_h^1$ if k is an integer $(k \geq 2)$, $\overline{G}_h^0 \neq \overline{G}_h^1$ if k is a noninteger.

Let $z^i(x, t)$, $(x, t) \in \overline{G}_h^i$, for $i = 1, 2$ be solutions of the difference schemes

$$\Lambda_{(10.5)} z^i(x, t) = f(x, t), \quad (x, t) \in G_h^i, \qquad (10.29a)$$

$$z^i(x, t) = \varphi(x, t), \quad (x, t) \in S_h^i, \quad i = 1, 2.$$

The solutions $z_{(10.29)}^1(x, t)$, $(x, t) \in \overline{G}_h^1$ and $z_{(10.29)}^2(x, t)$, $(x, t) \in \overline{G}_h^2$ admit expansions with respect to N^{-1} and N_0^{-1}, where the function $u(x, t)$, $(x, t) \in \overline{G}$, i.e., the solution of the boundary value problem, is the main term. Considering a linear combination of the grid solutions $z^1(x, t)$ and $z^2(x, t)$, we obtain on $\overline{G}_h^0$ more accurate solution than $z^1(x, t)$ and $z^2(x, t)$. Set

$$z^0(x, t) = \gamma\, z^1(x, t) + (1 - \gamma)\, z^2(x, t), \quad (x, t) \in \overline{G}_h^0, \qquad (10.29b)$$

$$\gamma = \gamma(k) = -(k - 1)^{-1}.$$

We say that the function $z_{(10.29)}^0(x, t)$, $(x, t) \in \overline{G}_h^0$, is the solution of the difference scheme (10.29), (10.28), i.e., *the scheme based on the Richardson extrapolation method on two embedded grids*, or briefly, *the Richardson scheme*. We call the functions $z_{(10.29)}^1(x, t)$, $(x, t) \in \overline{G}_h^1$ and $z_{(10.29)}^2(x, t)$, $(x, t) \in \overline{G}_h^2$, *the components of the solution* to the Richardson scheme. The coefficient γ in (10.29b) is chosen by such a way that the function $z^0(x, t)$, in regard to the expansion (10.30) with respect to the effective step-sizes N^{-1} and N_0^{-1} of the grid $\overline{G}_h^2$, does not contain the functions $u_1(x, t)$ and $u_0(x, t)$ (see the expansion (10.30)).

10.5.2. To justify convergence of the scheme (10.29), (10.28), we apply a technique similar to one used in [53, 197]). It is convenient to consider expansions of the functions $z^i(x, t)$, $(x, t) \in \overline{G}_h^i$, for $i = 1, 2$, with respect to the values N^{-1} and N_0^{-1}

$$z^i(x, t) = u(x, t) + k^{(i-1)} \left[N^{-1} u_1(x, t) + N_0^{-1} u_0(x, t) \right] + v^i(x, t), \quad (10.30)$$

$$(x, t) \in \overline{G}_h^i, \quad i = 1, 2,$$

where $v^i(x, t)$ is the remainder term and $k = k_{(10.28)}$. Here the function $u_0(x, t)$ is the solution of the problem

$$L_{(10.2)} u_0(x, t) = -2^{-1} T\, p(x, t) \frac{\partial^2}{\partial t^2} u(x, t), \quad (x, t) \in G,$$

$$u_0(x, t) = 0, \quad (x, t) \in S.$$

Write the function $u_1(x, t)$ as the sum of the functions

$$u_1(x, t) = u_{11}(x, t) + u_{12}(x, t) + u_{13}(x, t), \quad (x, t) \in \overline{G}.$$

Here $u_{1j}(x, t)$, $(x, t) \in \overline{G}$, are solutions of the problems

$$L_{(10.2)} \, u_{11}(x, t) = -\sigma \, b(x, t) \frac{\partial^2}{\partial x^2} V(x, t), \quad (x, t) \in G,$$

$$u_{11}(x, t) = 0, \quad (x, t) \in S;$$

$$L_{(10.2)} \, u_{12}(x, t) = -(d - \sigma) \, b(x, t) \frac{\partial^2}{\partial x^2} U(x, t), \quad (x, t) \in G,$$

$$u_{12}(x, t) = 0, \quad (x, t) \in S;$$

$$L_{(10.2)} \, u_{13}(x, t) = \left\{ \begin{array}{ll} (d - 2\sigma) \, b(x, t) \dfrac{\partial^2}{\partial x^2} \, U(x, t), & x < \sigma \\ 0, & x > \sigma \end{array} \right\}, \quad (x, t) \in G,$$

$$u_{13}(x, t) = 0, \quad (x, t) \in S.$$

Here $U(x, t) = U_{(10.14)}(x, t)$, $V(x, t) = V_{(10.14)}(x, t)$, $(x, t) \in \overline{G}$, and $\sigma = \sigma_{(10.28)}$.

The functions $u_0(x, t)$, $u_{11}(x, t)$, $u_{12}(x, t)$, $u_{13}(x, t)$, $(x, t) \in \overline{G}$, satisfy the estimates

$$|u_0(x, t)|, \ |u_{12}(x, t)| \leq M, \tag{10.31}$$

$$|u_{11}(x, t)| \leq M \ln N, \quad |u_{13}(x, t)| \leq M \, \sigma, \quad (x, t) \in \overline{G}.$$

The components $u_0(x, t)$, $u_{11}(x, t)$, and $u_{12}(x, t)$ are sufficiently smooth on $\overline{G}$, and the component $u_{13}(x, t)$ is sufficiently smooth on $\overline{G}$ on the sets $x \leq \sigma$ and $x \geq \sigma$. Taking into account the estimates of the components $u_1(x, t)$ and $u_0(x, t)$, one can find the estimate for the remainder term $v^i(x, t)$

$$|v^i(x, t)| \leq M \left[N^{-2} \ln^2 N + N_0^{-2} \right], \quad (x, t) \in \overline{G}_h^i, \quad i = 1, 2. \tag{10.32}$$

With regard to the expansion (10.30), the representation (10.29b) and the estimates (10.31) and (10.32), we obtain the following estimate for the solution $z^0(x, t)$ of the Richardson scheme:

$$|u(x, t) - z^0(x, t)| \leq M \left[N^{-2} \ln^2 N + N_0^{-2} \right], \quad (x, t) \in \overline{G}_h^0. \tag{10.33}$$

The interpolant $\overline{z}(x, t) = \overline{z}_{(10.27)}(x, t; z^0(\cdot))$, $(x, t) \in \overline{G}$, satisfies the estimate

$$|u(x, t) - \overline{z}(x, t)| \leq M \left[N^{-2} \ln^2 N + N_0^{-2} \right], \quad (x, t) \in \overline{G}. \tag{10.34}$$

Theorem 10.5.1 *Let the hypotheses of Theorem 10.4.1 be fulfilled. Then the solution of the Richardson difference scheme* (10.29), (10.28) *converges to the solution of the boundary value problem at the rate* $\mathcal{O}\left(N^{-2}\ln^2 N + N_0^{-2}\right)$, *i.e., ε-uniformly. For the discrete solution and the interpolant, the estimates* (10.33) *and* (10.34) *are valid.*

Remark 10.5.1 The constructed Richardson method allows us to construct schemes that converge ε-uniformly at the rate $\mathcal{O}\left(N^{-2}\ln^2 N + N_0^{-k_0}\right)$, where $k_0 \geq 2$ (a technique for a parabolic reaction-diffusion equation can be found, e.g., in [120]). ∎

10.6   Asymptotic constructs

In Sections 10.6–10.8, we consider the approach (c) based on asymptotic constructs that is used to construct a scheme with improved accuracy for the boundary value problem (10.2), (10.1).

For finite and not too small values of the parameter ε, when approximating the first-order derivative in x by the central difference derivative, a difference scheme on piecewise-uniform grids preserves monotonicity and ensures order of convergence for the solution with respect to the variable x close to two (see Section 10.7). Asymptotic constructs considered here, which use two terms in outer asymptotic expansion in ε, allow us to approximate the solution of the boundary value problem with accuracy of $\mathcal{O}\left(\varepsilon^2\right)$. Grid approximations based on these constructs lead to discrete solution that converges with the second order in N_0^{-1} and ε and close to second in N^{-1} (see Section 10.8) as $N, N_0 \to \infty$ and $\varepsilon \to 0$. A scheme that converges ε-uniformly with improved accuracy is considered in Section 10.8.

In this section we give the problem formulations for main terms in the asymptotics of the solution that are considered in a sufficiently small neighborhood of the output part of the boundary S^l and outside it. We shall use these problem formulations when constructing schemes for small values of the parameter.

Write the set $\overline{G}$ as the sum of two sets

$$\overline{G} = \overline{G}^{(1)} \cup \overline{G}^{(2)}, \quad S^{(k)} = \overline{G}^{(k)} \setminus G^{(k)}, \quad k = 1, 2; \tag{10.35}$$

$$G^{(1)} = (0, \sigma) \times (0, T], \quad G^{(2)} = [\sigma, d) \times (0, T];$$

the left "side" of the set $G^{(2)}$ belongs to this set. The condition imposed on the value of the parameter σ, which defines the domain decomposition (10.35), is given below.

Let us introduce the function $u^2(x,t)$, $(x,t) \in \overline{G}^{(2)}$, i.e., two first terms in the outer asymptotic expansion of the solution of the problem (10.2), (10.1). Set

$$u^2(x,t) = u_0^2(x,t) + \varepsilon\, u_1^2(x,t), \quad (x,t) \in \overline{G}^{(2)}.$$

The components $u_i^2(x,t)$ can be found by solving the problems

$$L^{(1)}\, u_0^2(x,t) = f(x,t), \quad (x,t) \in G^{(2)}, \tag{10.36a}$$

$$u_0^2(x,t) = \varphi(x,t), \quad (x,t) \in S^{(2)};$$

$$L^{(1)}\, u_1^2(x,t) = -a(x,t)\,\frac{\partial^2}{\partial x^2}\, u_0^2(x,t), \quad (x,t) \in G^{(2)}, \tag{10.36b}$$

$$u_1^2(x,t) = 0, \quad (x,t) \in S^{(2)},$$

where

$$L^{(1)}\, u(x,t) \equiv \left\{ b(x,t)\,\frac{\partial}{\partial x} - c(x,t) - p(x,t)\,\frac{\partial}{\partial t} \right\} u(x,t), \quad (x,t) \in G^{(2)}.$$

We say that the function $u^2(x,t)$, $(x,t) \in \overline{G}^{(2)}$, is the solution of problem (10.36a; 10.36b).

Next we find the function $u^1(x,t)$, $(x,t) \in \overline{G}^{(1)}$ (the main term in the inner asymptotic expansion of the solution of problem (10.2), (10.1)), by solving the problem

$$L_{(10.2)}\, u^1(x,t) = f(x,t), \quad (x,t) \in G^{(1)}, \tag{10.36c}$$

$$u^1(x,t) = \begin{cases} \varphi(x,t), & (x,t) \in S^{(1)} \cap S, \\ u^2(x,t), & (x,t) \in S^{(1)} \setminus S. \end{cases}$$

The function $u^1(x,t)$ is a solution of the parabolic equation, but in a small (in x) subdomain.

Using asymptotic expansions of the components in the representation of the solution to the boundary value problem (see (10.14)) provided that

$$\sigma = \sigma(\varepsilon,\delta) = \min\left[2^{-1}\, d,\, q\, m^{-1} \varepsilon\, \ln(1/\delta) \right], \tag{10.37a}$$

where $q \geq 1$, $m = m_{(10.19)}$, and δ is a sufficiently small value (it is chosen below), we find the estimate

$$|u(x,t) - u^k(x,t)| \leq M\left[\varepsilon^2 + \delta^q\right], \quad (x,t) \in \overline{G}^{(k)}, \quad k = 1,2. \tag{10.37b}$$

Theorem 10.6.1 *Let the hypotheses of Theorem 10.4.1 be fulfilled, and let also the condition (10.37a) hold. Then the functions $u^k(x,t)$, $(x,t) \in \overline{G}^{(k)}$, for $k = 1,2$, satisfy the estimate (10.37b).*

Remark 10.6.1 When the domain $\overline{G}$ is subdivided into subdomains (10.35), where $\sigma = \sigma_{(10.37a)}$, for $\delta \approx \varepsilon$ and $q \geq 2$, the solutions of the subproblems (10.36a; 10.36b), (10.36c) define on the subdomains $\overline{G}^{(k)}$ the functions $u^k(x,t)$, for $k = 1, 2$, which approximate the solution of the boundary value problem (10.2), (10.1) with accuracy $\mathcal{O}\left(\varepsilon^2\right)$. ∎

10.7 A scheme with improved convergence for finite values of ε

Here we give for the problem (10.2), (10.1) an ε-uniformly convergent finite difference scheme having (unlike the scheme (10.5), (10.11)) the rate of ε-uniform convergence of order close to two in x, but for not too small values of the parameter ε, and of the second order in t. The finite difference scheme is constructed by approximation of the problem (10.2), (10.1) with using the first central difference derivatives in x and defect correction for the residual of the implicit difference derivative in t (see also constructions in Section 10.4).

10.7.1. On the set $\overline{G}$, we introduce the grid

$$\overline{G}_h = \overline{G}_h^s = \overline{\omega}^* \times \overline{\omega}_0, \tag{10.38a}$$

where $\overline{\omega}^* = \overline{\omega}^*_{(10.11)}(\sigma)$ under the condition

$$\sigma = \sigma(\varepsilon, N, l = 2) = \min\left[2^{-1}d, \ 2\,m^{-1}\varepsilon \ln N\right], \quad m = m_{(10.19)}. \tag{10.38b}$$

On the grid $\overline{G}_h$ we consider the difference scheme

$$\Lambda^{(2)} z(x,t) = f(x,t), \quad (x,t) \in G_h, \tag{10.39}$$

$$z(x,t) = \varphi(x,t), \quad (x,t) \in S_h.$$

Here

$$\Lambda^{(2)} z(x,t) \equiv \{\Lambda_2 + \Lambda_1\}\, z(x,t),$$

$$\Lambda_2\, z(x,t) \equiv \{\varepsilon\, a(x,t)\, \delta_{\overline{x}\widehat{x}} + b(x,t)\, \delta_{\widehat{x}}\}\, z(x,t), \quad (x,t) \in G_h \cap G^{(1)};$$

$$\Lambda_2\, z(x,t) \equiv \begin{cases} \{\varepsilon\, a(x,t)\, \delta_{\overline{x}\widehat{x}} + b(x,t)\, \delta_{\widehat{x}}\}\, z(x,t), \\ \quad \text{for } \varepsilon\, N \geq M_0 \\ \{\varepsilon\, a(x,t)\, \delta_{\overline{x}\widehat{x}} + b(x,t)\, \delta_x\}\, z(x,t), \\ \quad \text{for } \varepsilon\, N < M_0 \end{cases}, \quad (x,t) \in G_h \cap G^{(2)};$$

$$\Lambda_1\, z(x,t) \equiv \{-p(x,t)\, \delta_{\overline{t}} - c(x,t)\}\, z(x,t), \quad (x,t) \in G_h;$$

$\overline{G}^{(k)} = \overline{G}^{(k)}_{(10.35)}(\sigma)$, $\sigma = \sigma_{(10.38)}$ and M_0 is any number satisfying the condition $M_0 \geq 2\,d\,\max_{\overline{G}}\left[a^{-1}(x,t)\,b(x,t)\right]$.

10.7.2. Scheme (10.39), (10.38) is monotone ε-uniformly. For solutions of the scheme (10.39), (10.38), using *a priori* estimates (10.19) and the maximum principle, we obtain the estimate

$$|u(x,t) - z(x,t)| \leq \begin{cases} M\left[N^{-1} + N_0^{-1}\right] & \text{if } \varepsilon N < M_0 \\ M\left[N^{-2}(\varepsilon + \ln^{-1}N)^{-2} + N_0^{-1}\right] & \text{if } \varepsilon N \geq M_0 \end{cases},$$
$$(10.40)$$
$$(x,t) \in \overline{G}_h,$$

where $M_0 = M_{0(10.39)}$. We have also the ε-uniform estimate

$$|u(x,t) - z(x,t)| \leq M\left[N^{-1} + N_0^{-1}\right], \quad (x,t) \in \overline{G}_h; \qquad (10.41)$$

thus, the scheme (10.39), (10.38) converges ε-uniformly with first-order accuracy.

The use of the first central difference derivative in x for not too small values of the parameter ε, precisely, for

$$\varepsilon N \geq M_{0(10.39)}, \qquad (10.42)$$

allowed us to obtain the order of the convergence rate in x close to two. According to (10.40), and under the condition (10.42), we have the estimate

$$|u(x,t) - z(x,t)| \leq M\left[N^{-2}\left(\varepsilon + \ln^{-1}N\right)^{-2} + N_0^{-1}\right], \quad (x,t) \in \overline{G}_h,$$

which is unimprovable with respect to N, N_0, ε.

10.7.3. We now give a difference scheme of improved convergence rate in t under the condition (10.42). For this we apply an approach based on the defect correction technique (see Section 10.4 and also, e.g., [46, 47, 50, 51, 52]); we use the scheme (10.39), (10.38) as the basic scheme.

For the problem (10.2), (10.1) we consider the difference scheme

$$\Lambda^{(2)}_{(10.39)}\, z^{(2)}(x,t) = f(x,t) + \psi^0(x,t), \quad (x,t) \in G_h, \qquad (10.43\text{a})$$

$$z^{(2)}(x,t) = \varphi(x,t), \qquad (x,t) \in S_h.$$

Here $\overline{G}_h = \overline{G}_{h(10.38)}$ and $\psi^0(x,t)$ is the correcting term

$$\psi^0(x,t) = \psi^0_{(10.22)}(x,t;\, z^{(1)}(x,t)), \quad (x,t) \in G_h, \qquad (10.43\text{b})$$

where $z^{(1)}(x,t)$, $(x,t) \in \overline{G}_h$, is the solution of the discrete problem (10.39), (10.38). We call the function $z^{(2)}(x,t)$, $(x,t) \in \overline{G}_h$, the solution of the difference scheme $\{(10.43), (10.39), (10.38)\}$, (10.42), i.e., *the improved difference scheme in x and t* $\{(10.43), (10.39), (10.38)\}$ *in the case of condition* (10.42).

Assume that the conditions (10.15) and (10.24) are satisfied. In this case, the solution of the difference scheme {(10.43), (10.39), (10.38)}, (10.42) satisfies the estimate (see its proof, e.g., in [50, 51, 52])

$$|u(x,t) - z^{(2)}(x,t)| \leq M\left[N^{-2}(\varepsilon + \ln^{-1} N)^{-2} + N_0^{-2}\right], \quad (10.44)$$
$$(x,t) \in \overline{G}_h, \quad \varepsilon N \geq M_0;$$

this estimate is unimprovable with respect to N, N_0, ε. We have also the ε-uniform estimate

$$|u(x,t) - z^{(2)}(x,t)| \leq M\left[N^{-2}\ln^2 N + N_0^{-2}\right], \; (x,t) \in \overline{G}_h, \; \varepsilon N \geq M_0. \; (10.45)$$

Thus, for not too small values of the parameter ε (for $\varepsilon N \geq M_{0(10.39)}$) the difference scheme {(10.43), (10.39), (10.38)}, (10.42) converges ε-uniformly with order of the convergence rate close to two in x and equal to two in t.

Theorem 10.7.1 *Let the hypotheses of Theorem 10.4.1 be fulfilled. Then the difference scheme {(10.43), (10.39), (10.38)}, (10.42) converges ε-uniformly. The discrete solutions satisfy the estimates (10.44), (10.45).*

10.8 Schemes based on asymptotic constructs

For small values of the parameter ε, we construct a higher-order accurate scheme by approximating the problems (10.36a; 10.36c). On the basis of this scheme and the scheme {(10.43), (10.39), (10.38)}, (10.42) we construct a scheme with improved accuracy that converges ε-uniformly.

10.8.1. The grid $\overline{G}_{h(10.38)}$ is decomposed into the sum of the grid sets

$$\overline{G}_h = \overline{G}_h^{(1)} \cup \overline{G}_h^{(2)}, \; \overline{G}_h^{(k)} = \overline{G}^{(k)} \cap \overline{G}_h, \; S_h^{(k)} = S^{(k)} \cap \overline{G}_h, \; k = 1, 2, \; (10.46)$$

where $\overline{G}^{(k)} = \overline{G}_{(10.35)}^{(k)}(\sigma)$ with $\sigma = \sigma_{(10.38)}$. On the grid $\overline{G}_h^{(2)}$, we approximate problem (10.36a; 10.36b) by the difference scheme

$$\Lambda^{(1)} z_0^2(x,t) = f(x,t), \quad (x,t) \in G_h^{(2)},$$
$$z_0^2(x,t) = \varphi(x,t), \quad (x,t) \in S_h^{(2)}; \tag{10.47a}$$

$$\Lambda^{(1)} z_1^2(x,t) = -a(x,t)\delta_{x\overline{x}} z_0^2(x,t) + \psi^1(x,t) + \psi^0(x,t), \; (x,t) \in G_h^{(2)},$$
$$z_1^2(x,t) = 0, \quad (x,t) \in S_h^{(2)}. \tag{10.47b}$$

Here

$$\Lambda^{(1)} z(x,t) \equiv \left\{b(x,t)\,\delta_x - c(x,t) - p(x,t)\,\delta_{\overline{t}}\right\} z(x,t), \quad (x,t) \in G_h^{(2)}; \tag{10.47c}$$

$$\psi^1(x,t) \equiv 2^{-1}\varepsilon^{-1}h^i\,b(x,t) \left\{ \begin{array}{l} \delta_{x\overline{x}}\,z_0^2(x,t),\ x > \sigma \\ \delta_{x\overline{x}}\,z_0^2(\widehat{x},t),\ x = \sigma \end{array} \right\},$$

$$\psi^0(x,t) \equiv 2^{-1}\varepsilon^{-1}\tau\,p(x,t) \left\{ \begin{array}{l} \dfrac{\partial^2}{\partial t^2}\,u(x,0),\ t = \tau \\ \delta_{2\overline{t}}\,z_0^2(x,t),\ t \geq 2\tau \end{array} \right\},$$

$$\delta_{x\overline{x}}\,z_0^2(\widehat{x},t) = \delta_{x\overline{x}}\,z_0^2(x^{i+1},t), \quad x = x^i;\quad h^i = x^{i+1} - x^i.$$

The function $z^2(x,t) = z_0^2(x,t) + \varepsilon\,z_1^2(x,t)$, $(x,t) \in \overline{G}_h^{(2)}$, is called the solution of the problem (10.47a; 10.47b).

The operator $\Lambda^{(1)}(\cdot)$ is monotone.

For the function $z^2(x,t)$, $(x,t) \in \overline{G}_h^{(2)}$, we obtain the estimate

$$|u^2(x,t) - z^2(x,t)| \leq M\,[\varepsilon^2 + N^{-2} + N_0^{-2}], \quad (x,t) \in \overline{G}_h^{(2)}. \tag{10.48}$$

Note that the function $z_{0(10.47a)}^2(x,t)$, $(x,t) \in \overline{G}_h^{(2)}$, approximates the function $u_{0(10.36a)}^2(x,t)$ only with the first-order accuracy. The use of the additional correcting terms

$$2^{-1}\,[h^i\,b(x,t)\,\delta_{x\overline{x}} + \tau\,p(x,t)\,\delta_{2\overline{t}}]\,z_0^2(x,t)$$

in the equation from (10.47b) to compute the function $z_1^2(x,t)$ allows us to improve the approximation of the function $u_{(10.36a;\ 10.36b)}^2(x,t)$ and accuracy of the numerical solution of the problem (10.2), (10.1) on $\overline{G}^{(2)}$ (see the estimate (10.48) on the set $\overline{G}_h^{(2)}$).

On the grid $\overline{G}_{h(10.46)}^{(1)}$, the problem (10.36c) is approximated by the difference scheme

$$\Lambda^1 z^1(x,t) \equiv \{\varepsilon\,a(x,t)\delta_{x\overline{x}} + b(x,t)\delta_{\widehat{x}} - c(x,t) - p(x,t)\delta_{\overline{t}}\}z^1(x,t) = f(x,t),$$

$$(x,t) \in G_h^{(1)}, \tag{10.47d}$$

$$z^1(x,t) = \left\{ \begin{array}{l} \varphi(x,t),\ (x,t) \in S_h^{(1)} \cap S, \\ z^2(x,t),\ (x,t) \in S_h^{(1)} \setminus S. \end{array} \right.$$

The operator Λ^1 is monotone on the grid set $G_h^{(1)}$.

The proximity of the solutions of the boundary value problem (10.36c) on the set $\overline{G}_h^{(1)}$ and its grid approximation (10.47d) follows from the *a priori* estimates for the solution of the differential problem and from the proximity of the functions $u^2(x,t)$ and $z^2(x,t)$ on the set $S_h^{(1)} \cap \overline{G}^{(2)}$. For the function $z^1(x,t)$, $(x,t) \in \overline{G}_h^{(1)}$, we obtain the estimate

$$|u^1(x,t) - z^1(x,t)| \leq M\,[\varepsilon^2 + \varepsilon\,N_0^{-1}\ln N + N^{-2}(\varepsilon + \ln^{-1} N)^{-2} + N_0^{-2}],$$

$$(x,t) \in \overline{G}_h^{(1)}. \tag{10.49}$$

10.8.2. On the grid $\overline{G}_{h(10.38)}$, we define the function $z(x,t)$ by the relation

$$z(x,t) = \{z^k(x,t), \quad (x,t) \in \overline{G}_h^{(k)}, \quad k = 1,2\}, \quad (x,t) \in \overline{G}_h, \qquad (10.50)$$

where $z^k(x,t)$, $(x,t) \in \overline{G}_h^{(k)}$, are the solutions of problems (10.47a; 10.47b) and (10.47c). The function $z_{(10.50)}(x,t)$, $(x,t) \in \overline{G}_h$, is called the solution of the difference scheme (10.47), (10.38), i.e., *the improved scheme in x, t, and ε (for $\varepsilon \to 0$) based on the asymptotics constructs.*

For the solution of the scheme (10.47), (10.38), by virtue of error bounds (10.48), (10.49), we have the estimate

$$|u(x,t) - z(x,t)| \le M\left[\varepsilon^2 + \varepsilon\,N_0^{-1}\ln N + N^{-2}(\varepsilon + \ln^{-1} N)^{-2} + N_0^{-2}\right],$$

$$(x,t) \in \overline{G}_h.$$

For sufficiently small values of the parameter ε (namely, for $\varepsilon \le M\,N^{-1}$), the scheme (10.47), (10.38) converges at the rate $\mathcal{O}\left(N^{-2}\ln^2 N + N_0^{-2}\right)$, i.e.,

$$|u(x,t) - z(x,t)| \le M\left[N^{-2}\ln^2 N + N_0^{-2}\right], \ (x,t) \in \overline{G}_h, \ \varepsilon \le M\,N^{-1}; \quad (10.51)$$

this estimate is unimprovable (with respect to N, N_0 for $\varepsilon \le M\,N^{-1}$).

10.8.3. The scheme with *improved ε-uniform order of convergence* for $\varepsilon \in (0,1]$ is constructed on the basis of schemes $\{(10.43), (10.39), (10.38)\}$, (10.42) and (10.47), (10.38).

Let

$$\varepsilon_0 = \varepsilon_0(N) = M^0_{(10.39)}N^{-1}. \qquad (10.52)$$

We use the scheme (i): (10.43), (10.39), (10.38) for $\varepsilon \ge \varepsilon_0$, , and we use the scheme (ii): (10.47), (10.38) for $\varepsilon < \varepsilon_0$. Then $z(x,t)$, $(x,t) \in \overline{G}_h$, is a solution of the schemes: (i) for $\varepsilon \ge \varepsilon_0$ and (ii) for $\varepsilon < \varepsilon_0$. We say that the function $z(x,t)$, $(x,t) \in \overline{G}_h$, is the solution of the difference scheme $\{(10.43), (10.39), (10.47), (10.52), (10.38)\}$, i.e., *the difference scheme with improved accuracy constructed on the basis of asymptotic constructs.*

From the estimates (10.44), (10.45), and (10.51) it follows that the solution of the scheme $\{(10.43), (10.39), (10.47), (10.52), (10.38)\}$ satisfies the estimate

$$|u(x,t) - z(x,t)| \le M \left\{ \begin{array}{ll} N^{-2}\ln^2 N + N_0^{-2}, & \varepsilon < \varepsilon_0 \\ N^{-2}\left(\varepsilon + \ln^{-1} N\right)^{-2} + N_0^{-2}, & \varepsilon \ge \varepsilon_0 \end{array} \right\}, \quad (x,t) \in \overline{G}_h,$$

$$(10.53a)$$

and also the ε-uniform estimate

$$|u(x,t) - z(x,t)| \le M \left[N^{-2}\ln^2 N + N_0^{-2}\right], \quad (x,t) \in \overline{G}_h. \qquad (10.53b)$$

The estimates (10.53a) and (10.53b) are unimprovable with respect to N, N_0, ε and N, N_0, respectively. The scheme $\{(10.43), (10.39), (10.47), (10.52),$

(10.38)} converges ε-uniformly with the second order up to a logarithmic factor in x and with the second order in t.

For the interpolant $\overline{z}(x,t) = \overline{z}_{(10.27)}(x, t; \, z(\cdot))$, $(x,t) \in \overline{G}$, one has the estimate

$$|u(x,t) - \overline{z}(x,t)| \leq M \left[N^{-2} \ln^2 N + N_0^{-2}\right], \quad (x,t) \in \overline{G}. \qquad (10.54)$$

Theorem 10.8.1 *Let the hypotheses of Theorem 10.4.1 hold. Then the solution of the difference scheme based on the asymptotic constructs {(10.43), (10.39), (10.47), (10.52), (10.38)} converges to the solution of the boundary value problem (10.2), (10.1) ε-uniformly at the rate $\mathcal{O}\left(N^{-2} \ln^2 N + N_0^{-2}\right)$ as N, $N_0 \to \infty$. The discrete solutions and the interpolant satisfy the estimates (10.53) and (10.54).*

Remark 10.8.1 For fixed values of the parameter ε, the difference scheme (10.43), {(10.39), (10.47), (10.52), (10.38)} converges with the second-order accuracy.

10.9 Boundary value problem for singularly perturbed parabolic convection-diffusion equation with piecewise-smooth initial data

In Sections 10.9–10.12, we consider a boundary value problem for a parabolic convection-diffusion equation with a piecewise-smooth initial data. Discontinuity of the first-order x-derivative of the initial function $\varphi(x,t)$ brings reduction of ε-uniform convergence rate; in this case a classical difference scheme on piecewise-uniform grid converges at the rate $\mathcal{O}(N^{-1/2}+N_0^{-1/2})$. The use of a *technique* of *additive splitting of a singularity* (generated by the discontinuity of the derivative) allows us to construct a difference scheme that converges ε-uniformly at the rate $\mathcal{O}\left(N^{-1} \ln N + N_0^{-1} \ln N_0\right)$. To improve the convergence rate, we combine the method of the additive splitting of singularities and the method either (a), or (b), or (c) (see the methods (a), (b), and (c) in Sections 10.4, 10.5, and 10.6–10.8, respectively) that bring us to a scheme which converges ε-uniformly with the convergence orders in x close to two and in t equal and more than two.

10.9.1 Problem (10.56) with piecewise-smooth initial data

In the domain $\overline{G}$ with boundary S, where

$$\overline{G} = G \cup S, \quad G = D \times (0, T], \quad D = \{x : x \in (-d, d)\}, \qquad (10.55)$$

we consider the Dirichlet problem for the singularly perturbed parabolic equation with constant coefficients and piecewise-smooth initial data

$$L_{(10.56)}\, u(x,t) = f(x,t), \quad (x,t) \in G, \quad u(x,t) = \varphi(x,t), \quad (x,t) \in S. \quad (10.56)$$

Here
$$L_{(10.56)}\, u(x,t) \equiv \varepsilon\, a\, \frac{\partial^2}{\partial x^2} + b\, \frac{\partial}{\partial x} - c - p\, \frac{\partial}{\partial t},$$

$a, b, p > 0$, $c \geq 0$, the right-hand side $f(x,t)$ is sufficiently smooth on $\overline{G}$; the parameter ε takes arbitrary values in $(0,1]$. The boundary function $\varphi(x,t)$ is continuous on S and sufficiently smooth on the lateral part $\overline{S}^L$ and the lower parts $\overline{S}^-$ and $\overline{S}^+$ of the boundary S; the first-order x-derivative of the function $\varphi(x,t)$ has a *discontinuity* of the first kind on the set $S^{(*)}$. Let the lower part of the boundary S be $S_0 = \overline{S}_0^- \cup \overline{S}_0^+$, where

$$\overline{S}_0^- = \{(x,t) : x \in [-d,0), t = 0\} \quad \text{and} \quad \overline{S}_0^+ = \{(x,t) : x \in [0,d), t = 0\};$$

$S^L = \Gamma \times (0,T]$, $\Gamma = \overline{D} \setminus D$; $S^{(*)} = \{(x,t) : x = t = 0\}$; $\overline{S}^+ = S \cap \{x > 0\}$, $S^- \cup S^+ = S^* = S \setminus S^{(*)}$.

By a solution of problem (10.56), we mean a function $u \in C(\overline{G}) \cap C^{2,1}(G)$ that satisfies the differential equation on G and the boundary condition on S.

For simplicity, we assume that compatibility conditions are fulfilled on the set of corner points $S_* = S_0 \cap \overline{S}^L$ that ensure the local smoothness of the solution for fixed ε [41]. Suppose that on the set $\overline{G}^\delta$, i.e., an δ-neighborhood of the set S_*, the following condition is valid:

$$u \in C^{l_0+\alpha,(l_0+\alpha)/2}(\overline{G}^\delta), \quad l_0 \geq 2, \quad \alpha \in (0,1). \quad (10.57)$$

The derivative $(\partial/\partial x)u(x,t)$ is continuous on $\overline{G}^*$, where $\overline{G}^* = \overline{G} \setminus S^{(*)}$, and it is bounded on $\overline{G}^*$ for fixed values of ε and is discontinuous on the set $S^{(*)}$.

10.9.2 The aim of the research

We are interested in an approximation of the solution $u(x,t)$, $(x,t) \in \overline{G}$. Let us specify the behaviour of the solution. Define the set $S^\gamma = \{(x,t) : x = \gamma(t), (x,t) \in \overline{G}\}$, where $x = \gamma(t)$, for $t \geq 0$, is the characteristic of the reduced equation passing through the point $(0,0)$. When the parameter ε tends to zero, in a neighborhood of the sets S^l (the left part of the lateral boundary) and S^γ there appear boundary and transient layers with the typical scales ε and $\varepsilon^{1/2}$, respectively; unlike the boundary layer, the transient layer is weak (the first x-derivative of the transient-layer function is bounded ε-uniformly; see estimates (10.63) in Section 10.10).

For simplicity, we assume that the characteristic S^γ does not meet the boundary S^l, i.e., the transient and boundary layers do not interact.

In the case of piecewise-uniform meshes that condense in the boundary layer, the classical finite difference scheme for the problem (10.56), (10.55) converges ε-uniformly at the rate $\mathcal{O}\left(N^{-1/2} + N_0^{-1/2}\right)$ (see the estimate (10.72) for the solutions of scheme (10.68), (10.71) in Section 10.11). This convergence rate is essentially lower than that for the problems with sufficiently smooth data.

Our aim for the problem (10.56), (10.55) is to construct an improved difference scheme that converges ε-uniformly with the convergence order close to one similar to the schemes for the problems with smooth solutions.

When constructing an improved difference scheme for the problem (10.56), (10.55), we use a method of additive splitting of singularities. We split the singular component, which is the main part of the transient-layer function generated by the discontinuity of the first-order derivative of the initial function. As a result, we come to a problem with an initial condition that has only a discontinuity of the second-order derivative. For such a problem, the solution of a scheme on piecewise-uniform meshes converges already ε-uniformly at the rate $\mathcal{O}\left(N^{-1}\ln N + N_0^{-1}\ln N_0\right)$ that implies convergence of the additive splitting scheme at the same rate.

10.10   *A priori* estimates for the boundary value problem (10.56) with piecewise-smooth initial data

We now give some estimates for the solution of the boundary value problem (10.56) and its derivatives. To derive the estimates, we apply a technique that can be found in [128, 138, 137, 184]. For simplicity, we assume that the functions $f(x,t)$ and $\varphi(x,t)$ are sufficiently smooth on the set $\overline{G}$ and on the sets $\overline{S}_0^-, \overline{S}_0^+, \overline{S}^L$, respectively, i.e.,

$$f \in C^{l,l}(\overline{G}), \quad \varphi \in C^l(\overline{S}_0^-) \cap C^l(\overline{S}_0^+) \cap C^l(\overline{S}^L) \cap C(S), \quad (10.58)$$

$$l = 3n + 1,$$

where $n > 0$. The solution of the problem satisfies condition (10.57), where $l_0 = n + 1$. Furthermore, let the following condition be fulfilled:

$$\frac{\partial^k}{\partial x^k}\varphi(x,t), \quad \frac{\partial^{k_0}}{\partial t^{k_0}}\varphi(x,t) = 0, \quad k + k_0 \leq 3n + 1, \quad (10.59)$$

$$\frac{\partial^{k+k_0}}{\partial x^k\,\partial t^{k_0}}f(x,t) = 0, \quad k + k_0 \leq 3n, \quad (x,t) \in S_*^{(r)}.$$

10.10.1. The set $\overline{G}$ is decomposed into the sum of overlapping sets

$$\overline{G} = \cup_j \overline{G}^j, \quad j = 1, 2, 3, \quad (10.60)$$

where

$$G^1 = G^1(m^1) = \{(x,t) : |x - \gamma(t)| < m^1, \ t \in (0,T]\},$$

$$G^2 = G^2(m^2) = \{(x,t) : x \in (-d, -d+m^2), \ t \in (0,T]\},$$

$$G^3 = G^3(m^3) = G \setminus \{G^1(m^3) \cup G^2(m^3)\}, \quad m^3 < m^1, m^2,$$

G^1 and G^2 are the neighborhoods of the transient and boundary layers, respectively; let $\overline{G}^1 \cap \overline{G}^2 = \emptyset$; G^3 is the domain where the solution is smooth. We denote the solution of the problem (10.56), (10.55) considered on the set $\overline{G}^j$ by $u^j(x,t)$, $j = 1, 2, 3$.

Using results from [128, 138, 137], we find the estimate

$$\left| \frac{\partial^{k+k_0}}{\partial x^k \partial t^{k_0}} u(x,t) \right| \leq M, \quad (x,t) \in \overline{G}^3, \ k + 2k_0 \leq K. \tag{10.61}$$

The value of K is defined by the problem data; $K = n + 1$.

10.10.2. Let us discuss the solution behaviour on the set $\overline{G}^1$ (details see in [184]).

Write the function $u(x,t)$, $(x,t) \in \overline{G}^1$, as the sum of the functions

$$u(x,t) = U^1(x,t) + W^1(x,t), \quad (x,t) \in \overline{G}^1, \tag{10.62a}$$

where $U^1(x,t)$ and $W^1(x,t)$ are the "regular" and singular parts of the solution, i.e.,

$$U^1(x,t) = U^1(x,t; i) = U(x,t) + \sum_{k=i+1}^{K-1} W_k(x,t), \quad (x,t) \in \overline{G}^1; \tag{10.62b}$$

$$W^1(x,t) = W^1(x,t; i) = \sum_{k=1}^{i} W_k(x,t), \quad (x,t) \in \overline{G},$$

where i takes one of its values, either 1 or 2. Here the functions $W_k(x,t)$, for $1 \leq k \leq K - 1$, are elementary functions of the transient layer generated by discontinuity of the k-order derivative of the initial function $\varphi(x,t)$. The function $U_{(10.62b)}(x,t)$, $(x,t) \in \overline{G}^1$, is the component of the solution to the inhomogeneous equation in (10.56), having ε-uniformly bounded derivatives in x and t up to the orders K and $2^{-1}K$, respectively. The functions $U_{(10.62b)}(x,t)$ on the set S^1 satisfy the condition

$$U(x,t) + \sum_{k=1}^{K-1} W_k(x,t) = \left\{ \begin{array}{ll} \varphi(x), & (x,t) \in S^1 \cap S_0 \\ u^3(x,t), & (x,t) \in S^1 \setminus S_0 \end{array} \right\}, \quad (x,t) \in S^1.$$

The functions $W_k(x,t)$, $(x,t) \in \mathbb{R} \times [0,T]$, are solutions of the Cauchy problems

$$L_{(10.56)} W_k(x,t) = 0, \quad (x,t) \in \mathbb{R} \times [0,T],$$

$$W_k(x,0) = 2^{-1} (k!)^{-1} \left[\frac{\partial^k}{\partial x^k} \varphi(x,0) \right] |x| x^{k-1}, \quad x \in \mathbb{R}, \ k = 1, \ldots, K-1.$$

Here $\left[\dfrac{\partial^k}{\partial x^k}\varphi(x,0)\right] = \dfrac{\partial^k}{\partial x^k}\varphi(+0,0) - \dfrac{\partial^k}{\partial x^k}\varphi(-0,0)$, $x=0$, $k \geq 1$, is the jump

of the derivative $\dfrac{\partial^k}{\partial x^k}\varphi(x,0)$ when passing through $x = 0$.

The function $W_1(x,t)$ is defined by the relation (details see in [184])

$$W_1(x,t) = 2^{-1}\left[\frac{\partial}{\partial x}\varphi(0,0)\right]\times \qquad (10.62c)$$

$$\times\left\{(x-\gamma(t))\,v\!\left(2^{-1}\varepsilon^{-1/2}a^{-1/2}p^{1/2}\,(x-\gamma(t))\,t^{-1/2}\right) + \right.$$

$$\left. +2\,\pi^{-1/2}\varepsilon^{1/2}a^{1/2}p^{-1/2}t^{1/2}\,\exp\left(-4^{-1}\varepsilon^{-1}a^{-1}p(x-\gamma(t))^2t^{-1}\right)\right\}\exp(-\alpha t),$$

$$v(\xi) = erf(\xi) = 2\pi^{-1/2}\int_0^\xi \exp(-\alpha^2)d\alpha, \quad \xi \in \mathbb{R}; \quad \gamma(t) = -bp^{-1}t, \quad \alpha = cp^{-1}.$$

For the components in the representation (10.62), taking into account the boundedness of the derivatives of the component $U(x,t)$ on the set $\overline{G}^1$ and the explicit form of the functions $W_k(x,t)$, for $k = 1,\ldots,K-1$, we find the estimates

$$\left|\frac{\partial^{k+k_0}}{\partial x^k \partial t^{k_0}}U^1(x,t)\right| \leq M\left[1 + \varepsilon^{(i+1-k-k_0)/2}\rho^{i+1-k-k_0} + \right. \qquad (10.63)$$

$$\left. + \varepsilon^{(i+1-k)/2}\rho^{i+1-k-2k_0}\right], \quad (x,t) \in \overline{G}^1,$$

$$\left|\frac{\partial^{k+k_0}}{\partial x^k \partial t^{k_0}}W^1(x,t)\right| \leq M\left[1 + \varepsilon^{(1-k-k_0)/2}\rho^{1-k-k_0} + \right.$$

$$\left. + \varepsilon^{(1-k)/2}\rho^{1-k-2k_0}\right]\exp(-m\varepsilon^{-1/2}|x-\gamma(t)|), \quad (x,t) \in \overline{G};$$

$$k + 2k_0 \leq K, \quad i = 1, 2,$$

where $\rho = \rho(x,t;\varepsilon) = \varepsilon^{-1/2}\,|x-\gamma(t)| + t^{1/2}$ and m is an arbitrary constant.

10.10.3. We now consider the solution of problem (10.56), (10.55) on the set $\overline{G}^2$. Write the solution as the sum of the functions

$$u(x,t) = U(x,t) + V(x,t), \quad (x,t) \in \overline{G}^2, \qquad (10.64)$$

where $U(x,t)$ and $V(x,t)$ are the regular and singular parts of the solution. The function $U(x,t)$ is the restriction to the set $\overline{G}^2$ of the function $U^e(x,t)$, $(x,t) \in \overline{G}^{2e}$, that is the solution of the problem

$$L_{(10.56)}\,U^e(x,t) = f^e(x,t), \quad (x,t) \in G^{2e},$$
$$U^e(x,t) = \varphi^e(x,t), \quad (x,t) \in S^{2e}. \qquad (10.65)$$

The domain G^{2e} is an extension of the domain G^2 beyond the boundary S^l. The right-hand side $f^e(x,t)$ of the equation in (10.65) is a smooth continuation of the function $f(x,t)$. The function $\varphi^e(x,t)$ is smooth on each piecewise-smooth part of the set S^{2e}, and it coincides with the functions $\varphi(x,t)$ and $u^1(x,t)$ on the sets $S^2 \cap S_0$ and $S^2 \cap G^1$, respectively. The function $V(x,t)$ is the solution of the problem

$$L_{(10.56)} V(x,t) = 0, \quad (x,t) \in G^2,$$

$$V(x,t) = \varphi(x,t) - U(x,t), \quad (x,t) \in S^l, \quad V(x,t) = 0, \quad (x,t) \in S^2 \setminus S^l.$$

For the functions $U(x,t)$ and $V(x,t)$, the following estimates are valid:

$$\left| \frac{\partial^{k+k_0}}{\partial x^k \partial t^{k_0}} U(x,t) \right| \leq M, \tag{10.66a}$$

$$\left| \frac{\partial^{k+k_0}}{\partial x^k \partial t^{k_0}} V(x,t) \right| \leq M \varepsilon^{-k} \exp\left(-m \varepsilon^{-1} r((x,t), \overline{S}^l) \right), \tag{10.66b}$$

$$(x,t) \in \overline{G}^2, \quad k + 2k_0 \leq K,$$

where $r((x,t), \overline{S}^l)$ is the distance from the point (x,t) to the set $\overline{S}^l$ and m is an arbitrary constant in the interval $(0, m_0)$, with $m_0 = a^{-1} b$.

The following theorem holds.

Theorem 10.10.1 *Let for the data of the boundary value problem (10.56), (10.55), the conditions (10.58), (10.59) for $n = K - 1$, where $K \geq 2$, be fulfilled, and let the solution of the problem satisfy the condition (10.57), where $l_0 = K$. Then the solution of the boundary value problem and its components in the representations (10.62) and (10.64) satisfy the estimates (10.61), (10.63) and (10.66).*

10.11 Classical finite difference approximations

In this section we construct a difference scheme that allows us to approximate the solution of the problem (10.56), (10.55) ε-uniformly.

10.11.1. We first consider a difference scheme based on classical approximations. On the set $\overline{G}_{(10.55)}$ we introduce the rectangular grid

$$\overline{G}_h = \overline{\omega} \times \overline{\omega}_0, \tag{10.67}$$

where $\overline{\omega}$ and $\overline{\omega}_0$ are meshes on the intervals $[-d, d]$ and $[0, T]$, respectively, the mesh $\overline{\omega}$ has an arbitrary distribution of its nodes satisfying only the condition $h \leq MN^{-1}$, where $h = \max_i h^i$, $h^i = x^{i+1} - x^i$, with $x^i, x^{i+1} \in \overline{\omega}$, and the mesh $\overline{\omega}_0 = \overline{\omega}_0^u$ is uniform with step-size $\tau = TN_0^{-1}$. Here $N + 1$ and $N_0 + 1$ are the numbers of nodes in the meshes $\overline{\omega}$ and $\overline{\omega}_0$, respectively.

We approximate the boundary value problem (10.56) by the finite difference scheme [108]

$$\Lambda_{(10.68)}\, z(x,t) = f(x,t), \quad (x,t) \in G_h, \quad z(x,t) = \varphi(x,t), \quad (x,t) \in S_h.$$
(10.68)

Here
$$\Lambda_{(10.68)} \equiv \varepsilon\, a\, \delta_{\bar{x}\hat{x}} + b\, \delta_x - c - p\, \delta_{\bar{t}}.$$

Consider the difference scheme (10.68) on the uniform grid

$$\overline{G}_h = \overline{\omega} \times \overline{\omega}_0.$$
(10.69)

Using the *a priori* estimates (10.61), (10.63), and (10.66) for the solutions of the problem (10.56), we find the estimate

$$|u(x,t) - z(x,t)| \le M \left[(\varepsilon + N^{-1})^{-1} N^{-1} + N^{-1/2} + N_0^{-1/2} \right],$$
(10.70)

$$(x,t) \in \overline{G}_h.$$

Thus, under the condition $N^{-1} = o(\varepsilon)$, the difference scheme (10.68), (10.69) converges; for fixed values of the parameter ε, the scheme converges at the rate $\mathcal{O}(N^{-1/2} + N_0^{-1/2})$.

10.11.2. We now construct a difference scheme that converges ε-uniformly in the case when the problem solution contains the boundary layer.

On the set $\overline{G}$ we construct a grid condensing in a neighborhood of the boundary layer, similar to that constructed in [33, 184]

$$\overline{G}_h = \overline{G}_h^s = \overline{\omega}^* \times \overline{\omega}_0,$$
(10.71a)

where $\overline{\omega}_0 = \overline{\omega}_{0(10.67)}^u$ and $\overline{\omega}^* = \overline{\omega}^*(\sigma)$ is a *piecewise-uniform mesh* on $[-d, d]$; here σ is a mesh parameter depending on ε and N. The value of σ is chosen to satisfy the condition

$$\sigma = \sigma(N, \varepsilon) = \min\left[\beta, 2\, m^{-1} \varepsilon \ln N \right],$$
(10.71b)

where β is an arbitrary number in the open-closed interval $(0, d]$, and $m = m_{(10.66)}$. The interval $[-d, d]$ is divided into two parts $[-d, -d + \sigma]$ and $[-d+\sigma, d]$; on each of these parts the mesh step-size is constant and equal to $h^{(1)} = 2\, d\, \sigma\, \beta^{-1} N^{-1}$ on $[-d, -d+\sigma]$ and $h^{(2)} = 2\, d\, (2\, d - \sigma)\, (2\, d - \beta)^{-1} N^{-1}$ on $[-d + \sigma, d]$, $\sigma \le d$.

The difference scheme (10.68), (10.71) converges ε-uniformly with the estimate

$$| u(x,t) - z(x,t) | \le M\, [\, N^{-1/2} + N_0^{-1/2}\,], \quad (x,t) \in \overline{G}_h.$$
(10.72)

For the interpolant $\overline{z}(x,t) = \overline{z}_{(10.27)}(x, t;\ z_{(10.68, 10.71)}(\cdot)),\ (x,t) \in \overline{G}$, one has the estimate

$$| u(x,t) - \overline{z}(x,t) | \le M\, [\, N^{-1/2} + N_0^{-1/2}\,], \quad (x,t) \in \overline{G}.$$
(10.73)

The following theorem holds.

Theorem 10.11.1 *Let the solution of the problem* (10.56), (10.55) *and its components in the representations* (10.62) *and* (10.64) *satisfy the estimates* (10.61), (10.63), (10.66) *for* $K = 4$. *Then the difference scheme* (10.68), (10.71) *converges ε-uniformly at the rate* $\mathcal{O}\left(N^{-1/2} + N_0^{-1/2}\right)$, *while the scheme* (10.68), (10.69) *converges under the condition* $N^{-1} = o(\varepsilon)$ *at the rate* $\mathcal{O}\big((\varepsilon + N^{-1})^{-1} N^{-1} + N^{-1/2} + N_0^{-1/2}\big)$. *The discrete solutions and the interpolant satisfy the estimates* (10.70), (10.72), *and* (10.73), *respectively.*

Remark 10.11.1 Let the component $W_1(x,t)$ be absent in the representation (10.62), i.e.,

$$W_1(x,t) = 0, \quad (x,t) \in \overline{G}^{\,1};$$

in this case the derivative $(\partial/\partial x)\,\varphi(x,t)$ is continuous on S_0 and satisfies on $S^{(*)}$ the relation

$$\left[\frac{\partial}{\partial x}\,\varphi(x,t)\right] = 0, \quad (x,t) \in S^{(*)}.$$

When using the scheme (10.68), (10.71), we obtain the estimate

$$|u(x,t) - z(x,t)| \le M\left[N^{-1}\ln N + N_0^{-1}\ln N_0\right], \quad (x,t) \in \overline{G}_h.$$

∎

10.12   Improved finite difference scheme

To construct an improved difference scheme for the problem (10.56), (10.55) (in comparison with the scheme (10.68), (10.71)), we apply the method of additive splitting of singularities (see [80] in the case of regular problems and [64, 184, 71] in the case of singular problems and the bibliography therein).

10.12.1. Decompose the problem solution as the sum of the functions

$$u(x,\,t) = u_1(x,\,t) + u_2(x,\,t), \quad (x,\,t) \in \overline{G}, \tag{10.74a}$$

where the singular component $u_2(x,\,t)$ is given in the explicit form

$$u_2(x,\,t) = W^{\,1}(x,t) = W^{\,1}_{(10.62)}(x,t;\,i), \quad (x,\,t) \in \overline{G}, \quad i = 1. \tag{10.74b}$$

The function $u_1(x,\,t)$ is the solution of the problem

$$\begin{aligned} L_{(10.56)}\,u_1(x,\,t) &= f(x,\,t), & (x,\,t) \in G, \\ u_1(x,\,t) &= \varphi_1(x,\,t), & (x,\,t) \in S, \end{aligned} \tag{10.74c}$$

where

$$\varphi_1(x,\,t) = \varphi(x,\,t) - W^{\,1}(x,t), \quad (x,\,t) \in S;$$

the function $\varphi_1(x, t)$ and its first derivative in x are continuous on S_0.

We approximate the problem (10.74c) on the grid (10.67) by the difference scheme

$$\Lambda_{(10.68)}\, z_1(x, t) = f(x, t), \qquad (x, t) \in G_h,$$
$$z_1(x, t) = \varphi_1(x, t), \qquad (x, t) \in S_h.$$

(10.75a)

On the set $\overline{G}$ we construct the interpolant

$$\overline{z}_1(x, t) = \overline{z}_{(10.27)}(x, t;\, z_1(\cdot)) \qquad (x, t) \in \overline{G},$$

of the function $z_1(x, t)$, $(x, t) \in \overline{G}_h$. Next, we construct the function

$$u_0^h(x, t) = \overline{z}_1(x, t) + u_2(x, t), \qquad (x, t) \in \overline{G},$$

(10.75b)

where

$$u_2(x, t) = u_{2(10.74b)}(x, t), \qquad (x, t) \in \overline{G}.$$

(10.75c)

We say that the function $u_0^h(x, t)$, $(x, t) \in \overline{G}$, is the solution of the difference scheme (10.75), (10.67) based on *the method of the additive splitting of singularities* (namely, the main term of the transient-layer function is split off).

10.12.2. Consider the difference scheme (10.75), (10.71).

Taking the estimates (10.63) into account, we have

$$|u_1(x, t) - z_1(x, t)| \le M \left[N^{-1} \ln N + N_0^{-1} \ln N_0\right], \qquad (x, t) \in \overline{G}_h.$$

For the function $u_0^h(x, t)$, $(x, t) \in \overline{G}$, we obtain the estimate

$$|u(x, t) - u_0^h(x, t)| \le M \left[N^{-1} \ln N + N_0^{-1} \ln N_0\right], \qquad (x, t) \in \overline{G}, \quad (10.76)$$

which is better than (10.73).

Thus, the difference scheme (10.75), (10.71) is *the improved scheme* that converges ε-uniformly at the rate $\mathcal{O}\left(N^{-1} \ln N + N_0^{-1} \ln N_0\right)$.

Theorem 10.12.1 *Let the hypothesis of Theorem 10.11.1 be fulfilled. Then the difference scheme* (10.75), (10.71) *converges ε-uniformly with the estimate* (10.76).

Remark 10.12.1 In the case of the problem (10.56), (10.55), using more terms in the component $u_{1(10.74a)}(x, t)$ and approximating problem (10.74c) by a scheme of improved accuracy, it is possible to construct a scheme that converges ε-uniformly at the rate $\mathcal{O}\left(N^{-2} \ln^2 N + N_0^{-2}\right)$.

For example, if we use the Richardson extrapolation method to improve accuracy of the scheme (10.75), (10.71), under the condition

$$u_2(x, t) = W_{(10.62)}^1(x, t;\, i), \qquad (x, t) \in \overline{G}, \quad i = 2,$$

then the improved scheme converges ε-uniformly at the rate $\mathcal{O}\big(N^{-(3-\nu)/2} + N_0^{-(3-\nu)/2}\big)$, where ν is an arbitrary constant in $(0, 1)$, and for $i \ge 4$, for corresponding problem data, one obtains the convergence rate $\mathcal{O}\left(N^{-2} \ln^2 N + N_0^{-2}\right)$.

∎

Chapter 11

A finite difference scheme on *a priori* adapted grids for a singularly perturbed parabolic convection-diffusion equation

A boundary value problem is considered for a singularly perturbed parabolic convection-diffusion equation. A finite difference scheme on *a priori* (sequentially) *adapted* grids is constructed and its convergence is studied. The scheme is constructed using a majorant function for the singular component of the discrete solution that allows us to find *a priori* a subdomain where the computed solution requires a further improvement. This subdomain is defined by the perturbation parameter ε, the step-size of a uniform mesh in x, and also by the required accuracy of the discrete solution and the prescribed number K of refinement iterations for the improvement of the solution. When we solve the discrete problems to improve the solution, we use uniform meshes on the subdomains. The error of the numerical solution depends weakly on the parameter ε. The scheme converges almost ε-uniformly. The *advantage* of this approach consists in the *uniform meshes* used.

11.1   Introduction

At present, methods for constructing ε-uniformly convergent schemes on special grids that are *a priori* adapted in a boundary layer region and do not change in the computational process, or, briefly, schemes on meshes condensing in boundary layers *a priori*, are well developed (see, e.g., [15, 138, 87, 102, 33] for partial differential equations and [75] for ordinary differential equations). Methods based on piecewise-uniform meshes that condense in boundary layers are used fairly widely owing to their simplicity and convenience in application (see, e.g., [138, 87, 102, 33] and the references therein). But necessity to solve the difference equations on meshes whose step-size changes sharply in a neighborhood of the boundary layer can be considered as a weakness of the numerical methods on *a priori adapted meshes*.

Another alternative approach to the construction of numerical methods for singularly perturbed boundary value problems developed, e.g., in [166, 165, 203, 185], leads to methods on sequentially *a posteriori* adapted meshes that are fitted (refined) in the computational process depending on the computed solution, or briefly, methods on *a posteriori adapted* meshes. In this approach, classical finite difference approximations of the boundary value problem are used; the discrete solution is corrected on a finer mesh in that subdomain where errors in the solution turn to be inadmissibly large. The subdomain in which the solution should be locally improved is determined using an indicator that is a functional of the solution (e.g., the solution gradient) of the discrete problem. In these methods, the discrete problems are solved on uniform meshes in the subdomains where the solution is *a posteriori* improved.

In this respect, it would be of interest to consider such numerical methods on *a priori* adapted meshes in which the discrete problems are solved also on uniform meshes in the subdomains where the computed solution is *a priori* corrected. Methods of this kind are unknown in the literature. This chapter is based on the paper [196] where this approach is applied for the first time.

In this chapter, a Dirichlet problem is considered for a parabolic convection-diffusion equation with a small parameter ε multiplying the highest-order derivative. For the boundary value problem, finite difference schemes on *locally uniform* grids (namely, *uniform meshes on the subdomains* where the solution should be improved) are constructed. These schemes are *adapted a priori*, and their convergence is studied. To construct the schemes, a standard finite difference approximation of the differential equation is used. Note that the scheme on *a priori condensing* (in the layer) *piecewise-uniform meshes* converges ε-uniformly. The standard scheme on *uniform meshes* converges only under the condition $N^{-1} \ll \varepsilon$, where the value N defines the number of mesh points in x.

For the scheme on *a priori* adapted meshes, boundaries of the subdomains, where it requires to improve the solution, are determined by a *majorant for the singular component of the discrete solution*. This majorant is specified by the perturbation parameter ε, the step-size in a mesh used in x, and also by the required accuracy of the discrete solution. On the meshes adapted in the majorant function of the discrete solution, a sufficiently simple finite difference scheme is constructed for which the solution error depends weakly on the parameter ε. The scheme constructed on *a priori* adapted grids converges *almost ε-uniformly*, precisely, under the condition $N^{-1} \ll \varepsilon^{\nu}$, where the value ν defining the scheme (the number of refinement iterations required for the discrete solution to be improved) can be chosen arbitrarily in $(0, 1]$.

11.2   Problem formulation. The aim of the research

11.2.1. Problem formulation

We consider the boundary value problem (10.2), (10.1) for the singularly perturbed parabolic convection-diffusion equation. Problem formulation is given in Section 10.2 of Chapter 10. In the same place, one can find:

(i) the rectangular grid $\overline{G}_{h(10.4)}$ with an arbitrary distribution of nodes, and the finite difference scheme (10.5) on $\overline{G}_h$;

(ii) the uniform grid $\overline{G}_{h(10.7)}^{\,u}$, and estimates of the solution to the difference scheme (10.5) on the grid (10.7);

(iii) the special piecewise-uniform grid $\overline{G}_{h(10.11)}^{\,s}$ and ε-uniform estimates of the solution to the difference scheme (10.5) on the grid (10.11).

The following convergence theorem holds.

Theorem 11.2.1 *Let the components of the solution $u(x,t)$ of the boundary value problem (10.2), (10.1) in the representation (11.29) satisfy the estimates of Theorem 11.6.1. Then the difference scheme (10.5), (10.11) (the scheme (10.5), (10.7)) converges ε-uniformly (converges under the condition (10.10)). For the discrete solutions and their interpolants, the estimates (10.8), (10.12) and (10.9), (10.13) are valid, respectively.*

Remark 11.2.1 For the mesh $\overline{G}_{h(10.11)}$, the ratio of $h^{(2)}$ and $h^{(1)}$, i.e., the mesh sizes in x on the mesh intervals with constant step-size, is of order $\mathcal{O}\left(\varepsilon^{-1}\ln^{-1}N\right)$.

11.2.2. The aim of the research

As a rule, for boundary value problems of type (10.2), (10.1), we are interested in numerical methods whose solutions converge ε-uniformly in the maximum discrete norm. But in the case of singularly perturbed problems, the ε-uniform convergence of the numerical solution $z(x,t)$ at the points of the mesh $\overline{G}_h$ is, in general, inadequate to give a representation about ε-uniform convergence of an approximation constructed on the whole set $\overline{G}$. For example, the solution of a difference scheme obtained by the classical approximation of problem (10.2), (10.1) on a uniform grid like $\overline{G}_{h(10.7)} = \overline{G}_h^{\,u}$ converges on the grid $\overline{G}_h$ when $\varepsilon^{-1}h \to \infty$ for $h \to 0$, i.e., when the typical width of the boundary layer defined by ε is much less than the mesh size in x. But, even the simplest interpolant (10.6) does not converge on $\overline{G}$. Under the given requirements on the grid $\overline{G}_{h(10.7)}$, the interpolant

$$\overline{u}^h(x,t) = \overline{z}_{(10.6)}\big(x,t;\ u^h(\cdot),\ \overline{G}_h\big), \quad (x,t) \in \overline{G},$$

constructed using the discrete function

$$u^h(x,t) = u(x,t), \quad (x,t) \in \overline{G}_h,$$

where $u(x,t)$ is the solution of the problem (10.2), (10.1), does not converge on $\overline{G}$ as well.

We now give some *definitions*.

Solvability of difference scheme. In the case when the interpolant $\overline{z}_{(10.6)}(x,t)$, $(x,t) \in \overline{G}$, converges on $\overline{G}$, we say that the difference scheme *resolves* the

boundary value problem (*converges on* $\overline{G}$); otherwise, we say that the difference scheme *does not resolve* the boundary value problem.

$\mathcal{E}$-uniform solvability of difference scheme.   In that case when the inter-
polant $\overline{z}(x,t)$, $(x,t) \in \overline{G}$, converges on $\overline{G}$ ε-uniformly, we say that the difference scheme *converges* (*resolves* the boundary value problem) ε-*uniformly*.

We are interested in *difference schemes* that *resolve the boundary value
problem on* $\overline{G}$ ε-*uniformly*, or in *difference schemes* that are *close to ε-uni-
formly convergent schemes on* $\overline{G}$.

The estimate (10.8) for the discrete solution in Section 10.2, Chapter 10
implies that the solution of the classical difference scheme (10.5) on the uniform grid (10.7) converges under the rather restrictive condition $(h \ll \varepsilon)$
$\varepsilon^{-1} = \mathcal{O}(N)$, where $N+1$ is the number of mesh points in x. If this condition is violated, for example, for $\varepsilon^{-1} = \mathcal{O}(N)$, then, in general, the solution
of the difference scheme (10.5), (10.7) does not converge to the solution of
the problem (10.2), (10.1) as $N, N_0 \to \infty$; here $N_0 + 1$ is the number of mesh
points in t.

$\mathcal{E}$-uniform convergence of difference scheme.   Let $z(x,t)$, $(x,t) \in \overline{G}_h$, be
a solution of some difference scheme, and let the function $z(x,t)$ satisfy the
estimate [166]

$$|u(x,t) - z(x,t)| \leq M\lambda \left(\varepsilon^{-\nu} N^{-1}, N_0^{-1}\right), \quad (x,t) \in \overline{G}_h, \qquad (11.1)$$

where $\lambda(\xi_1, \xi_2) \to 0$ as $\xi_1, \xi_2 \to 0$, uniformly in the parameter ε; and $\nu \geq 0$.
The solution of the difference scheme converges on the set $\overline{G}_h$ *uniformly with
respect to the parameter ε* (or, briefly, ε-*uniformly*) if $\nu = 0$ in the estimate
(11.1). Otherwise, we say that the difference scheme does not converge ε-
uniformly on $\overline{G}_h$.

Convergence of difference scheme with defect ν.   If the difference scheme
converges for $N^{-1} = o(\varepsilon^{\nu})$, where the constant M in the estimate (11.1), in
general, depends on ν, but, in general, there is no convergence for $N^{-1} =
\mathcal{O}(\varepsilon^{\nu})$, we say that the difference scheme converges with *defect ν* of ε-uniform
convergence.

Almost ε-uniform convergence of difference scheme.   In the case when the
value ν can be chosen arbitrarily small, and also the solution of the difference
scheme *controlled by the value ν* satisfies the estimate (11.1), we say that
the difference scheme converges on $\overline{G}_h$ *almost ε-uniformly with defect ν* (or,
briefly, *almost ε-uniformly*).

In a similar way, the convergence defect of the scheme on the set $\overline{G}$ can be
defined.

Defect of the scheme (10.5), (10.7) equals 1.

For the problem (10.2), (10.1), the difference scheme from [138] in the case
$n = 1$ (that is, the scheme on *a priori* adapted piecewise-uniform grid with
a single *transition point* at which the mesh changes its step-size) converges
ε-uniformly. Note that, in schemes on piecewise-uniform meshes (see, e.g.,
[178, 87, 102, 33] and the references therein), the mesh size changes sharply

at the transition points where the mesh switches from coarse to fine (the ratio of the mesh sizes is not ε-uniformly bounded; see Remark 11.2.1 in Section 10.2). This, in general, can lead to restrictions in using efficient numerical approaches to the computation of discrete solutions and improvement of their accuracy (see, e.g., [108, 110, 79, 80, 16] and the references therein).

Schemes on *a posteriori* adapted grids that converge almost ε-uniformly were considered in [166, 165, 185]. An adapted mesh is constructed using meshes that are uniform on the subdomains in which the computed solution is corrected. The advantage of this scheme is that its solution is "synthesized" using the components of solutions of auxiliary intermediate problems that are solved on the corresponding subdomains having uniform meshes with the same numbers of mesh point in x, t on each subdomain.

It should be noted that in order to solve discrete problems on uniform meshes, highly efficient numerical methods have been developed, which require the number of operations for computing the solution of the same order as the number of mesh points (see, e.g., [108, 110, 79] and the references therein). Owing to this, it would be of interest to construct and examine almost ε-uniformly convergent schemes on *a priori* adapted grids based on locally uniform grids, that is, uniform grids on each of the subdomains.

Our aim for the boundary value problem (10.2), (10.1) is to construct a scheme, on *a priori* adapted locally uniform grids, which converges almost ε-uniformly.

11.3 Grid approximations on locally refined grids that are uniform in subdomains

In this section, we present an algorithm for constructing a locally refined grid (adapted in the boundary layer) and a grid solution on it. In each of subdomains subjected to grid refinement, this algorithm uses uniform meshes in space and time (the temporal mesh is not refined).

11.3.1. A formal iterative algorithm

First, we describe a *formal iterative algorithm* for constructing approximate solutions for the boundary value problem (10.2), (10.1) [108].

On the set $\overline{G}$, we introduce the coarsened (initial) grid

$$\overline{G}_{1h} = \overline{\omega}_1 \times \overline{\omega}_0, \tag{11.2a}$$

where $\overline{\omega}_1$ and $\overline{\omega}_0$ are uniform meshes, $\overline{\omega}_0 = \overline{\omega}^u_{0(10.7)}$; the step-size in $\overline{\omega}_1$ is $h_1 = d\,N^{-1}$, and $N + 1$ is the number of nodes. We denote the solution of problem (10.5), (11.2a) by $z_1(x, t)$, $(x, t) \in \overline{G}_{1h}$, where $\overline{G}_{1h} = \overline{G}_{1h(11.2)}$. Note that $\overline{G}_{1h(11.2)} = \overline{G}_{h(10.7)}$.

Let the value $d_1 \in \overline{\omega}_1$ be found in such a way that for $x \geq d_1$, the discrete solution $z_1(x,t)$, $(x,t) \in \overline{G}_{1h}$, is a good approximation of the solution of problem (10.2), (10.1), moreover,

$$|u(x,t) - z_1(x,t)| \leq M\,\delta, \quad (x,t) \in \overline{G}_{1h}, \quad x \geq d_1, \tag{11.3a}$$

where $\delta > 0$ is an arbitrary sufficiently small number specifying the required accuracy of the discrete solution, and M is a constant independent of δ, and $d_1 \in [0, d)$.

If it turns out that $d_1 > 0$, then we define the subdomain

$$\overline{G}_{(2)} = G_{(2)} \cup S_{(2)}, \quad G_{(2)} = G_{(2)}(d_1), \quad G_{(2)} = D_{(2)} \times (0, T], \quad D_{(2)} = (0, d_1),$$

where we shall refine the grid. On the subdomain $\overline{G}_{(2)}$ we introduce the grid

$$\overline{G}_{(2)h} = \overline{\omega}_{(2)} \times \overline{\omega}_0,$$

where $\overline{\omega}_{(2)}$ is a uniform mesh with $N + 1$ nodes and step-size $h_{(2)}$.

On the set $\overline{G}_{(2)h}$ we find the solution $z_{(2)}(x,t)$ of the discrete problem

$$\Lambda_{(10.5)}\, z_{(2)}(x,t) = f(x,t), \qquad (x,t) \in G_{(2)h},$$

$$z_{(2)}(x,t) = \begin{cases} z_1(x,t), & (x,t) \in S_{(2)h} \setminus S, \\ \varphi(x,t), & (x,t) \in S_{(2)h} \cap S, \end{cases}$$

where

$$G_{(2)h} = G_{(2)} \cap \overline{G}_{(2)h},\ S_{(2)h} = S_{(2)} \cap \overline{G}_{(2)h}.$$

The grid set $\overline{G}_{2h}$ on $\overline{G}$ and the function $z_2(x,t)$, $(x,t) \in \overline{G}_{2h}$, are defined by

$$\overline{G}_{2h} = \overline{G}_{(2)h} \cup \{\overline{G}_{1h} \setminus \overline{G}_{(2)}\}, \quad z_2(x,t) = \begin{cases} z_{(2)}(x,t), & (x,t) \in \overline{G}_{(2)h}, \\ z_1(x,t), & (x,t) \in \overline{G}_{1h} \setminus \overline{G}_{(2)}. \end{cases}$$

For $k \geq 3$ at the $(k-1)$th iteration, assume that the grid set $\overline{G}_{k-1,h}$ and the grid function $z_{k-1}(x,t)$ on this set have already been constructed. Furthermore, let the value $d_{k-1} \in \omega_{k-1}$ be found in such a way that for $x \geq d_{k-1}$ the discrete solution $z_{k-1}(x,t)$, $(x,t) \in \overline{G}_{k-1,h}$, is a good approximation of the solution of the problem (10.2), (10.1), i.e., the following estimate holds:

$$|u(x,t) - z_{k-1}(x,t)| \leq M\,\delta, \quad (x,t) \in \overline{G}_{k-1,h}, \quad x \geq d_{k-1}. \tag{11.3b}$$

The constant M depends on k, i.e., $M_{(11.3b)} = M_{(11.3b)}(k-1)$, where $M(k) = M^* k$.[1] Here

$$\overline{G}_{k-1,h} = \overline{\omega}_{k-1} \times \overline{\omega}_0,$$

---

[1] Here and in what follows M^* denote constants independent of k.

where $\overline{\omega}_{k-1}$ is a mesh generating the grid $\overline{G}_{k-1,h}$; $N_k + 1$ is the number of nodes in the mesh $\overline{\omega}_{k-1}$, for $k \geq 2$; and $N_1 = N$.

If it happens that $d_{k-1} > 0$, then we define the subdomain

$$\overline{G}_{(k)} = G_{(k)} \cup S_{(k)}, \quad G_{(k)} = G_{(k)}(d_{k-1}),$$
$$G_{(k)} = D_{(k)} \times (0, T], \quad D_{(k)} = (0, d_{k-1}). \tag{11.2b}$$

On the set $\overline{G}_{(k)}$, we introduce the grid

$$\overline{G}_{(k)h} = \overline{\omega}_{(k)} \times \overline{\omega}_0, \tag{11.2c}$$

where $\overline{\omega}_{(k)}$ is a uniform mesh with step-size $h_{(k)}$. Let $N + 1$ be the number of nodes in the mesh $\overline{\omega}_{(k)}$ and $h_{(k)} = d_{k-1} N^{-1}$, where $d_{k-1} = d$ for $k = 1$. Let $z_{(k)}(x, t)$, $(x, t) \in \overline{G}_{(k)h}$, be the solution of the grid problem

$$\Lambda_{(10.5)} z_{(k)}(x, t) = f(x, t), \qquad (x, t) \in G_{(k)h},$$
$$z_{(k)}(x, t) = \begin{cases} z_{k-1}(x, t), & (x, t) \in S_{(k)h} \setminus S, \\ \varphi(x, t), & (x, t) \in S_{(k)h} \cap S. \end{cases} \tag{11.2d}$$

We set

$$\overline{G}_{kh} = \overline{G}_{(k)h} \cup \{\overline{G}_{k-1,h} \setminus \overline{G}_{(k)}\},$$
$$z_k(x, t) = \begin{cases} z_{(k)}(x, t), & (x, t) \in \overline{G}_{(k)h}, \\ z_{k-1}(x, t), & (x, t) \in \overline{G}_{k-1,h} \setminus \overline{G}_{(k)}. \end{cases}$$

If for some value $k = K_0$ it turns out that $d_{K_0} = 0$, then we set $d_k = 0$ for $k \geq K_0$. For $k \geq K_0 + 1$, the sets $\overline{G}_{(k)}$ are assumed to be empty, and we do not compute the functions $z_{(k)}(x, t)$. For example, for $k \geq K_0$ we have $z_k(x, t) = z_{K_0}(x, t)$, $\overline{G}_{kh} = \overline{G}_{K_0 h}$.

For $k = K$, where K is a given fixed number (the number of iterations for improving the grid solution), $K \geq 1$, we assume

$$\overline{G}_h^K = \overline{G}_{Kh} \equiv \overline{G}_h, \quad z^K(x, t) = z_K(x, t) \equiv z(x, t). \tag{11.2e}$$

The grid $\overline{G}_h$ and the function $z(x, t)$ in (11.2e) are constructed using the grid sets $\overline{G}_{(k)h}$ and the functions $z_{(k)}(x, t)$, $(x, t) \in \overline{G}_{(k)h}$, $k = 1, \dots, K$.

We say that the function $z_{(11.2)}(x, t)$, $(x, t) \in \overline{G}_{h(11.2)}$ is the solution of the scheme (10.5), (11.2) (of the scheme on refined grids being uniform on local subdomains). The functions $z_k(x, t)$, $(x, t) \in \overline{G}_{kh}$, $k = 1, \dots, K$ are called the components of the solution of the difference scheme.

Let the value $d^K \in \overline{\omega}_K$, $d^K = d_K$, be found so that for $x \geq d_K$ the solution $z_K(x, t)$ approximates the solution of problem (10.2), (10.1); in this case we have

$$|u(x, t) - z(x, t)| \leq M\delta, \quad (x, t) \in \overline{G}_h, \quad x \geq d^K, \tag{11.3c}$$

where $z(x, t) = z_{(11.2)}(x, t)$, $\overline{G}_h = \overline{G}_{h(11.2)}$.

The difference scheme (10.5), (11.2) is a *difference scheme on locally refined grids* that are uniform on the subdomains $\overline{G}_{(k)h}$ in which the computed solution is corrected.

In the case when the values d_k are determined in the process of the numerical solution of the problem (10.5), (11.2) depending on the results of computations, the scheme (10.5), (11.2) is a *difference scheme on a posteriori adapted grids* (or briefly, *a posteriori* adapted scheme).

If the values d_k are given before the start of computations regardless of the computational results obtained, the scheme (10.5), (11.2) is a *difference scheme on a priori adapted grids* (or briefly, *a priori* adapted scheme).

11.3.2. Maximum principle for the algorithm $A_{(11.2)}$

The given algorithm (we call it $A_{(11.2)}$) allows us to construct the solution of the problem (10.5), (11.2) based on the sequence of values d_k, $k = 1, ..., K$. The value $N_K + 1$ is the number of nodes in the mesh $\overline{\omega}^K = \overline{\omega}_K$ used for the construction of the function $z^K(x,t)$. For the value N_K, we have the estimate

$$N_K \leq K\,(N - 1) + 1 \leq K\,N.$$

The ratio of $h_{(k)}$ and $h_{(k+1)}$, i.e., the mesh step-sizes in x in neighboring subregions of the adaptive grid, does not exceed the value N.

In the scheme (10.5), (11.2) when solving the intermediate problems (11.2d), an interpolation is not required to find values of the functions $z_{(k)}(x,t)$ on the boundary $S_{(k)h}$.

For the scheme (10.5), (11.2), the *maximum principle* holds; the following comparison theorem is valid.

Theorem 11.3.1 *Let the functions $z_k^1(x,t)$ and $z_k^2(x,t)$, $(x,t) \in \overline{G}_{kh}$, $\overline{G}_{kh} = \overline{G}_{kh(11.2)}$, $k = 1, 2, \ldots, K$, satisfy the conditions*

$$\Lambda\, z_1^1(x,t) \leq \Lambda\, z_1^2(x,t), \quad (x,t) \in G_{(1)h},$$

$$z_1^1(x,t) \geq z_1^2(x,t), \quad (x,t) \in S_{(1)h};$$

$$\Lambda\, z_k^1(x,t) \leq \Lambda\, z_k^2(x,t), \quad (x,t) \in G_{(k)h},$$

$$z_k^1(x,t) \geq z_k^2(x,t), \quad (x,t) \in S_{(k)h} \cap S,$$

$$z_k^1(x,t) \geq z_{k-1}^1(x,t), \quad z_{k-1}^2(x,t) \geq z_k^2(x,t),$$

$$(x,t) \in \overline{G}_{kh} \setminus \{G_k \cup \{S_{(k)} \cap S\}\}, \quad k = 2, ..., K.$$

Then $z_K^1(x,t) \geq z_K^2(x,t)$, $(x,t) \in \overline{G}_{Kh}$.

The theorem is proved by the induction with respect to k, where k is the number of the current iteration in the iterative process.

The grids $\overline{G}_{kh}$ obtained by the algorithm $A_{(11.2)}$, are defined by the choice of the values d_k, for $k = 1, 2, ..., K$, and also by the values K and N, N_0. The values d_k are determined regardless of the results obtained in the computational process, i.e., the grids $\overline{G}_{kh}$ belong to *a priori* condensing grids.

Note that there exist no schemes in this class of difference schemes whose solutions converge ε-uniformly to the solution of the boundary value problem (10.2), (10.1).

11.4 Difference scheme on *a priori* adapted grid

In this section, we consider a difference scheme on *a priori* adapted grids constructed using a majorant function for the singular component of the grid solution.

11.4.1. Auxiliary constructions

We present some auxiliary constructions. For the differential and the difference problems, we introduce the width of the boundary layer specified by majorant functions for the singular components of their solutions.

The function

$$W^c(x) = W^c(x; \varepsilon) = \exp(-m^0 \varepsilon^{-1} x), \quad x \in \overline{D}^\infty, \tag{11.4a}$$

is a majorant (up to a constant factor) for the singular component $V(x,t)$ in the representation (11.29) of the solution to the boundary value problem (10.2), (10.1); here $m^0 = \min_{\overline{G}}[a^{-1}(x,t)\,b(x,t)]$, and

$$\overline{D}^\infty = [0, \infty). \tag{11.4b}$$

Based on the function $W^c(x)$, we introduce the *width of the boundary layer* for problem (10.2), (10.1). Let $\delta > 0$ be a sufficiently small number. We say that the value

$$\eta^c = \eta^c(\delta; \varepsilon) \tag{11.5a}$$

is the *width of the boundary layer* (defined by the majorant function for the singular component $V(x,t)$) with a *threshold* of order δ (or, briefly, the width of the boundary layer defined by the majorant function), if η^c is the minimum value of η^0 for which the following estimate is fulfilled:

$$W^c(x; \varepsilon) \le \delta, \quad x \in \overline{D}^\infty, \quad r(x, \Gamma_1) \ge \eta^0, \tag{11.5b}$$

where Γ_1 is the boundary of the set $\overline{D}^\infty$ with $\overline{D}^\infty = D^\infty \bigcup \Gamma$ and $\Gamma = \Gamma_1$. The value η^c may take magnitudes exceeding $d_{(10.1)}$ (for sufficiently small values δ such that $\delta \le \delta(\varepsilon)$); η^c is defined by the formula

$$\eta^c = (m^0)^{-1} \varepsilon \ln \delta^{-1}. \tag{11.5c}$$

Next, we introduce the *width of the discrete boundary layer* defined on the basis of a majorant function for the discrete singular component. By $z_v(x,t)$, $(x,t) \in \overline{G}$, we denote the solution of the difference scheme

$$\Lambda_{(10.5)} z(x,t) = L_{(10.2)} v(x,t), \quad (x,t) \in G_h, \quad z(x,t) = v(x,t), \quad (x,t) \in S_h,$$

where $v \in C^{2,1}(G) \cap C(\overline{G})$. Write the solution of the problem (10.5), (10.4) as the sum of the functions

$$z(x,t) = z_U(x,t) + z_V(x,t), \quad (x,t) \in \overline{G}_h, \tag{11.6}$$

where $z_U(x,t)$ and $z_V(x,t)$ are grid functions that approximate the components $U(x,t)$ and $V(x,t)$ in the representation (11.29), and $z_V(x,t)$ is the function of the discrete boundary layer.

Let

$$\overline{D}_h^\infty \tag{11.7a}$$

be a uniform grid on the semi-axis $\overline{D}_{(11.4)}^\infty$ with step-size h. The function

$$W(x) = W(x; \varepsilon, h) = (1 + m^0 \varepsilon^{-1} h)^{-n}, \quad x = x^n \in \overline{D}_h^\infty, \quad x^n = nh, \tag{11.7b}$$

is a *majorant* (up to a constant factor) for the singular component $z_V(x,t)$ in the representation (11.6) of the solution to the difference scheme (10.5) on the grid (10.7), where $h_{(10.7)} = h_{(11.7)}$ and $m^0 = m_{(11.4)}^0$. We say that the value

$$\eta = \eta(\delta; \varepsilon, h), \tag{11.8a}$$

where $\delta > 0$ is a sufficiently small number, is *the width of the discrete boundary layer* (defined by the majorant function $W(x)$ for the singular component $z_V(x,t)$) with a *threshold* of order δ (or, briefly, the width of the discrete boundary layer defined by the majorant function) if η is the minimum value of η_0 for which the estimate

$$W(x; \varepsilon, h) \leq \delta, \quad x \in \overline{D}_h^\infty, \quad r(x, \Gamma_1) \geq \eta_0 \tag{11.8b}$$

holds. The quantity η may take values exceeding $d_{(10.1)}$, and η is defined by the formula

$$\tag{11.8c}$$
$$\eta = \eta(\delta; \varepsilon, h) = \begin{cases} h \ln \delta^{-1} \ln^{-1}(1 + m^0 \varepsilon^{-1} h) \quad \text{for} \\ \left[\ln \delta^{-1} \ln^{-1}(1 + m^0 \varepsilon^{-1} h) \right]^e = \ln \delta^{-1} \ln^{-1}(1 + m^0 \varepsilon^{-1} h), \\ h \left\{ \left[\ln \delta^{-1} \ln^{-1}(1 + m^0 \varepsilon^{-1} h) \right]^e + 1 \right\} \quad \text{for} \\ \left[\ln \delta^{-1} \ln^{-1}(1 + m^0 \varepsilon^{-1} h) \right]^e < \ln \delta^{-1} \ln^{-1}(1 + m^0 \varepsilon^{-1} h), \end{cases}$$

with $\delta \in (0,1)$ and $\varepsilon \in (0,1]$; here $h = h_{(11.7)}$, and $[a]^e$ is the integer part of a number a.

It is convenient to use the following notation. On the uniform grid $\overline{D}_{h\,(11.7)}^{\infty}$ with step-size h, we associate the value $a \geq 0$ with the value $\{a; h\}^e$ defined by the relation

$$\{a; h\}^e = \begin{cases} a & for \quad h\left[h^{-1}a\right]^e = a, \\ h\left\{h\left[h^{-1}a\right]^e + 1\right\} & for \quad h\left[h^{-1}a\right]^e < a, \end{cases}$$

where $[a]^e = [a]^e_{(11.8)}$. Write the value η in the form

$$\eta = \eta(\delta; \varepsilon, h) = \{a; h\}^e, \tag{11.8d}$$

where $a = h \ln \delta^{-1} \ln^{-1}(1 + m^0 \varepsilon^{-1} h)$.

On the set $\overline{G}_{(k)}$, for $k \geq 1$, we define the grid $\overline{G}_{(k)h}$ with step-size $h_{(k)}$ in x. Define the values d_k in (11.2) by the relation

$$d_k = d_k(\delta; \varepsilon, N) \equiv \min\left[\eta(\delta; \varepsilon, h_{(k)}), d\right], \quad k = 1, \ldots, K, \tag{11.9a}$$

where $h_{(1)} = dN^{-1}$ and $h_{(k)} = d_{k-1}N^{-1}$, for $k \geq 2$. Set

$$\delta = \delta(N) \to 0 \quad for \quad N \to \infty. \tag{11.9b}$$

The difference scheme (10.5), (11.2), (11.9) is a scheme on *a priori* adapted grids. The values d_k are computed using an *indicator* based on the *majorant function* of the discrete boundary layer controlled by the parameters δ, ε, and h.

11.4.2. A nonconstructive estimate
For the solution of the difference scheme (10.5), (11.2), (11.9), using the maximum principle, we establish the estimate

$$|u(x,t) - z(x,t)| \leq \tag{11.10}$$

$$\leq \begin{cases} M\left[\delta(N) + N^{-1} + N_0^{-1}\right], & (x,t) \in \overline{G}_h, \quad r(x, \Gamma_1) \geq d_K, \\ M\left[(\varepsilon + d_{K-1}N^{-1})^{-1} d_{K-1}N^{-1} + \delta(N) + N^{-1} + N_0^{-1}\right], \end{cases} \quad (x,t) \in \overline{G}_h.$$

Thus, the difference scheme (10.5), (11.2), (11.9) converges ε-uniformly outside an d_K-neighborhood of the boundary S_1^L, and also on the whole set $\overline{G}_h$ under the condition $h_{(K)} \ll \varepsilon$, or more precisely,

$$\varepsilon^{-1} = o\left(d_{K-1}^{-1} N\right),$$

which is more weaker in comparison with the convergence condition (10.10).

The estimate (11.10) is *nonconstructive* since the values $d_{K-1\,(11.9)}$ and $d_{K\,(11.9)}$ depend on ε, N, and K implicitly that complicates the study of the scheme (10.5), (11.2), (11.9) depending on the values of N, ε, and K.

Theorem 11.4.1 *Let the solution of the problem* (10.2), (10.1) *satisfy the hypothesis of Theorem* 11.2.1. *Then the solution of the difference scheme* (10.5), (11.2), (11.9) *satisfies the bound* (11.10).

11.4.3. A difference scheme on *a priori* adapted grids

We now consider a difference scheme on *a priori* adapted grids that allows us to write out efficient estimates for $\eta(\delta; \varepsilon, h_{(k)})$, i.e., the width of the discrete boundary layer. These estimates make it possible to study convergence properties of the scheme on *a priori* adapted grids.

Note some properties of the value $\eta_{(11.8)}$ implied by its explicit form. The function $\eta(\delta; \varepsilon, h)$ for fixed values of δ and h is a piecewise-constant nondecreasing function with respect to the variable ε.

We assume that the following condition is fulfilled:

$$\delta = N^{-\alpha}, \quad \alpha \in (0, 1]. \tag{11.11}$$

For the value η, we have the estimate

$$\eta(\delta; \varepsilon, h_1) > \eta^c(\delta; \varepsilon),$$

where $h_1 = h_{1(11.2a)}$. However,

$$\eta(\delta; \varepsilon, h_1) \le M_1 \eta^c(\delta; \varepsilon) \tag{11.12a}$$

provided that the condition

$$h_1 \le m_1 (m^0)^{-1} \varepsilon; \quad M_1 = M_1(m_1), \quad m^0 = m^0_{(11.4)}, \tag{11.12b}$$

where $M_1(m_1)$ is evaluated by the inequality

$$\alpha_1 \ln^{-1}(1 + \alpha_1) \le M_1 \quad for \quad \alpha_1 \le m_1; \quad e.g., \quad M_1(m_1 = 1) = 2. \tag{11.12c}$$

In the case of the condition

$$\varepsilon \ge \varepsilon^{(0)},$$

the width $\eta(\delta; \varepsilon, h)$ of the discrete boundary layer satisfies the lower bound

$$\eta(\delta; \varepsilon, h_1) \ge h_1;$$

furthermore, under the condition

$$\varepsilon \le \varepsilon^{(1)},$$

for $\eta(\delta; \varepsilon, h)$ we have the upper bound

$$\eta(\delta; \varepsilon, h_1) \le h_1.$$

Here the values $\varepsilon^{(j)}$ are defined by the relations

$$\varepsilon^{(j)} = \varepsilon^{(j)}(\delta, N) = \varepsilon^{(j)}(\delta, N; d), \quad j \geq -1; \tag{11.13}$$

$$\varepsilon^{(-1)} = M_2 m^0 d \, \ln^{-1} \delta^{-1}, \quad \varepsilon^{(0)} = M_1 m^0 d \, N^{-1},$$

$$\varepsilon^{(j)} = m^0 d \, \delta (1 - \delta)^{-1} N^{-j}, \qquad j \geq 1,$$

where $d = d_{(10.1)}$, $N = N_{(11.2a)}$, $m^0 = m^0_{(11.4)}$, $M_1 = M_{1(11.12)}$, and M_2 is an arbitrary constant satisfying the inequality

$$M_2 \leq M_1^{-1},$$

$j \geq -1$ is an integer. By this choice of the constants M_1, M_2, we have $\eta(\delta; \varepsilon, h_1) \leq d$ for $\delta = \delta_{(11.11)}$ and $\varepsilon \leq \varepsilon^{(-1)}$.

We describe a rule for determining the values $d_{k(11.2)}$ in the grid construction (10.5), (11.2) for the given values of K and ε. Assume that the parameter ε belongs to the prescribed fixed intervals which are defined by the values $\varepsilon^{(j)}$. To construct the scheme on adapted grids for given K, it is necessary to prescribe the values d_k for $k \leq K - 1$. But when studying the schemes, we need the values d_k for $k \leq K$.

Let the parameter ε belong to one of the following intervals defined by the value j

$$\left\{ \begin{array}{ll} \varepsilon \in \left[\varepsilon^{(j)}, 1 \right] & for \quad j = -1 \\ \varepsilon \in \left[\varepsilon^{(j)}, \varepsilon^{(j-1)} \right) & for \quad j \geq 0 \end{array} \right\}, \quad \begin{array}{l} \varepsilon^{(j)} = \varepsilon^{(j)}_{(11.13)}(\delta, N), \\ j \geq -1. \end{array} \tag{11.14a}$$

The value d_k depends on K, j and also on δ, ε, k, and it is chosen in the set $\overline{G}_{(k)h}$ so that the value of $\eta(\delta; \varepsilon, h_{(k)})$, i.e., the width of the discrete boundary layer, satisfies the estimate

$$\eta(\delta; \varepsilon, h_{(k)}) \leq d_k \quad for \quad 1 \leq k \leq K$$

in the case when the parameter ε belongs to one of the intervals in (11.14a).

Consider the case when the following relation is fulfilled:

$$K = K_{(11.14b)}(j) \equiv j + 2, \quad j \geq -1, \tag{11.14b}$$

where $j = j_{(11.14a)}$ defines the interval of varying the parameter ε. Let $\varepsilon \in [\varepsilon^{(j)}, 1]$ for $j = -1$. In this case $K = 1$, and we set

$$d_1 = \min \left[\{ M_1 (m^0)^{-1} \varepsilon \, \ln \delta^{-1}; \, h_{(1)} \}^e, \, d \right]. \tag{11.14c}$$

Let $\varepsilon \in \left[\varepsilon^{(j)}, \varepsilon^{(j-1)}\right)$, for $j \geq 0$. Set

$$d_1 = d_2 = \left\{M_1(m^0)^{-1}\varepsilon \ln \delta^{-1}; \; h_{(1)}\right\}^e \quad \text{if} \; j = 0; \tag{11.14d}$$

$$d_1 = \left\{h_{(1)} \ln \delta^{-1} \ln^{-1}(1 + m^0 \varepsilon^{-1} h_{(1)}); \; h_{(1)}\right\}^e,$$

$$d_2 = d_3 = \left\{M_1(m^0)^{-1}\varepsilon \ln \delta^{-1}; \; h_{(2)}\right\}^e \quad \text{if} \; j = 1;$$

$$d_1 = h_{(1)}, \dots, d_k = h_{(k)}, \quad k \leq j - 1,$$

$$d_k = \left\{h_{(k)} \ln \delta^{-1} \ln^{-1}(1 + m^0 \varepsilon^{-1} h_{(k)}); \; h_{(k)}\right\}^e, \quad k = j,$$

$$d_k = d_{k+1} = \left\{M_1(m^0)^{-1}\varepsilon \ln \delta^{-1}; \; h_{(k)}\right\}^e, \quad k = j+1 \quad \text{if} \; j \geq 2.$$

Here $h_{(i)} = d_{i-1}N^{-1}$, $i = 1, \dots, j+1$, $d_0 = d_{(10.1)}$, $h_{(1)} = h_{1(11.2)}$, $m^0 = m^0_{(11.4)}$, and $M_1 = M_{1(11.12)}$.

The relations (11.14 b, c, d) prescribe the values d_k depending on the values δ, ε, $h_{(k)}$ and on the ratio between j and k for $k \leq K$, $K = j + 2$.

In the case when

$$K > j + 2, \quad j \geq -1, \tag{11.14e}$$

we set

$$d_k = d_{k(11.14d)} \quad \text{for} \quad k \leq j + 2,$$
$$d_k = d_{j+2\,(11.14d)} \quad \text{for} \; j + 2 < k \leq K, \quad j \geq -1; \tag{11.14f}$$

here $K > K_{(11.14b)}(j)$. But if

$$K \leq j + 1, \quad K \geq 1, \quad j \geq 0, \tag{11.14g}$$

then we set

$$d_k = d_{k(11.14d)} \quad \text{for} \; 1 \leq k \leq K; \tag{11.14h}$$

here $K < K_{(11.14b)}(j)$.

Thus, for the parameter ε chosen in one of the intervals in (11.14a) and for given K, formulae (11.14) give us the set of the values $d_k = d_k(\delta; \varepsilon, h_{(k)})$ depending on the relation between K and $j = j_{(11.14a)}$.

As follows from (11.14b–h), the values d_k, by virtue of the relation $h_{(k)} = d_{k-1}N^{-1}$, are defined only by the parameters j, k and δ, ε, N; we have

$$d_k = d_{k(11.14)}(\delta; \varepsilon, N) = d_k^j(\delta; \varepsilon, N), \quad 1 \leq k \leq K, \quad j \geq -1. \tag{11.14i}$$

The difference scheme (10.5), (11.2), (11.14) is a scheme on *a priori* adapted grids refined sequentially in a neighborhood of the boundary layer. When choosing the values d_k, we use, as an indicator, the majorant function of the discrete boundary layer controlled by the parameters δ, ε, h, taking into account that the parameter ε belongs to the prescribed intervals from (11.14a); $\varepsilon \in (0, 1]$.

11.4.4. Some estimates for the width η

Under the above choice of the values $d_{k(11.14)}$, taking into account the explicit form of the width of the discrete boundary layer $\eta(\delta; \varepsilon, h)$, we find the estimates

$$\eta(\delta; \varepsilon, h_{(1)}) \geq m \quad for \quad \varepsilon \in \left[\varepsilon^{(-1)}, 1\right]; \tag{11.15}$$

$$\eta(\delta; \varepsilon, h_{(k)}) \leq d_k, \quad 1 \leq k \leq K,$$

$$\eta(\delta; \varepsilon, h_{(k)}) \geq m\, d_k, \quad j+1 \leq k \leq K \quad for \quad \varepsilon \in \left[\varepsilon^{(j)}, \varepsilon^{(j-1)}\right), \quad j \geq 0,$$

where $h_{(k)} = d_{k-1} N^{-1}$. The smallest step-size attained in this process is not less than dN^{-K}.

Lemma 11.4.1 *In the case of the difference scheme* (10.5), (11.2), (11.14), *the estimates* (11.15) *hold for the values* $\eta(\delta; \varepsilon, h_{(k)})$ *and* $d_{k(11.14i)}$.

Lemma 11.4.2 *In the case of the difference scheme* (10.5), (11.2), (11.9), *the values* $d_{k(11.9a)}$ *and* $d^j_{k(11.14i)}$ *satisfy the estimate*

$$d_{k\,(10.5,\,11.2,\,11.9a)} \leq d^j_{k\,(10.5,\,11.2,\,11.14i)}, \quad 1 \leq k \leq K, \tag{11.16}$$

where $j = j_{(11.14a)}$ *defines the interval in* (11.14a) *to which the parameter* ε *belongs.*

11.5 Convergence of the difference scheme on *a priori* adapted grid

We consider the difference scheme (10.5), (11.2), (11.14) provided that the following condition is fulfilled:

$$\delta = N^{-1}. \tag{11.17}$$

11.5.1. Estimates of solutions on subdomains

Let $z_{[k]}(x,t)$, $(x,t) \in \overline{G}_{(k)h}$, be a solution of the difference scheme (10.5), (11.2c) that approximates the boundary value problem

$$L_{(10.2)}\, u(x,t) = f(x,t), \quad (x,t) \in G_{(k)},$$
$$u(x,t) = \varphi(x,t), \quad (x,t) \in S_{(k)}, \tag{11.18}$$

where $\overline{G}_{(k)} = \overline{G}_{(k)(11.2b)}$, $\overline{G}_{(k)h} = \overline{G}_{(k)h(11.2c)}$, $k \geq 1$. For the solution $z_{[k]}(x,t)$, we have the estimate

$$|u(x,t) - z_{[k]}(x,t)| \leq \tag{11.19}$$

$$\leq \begin{cases} M\left[h_{(1)}(\varepsilon + h_{(1)})^{-1} + N^{-1} + N_0^{-1}\right], & k \geq 1, \ j = -1, 0; \\ M\left[h_{(k)}(\varepsilon + h_{(k)})^{-1} + N^{-1} + N_0^{-1}\right], & k, j \geq 1; \end{cases} \quad (x,t) \in \overline{G}_{(k)h},$$

where the parameter ε belongs to one of the intervals in (11.14a), and

$$h_{(k)} = h_{(k)}(j) = h_{(k)}(j;\, N) \leq \begin{cases} MN^{-1} & \text{for } j = -1, 0, \quad k \geq 1; \\ MN^{-j-1}\ln N, & j \leq k-1, \\ MN^{-k}, & k \leq j \text{ for } j \geq 1, \quad k \geq 1. \end{cases}$$

For $k \geq j+2$, the function $z_{[k]}(x,t)$ satisfies the estimate

$$|u(x,t) - z_{[k]}(x,t)| \leq M\left[N^{-1}\ln N + N_0^{-1}\right], \quad (x,t) \in \overline{G}_{(k)h}, \qquad (11.20)$$

$$k \geq j+2, \quad j \geq -1.$$

Outside an σ_k^j-neighborhood of the boundary S_1^L, the following estimate holds for $z_{[k]}(x,t)$:

$$|u(x,t) - z_{[k]}(x,t)| \leq M\left[N^{-1} + N_0^{-1}\right], \quad (x,t) \in \overline{G}_{(k)h}, \qquad (11.21)$$

$$r(x, \Gamma_1) \geq \sigma_k^j, \quad k \geq 1, \quad j \geq 0,$$

where

$$\sigma_k^j = d_k^j, \ 1 \leq k \leq j+1; \quad \sigma_k^j = d_{j+2}^j, \ k \geq j+2; \quad d_k^j = d_{k(11.14i)}^j.$$

Lemma 11.5.1 *Let the hypothesis of Theorem 11.2.1 be fulfilled. Then the function $z_{[k]}(x,t)$, $(x,t) \in \overline{G}_{(k)h(11.2c)}$, i.e., the solution of the difference scheme (10.5), (11.2c) that approximates the boundary value problem (11.18), satisfies the estimates (11.19)–(11.21).*

Remark 11.5.1 The interpolant $\overline{z}_{[k]}(x,t)$ that is constructed on $\overline{G}_{(k)}$ using the function $z_{[k]}(x,t)$, under the hypothesis of Theorem 11.2.1, satisfies the estimates (11.19)–(11.21), where $z_{[k]}(x,t)$ and $\overline{G}_{(k)h}$ are $\overline{z}_{[k]}(x,t)$ and $\overline{G}_{(k)}$, respectively.

11.5.2. Main convergence results

First, we consider the difference scheme (10.5), (11.2), (11.14), (11.17).

Taking into account the estimates (11.19)–(11.21), for the solution of the difference scheme (10.5), (11.2), (11.14), (11.17) for $\varepsilon \in (0,1]$, we obtain the estimate

$$|u(x,t) - z(x,t)| \leq \qquad\qquad\qquad\qquad (11.22)$$

$$\leq \begin{cases} M\left\{\min\left[\varepsilon^{-1}N^{-1},\, 1\right] + N_0^{-1}\right\}, & K = 1 \\ M\left\{\min\left[\varepsilon^{-1}N^{-K}\ln N,\, 1\right] + N^{-1}\ln N + N_0^{-1}\right\}, & K \geq 2 \end{cases},$$

$$(x,t) \in \overline{G}_h, \quad K \geq 1, \quad \varepsilon \in (0,1].$$

Thus, the *scheme converges* on $\overline{G}_h$ under the condition $N^{-K} \ln N \ll \varepsilon$, or more precisely,

$$\varepsilon^{-1} = o\left(N^K \ln^{-1} N\right) \quad \text{for } K \geq 2, \ \varepsilon \in (0, 1].$$

Let the parameter ε satisfy the condition

$$\varepsilon \in \left(0, \varepsilon^{(j)}\right], \quad j \geq 2, \quad \varepsilon^{(j)} = \varepsilon^{(j)}_{(11.13)}. \tag{11.23}$$

For the error of the solution of the boundary value problem (10.2), (10.1) outside an σ_K-neighborhood of the set S_1^L, we obtain the estimate

$$|u(x, t) - z(x, t)| \leq M\left[N^{-1} + N_0^{-1}\right], \quad (x, t) \in \overline{G}_h, \ r(x, \Gamma_1) \geq \sigma_K, \tag{11.24a}$$

where

$$\sigma_K = d_K^j, \quad d_K^j = d_{K(11.14i)}^j, \quad 1 \leq K \leq j - 1, \quad j = j_{(11.23)}. \tag{11.24b}$$

The value σ_K satisfies the relation

$$\sigma_K = d\, N^{-K}. \tag{11.24c}$$

Thus, in the case of the condition (11.23), the solution of the *difference scheme* converges ε-*uniformly* with the first order of accuracy in x and t outside an σ_K-neighborhood of the boundary S_1^L, where σ_K tends to zero at the rate of order $\mathcal{O}\left(N^{-K}\right)$.

Let the parameter ε belong to one of the intervals in (11.14a). In this case, depending on the relation between K and j, we obtain the estimate

$$|u(x, t) - z(x, t)| \leq \tag{11.25}$$

$$\leq \begin{cases} M\left\{\min\left[\varepsilon^{-1}N^{-1}, 1\right] + N_0^{-1}\right\}, & K = 1 \\ M\left\{\min\left[\varepsilon^{-1}N^{-K}\ln N, 1\right] + N^{-1}\ln N + N_0^{-1}\right\}, & K \geq 2 \\ M\left[N^{-1}\ln N + N_0^{-1}\right], & K \geq j + 2, \end{cases} \Bigg\}, K = j+1,$$

$$(x, t) \in \overline{G}_h, \quad K \geq j + 1, \quad j \geq -1.$$

Outside an σ_K^j-neighborhood of the set S_1^L, we have the estimate

$$|u(x, t) - z(x, t)| \leq M\left[N^{-1} + N_0^{-1}\right], \quad (x, t) \in \overline{G}_h, \tag{11.26a}$$

$$r(x, \Gamma_1) \geq \sigma_K^j, \quad K \geq 1, \ j \geq 0;$$

the value σ_K^j, where

$$\sigma_K^j = d_K^j, \quad 1 \leq K \leq j + 1; \quad \sigma_K^j = d_{j+2}^j, \quad K \geq j + 2; \quad j \geq 0, \tag{11.26b}$$

satisfies the estimate

$$\sigma_K^j \leq \begin{cases} M\,\varepsilon\,\ln N, & K \geq j+1,\ \ j \geq 0, \\ M\,N^{-K}\,\ln N, & K = j, \qquad j \geq 1, \end{cases} \tag{11.26c}$$

and the relation

$$\sigma_K^j = d\,N^{-K}, \quad K \leq j-1, \quad j \geq 2. \tag{11.26d}$$

Thus, in the case when the parameter ε belongs to one of the intervals in (11.14a), the *convergence rate* of the scheme on the set $\overline{G}_h$, as well as the size of the neighborhood of the set S_1^L outside which the scheme converges at the rate $\mathcal{O}\left(N^{-1} + N_0^{-1}\right)$, *depends essentially on the parameters K and j.*

According to the estimate (11.25), it requires K iterations, where $K = j+2$, in order to obtain the solution of the difference scheme (10.5), (11.2), (11.14), (11.17) with the estimate

$$|u(x,t) - z(x,t)| \leq M\left[N^{-1}\ln N + N_0^{-1}\right], \quad (x,t) \in \overline{G}_h,$$

provided that the parameter ε belongs to one of the intervals in (11.14a).

By virtue of estimate (11.22), the difference scheme (10.5), (11.2), (11.14), (11.17) converges on $\overline{G}_h$ under the conditions $N^{-1} \ll \varepsilon$ for $K = 1$ and $N^{-K}\ln N \ll \varepsilon$ for $K \geq 2$, or more precisely,

$$\left.\begin{array}{ll} \varepsilon^{-1} = o(N) & \text{for } K = 1 \\ \varepsilon^{-1} = o\left(N^K \ln^{-1} N\right) & \text{for } K \geq 2 \end{array}\right\}, \quad N \to \infty,\ \varepsilon \in (0,1]. \tag{11.27}$$

In order that the difference scheme be convergent almost ε-uniformly with the convergence defect no greater than the value $\nu_{(11.1)}$, it is sufficient to choose the value K satisfying the condition

$$K > K(\nu), \quad K(\nu) = \nu^{-1}. \tag{11.28}$$

Thus, the difference scheme (10.5), (11.2), (11.14), (11.17), (11.28) converges almost ε-uniformly with defect ν.

Theorem 11.5.1 *Let for the solution of the boundary value problem (10.2), (10.1) the hypothesis of Theorem 11.2.1 be satisfied. Then the difference scheme (10.5), (11.2), (11.14), (11.17) converges on $\overline{G}_h$ under the condition (11.27); under the condition (11.28), the scheme converges almost ε-uniformly with defect ν. The discrete solution satisfies the estimate (11.22) and, in the case of conditions (11.23) and (11.14a), it satisfies the estimates (11.24) and (11.25), (11.26), respectively.*

We now consider the difference scheme (10.5), (11.2), (11.9), (11.17). Taking into account the estimate (11.16), we establish the following theorem.

Theorem 11.5.2 *Let for the solution of the boundary value problem (10.2), (10.1) the hypothesis of Theorem 11.2.1 be satisfied. Then the difference scheme (10.5), (11.2), (11.9), (11.17) converges on $\overline{G}_h$ under the condition (11.27); under the condition (11.28), the scheme converges almost ε-uniformly with defect ν. For the discrete solution, the estimate (11.22) holds and, in the case of conditions (11.23) and (11.14a), the estimates (11.24) and (11.25), (11.26) are valid, respectively, where*
in (11.24), provided that $\varepsilon^{-1} h_{(K)} \geq (m^0)^{-1} N$,

$$\sigma_K = d_{K(11.9)}(\delta; \varepsilon, N), \quad \delta = \delta_{(11.17)}, \quad \varepsilon = \varepsilon_{(11.23)},$$

and in (11.26),

$$\sigma_K^j = d_{K(11.9)}(\delta; \varepsilon, N), \quad \delta = \delta_{(11.17)}, \quad \varepsilon = \varepsilon_{(11.14a)}, \quad j = j_{(11.14a)}.$$

Remark 11.5.2 Let the hypothesis of Theorem 11.5.1 (Theorem 11.5.2) be fulfilled. Then for the interpolants $\overline{z}(x,t)$ constructed on $\overline{G}$ using the functions $z(x,t)$, $(x,t) \in \overline{G}_h$, the estimates of Theorem 11.5.1 (Theorem 11.5.2) remain valid, where $z(x,t)$ and $\overline{G}_h$ are $\overline{z}(x,t)$ and $\overline{G}$, respectively.

11.6   Appendix

In this section, we present estimates for the solution of the boundary value problem and its derivatives. These estimates can be derived similar to that in [47, 52, 182]. We write the solution of the problem (10.2) as the decomposition

$$u(x,t) = U(x,t) + V(x,t), \quad (x,t) \in \overline{G}, \tag{11.29}$$

where $U(x,t)$ and $V(x,t)$ are the regular and singular components of the solution.

The functions $U(x,t)$, $V(x,t)$ satisfy the estimates

$$\left| \frac{\partial^{k+k_0}}{\partial x^k \partial t^{k_0}} U(x,t) \right| \leq M\left[1 + \varepsilon^{2-k} \right], \tag{11.30}$$

$$\left| \frac{\partial^{k+k_0}}{\partial x^k \partial t^{k_0}} V(x,t) \right| \leq M\varepsilon^{-k} \exp\left(-m\varepsilon^{-1} r(x, \Gamma_1) \right),$$

$$(x,t) \in \overline{G}, \quad k + 2k_0 \leq 4, \quad k \leq 3,$$

where m is an arbitrary number in the interval $(0, m_0)$,

$$m_0 = \min_{\overline{G}} \left[a^{-1}(x,t) b(x,t) \right];$$

and $r(x, \Gamma_1)$ is the distance between the point x and the left boundary Γ_1 of the set D.

Theorem 11.6.1 *Let the data of the boundary value problem (10.2), (10.1) satisfy condition (10.3), the condition a, b, c, p, f $\in C^{6+\alpha}(\overline{G})$, $\varphi \in C^{6+\alpha}(S)$, $\alpha > 0$, and also the condition*

$$\varphi(x,t) = 0, \quad (x,t) \in S_0;$$

$$\frac{\partial^{k_0}}{\partial t^{k_0}} \varphi(x,t) = 0, \quad \frac{\partial^{k+k_0}}{\partial x^k \, \partial t^{k_0}} f(x,t) = 0, \quad (x,t) \in S^c,$$

where k, $k_0 \le 6$, $S^c = \overline{S}^L \cap S_0$. Then the components in representation (11.29) of the solution of the boundary value problem satisfy estimates (11.30).

Chapter 12

On conditioning of difference schemes and their matrices for singularly perturbed problems

In this chapter a new approach is discussed for studying ε-*uniform well conditioning* of difference schemes. For this, a model boundary value problem for a linear ordinary convection-diffusion differential equation is considered. Some results are also given on the investigation of conditioning of difference schemes and their matrices to a boundary value problem for a singularly perturbed parabolic convection-diffusion equation.

12.1 Introduction

In ε-uniformly convergent difference schemes based on the condensing mesh method, essentially nonuniform meshes are used. Here a mesh step-size in a neighborhood of a boundary layer is much less than the value of the perturbation parameter ε that defines typical width of the boundary layer, where $\varepsilon \in (0, 1]$, (see, e.g., Remark 11.2.1 in Section 11.2). When solving numerically even regular problems, the use of meshes whose step-size may be extremely small leads to a loss of well conditioning of matrices to discrete problems that, in general, causes increased sensitivity of solutions to perturbations which arise under the numerical solving (see, e.g., [16, 10], where conditioning of a linear system of discrete equations (a difference scheme) and a matrix related to the system (a matrix associated with the difference scheme) was discussed).

Conditioning of a *numerical method* (or a difference scheme) that depends on the value of the parameter ε shows that *correctness* of the discrete problem depends also on the value of the parameter ε and deteriorates (in general, up to loss of correctness) as $\varepsilon \to 0$. When conditioning of a numerical method is independent of the parameter ε, we say that such a numerical method is ε-*uniform well-conditioned.* ε-*uniform correctness* of a numerical method is defined in a similar way. Definition of *correctness* (or *well-posedness*) to numerical methods for regular problems is given, e.g., in [108].

The question of ill-conditioning (in the classical sense) of schemes on nonuniform meshes has been touched in the literature; we mention [104], where the technique of "classical preconditioning" for special difference schemes was discussed. But conditioning of a discrete problem itself was not considered in [104].

In the case of singularly perturbed boundary value problems for a one-dimensional convection-diffusion equation, ε-uniform well-conditioning of difference schemes was established in [162, 174] (for a scheme on piecewise-uniform meshes in [162] and for a domain decomposition scheme based on a scheme on piecewise-uniform meshes in [174]).

For a singularly perturbed parabolic convection-diffusion equation, a new approach was developed in [192] for studying of ε-uniform well-conditioning of difference schemes on piecewise-uniform meshes.

In the case of regular problems, conditioning of a grid method is defined by conditioning of a matrix of a grid problem, namely, by a conditioning number of the matrix that is specified by the value N, where $N + 1$ is the number of mesh points used.

In the case of singularly perturbed problems the conditioning number of a matrix of a grid problem, as well as a domain of convergence of the grid problem, depend on both the value N and the value of the perturbation parameter ε. Conditioning of the matrix of the grid problem is inadequate to describe conditioning of the grid problem itself. Therefore, their conditioning needs special studying.

In the present chapter the new *approach* based on results of [192] is shown for the investigation of ε-*uniform well-conditioning of difference schemes* using a model boundary value problem for a linear ordinary convection-diffusion differential equation. When analyzing conditioning of a grid problem, the variables δ and ε are used instead of N and ε, where the value δ is an error of the grid problem solution, and the conditioning number of the grid problem itself is considered instead of the conditioning number of the matrix to the grid problem. Such an approach to study conditioning of the grid problem turns out to be efficient. For the model problem, it is shown that the estimate of the conditioning number to a difference scheme on a piecewise-uniform grid is independent of the parameter ε and is defined by only the value N, i.e., such a *grid problem is ε-uniformly well conditioned*. On the other hand, a difference scheme on a uniform mesh (under the imposed condition on step-sizes of the mesh and the parameter ε guaranteeing convergence of the scheme) is not ε-uniformly well conditioned; for a fixed value of the value δ, *the conditioning number* of this *grid problem grows unboundedly as $\varepsilon \to 0$*. Neither the matrix of the ε-uniformly convergent scheme nor the matrix of the scheme on a uniform grid that converges under the condition are ε-uniformly well conditioned.

The efficiency of this approach to study conditioning of difference schemes for a singularly perturbed parabolic convection-diffusion equation is discussed in Section 12.2. Here, the ε-uniform estimate of the conditioning number to the grid problem is exposed for a difference scheme on a piecewise-uniform mesh

12.2   Conditioning of matrices to difference schemes on piecewise-uniform and uniform meshes. Model problem for ODE

When conditioning of difference schemes that approximate a boundary value problem is studied, difficulties appear. It is convenient to discuss these difficulties using a model boundary value problem for a linear ordinary differential convection-diffusion equation. In order to reveal an influence of the perturbation parameter ε on sensitivity of solutions of difference schemes to perturbations in the data of the grid problem, we consider conditioning of matrices of the difference schemes.

12.2.1. We consider the boundary value problem for the singularly perturbed ordinary differential equation

$$L_{(12.1)}u(x) \equiv \left\{\varepsilon\, a(x)\, \frac{d^2}{dx^2} + b(x)\, \frac{d}{dx}\right\} u(x) = f(x), \qquad x \in D, \quad (12.1)$$

$$u(x) = \varphi(x), \qquad x \in \Gamma.$$

Here

$$\overline{D} = D \cup \Gamma, \quad D = (0, d), \tag{12.2}$$

the coefficients $a(x)$, $b(x)$ and the right-hand side $f(x)$ are sufficiently smooth functions on $\overline{D}$, moreover

$$a(x) \geq a_0, \quad b(x) \geq b_0, \quad x \in \overline{D}, \quad a_0, b_0 > 0;$$

the parameter ε takes arbitrary values in the open-closed interval $(0, 1]$.

On $\overline{D}$ we introduce the grid

$$\overline{D}_h = \overline{\omega}, \tag{12.3}$$

where $\overline{\omega} = \overline{\omega}_{(12.3)}$ is a mesh with an arbitrary distribution of nodes. Let $N+1$ be the number of nodes in the mesh $\overline{\omega}$. We approximate problem (12.1), (12.2) by the classical [108] finite difference scheme

$$\Lambda z(x) \equiv \{\varepsilon\, a(x)\, \delta_{\widehat{x}\widehat{x}} + b(x)\, \delta_x\}\, z(x) = f(x), \quad x \in D_h,$$
$$z(x) = \varphi(x), \quad x \in \Gamma_h. \tag{12.4}$$

The maximum principle holds for the finite difference scheme (12.4), (12.3).

In the case of the uniform grid

$$\overline{D}_h = \overline{D}_h^{\,u} \equiv \overline{\omega}_{(12.5)}^{\,u} \tag{12.5}$$

with step-size $h = dN^{-1}$, for the solution of the difference scheme one has

$$|u(x) - z(x)| \leq M\,(\varepsilon + N^{-1})^{-1}\, N^{-1}, \quad x \in \overline{D}_h. \tag{12.6}$$

On $\overline{D}$ we introduce the piecewise-uniform grid

$$\overline{D}_h = \overline{D}_h^s \equiv \overline{\omega}_{(10.11a)}^s. \qquad (12.7)$$

The parameter σ in this mesh is specified by

$$\sigma = \sigma(\varepsilon,\, N) = \min \left[2^{-1} d,\ m^{-1} \varepsilon \ln N \right].$$

Here m is an arbitrary number in $(0, m_0)$, where

$$m_0 = \min_{\overline{D}} \left[a^{-1}(x)\, b(x) \right],$$

i.e., $\overline{\omega}_{(10.11a)}^s = \overline{\omega}_{(10.11a)}^s (m_0)$. In the case when the parameters ε and N satisfy the condition

$$m^{-1} \varepsilon \ln N \geq 2^{-1} d, \qquad (12.8)$$

the grid $\overline{G}_{h(12.7)}$ becomes uniform.

For the solution of the difference scheme (12.4), (12.3) we obtain the ε-dependent estimate

$$|u(x) - z(x)| \leq M\, N^{-1} \left(\varepsilon^{-1} + \ln^{-1} N \right)^{-1} \approx \qquad (12.9a)$$

$$\approx M\, N^{-1} \min[\varepsilon^{-1},\, \ln N], \quad x \in \overline{D}_h,$$

which is unimprovable with respect to ε, N, and the ε-uniform estimate

$$|u(x) - z(x)| \leq M\, N^{-1} \ln N, \quad x \in \overline{D}_h, \qquad (12.9b)$$

which is unimprovable with respect to N.

12.2.2. We now consider conditioning of matrices of difference schemes for the problem (12.1), (12.2).

We write the difference scheme (12.4), (12.3) as a system of algebraic equations. Let an $(N+1)$-dimensional vector y correspond to the $N+1$ components of the function $z(x)$, for $x \in \overline{D}_h$. After the ordering of the elements $z(x)$, $x \in \overline{D}_h$, in the scheme (12.4), we come to the matrix system

$$A\, y = b. \qquad (12.10)$$

Here A is a tridiagonal $(N+1) \times (N+1)$-matrix (a_{ij}) [108], and b is an $(N+1)$-dimensional vector. The components of the vector b that correspond to the nodes $x \in D_h$ and $x \in \Gamma_h$, are the values of $-f(x)$ and $\varphi(x)$, respectively; let the first component of the vector b correspond to the point $x^0 = 0$. The matrix A is an M-matrix with the properties of nonstrict diagonal dominance and strict dominance with respect to the first and last rows. Let y and b be vectors from the normed spaces Y and B endowed with the maximum vector-norm $\| \cdot \|$. The notation $y \geq 0$ means that $y_i \geq 0$ for all i.

Thus, the system (12.10) corresponds to canonical form of the difference scheme (12.4), (12.3) [108].

The operator $A_{(12.10)}$ satisfies the *monotonicity principle*

$$\text{the condition } A\,y^1 \geq A\,y^2 \text{ implies } y^1 \geq y^2.$$

Using the majorant function technique (see, e.g., [138, 87, 108]) applied for the discrete problem (12.4), (12.7), we establish ε-uniform boundedness for the norm of the inverse matrix A^{-1}:

$$\|A^{-1}\| \leq M.$$

Here $\|A^{-1}\|$ is the matrix norm induced by the maximum vector-norm $\|\cdot\|$. We define the conditioning number $\math'(A)$ for the matrix $A_{(12.10)}$

$$æ(A) = æ_M(A) = \|A\|\,\|A^{-1}\|.$$

For the norm of the matrix $A_{(12.10)}$ and its conditioning number $æ(A)$ on the piecewise-uniform grid $\overline{D}_{h(12.7)}$, we obtain the estimate [162, 166]

$$\|A(\overline{D}_{h(12.7)})\|, \ æ_M(A; \overline{D}^{\,s}_{h(12.7)}) \leq M\,\varepsilon^{-1} N^2 \left(\varepsilon + \ln^{-1} N\right)^2, \qquad (12.11)$$

which is unimprovable with respect to the values ε, N; the conditioning number $æ(A)$ is not ε-uniformly bounded and it grows exponentially as N increases.

In the case of uniform grid (12.5) we obtain the estimate

$$\|A(\overline{D}_{h(12.5)})\|, \ æ_M(A; \overline{D}^{\,u}_{h(12.5)}) \leq M\,N\,(1 + \varepsilon\,N), \qquad (12.12)$$

which is unimprovable with respect to the values ε, N; the conditioning number $æ(A)$ in this case is ε-uniformly bounded.

By virtue of the unimprovability of the estimates (12.11) and (12.12), one has

$$æ_M\left(A; \overline{D}^{\,u}_{h(12.5)}\right) = o\left(æ_M(A; \overline{D}^{\,s}_{h(12.7)})\right) \ \text{ for } \ \varepsilon = o(\ln^{-1} N), \qquad (12.13a)$$

$$æ_M\left(A; \overline{D}^{\,u}_{h(12.5)}\right) \approx æ_M\left(A; \overline{D}^{\,s}_{h(12.7)}\right) \qquad \text{ for } \ \ln^{-1} N = \mathcal{O}\,(\varepsilon); \qquad (12.13b)$$

$$N \to \infty, \quad \varepsilon \in (0, 1].$$

The conditioning number of the matrix A on the uniform grid $\overline{D}_{h(12.5)}$ (unlike the conditioning number of the matrix on the piecewise-uniform grid $\overline{D}_{h(12.7)}$) turns out to be ε-uniformly bounded, moreover, it grows as N increases. But unlike the ε-uniformly convergent scheme (12.4), (12.7), the difference scheme (12.4), (12.5) converges only under the condition $N^{-1} \ll \varepsilon$, or more precisely:

$$\varepsilon^{-1} = o(N), \quad N \to \infty, \quad \varepsilon \in (0, 1]. \qquad (12.14)$$

Furthermore, it converges at the rate $\mathcal{O}\left((\varepsilon + N^{-1})^{-1}N^{-1}\right)$; the convergence condition and the convergence order are unimprovable. Under the condition (12.14) we have the estimate (12.13b) and the following estimate similar to (12.13a):

$$\text{æ}_M\left(A; \overline{D}^u_{h(12.5)}\right) = o\left(\text{æ}_M\left(A; \overline{D}^s_{h(12.7)}\right)\right) \qquad (12.13c)$$

$$\text{for} \ \ N^{-1} = \mathcal{O}\left(\varepsilon\right), \ \ \varepsilon = o(\ln^{-1} N).$$

Theorem 12.2.1 *Let the data of the boundary value problem* (12.1), (12.2) *satisfy the conditions* $a, b, f \in C^1(\overline{D})$, $a(x), b(x) \geq m$, $x \in \overline{D}$. *Then the conditioning number* $\text{æ}(A)$ *of the matrices A related to the schemes* (12.4), (12.5) *and* (12.4), (12.7) *satisfy the estimates* (12.11), (12.12), *and also either* {(12.13a, 12.13b)} *and* {(12.13b), (12.13c)} *under the condition* (12.14).

Remark 12.2.1 In the case of a *regular* boundary value problem (the problem (12.1), (12.2) for $\varepsilon \approx 1$), from the estimates (12.11), (12.12), (12.13) it follows that in the variable N *the matrices of the scheme* (12.4) on the grids (12.5) and (12.7) *are of the same order* and *well conditioned.* These schemes converge with the first convergence order. ∎

Remark 12.2.2 From the estimates (12.11), (12.12), (12.13) it follows that in the variables ε and N the conditioning of the matrix to the difference scheme (12.4) on the grid (12.5)

  (i) *is not worse* than on the grid (12.7),

at that time as it

  (ii) is *much better* than on the grid (12.7) *under the condition* $\varepsilon \ln N = o(1)$.
However, the convergence domains and the convergence orders of the difference schemes (12.4), (12.5) and (12.4), (12.7) in the variables ε and N are rather different.

Thus, unlike regular boundary value problems, the comparison in the variables ε and N of the conditioning numbers of the matrix A and the convergence orders of the difference schemes on the grids $\overline{D}_{h(12.5, \, 12.14)}$ and $\overline{D}_{h(12.7)}$ does not allow us to define which of the difference schemes is better either (12.4), (12.5), (12.14) and (12.4), (12.7). ∎

12.2.3. When we solve a problem numerically, quality of the obtained solution and expenditure for the solving of the discrete problem are defined by the solution accuracy and the number of nodes used in the grid domain, respectively. We study dependence of the conditioning number of the matrix A on the accuracy of the discrete solution and the value of the parameter ε.

We give some *definitions.* Let δ be an error of a discrete problem, moreover, in general, $\delta = \delta(N, \varepsilon)$. Let A be a matrix related to a scheme for the problem (12.2), (12.1) that converges at the rate $\mathcal{O}(\delta)$. We say that the *matrix A is* *well conditioned for fixed values of the parameter ε if its conditioning number $\mathit{æ}(A)$ satisfies the estimate*

$$\mathit{æ}(A) \leq Q(\varepsilon)\,\mu(\delta^{-1}),$$

where $\mu(\delta^{-1})$ grows (exponentially with respect to δ^{-1}) as $\delta \to 0$.

Let for $\mathit{æ}(A)$ the following estimate hold

$$\mathit{æ}(A) \leq M\,\mu(\varepsilon^{-\nu_0}\,\delta^{-1}), \tag{12.15}$$

where $\nu_0 \geq 0$. We say that the *matrix A is ε-uniformly well conditioned* if $\nu_0 = 0$ in the estimate (12.15). Otherwise, we say that the *matrix A is not ε-uniformly well conditioned*.

When

(i) $\nu_0 > 0$ in the estimate (12.15),
(ii) the constant M, in general, depends on ν_0,
(iii) the estimate $\mathit{æ}(A) \leq M\,\mu(\varepsilon^{-\nu}\,\delta)$ for $\nu < \nu_0$ is violated,

we say that the *matrix A is well conditioned with defect ν_0 (of the ε-uniform well conditioning)*.

By virtue of the estimates (12.9), (12.11) in the case of the scheme (12.4), (12.7), the following unimprovable estimate for the norm of the matrix A and its conditioning number is valid:

$$\|A(\overline{D}_{h(12.7)})\|,\ \mathit{æ}_M(A;\ \overline{D}^{\,s}_{h(12.7)}) \leq M\,\varepsilon^{-1}\delta^{-2}, \tag{12.16}$$

where

$$\delta = N^{-1}\,\min\left[\varepsilon^{-1},\ \ln N\right] \leq N^{-1}\,\ln N.$$

In the case of the convergent difference scheme (12.4), (12.5), i.e., the scheme (12.4), (12.5), (12.14), by virtue of (12.6), (12.12), one has the estimate

$$\|A(\overline{D}_{h(12.5,\,12.14)})\|,\ \mathit{æ}_M(A;\ \overline{D}^{\,u}_{h(12.5,\,12.14)}) \leq M\,\varepsilon^{-1}\delta^{-2}, \tag{12.17}$$

where

$$\delta = \left(\varepsilon + N^{-1}\right)^{-1} N^{-1}.$$

The estimate (12.17) is unimprovable with respect to the values δ and ε.

We say that the *variable δ* appeared in the estimates (12.16), (12.17) for conditioning numbers of matrices is *effective*, and the *variable N* is *primary*. We refer the parameter ε to both primary and effective variables.

Theorem 12.2.2 *Let the data of the boundary value problem (12.1), (12.2) satisfy the hypotheses of Theorem 12.2.1. Then the conditioning numbers $\mathit{æ}(A)$ of the matrices A, related to the convergent schemes*

(i) *(12.4) on the uniform grid (12.5) under the condition (12.14)*
and

(ii) *(12.4) on the piecewise-uniform grid (12.7),*
are ε-uniformly well conditioned; the conditioning numbers $\mathit{æ}(A)$ satisfy the estimates (12.16), (12.17).

Remark 12.2.3 The unimprovable estimates (12.16), (12.17) of the conditioning numbers of the matrix A in the variables ε and δ allow us to compare conditioning of the matrices. For the model problem (12.1), (12.2), conditioning of the matrix $A_{(12.10)}$ in the case of the grids (12.7) and (12.5), (12.14) is of the same order. The matrix A

(i) is *well conditioned for fixed values of the parameter* ε,

but

(ii) is *not* ε-*uniformly well conditioned*;

(iii) in the case of the grids (12.7) and (12.5), (12.14) is *well conditioned with the same defect* 2^{-1}.

However, the estimates (12.16) and (12.17) of the matrix A do not allow us to clarify differences in sensitivity of solutions of the difference schemes (12.4), (12.7) and (12.4), (12.5), (12.14) to perturbations in the data of the grid problem.   ∎

12.3 Conditioning of difference schemes on uniform and piecewise-uniform grids for the model problem

For the model problem (12.1), (12.2) in the case of the difference scheme (12.4), (12.3), we consider conditioning of the scheme on the piecewise-uniform grid (12.7) and on the uniform grid (12.5), (12.14).

12.3.1. Let in the matrix notation of the difference scheme (12.4), (12.3) (see (12.10)) the vector z in nodes of $\overline{D}_h$ undergo a perturbation (e.g., by virtue of a perturbation in the problem data or in a computational process). We call it the *primary perturbation*. The perturbations of components of z that are placed in different lines and columns are not related to each other. The primary perturbations of the components of z appear as perturbations in the right-hand side of the difference scheme (12.4), (12.3) and bring to a perturbation of its solution. We consider the perturbation of the solution to the scheme (12.4), (12.3) as a *secondary perturbation* of the vector z.

In the case of the piecewise-uniform grid (12.7), by virtue of its structure, the perturbation appeared in the right-hand side of the system of grid equations that arise in nodes from the sets $(0, \sigma)$, $\{\sigma\}$, $(\sigma, d) \subset D$ belongs to three, in general, different intervals which are defined by the intervals of the perturbations of the components of the vector z and parameters of the stencil to the difference scheme. In the case of the difference scheme (12.4) on the uniform grid (12.5), the perturbations in the right-hand side in $(0, d)$ belong only to one interval.

In the case of the problem (12.4), (12.7), we define a class $Q = Q(D_h)$ of "piecewise-bounded" grid functions $f^*(x)$, $x \in D_h$ bounded on each of the subsets $\{D_h \cap (0, \sigma)\}$, $\{D_h \cap \sigma\}$, and $\{D_h \cap (\sigma, d)\}$ by constants that are not related to each other. These constants may be dependent on ε and N.

We now estimate the solution of the problem (12.4), (12.7) in the case when its right-hand side $f(x) = f(x; \varepsilon, N)$, $x \in D_h$, belongs to the class Q, i.e.,

$$|f(x)| \le M\eta_{(12.18)}(x), \quad x \in D_h, \tag{12.18a}$$

where $M = M(d)$. In the case when the condition (12.8) is violated, i.e., when the grid (12.7) is nonuniform, the function $\eta_{(12.18)}(x)$, $\eta \in Q(D_h)$, is defined by the relation

$$\eta_{(12.18)}(x) = \eta_{(12.18)}(x; \varepsilon, N) = \tag{12.18b}$$

$$= \begin{cases} \varepsilon^{-1} \left(\varepsilon + \ln^{-1} N \right), & x < \sigma \\ N, & x = \sigma \\ 1, & x > \sigma \end{cases}, \quad x \in D_h.$$

Under the condition (12.8), i.e., when the grid (12.7) is uniform, we set

$$\eta_{(12.18)}(x) = 1, \quad x \in D_h. \tag{12.18c}$$

Then for the solution of the difference scheme (12.4), (12.7), (12.18) we have the estimate

$$|z(x)| \le M \left[\max_{D_h} \left[(\eta_{(12.18)}(x))^{-1} |f(x)| \right] + \max_{\Gamma_h} |\varphi(x)| \right], \quad x \in \overline{D}_h; \tag{12.19}$$

here $\eta_{(12.18)}(x) = \eta_{(12.18)}(x; \varepsilon, N)$, furthermore, $\eta_{(12.18)}(x) \ge 1$, $x \in D_h$. This estimate is unimprovable with respect to the values ε and N, keeping in mind dependence of the function $\eta(x) = \eta_{(12.18)}(x; \varepsilon, N)$ on the values ε and N.

12.3.2. Write the right-hand part of the grid equations in the matrix notation of the scheme (12.4), (12.7):

$$F(x) = F(x; f(\cdot), \varphi(\cdot)) \equiv \{f(x), \varphi(x)\}, \quad x \in \overline{D}_h, \tag{12.20a}$$

where $F(x) = f(x)$, $x \in D_h$, $F(x) = \varphi(x)$, $x \in \Gamma_h$. When estimating the solution of the difference scheme (12.4), (12.7), (12.18), it is convenient to measure the function $F(x)$ in a special *weight maximum norm* $\| \cdot \|^S$ with a *weight factor* $\eta_{(12.18)}(x)$, $x \in D_h$:

$$\|F\|^S = \|F\|_{\overline{D}_h}^S = \max \left\{ \max_{D_h} \left[(\eta_{(12.18)}(x))^{-1} |f(x)| \right], \max_{\Gamma_h} |\varphi(x)| \right\}.$$
$$\tag{12.20b}$$

The estimate (12.19) in this case takes the form

$$|z(x)| \le M \|F\|^S, \quad x \in \overline{D}_h. \tag{12.20c}$$

Thus, the difference scheme (12.4), (12.7), (12.18) can be considered as a transformation that brings a grid space with the norm $\| \cdot \|$ (the discrete maximum norm) to a grid space with the special norm $\| \cdot \|_{(12.20)}^S$ (the weight discrete maximum norm). It is convenient to apply such an approach in order to analyze conditioning numbers of the difference scheme (12.4), (12.7).

The operator $\Lambda_{(12.4)}^e$ related to the matrix notation of the scheme (12.4), by virtue of (12.20c), satisfies the estimate

$$\left\| \left(\Lambda_{(12.4,\,12.7)}^e \right)^{-1} \right\|^* \le M,$$

where $\left\| \left(\Lambda_{(12.4,\,12.7)}^e \right)^{-1} \right\|^*$ is the matrix norm induced by the maximum norms $\| \cdot \|$ and $\| \cdot \|_{(12.20)}^S$.

12.3.3. We are interested in the estimate of the value $\mathrm{æ}^P(\Lambda_{(12.4)}; \overline{D}_{h(12.7)})$, i.e., the *conditioning number of the difference scheme* (12.4), (12.7).

We define the conditioning number of the difference scheme (12.4), (12.3) by the relation

$$\mathrm{æ}^P(\Lambda_{(12.4)}) = \mathrm{æ}^P(\Lambda_{(12.4)}; \overline{D}_{h(12.3)}) = \max_{|\omega(\cdot)| \le 1,\ \overline{D}_h} |z(x; \omega(\cdot))|. \tag{12.21a}$$

Here $z(x) = z(x; \omega(\cdot))$ is the solution of the difference scheme

$$\begin{aligned} \Lambda_{(12.4)}\, z(x) &= \Lambda_{(12.21)}^y\, \omega(x; y), \quad x \in D_h, \quad y \in \overline{D}_h, \\ z(x) &= \omega(x; x), \quad x \in \Gamma_h, \end{aligned} \tag{12.21b}$$

where

$$\Lambda_{(12.21)}^y\, v(x; y) = \Lambda_{(12.21,\,12.4)}^y\, v(x; y) \equiv \varepsilon\, a(x)\, \delta_{\overline{y}\widehat{y}}\, v(x; y) + b(x)\, \delta_y\, v(x; y),$$

$$x \in D_h, \quad y \in \overline{D}_h, \quad y = x;$$

here $\omega(x; y)$, $x, y \in \overline{D}_h$ is a function different from zero for $y = y^i$, y^{i-1}, y^{i+1}, where $y^i = x$, $x \in D_h$, and for $y = x$, $x \in \Gamma_h$.

Perturbation in the data of the grid problem is clarified by means a function $\omega(x; y)$. The function $\omega(x; y)$ in (12.21) is the normed perturbation (unit in the maximum norm) of the elements $z(x)$ in the matrix notation of the difference scheme (12.4), (12.3). The value $\mathrm{æ}^P(\Lambda_{(12.4)})$ is a perturbation of the solution, maximal on $\overline{D}_h$, that is caused by various admissible perturbations, i.e., the function $\omega(x; y)$. By virtue of the choice of $\omega(x; y)$, the value $\mathrm{æ}^P(\Lambda_{(12.4)})$ is the maximum ratio of the solution perturbation to an admissible perturbation of the elements $z(x)$.

In the case of the difference scheme (12.4) on the piecewise-uniform grid (12.7), we obtain for the value $\mathrm{æ}^P(\Lambda_{(12.4)})$ the estimate (*the equivalence relation*)

$$m\mathrm{æ}^S(\Lambda_{(12.4)}; \overline{D}_{h(12.7)}) \le \mathrm{æ}^P(\Lambda_{(12.4)}; \overline{D}_{h(12.7)}) \le \tag{12.22}$$

$$\le M\mathrm{æ}^S(\Lambda_{(12.4)}; \overline{D}_{h(12.7)}),$$

where

$$æ^S(\Lambda_{(12.4)}) = æ^S(\Lambda_{(12.4)}; \overline{D}_{h(12.3)}) = \qquad (12.23)$$

$$= \max \left\{ \max_{|\omega(\cdot)|\le 1, D_h} \left[(\eta_{(12.18)}(x))^{-1} \left| \Lambda^y_{(12.21)} \omega(x;y) \right| \right], \max_{|\omega(\cdot)|\le 1, \Gamma_h} [\omega(x;x)] \right\},$$

$$\omega(x;y) = \omega_{(12.21)}(x;y), \quad x,y \in \overline{D}_h; \quad \overline{D}_h = \overline{D}_{h(12.3)}.$$

We call the value $æ^S_{(12.23)}(\Lambda_{(12.4)}; \overline{D}_{h(12.7)})$ the *modified conditioning number of the difference scheme* (12.4), (12.7). Because of the equivalence relation (12.22), we shall also call this value *the conditioning number of the difference scheme* (12.4), (12.7).

We are interested in the estimate of $æ^S(\Lambda_{(12.4)}; \overline{D}_{h(12.7)})$ in the values (effective variables) δ and ε, where $\delta = \delta(N, \varepsilon)$ is the error of the solution to the difference scheme(12.4), (12.7).

Using the estimate (12.19) (unimprovable with respect to the primary variables ε and N) and the estimate

$$\max_{|\omega(\cdot)|\le 1, D_h^k} \left| \Lambda^y_{(12.21)} \omega(x;y) \right| \le M\,\eta(x), \quad x \in D_h^k, \quad k = 1,2,3, \qquad (12.24)$$

where $\eta(x)$ is the piecewise-uniform function, i.e.,

$$\eta(x) = \eta(x; \varepsilon, N) = \left\{ \begin{array}{ll} \varepsilon^{-1}N^2\left(\varepsilon^2 + \ln^{-2} N\right), & x \in D_h^1 \\ N^2\left(\varepsilon + \ln^{-1} N\right), & x \in D_h^2 \\ N(1 + \varepsilon N), & x \in D_h^3 \end{array} \right\}, \quad x \in D_h,$$

$$D_h^k = \left\{ \begin{array}{ll} D_h \cap (0, \sigma), & k = 1, \\ D_h \cap \{\sigma\}, & k = 2, \\ D_h \cap (\sigma, d), & k = 3, \end{array} \right\}, \quad k = 1, 2, 3,$$

we obtain the following estimate for the value $æ^S(\Lambda_{(12.4)})$:

$$æ^S\left(\Lambda_{(12.4)}\right) = æ^S_{(12.23)}\left(\Lambda_{(12.4)}; \overline{D}^s_{h(12.7)}\right) \le M\,\eta^1; \qquad (12.25a)$$

$$\eta^1 = \eta^1(\varepsilon, N) = N^2\left(\varepsilon + \ln^{-1} N\right).$$

The estimate $æ^S$ is unimprovable with respect to the values ε and N. One has also the ε-uniform estimate

$$æ^S\left(\Lambda_{(12.4)}\right) = æ^S\left(\Lambda_{(12.4)}; \overline{D}^s_{h(12.7)}\right) \le M\,N^2. \qquad (12.25b)$$

12.3.4. In the case of the uniform grid (12.5) the special norm $\|\cdot\|^S$ transforms into the maximum norm

$$\|F\|^S_{\overline{D}_{h(12.5)}} = \max\left\{ d\max_{D_h}|f(x)|, \max_{\Gamma_h}|\varphi(x)| \right\}, \quad d = d_{(12.2)}.$$

Thus we have $\left\| \left(\Lambda^e_{(12.4,\,12.5)} \right)^{-1} \right\|^* \le M$.

For the value $\text{æ}^S(\Lambda_{(12.4)}; \overline{D}_{h(12.5)})$, i.e., the modified conditioning number of the difference scheme (12.4) on the uniform grid (12.5), where

$$\text{æ}^S\left(\Lambda_{(12.4)}; \overline{D}^u_{h(12.5)} \right) = \tag{12.26}$$

$$= \max \left\{ \max_{|\omega(\cdot)| \le 1,\, D_h} \left[d\left| \Lambda^y_{(12.21)}\, \omega(x; y) \right| \right],\ \max_{|\omega(\cdot)| \le 1,\, \Gamma_h} [\omega(x; x)] \right\},$$

we obtain the estimate (unimprovable with respect to ε and N)

$$\text{æ}^S_{(12.26)}\left(\Lambda_{(12.4)}; \overline{D}^u_{h(12.5)} \right) \le M\,N\,(1 + \varepsilon\,N). \tag{12.27}$$

For the conditioning number $\text{æ}^P\left(\Lambda_{(12.4)}; \overline{D}_{h(12.5)} \right)$ of the difference scheme (12.4), (12.5), one has the estimate

$$m\text{æ}^S\left(\Lambda_{(12.4)}; \overline{D}_{h(12.5)} \right) \le \text{æ}^P\left(\Lambda_{(12.4)}; \overline{D}_{h(12.5)} \right) \le M\text{æ}^S\left(\Lambda_{(12.4)}; \overline{D}_{h(12.5)} \right).$$

Next, because the values $\text{æ}^S\left(\Lambda_{(12.4)} \right)$ and $\text{æ}^P\left(\Lambda_{(12.4)} \right)$ are equivalent and the value $\text{æ}^S_{(12.26)}\left(\Lambda_{(12.4)} \right)$ is constructive, we shall study the value $\text{æ}^S\left(\Lambda_{(12.4)} \right)$.

By virtue of the unimprovability of the estimates (12.25), (12.27), in the primary variables ε, N we have the estimates

$$\text{æ}^S(\Lambda_{(12.4)}; \overline{D}^u_{h(12.5)}) = o\big(\text{æ}^S(\Lambda_{(12.4)}; \overline{D}^s_{h(12.7)}) \big) \ \text{for } \varepsilon = o(\ln^{-1} N), \tag{12.28a}$$

$$\text{æ}^S(\Lambda_{(12.4)}; \overline{D}^u_{h(12.5)}) \approx \text{æ}^S(\Lambda_{(12.4)}; \overline{D}^s_{h(12.7)}) \ \text{for } \ln^{-1} N = \mathcal{O}(\varepsilon), \tag{12.28b}$$

$$N \to \infty, \ \varepsilon \in (0, 1].$$

Under the condition (12.14), in the primary variables ε, N we have (12.28b) and the following estimate similar to (12.28a):

$$\text{æ}^S(\Lambda_{(12.4)}; \overline{D}^u_{h(12.5)}) = o\big(\text{æ}^S(\Lambda_{(12.4)}; \overline{D}^s_{h(12.7)}) \big) \tag{12.28c}$$

$$\text{for } N^{-1} = \mathcal{O}(\varepsilon), \varepsilon = o(\ln^{-1} N).$$

Theorem 12.3.1 *Let the data of the boundary value problem* (12.1), (12.2) *satisfy the hypotheses of Theorem 12.2.1. Then for the conditioning number* $\text{æ}(\Lambda_{(12.4)})$ *of the difference schemes* (12.4), (12.5) *and* (12.4), (12.7), *the estimates* (12.25), (12.27) *are valid and also either* (12.28a), (12.28b) *and* (12.28b), (12.28c) *under the condition* (12.14).

Remark 12.3.1 From the estimates (12.25), (12.27), and (12.28) it follows that the conditioning numbers of the difference schemes (12.4), (12.7) and (12.4), (12.5), (12.14) (as well as the conditioning numbers of the matrices of those schemes) written in the variables ε and N are "little informative".

Thus, in the case of singularly perturbed problems, the variables ε, N are inadequate to describe the conditioning of both difference schemes and their matrices.   ∎

Remark 12.3.2 In the case of a *regular* boundary value problem (the problem (12.1), (12.2) for $\varepsilon \approx 1$) from the estimates (12.11), (12.12), (12.13) it follows that in *the variables* either δ or N *matrices of the scheme* (12.4) *on the grids* (12.5) *and* (12.7) *are of the same order* and *well conditioned*. Hence, *the variables N and δ are adequate* to describe *conditioning of matrices of the difference schemes*.   ∎

12.3.5. We now give estimates of conditioning numbers of difference schemes in the effective variables δ and ε.

Well-conditioning (either *for fixed values of the parameter ε*, or *ε-uniform*, or *with defect ν_0*) *of difference schemes* is defined similar to that for matrices of difference schemes.

Under the condition (12.14) by virtue of (12.6), (12.27), the conditioning number of the difference scheme (12.4) on the uniform grid (12.5) satisfies the estimate

$$\mathscr{æ}^S \left(\Lambda_{(12.4)}; \overline{D}^{u}_{h(12.5,\, 12.14)} \right) \le M \, \varepsilon^{-1} \delta^{-2}, \tag{12.29}$$

where $\delta = \delta_{(12.17)}$; the estimate is unimprovable with respect to δ, ε. The scheme (12.4), (12.5), (12.14) is not well conditioned ε-uniformly; the conditioning defect of the scheme equals 2^{-1}.

In the case of the scheme (12.4) on the piecewise-uniform grid (12.7), by virtue of (12.9), (12.25) we have (unimprovable) the ε-dependent estimate

$$\mathscr{æ}^S \left(\Lambda_{(12.4)}; \overline{D}^{s}_{h(12.7)} \right) \le M \, \delta^{-2} \left(\varepsilon + \ln^{-1} \delta^{-1} \right)^{-1} \tag{12.30a}$$

and the ε-uniform estimate

$$\mathscr{æ}^S \left(\Lambda_{(12.4)}; \overline{D}^{s}_{h(12.7)} \right) \le M \, \delta^{-2} \ln \delta^{-1}, \tag{12.30b}$$

where $\delta = \delta_{(12.16)}$; the difference scheme (12.4) on the piecewise-uniform grid (12.7) is well conditioned ε-uniformly.

Theorem 12.3.2 *Let the data of the boundary value problem* (12.1), (12.2) *satisfy the hypotheses of Theorem 12.2.1. Then the difference scheme* (12.4), (12.7) *is ε-uniformly well conditioned while the scheme* (12.4), (12.5), (12.14) *is well conditioned with defect $\nu_0 = 2^{-1}$; the conditioning numbers satisfy the estimates* (12.29), (12.30).

Remark 12.3.3 In the case of the model problem (12.1), (12.2) the *effective variables δ, ε* allow us to establish
  (i) ε-uniform well conditioning of the scheme (12.4), (12.7),

(ii) well conditioning with defect $\nu_0 = 2^{-1}$ of the scheme (12.4), (12.5), (12.14).

Thus, the effective variables δ, ε are *adequate to analyze the conditioning of the special difference scheme* (12.4), (12.7) (the scheme of the condensing mesh method) and *the classical difference scheme* (12.4), (12.5) that approximate the singularly perturbed boundary value problem (12.1), (12.2). ∎

Remark 12.3.4 For the singularly perturbed problem (12.1), (12.2), by virtue of the estimates (12.29), (12.30), the ε-uniform estimate of the conditioning number to the special difference scheme (12.4), (12.7) is, up to a logarithmic factor, the same as that for the classical difference scheme (12.4), (12.5) but already in the case of the regular problem (12.1), (12.2) for $\varepsilon = 1$.

Thus, the special piecewise-uniform grid (12.7) condensing in a neighborhood of the boundary layer appears to be a "regularizator" that guarantees ε-uniform boundedness of the conditioning number of the ε-uniformly convergent difference scheme (12.4). ∎

Remark 12.3.5 In the case of the boundary value problem (12.1), (12.2) *for the scheme* (12.4) *on the uniform grid* (12.5), by virtue of the estimates (12.12), (12.27) and (12.17), (12.29), *the conditioning numbers of both matrix and scheme* are *of the same order in both variables* ε, N *and* ε, δ. ∎

Remark 12.3.6 In the *effective variables* ε, δ, *the matrix of the difference scheme* (12.4), (12.7) *is not ε-uniformly well conditioned*, at that time as *the difference scheme itself is ε-uniformly well conditioned*.

Thus, *the classical difference scheme* (12.4) *on the piecewise-uniform grid* (12.7), which is

(i) ε-uniformly stable,

(ii) ε-uniformly well-conditioned,

is *ε-uniformly correct*. Definition of *correctness* (or *well-posedness*) to numerical methods for regular problems can be found in [108].

But *the classical difference scheme* (12.4) that converges *on the uniform grid* (12.5) *under the condition* (12.14), which is

(i) ε-uniformly stable,

(ii) not ε-uniformly well-conditioned,

is *not ε-uniformly correct*.

At the same time, for *regular boundary value problems, the schemes* (12.4), (12.7) and (12.4), (12.5), which are stable and well conditioned, are *correct*. ∎

12.4 On conditioning of difference schemes and their matrices for a parabolic problem

In this section we show some results on conditioning of difference schemes and their matrices for the boundary value problem (10.2), (10.1) for the singularly perturbed parabolic convection-diffusion equation. More detailed presentation and justification of these results can be found in [192]. Problem formulation is given in Section 10.2 of Chapter 10. In the same place, one can find:

(i) the rectangular grid $\overline{G}_{h(10.4)}$ with an arbitrary distribution of nodes, and the finite difference scheme (10.5) on $\overline{G}_h$;

(ii) the uniform grid $\overline{G}^u_{h(10.7)}$, and estimates of the solution to the difference scheme (10.5) on the grid (10.7);

(iii) the special piecewise-uniform grid $\overline{G}^s_{h(10.11)}$ and ε-uniform estimates of the solution to the difference scheme (10.5) on the grid (10.11).

The convergence Theorem 11.2.1 and *a priori* estimates from Chapter 11 are satisfied.

12.4.1. First, we consider grid approximations of the boundary value problem (10.2), (10.1). Let for the solution of the problem (10.2), (10.1) the hypotheses of Theorem 11.2.1 be fulfilled. Then the conditioning number of the matrix A related to the ε-uniformly convergent difference scheme (10.5), (10.11) satisfy the estimate

$$\ae_M(A; \overline{G}^s_{h(10.11)}) \le M \left[\varepsilon^{-1} \delta_1^{-2} + \delta_0^{-1} \right], \qquad (12.31a)$$

where δ_1 and δ_0 are the components of the error δ of the discrete solution (up to a constant factor) that appear because of an approximation of the boundary value problem (10.2), (10.1), respectively, in x and t, i.e.,

$$\delta_1 = N^{-1} \min \left[\varepsilon^{-1}, \ln N \right] \approx N^{-1} \left(\varepsilon + \ln^{-1} N \right)^{-1}, \qquad \delta_0 = N_0^{-1}. \quad (12.31b)$$

The estimate is unimprovable with respect to δ_1, δ_0, and ε.

For the matrix related to the difference scheme (10.5) that converges on the uniform grid (10.7) under the condition (10.10) ($N^{-1} \ll \varepsilon$, see Section 10.2), we obtain the (unimprovable) estimate

$$\ae_M(A; \overline{G}^u_{h(10.7, 10.10)}) \le M \left[\varepsilon^{-1} \delta_1^{-2} + \delta_0^{-1} \right], \qquad (12.32)$$

where

$$\delta_1 = (\varepsilon + N^{-1})^{-1} N^{-1}, \qquad \delta_0 = \delta_{0(12.31)}.$$

Conditioning defect of the matrices to the difference schemes for the boundary value problem (10.2), (10.1) is defined similar to that in the case of the problem (12.1), (12.2).

Under the condition (10.10), taking into account the unimprovable estimates (12.31a) and (12.32), one has

$$\ae_M\left(A; \overline{G}_{h(10.7),(10.10)}^u\right) \approx \ae_M\left(A; \overline{G}_{h(10.11)}^s\right). \tag{12.33}$$

The following theorem holds (see [192]).

Theorem 12.4.1 *Let for the solution of the boundary value problem (10.2), (10.1) the hypotheses of Theorem 11.2.1 be fulfilled. Then the matrices A related to the convergent schemes*
(i) (10.5) on the uniform grid (10.7) under the condition (10.10)
and
(ii) (10.5) on the piecewise-uniform grid (10.11),
are ε-uniformly well conditioned in t and well conditioned in x with defect $\nu_0 = 2^{-1}$; the conditioning numbers $\ae(A)$ satisfy the estimates (12.31a), (12.32), and (12.33).

Remark 12.4.1 In the case of a *regular* boundary value problem (the problem (12.1), (12.2) for $\varepsilon \approx 1$), the matrices of the difference schemes (10.5), (10.7), (10.10) and (10.5), (10.11) are well conditioned with the conditioning numbers of order $\mathcal{O}\left(\delta_1^{-2} + \delta_0^{-1}\right)$.

12.4.2. Next, we give the conditioning numbers $\ae^S\left(\Lambda_{(10.5)}; \overline{G}_{h(10.7, 10.10)}^u\right)$ and $\ae^S\left(\Lambda_{(10.5)}; \overline{G}_{h(10.11)}^s\right)$ of the difference schemes (10.5), (10.7), (10.10) and (10.5), (10.11), respectively. Strict definition of the conditioning numbers $\ae^S\left(\Lambda; \overline{G}_h\right)$ for the difference schemes on uniform and piecewise-uniform grids is given in [192].
The conditioning numbers of the difference scheme (10.5), (10.7), (10.10) satisfy the estimate

$$\ae^S\left(\Lambda_{(10.5)}; \overline{G}_{h(10.7, 10.10)}^u\right) \le M\left[\varepsilon^{-1}\delta_1^{-2} + \delta_0^{-1}\right], \tag{12.34}$$

where

$$\delta_1 = \delta_{1(12.32)} = (\varepsilon + N^{-1})^{-1}N^{-1}, \quad \delta_0 = \delta_{0(12.31b)} = N_0^{-1};$$

the estimate (12.34) is unimprovable with respect to δ_1, δ_0, and ε. The difference scheme (10.5), (10.7), (10.10) is well conditioned with defect $\nu_1 = 2^{-1}$ in x and ε-uniformly well conditioned in t.
On the grid (10.11) the following estimate holds:

$$\ae^S\left(\Lambda_{(10.5)}; \overline{G}_{h(10.11)}^s\right) \le M\left[\delta_1^{-2}\left(\varepsilon + \ln^{-1}\delta_1^{-1}\right)^{-1} + \delta_0^{-1}\right], \tag{12.35a}$$

where

$$\delta_1 = \delta_{1(12.01b)} = N^{-1}\left(\varepsilon + \ln^{-1} N\right)^{-1}, \quad \delta_0 = \delta_{0(12.31b)} = N_0^{-1};$$

the estimate (12.35a) is unimprovable with respect to δ_1, δ_0, ε. One also has the ε-uniform estimate

$$\text{æ}^S \left(\Lambda_{(10.5)}; \overline{G}^s_{h(10.11)} \right) \leq M \left[\delta_1^{-2} \ln \delta_1^{-1} + \delta_0^{-1} \right]. \tag{12.35b}$$

Thus, the difference scheme (10.5), (10.11) is ε-uniformly well conditioned.
    The following theorem holds (see [192]).

Theorem 12.4.2 *Let for the solution of the boundary value problem* (10.2), (10.1) *the hypotheses of Theorem* 11.2.1 *be fulfilled. Then the difference scheme* (10.5), (10.11) *is ε-uniformly well conditioned; the difference scheme* (10.5), (10.7), (10.10) *is well conditioned with defect* $\nu_1 = 2^{-1}$ *in* x, *and it is ε-uniformly well conditioned in* t. *For the conditioning numbers the estimates* (12.34) *and* (12.35) *hold.*

Remark 12.4.2 In the case of the singularly perturbed problem (10.2), (10.1), by virtue of the estimates (12.34), (12.35), ε-uniform estimate of the conditioning number to the special difference scheme (10.5), (10.11) is the same (up to a logarithmic factor of the variable δ_1^{-2}) as that for the classical difference scheme (10.5), (10.7) but already in the case of the regular problem (10.2), (10.1) for $\varepsilon \approx 1$. ∎

Remark 12.4.3 In the case of the difference scheme (10.5) on the uniform grid (10.7), the conditioning numbers of the matrix and the difference scheme (by virtue of the estimates (12.32), (12.34)) are of the same order. ∎

Chapter 13

Approximation of systems of singularly perturbed
elliptic reaction-diffusion equations with two parameters

The grid approximations of a Dirichlet problem are considered for a system of
two singularly perturbed elliptic reaction-diffusion equations with two perturbation parameters on a rectangle. For the data of the boundary value problem,
compatibility conditions are given to ensure sufficient smoothness of the solution to the problem that is required for the construction and justification of
schemes convergent uniformly with respect to the perturbation parameters
ε_i^2, $i = 1, 2$. A priori estimates are constructed for the problem solution.
Using these estimates, the condensing mesh technique and classical finite difference approximations of the problem, special finite difference schemes are
constructed that converge $\varepsilon_1^2, \varepsilon_2^2$-uniformly with the convergence order close
to 2.

13.1   Introduction

Grid approximations of boundary value problems for a system of singularly
perturbed equations on a strip were considered in [143, 152, 158] (in [143] for
reaction-diffusion equations and in [152, 158] for convection-diffusion equations). In these papers special finite difference schemes were constructed that
converge uniformly with respect to the perturbation parameters ε_i^2, $i = 1, 2$,
which are components of the vector parameter ε. The technique for studying
the special finite difference schemes that was introduced in [143, 152, 158] allowed to obtain only sufficiently lower order of ε-uniform rate of the schemes
constructed. Here a Dirichlet problem is considered for a system of two singularly perturbed elliptic reaction-diffusion equations on a rectangle. The
highest-order derivatives in the i-th differential equations are multiplied by
the perturbation parameter ε_i^2, for $i = 1, 2$; the parameters ε_i take arbitrary
values in the open-closed interval $(0, 1]$. When the component-parameters ε_1
and/or ε_2 tend to zero, a double boundary layer appears in a neighborhood
of the boundary. The characteristic width of the boundary layers is of the
order of ε_1 and ε_2. The problem under consideration is more complicated
compared with those in [143, 152, 158], because here we study the problem
in a *domain with a piecewise-smooth boundary*. Such a problem is considered

in the paper [187]. In this chapter, in order to demonstrate a technique of constructing special finite difference schemes for problems of such type, some results of this papers are given:

(i) the derivation of compatibility conditions that are necessary to guarantee required smoothness of the solution components;

(ii) the derivation of *a priori* estimates for the regular and singular components of the solution that are necessary to construct and justify schemes convergent $\bar{\varepsilon}$-uniformly;

(iii) Using such estimates and the condensing mesh technique and classical finite difference approximations of the boundary value problem, special finite difference schemes are constructed that converge $\varepsilon_1^2, \varepsilon_2^2$-uniformly at the rate $\mathcal{O}\left(N^{-2} \ln^2 N\right)$, where $N = \min_s N_s$ and $N_s + 1$ is the number of mesh points on the x_s-axis with $s = 1, 2$.

Note that an initial-boundary value problem is considered in [199, 198] for a system of two parabolic reaction-diffusion equations with a scalar perturbation parameter ε on a rectangle. For such a problem, a difference scheme on piecewise-uniform grids is constructed that converges ε-uniformly.

13.2   Problem formulation. The aim of the research

13.2.1. On the rectangle $\overline{D}$, where

$$\overline{D} = D \bigcup \Gamma, \quad D = D_{(13.1)} = \{x : \ 0 < x_s < d_s, \ \ s = 1, 2\}, \qquad (13.1)$$

we consider the Dirichlet problem for the system of singularly perturbed elliptic equations

$$L\,\mathbf{u}(x) = \mathbf{f}(x), \quad x \in D, \qquad (13.2a)$$

$$\mathbf{u}(x) = \boldsymbol{\varphi}(x), \quad x \in \Gamma. \qquad (13.2b)$$

Here

$$L = L_{(13.2)}(\varepsilon_1, \varepsilon_2) \equiv \begin{pmatrix} L_0^1 & 0 \\ 0 & L_0^2 \end{pmatrix} - C(x) \equiv A(\varepsilon_1, \varepsilon_2) \sum_{s=1,2} \frac{\partial^2}{\partial x_s^2} - C(x),$$

$$A(\varepsilon_1, \varepsilon_2) = \begin{pmatrix} \varepsilon_1^2 & 0 \\ 0 & \varepsilon_2^2 \end{pmatrix}, \qquad C(x) = \begin{pmatrix} c^{11}(x) & c^{12}(x) \\ c^{21}(x) & c^{22}(x) \end{pmatrix},$$

and $\mathbf{u}(x)$, $\mathbf{f}(x)$, and $\boldsymbol{\varphi}(x)$ are vector functions, e.g., $\mathbf{u}(x) = (u^1(x), u^2(x))^T$, $x \in \overline{D}$. Along the vector form we shall use the scalar form

$$L^i\,\mathbf{u}(x) = f^i(x), \quad x \in D, \qquad u^i(x) = \varphi^i(x), \quad x \in \Gamma, \quad i = 1, 2; \qquad (13.2c)$$

here the operator $L^i = L^i_{(13.2)}$ is defined by the relation

$$L^i\,\mathbf{u}(x) = L^i_0\,u^i(x) - \sum_{j=1,2} c^{ij}(x)\,u^j(x), \qquad L^i_0 = L^i_{0(13.2)}(\varepsilon_i) \equiv \varepsilon_i^2 \sum_{s=1,2} \frac{\partial^2}{\partial x_s^2}.$$

The functions $c^{ij}(x)$, $f^i(x)$, and $\varphi^i(x)$ are assumed to be sufficiently smooth on the set $\overline{D}$ and on the boundary Γ, respectively. Assume also that the following conditions are satisfied:

$$
\begin{aligned}
c^{ii}(x) \geq c_0, \quad mc^{ii}(x) \geq |c^{ij}(x)|, \quad x \in \overline{D}, \\
i,j = 1,2, \quad i \neq j, \quad c_0 > 0, \quad m = m_{(2.3)} < 1.
\end{aligned}
\tag{13.3}
$$

The parameters ε_1 and ε_2, i.e., the components of the vector parameter $\boldsymbol{\varepsilon} = (\varepsilon_1, \varepsilon_2)$, take arbitrary values in the open-closed interval $(0, 1]$.

When the components-parameters (or one of them) tend to zero, a double boundary layer appears in a neighborhood of the set Γ.

By a solution of the problem (13.2), we mean a function $\mathbf{u} \in C^2(D)$ which is continuous on $\overline{D}$ and satisfies the differential equation (13.2a) on D and the boundary condition (13.2b) on Γ.

Assume that for fixed values of the parameter $\boldsymbol{\varepsilon}$, the solution of the problem is sufficiently smooth. Conditions sufficient for existence of a smooth solution of a boundary value problem in a domain with smooth boundaries are considered in[68]; for a scalar equation on a rectangle, such conditions are given in [212].

We denote by Γ_j, $\Gamma = \bigcup \Gamma_j$, for $j = 1, 2, 3, 4$, the sides of the rectangle D; the sides Γ_s and Γ_{s+2} are orthogonal to the x_s-axis, for $s = 1, 2$; the sides Γ_1 and Γ_2 pass through the point $(0, 0)$, moreover, the Γ_j are closed sets; and Γ^c is the set of corner points.

We shall consider the boundary value problem (13.2) subject to each of the conditions:

$$\varepsilon_1 = \varepsilon_2 = \varepsilon, \qquad \varepsilon \in (0, 1], \tag{13.4}$$

$$\varepsilon_1 = \varepsilon, \quad \varepsilon_2 = 1, \qquad \varepsilon \in (0, 1], \tag{13.5}$$

$$\varepsilon_1 = \varepsilon, \quad \varepsilon_2 = \mu, \quad \varepsilon, \mu \in (0, 1]. \tag{13.6}$$

For definiteness, we assume that $\varepsilon_1 \leq \varepsilon_2$.

More simple singular components of the solution of problem (13.2) subject to conditions (13.4) and (13.5) (compared to (13.6)) allow us to establish the required *a priori* estimates of the solutions under less strict conditions imposed on the data of the boundary value problem.

Our aim is for the boundary value problem (13.2) to derive *a priori* estimates of the regular and singular components to the solution for various sets of the parameters (conditions (13.4)–(13.6)). These estimates will be used to construct and justify a difference scheme that converges ε-uniformly with respect to the perturbation component-parameters ε_1, ε_2 (or, briefly, convergent $(\varepsilon_1, \varepsilon_2)$-uniformly, or ε-uniformly) with an ε-uniform convergence order close to 2.

13.3 Compatibility conditions. Some *a priori* estimates

13.3.1. We now give conditions imposed on the data of the problem (13.2), that guarantee sufficient smoothness of the solution required for constructions.

For the particular Dirichlet boundary value problem of type (13.2)

$$\triangle \mathbf{w}(x) = \mathbf{F}(x), \quad x \in D, \quad \mathbf{w}(x) = \mathbf{\Phi}(x), \quad x \in \Gamma, \tag{13.7}$$

in [212, 68, 80] (see also [180]), necessary and sufficient conditions are given in order that $\mathbf{w} \in C^{l+2+\alpha}(\overline{D})$ for $l \geq 0$ and $\alpha \in (0, 1)$. For example, for

$$\mathbf{F} \in C^{l+\alpha}(\overline{D}); \quad \mathbf{\Phi}(x) = \mathbf{0}, \quad x \in \Gamma$$

such a condition takes the form

$$\sum_{k=0}^{q} (-1)^k \frac{\partial^{2q}}{\partial x_1^{2(q-k)} \partial x_2^{2k}} \mathbf{F}(x) = \mathbf{0}, \quad x \in \Gamma^c, \quad 0 \leq q \leq [l/2], \tag{13.8}$$

where $[l/2] = [l/2]_{(13.8)}$ is the integer part of the number $l/2$ and Γ^c is the set of corner points of the boundary Γ. In the problem (13.7), (13.1), using a standard transformation of variables, the condition $\varphi(x) \neq \mathbf{0}$ can be brought to a homogeneous boundary condition, however, assuming that $\varphi \in C^{l+2+\alpha}(\Gamma)$.

Let the data of the problem (13.2) satisfy the conditions

$$C_{(13.2)}, \mathbf{f} \in C^{l+\alpha}(\overline{D}), \quad \varphi \in C^{l+2+\alpha}(\Gamma_j), \quad \varphi \in C(\Gamma),$$
$$j = 1, 2, 3, 4, \quad l \geq 0, \quad \alpha \in (0, 1). \tag{13.9a}$$

Then for its solution one has

$$\mathbf{u} \in C^{l+2+\alpha}(\overline{D} \setminus D^{\delta/2}), \tag{13.10}$$

where D^δ is an δ-neighborhood of the set Γ^c (the value δ is independent of ε). Taking into account (13.10) and assuming that

$$\frac{\partial^k}{\partial x_1^{k_1} \partial x_2^{k_2}} C(x) = 0, \quad x \in \Gamma^c, \quad 1 \leq k \leq 2([l/2] - 1) \quad for \quad l \geq 4;$$
$$for \quad l \leq 3 \quad (l \geq 0) \quad restrictions \quad on \quad the \quad derivatives \tag{13.9b}$$

of the matrix function $C(x)$, $x \in \Gamma^c$ are not imposed.

where $[l/2] = [l/2]_{(13.8)}$, one can show that the condition (13.8) is necessary and sufficient for

$$\mathbf{u} \in C^{l+2+\alpha}(\overline{D}). \qquad (13.11)$$

In this case, the function $\mathbf{F}(x)$ in (2.8) is defined by the relation

$$\mathbf{F}(x) = \begin{pmatrix} \varepsilon_1^{-2} & 0 \\ 0 & \varepsilon_2^{-2} \end{pmatrix} [\mathbf{f}(x) - \mathbf{f}_{\Phi}(x) + C(x)\,\mathbf{v}(x)], \quad x \in \overline{D}^{\delta}, \qquad (13.12)$$

$$\mathbf{v}(x) = \mathbf{u}(x) - \mathbf{\Phi}(x), \quad \mathbf{f}_{\Phi}(x) = L_{(13.2)}\mathbf{\Phi}(x), \quad \mathbf{\Phi}(x) = \varphi_i(x^i) + \varphi_j(x^j) - \varphi_{ij},$$

where $\varphi_j(x) = \varphi(x), \ x \in \Gamma_j; \ x^j = x^j(x), \ x \in \overline{D}$ is the projection of the point x to Γ_j; $\varphi_{ij} = \varphi(x^{ij}), \ x^{ij} = (x_1^{ij}, x_2^{ij}) \in \Gamma_{ij}^c$, where $\Gamma_{ij}^c = \Gamma_i \cap \Gamma_j$ for $\Gamma_{ij}^c \subset \overline{D}^{\delta}$.

Under the assumptions (13.9a), (13.9b), we have the following sufficient condition to ensure (13.11):

$$\frac{\partial^k}{\partial x_1^{k_1} \partial x_2^{k_2}} \mathbf{f}(x) = \mathbf{0}, \quad k \le l,$$

$$\frac{\partial^{k_1}}{\partial x_1^{k_1}} \varphi(x) = \frac{\partial^{k_2}}{\partial x_2^{k_2}} \varphi(x) = \mathbf{0}, \quad k_1, k_2 \le l+2, \quad x \in \Gamma^c. \qquad (13.9c)$$

This condition is simpler than the necessary and sufficient compatibility condition (13.8), (13.12).

For simplicity, we assume that the following condition is valid (we shall call this the condition (13.13)):

> *The data of the problem (13.2) satisfy the condition (13.9) that ensures the smoothness of the solution of the boundary value problem. If the perturbation parameters satisfy the conditions either (13.4) or (13.6), where $\mu^2 = \mathcal{O}(\varepsilon)$ and $\varepsilon \le \mu$, then it is assumed that the following condition is valid:*

$$c^{iq}, f^i \in C^{l_1+\alpha}(\overline{D}), \quad \varphi^i \in C^{l_1+\alpha}(\Gamma_j), \quad \varphi^i \in C(\Gamma); \qquad (13.13a)$$

$$\frac{\partial^k}{\partial x_1^{k_1} \partial x_2^{k_2}} f^i(x) = 0, \quad x \in \Gamma^c, \quad k = k_1 + k_2, \quad k \le l_1;$$

$$\frac{\partial^{k_1}}{\partial x_1^{k_1}} \varphi^i(x) = \frac{\partial^{k_2}}{\partial x_2^{k_2}} \varphi^i(x) = 0, \quad x \in \Gamma^c,$$

$$k_1, k_2 \le l_1; \quad i, q = 1, 2; \quad l_1 \ge l.$$

> *But if the perturbation parameters satisfy the conditions either (13.5) or (13.6), where $\varepsilon = \mathcal{O}(\mu^2)$, it is assumed that the fol-*

lowing condition is fulfilled:

$$c^{1q},\ f^1 \in C^{l_1+2+\alpha}(\overline{D}),\quad c^{2q},\ f^2 \in C^{l_1+\alpha}(\overline{D}), \tag{13.13b}$$

$$\varphi^i \in C^{l_1+2+\alpha}(\Gamma_j),\quad \varphi^i \in C(\Gamma);$$

$$\frac{\partial^k}{\partial x_1^{k_1} \partial x_2^{k_2}}\, f^1(x) = 0,\quad k \leq l_1 + 2,$$

$$\frac{\partial^k}{\partial x_1^{k_1} \partial x_2^{k_2}}\, f^2(x) = 0,\quad k \leq l_1,\quad k = k_1 + k_2,\quad x \in \Gamma^c;$$

$$\frac{\partial^{k_1}}{\partial x_1^{k_1}}\, \varphi^i(x) = \frac{\partial^{k_2}}{\partial x_2^{k_2}}\, \varphi^i(x) = 0,\quad x \in \Gamma^c,$$

$$k_1,\, k_2 \leq l_1 + 2;\quad i, q = 1, 2,\quad l_1 \geq l.$$

The conditions (13.13a) and (13.13b) guarantee the smoothness of the regular and singular components of the solution (that are given in Section 13.4).

The actual values of l and l_1 are specified where it is required. The fulfillment of other conditions in addition to (13.9) and (13.13) is not assumed.

13.3.2. When constructing and studying classical and special finite difference schemes, we shall need bounds on the solutions and on their derivatives for the boundary value problem (13.2) under the conditions (13.4)–(13.6).

Introducing new variables $\tilde{x}_i = \varepsilon^{-1} x_i$, for $i = 1, 2$ (where $\varepsilon = \min[\varepsilon_1, \varepsilon_2]$), we bring the problem (13.2) to a form, in which the coefficients of the higher-order derivatives become of the same order. In this case, the derivatives of the function $\tilde{u}(\tilde{x}) = u(x(\tilde{x}))$ in the new variables are of order 1. Returning to the origin variables, in the case of the condition (13.13), where

$$l \geq K - 2, \tag{13.14}$$

we obtain the estimates

$$|\mathbf{u}(x)| \leq M,\quad \left| \frac{\partial^k}{\partial x_1^{k_1} \partial x_2^{k_2}}\, \mathbf{u}(x) \right| \leq M \left[\varepsilon_1^{-k} + \varepsilon_2^{-k} \right],\quad x \in \overline{D},\quad k \leq K, \tag{13.15}$$

where $|\mathbf{u}(x)| = \max_{\overline{D}} |\mathbf{u}(x)| = \max_{\overline{D},i} |u^i(x)|.$

Theorem 13.3.1 *Let the data of the boundary value problem (13.2), (13.4)–(13.6) satisfy the conditions (13.13), (13.14) for $K \geq 2$. Then the problem solution satisfies the estimates (13.15).*

13.4   Derivation of *a priori* estimates for the problem (13.2) under the condition (13.5)

Let us give some estimates that are obtained using main terms in the asymptotic expansion of the solution (see, e.g., [187]). The problem (13.2) for $\varepsilon_1 = 0$ and $\varepsilon_2 = 1$ is a differential-algebraic problem. This feature of the reduced problem is used for constructing asymptotics.

13.4.1. For the problem (13.2) subject to the condition (13.5), we construct a "preliminary" *decomposition* of the problem solution as the sum

$$\mathbf{u}(x) = \mathbf{u}_{(13.2,\,13.5)}(x) = \mathbf{U}(x) + \mathbf{V}(x) + \mathbf{u}^{(1)}(x) \equiv \mathbf{U}(x) + \mathbf{P}(x), \quad (13.16)$$

$$x \in \overline{D},$$

where $\mathbf{U}(x)$ and $\mathbf{V}(x)$ are the regular and singular components of the solution, $\mathbf{u}^{(1)}(x)$ is the "remainder" term; the function $\mathbf{u}^{(1)}(x)$ is considered in Section 13.4.4. The function $\mathbf{U}(x) = (U^1(x),\, U^2(x))^T$ is the solution of the problem

$$L_{(13.2)}\,\mathbf{U}(x) = L_{(13.2)}(\varepsilon_1 = \varepsilon,\, \varepsilon_2 = 1)\,\mathbf{U}(x) = \mathbf{f}(x), \quad x \in D, \quad (13.17\text{a})$$

its second component $U^2(x)$ on the boundary Γ satisfies the condition

$$U^2(x) = \varphi^2(x), \quad x \in \Gamma; \quad (13.17\text{b})$$

it is required that the derivatives up to order K of the component $U^1(x)$ to be ε-uniformly bounded on $\overline{D}$:

$$\left| \frac{\partial^k}{\partial x_1^{k_1} \partial x_2^{k_2}} U^1(x) \right| \leq M, \quad x \in \overline{D}, \quad k \leq K. \quad (13.17\text{c})$$

The function $\mathbf{P}(x)$ is the solution of the problem

$$L_{(13.2)}\,\mathbf{P}(x) = \mathbf{0}, \quad x \in D, \quad \mathbf{P}(x) = \boldsymbol{\varphi}(x) - \mathbf{U}(x) \equiv \boldsymbol{\varphi}_{\mathbf{V}}(x), \quad x \in \Gamma. \quad (13.18)$$

We represent the function $\mathbf{U}(x)$ as an expansion in the parameter ε^2

$$\mathbf{U}(x) = \sum_{k=0}^{n} \varepsilon^{2k} \mathbf{U}_k(x) + \mathbf{v}_{\mathbf{U}}^n(x) \equiv \mathbf{U}^n(x) + \mathbf{v}_{\mathbf{U}}^n(x), \quad x \in \overline{D}. \quad (13.19)$$

The function $\mathbf{U}_k(x)$, $x \in \overline{D}$, for $k = 0, \ldots, n$, are solutions of the problem

$$L_{(13.20)}\,\mathbf{U}_0(x) = \mathbf{f}(x), \quad x \in D, \quad U_0^2(x) = \varphi_0^2(x), \quad x \in \Gamma; \quad (13.20)$$

$$L_{(13.20)}\,\mathbf{U}_k(x) = \varepsilon^{-2}\left\{ L_{(13.20)} - L_{(13.2)} \right\}\mathbf{U}_{k-1}(x), \quad x \in D,$$

$$U_k^2(x) = \varphi_k^2(x), \quad x \in \Gamma, \quad k \geq 1,$$

where

$$L_{(13.20)} = L_{(13.2)}(\varepsilon_1 = 0, \varepsilon_2 = 1), \quad \varphi_0^2(x) = \varphi^2(x), \quad \varphi_k^2(x) = 0, \quad x \in \Gamma, \quad k \geq 1.$$

In the case of condition (13.13), where

$$l \geq K - 2, \quad l_1 \geq K + 2n \quad for \quad n = [(K+1)/2]_{(13.8)} - 1, \quad K \geq 2, \qquad (13.21)$$

one has $\mathbf{U} \in C^{K+\alpha}(\overline{D})$. For the function $\mathbf{U}_{(13.19)}(x)$ we have the estimate

$$\left| \frac{\partial^k}{\partial x_1^{k_1} \partial x_2^{k_2}} \mathbf{U}(x) \right| \leq M \left[1 + \varepsilon^{K-k} \right], \quad x \in \overline{D}, \quad k \leq K; \qquad (13.22a)$$

moreover, for the component $\mathbf{U}^n(x)$ and the remainder term $\mathbf{v}_{\mathbf{U}}^n(x)$ in (13.19), we have the estimates

$$\left| \frac{\partial^k}{\partial x_1^{k_1} \partial x_2^{k_2}} \mathbf{U}^n(x) \right| \leq M, \qquad (13.22b)$$

$$\left| \frac{\partial^k}{\partial x_1^{k_1} \partial x_2^{k_2}} \mathbf{v}_{\mathbf{U}}^n(x) \right| \leq M \varepsilon^{K-k}, \quad x \in \overline{D}, \quad k \leq K. \qquad (13.22c)$$

Remark 13.4.1 In the problem (13.20), the components $U_k^2(x)$, for $k \geq 1$, vanish on the boundary Γ but the components $U_k^1(x)$, for $k \geq 0$, are, in general, not equal to zero. For the function $\varphi_{\mathbf{U}}(x) = \mathbf{U}(x)$, $x \in \Gamma$, we have the representation $\varphi_{\mathbf{U}}(x) = \sum_{k=0}^{n} \varepsilon^{2k} \varphi_{k\,\mathbf{U}}(x)$, $x \in \Gamma$, where $\varphi_{k\,\mathbf{U}}^2(x) = \varphi_{k(13.20)}^2(x)$, $\varphi_{k\,\mathbf{U}}^1(x) = U_k^1(x) \neq 0$, $x \in \Gamma$, $k = 0, 1, \ldots, n$. ∎

Remark 13.4.2 In the case when the data of the problem (13.20) satisfy the condition

$$\mathbf{f}(x) = \mathbf{0}, \quad x \in \overline{D}, \qquad \varphi_i^2(x) = 0, \quad x \in \Gamma, \quad i < k_0,$$

$$\frac{\partial^{k_1}}{\partial x_1^{k_1}} \varphi_i^2(x) = \frac{\partial^{k_2}}{\partial x_2^{k_2}} \varphi_i^2(x) = 0, \quad x \in \Gamma^c, \quad i \geq k_0,$$

where

$$k_1, k_2 \leq l + 2 - 2k_0 \quad for \quad k_0 = 0,$$

$$k_1, k_2 \leq l + 4 - 2k_0 \quad for \quad k_0 \geq 1; \quad i = 0, 1, \ldots, n; \quad 0 \leq k_0 \leq n,$$

the component $\mathbf{U}^n(x)$ satisfies the estimate

$$\left| \frac{\partial^k}{\partial x_1^{k_1} \partial x_2^{k_2}} \mathbf{U}^n(x) \right| \leq M \varepsilon^{2k_0}, \quad x \in \overline{D}, \quad k \leq K;$$

for the component $\mathbf{v}_{\mathbf{U}}^n(x)$, the estimate (13.22c) remains true. ∎

13.4.2. Consider a *decomposition* of the singular part of the solution to the boundary value problem. Write the function $\mathbf{V}(x)$ as the sum

$$\mathbf{V}(x) = \sum_{j=1}^{4} \left[\mathbf{V}_{(j)}(x) + \mathbf{V}_{(j,\,j+1)}(x) \right], \quad x \in \overline{D}. \tag{13.23}$$

The functions $\mathbf{V}_{(j)}(x)$ and $\mathbf{V}_{(j,j+1)}(x)$, $x \in \overline{D}$ are restrictions to the set $\overline{D}$ of the functions $\mathbf{V}^0_{(j)}(x)$, $x \in \overline{D}_{(j)}$ and $\mathbf{V}^0_{(j,j+1)}(x)$, $x \in \overline{D}_{(j,j+1)}$, that are the regular and corner boundary layers. It is convenient to choose the sets $\overline{D}_{(j)}$ and $\overline{D}_{(j,j+1)}$ as a half-plane and a quarter-plane, respectively. Here $\Gamma_j \subset \Gamma_{(j)}$ and $\Gamma_j \bigcup \Gamma_{j+1} \subset \Gamma_{(j,j+1)}$. The functions $\mathbf{V}^0_{(j)}(x)$ and $\mathbf{V}^0_{(j,j+1)}(x)$ are solutions of the problems

$$L^0_{(13.2)} \mathbf{V}^0_{(j)}(x) = \mathbf{0}, \quad x \in D_{(j)}, \quad V^{10}_{(j)}(x) = \varphi^{10}_{\mathbf{V}_{(j)}}(x), \quad x \in \Gamma_{(j)}; \tag{13.24}$$

$$L^0_{(13.2)} \mathbf{V}^0_{(j,\,j+1)}(x) = \mathbf{0}, \quad x \in D_{(j,\,j+1)}, \tag{13.25}$$

$$V^{10}_{(j,\,j+1)}(x) = \varphi^{10}_{\mathbf{V}_{(j,\,j+1)}}(x), \quad x \in \Gamma_{(j,\,j+1)}, \quad j = 1, 2, 3, 4.$$

The second components $V^{20}_{(j)}(x)$ and $V^{20}_{(j,j+1)}(x)$ (of the functions $\mathbf{V}^0_{(j)}(x)$ and $\mathbf{V}^0_{(j,j+1)}(x)$) on the boundaries $\Gamma_{(j)}$ and $\Gamma_{(j,j+1)}$ are not specified. The functions $\mathbf{V}^0_{(j)}(x)$ and $\mathbf{V}^0_{(j,j+1)}(x)$ decrease exponentially, as $r(x, \Gamma_{(j)})$ and $r(x, \Gamma_j \cap \Gamma_{j+1})$ grow, at the rate

$$\mathcal{O}\left(\exp\left(-m_* \varepsilon^{-1} r(x, \Gamma_{(j)}) \right) \right) \quad and \quad \mathcal{O}\left(\exp\left(-m_* \varepsilon^{-1} r(x, \Gamma_j \cap \Gamma_{j+1}) \right) \right)$$

respectively, where m_* is sufficiently small. The functions $\varphi^{10}_{\mathbf{V}_{(j)}}(x)$, $x \in \Gamma_{(j)}$ and $\varphi^{10}_{\mathbf{V}_{(j,j+1)}}(x)$, $x \in \Gamma_{(j,j+1)}$ are sufficiently smooth and satisfy the condition

$$\varphi^{10}_{\mathbf{V}_{(j)}}(x) = \varphi^1(x) - U^1(x), \quad x \in \Gamma_j,$$

$$\varphi^{10}_{\mathbf{V}_{(j,\,j+1)}}(x) = \varphi^1(x) - \left[U^1(x) + V^1_{(j)}(x) + V^1_{(j+1)}(x) \right], \quad x \in \Gamma_{j,\,j+1}.$$

The function $\mathbf{V}^0_{(j)}(x)$, $x \in \overline{D}_{(j)}$ is represented as the expansion

$$\mathbf{V}^0_{(j)}(x) = \sum_{k=0}^{n} \varepsilon^{2k} \mathbf{V}^0_{k(j)}(x) + \mathbf{v}^{n\,0}_{\mathbf{V}_{(j)}}(x) \equiv \mathbf{V}^{n\,0}_{(j)}(x) + \mathbf{v}^{n\,0}_{\mathbf{V}_{(j)}}(x), \tag{13.26}$$

$$x \in \overline{D}_{(j)},$$

which is associated with the representation of the function $\mathbf{V}_{(j)}(x)$ in (13.23):

$$\mathbf{V}_{(j)}(x) = \sum_{k=0}^{n} \varepsilon^{2k} \mathbf{V}_{k(j)}(x) + \mathbf{v}^n_{\mathbf{V}_{(j)}}(x) \equiv \mathbf{V}^n_{(j)}(x) + \mathbf{v}^n_{\mathbf{V}_{(j)}}(x), \quad x \in \overline{D}. \tag{13.27}$$

The components $V_{k(j)}^{10}(x)$ and $V_{k(j)}^{20}(x)$, $x \in \overline{D}_{(j)}$, of the function $\mathbf{V}_{k(j)}^{0}(x)$ in (13.26) are solutions of the following problems for ordinary differential equations with respect to the variable, which is orthogonal to the boundary $\Gamma_{(j)}$:

$$L_{(13.28)}^{10} V_{0(j)}^{10}(x) = 0, \quad x \in D_{(j)}, \qquad V_{0(j)}^{10}(x) = \varphi_{0\mathbf{V}_{(j)}}^{10}(x), \quad x \in \Gamma_{(j)};$$

$$L_{(13.28)}^{20} V_{1(j)}^{20}(x) = c^{21,0}(x) V_{0(j)}^{10}(x), \quad x \in D_{(j)};$$

$$L_{(13.28)}^{10} V_{1(j)}^{10}(x) = -\frac{\partial^2}{\partial x_{3-s}^2} V_{0(j)}^{10}(x) + c^{12,0}(x) V_{1(j)}^{20}(x), \quad x \in D_{(j)},$$

$$V_{1(j)}^{10}(x) = \varphi_{1\mathbf{V}_{(j)}}^{10}(x), \quad x \in \Gamma_{(j)};$$

$$L_{(13.28)}^{20} V_{k(j)}^{20}(x) = -\left(\frac{\partial^2}{\partial x_{3-s}^2} - c^{22,0}(x)\right) V_{k-1,(j)}^{20}(x) + c^{21,0}(x) V_{k-1,(j)}^{10}(x),$$

$$x \in D_{(j)};$$

$$L_{(13.28)}^{10} V_{k(j)}^{10}(x) = -\frac{\partial^2}{\partial x_{3-s}^2} V_{k-1,(j)}^{10}(x) + c^{12,0}(x) V_{k-1,(j)}^{20}(x), \quad x \in D_{(j)},$$

$$V_{k(j)}^{10}(x) = \varphi_{k\mathbf{V}_{(j)}}^{10}(x), \quad x \in \Gamma_{(j)}, \quad k \geq 2.$$

Here

$$L_{(13.28)}^{10} \equiv \varepsilon^2 \frac{\partial^2}{\partial x_s^2} - c^{11,0}(x), \quad L_{(13.28)}^{20} \equiv \varepsilon^2 \frac{\partial^2}{\partial x_s^2}, \quad s = s(j), \qquad (13.28\text{b})$$

$s(j) = 1$ for $j = 1,3$ while $s(j) = 2$ for $j = 2,4$; and $c^{ij,0}(x)$ are the components of the matrix $C^0(x)$ that appears in the operator $L_{(13.2)}^0$, which is extended to $\mathbb{R}^2$, preserving the properties (13.3).

The components $V_{k(j)}^{20}(x)$, for $k \geq 1$, are not specified on the boundary $\Gamma_{(j)}$, but the following additional condition holds: the components $V_{k(j)}^{10}(x)$, for $k \geq 0$, and $V_{k(j)}^{20}(x)$, for $k \geq 1$, decrease exponentially, as $r(x, \Gamma_{(j)})$ grows, at the rate $\mathcal{O}\left(\exp\left(-m_* \varepsilon^{-1} r(x, \Gamma_{(j)})\right)\right)$. Set $V_{0(j)}^{20}(x) = 0$, $x \in \overline{D}_{(j)}$. The functions $\varphi_{k\mathbf{V}_{(j)}}^{10}(x)$, $x \in \Gamma_{(j)}$ are sufficiently smooth and satisfy the condition

$$\begin{aligned}
\varphi_{k\mathbf{V}_{(j)}}^{10}(x) &= \varphi^1(x) - U_0^1(x), \quad k = 0, \\
\varphi_{k\mathbf{V}_{(j)}}^{10}(x) &= -U_k^1(x), \qquad\qquad k \geq 1, \quad x \in \Gamma_j.
\end{aligned} \qquad (13.28\text{c})$$

Let the condition (13.13) be satisfied, where the values l and l_1 satisfy the condition (13.21)

$$l \geq K - 2, \quad l_1 > 2K - 2, \quad \text{for } 2n = K, \quad K \geq 2, \qquad (13.29)$$

which is somewhat stronger than the condition (13.21). Then the components $\mathbf{V}^n_{(j)}(x)$, $\mathbf{v}^n_{\mathbf{V}_{(j)}}(x)$, $x \in \overline{D}$, that are restrictions of the corresponding components in the representation (13.26) of the regular boundary layer $\mathbf{V}_{(j)}(x)$, satisfy the estimates

$$\left| \frac{\partial^k}{\partial x_1^{k_1} \partial x_2^{k_2}} V^{ni}_{(j)}(x) \right| \le M\varepsilon^{2(i-1)-k_{(j)}} \exp\left(-m\varepsilon^{-1} r(x, \Gamma_j)\right), \quad i = 1, 2,$$

$$\tag{13.30}$$

$$\left| \frac{\partial^k}{\partial x_1^{k_1} \partial x_2^{k_2}} \mathbf{v}^n_{\mathbf{V}_{(j)}}(x) \right| \le M\varepsilon^{K-k}, \quad x \in \overline{D}, \quad k \le K, \quad j = 1, 2, 3, 4.$$

Here $k_{(j)} = k_1$ for $j = 1, 3$ while $k_{(j)} = k_2$ for $j = 2, 4$; m is an arbitrary constant in the interval $(0, m_0)$, where $m_0 = c_0^{1/2}(1-m)^{1/2}$ with $c_0 = c_{0(13.3)}$ and $m = m_{(13.3)}$.

Remark 13.4.3 In the function $\mathbf{V}^n_{(j)(13.27)}(x)$, the components $V^2_{k(j)}(x)$, for $k \ge 1$, on Γ_j, are, in general, not equal to zero; the components $V^i_{k(j)}(x)$, for $k \ge 0$ and $i = 1, 2$, satisfy the relations

$$U_0^1(x) + V_{0(j)}^1(x) = \varphi^1(x), \quad U_k^1(x) + V_{k(j)}^1(x) = 0,$$

$$U_k^2(x) + V_{k(j)}^2(x) \ne 0, \quad k \ge 1, \quad x \in \Gamma_j.$$

For the function $\varphi_{\mathbf{V}_{(j)}}(x) = \mathbf{V}^0_{(j)}(x)$, $x \in \Gamma_{(j)}$, we have the representation

$$\varphi_{\mathbf{V}_{(j)}}(x) = \sum_{k=0}^n \varepsilon^{2k} \varphi_{k\mathbf{V}_{(j)}}(x), \quad x \in \Gamma_{(j)},$$

where

$$\varphi^1_{k\mathbf{V}_{(j)}}(x) = \varphi^{10}_{k\mathbf{V}_{(j)}(13.28)}(x), \quad \varphi^2_{k\mathbf{V}_{(j)}}(x) = V^{20}_{k(j)}(x), \quad x \in \Gamma_{(j)},$$

$$\varphi^2_{k\mathbf{V}_{(j)}}(x) \ne 0, \quad x \in \Gamma_j, \quad k = 0, 1, \dots, n.$$

∎

Remark 13.4.4 In the case when the data of the problem (13.28) satisfy the condition

$$\varphi^1_{k\mathbf{V}_{(j)}}(x) = 0, \quad x \in \Gamma_j, \quad k < k_0, \tag{13.31}$$

$$\frac{\partial^{k_s}}{\partial x_s^{k_s}} \varphi^1_{k\mathbf{V}_{(j)}}(x) = 0, \quad x \in \Gamma^c, \quad k \ge k_0,$$

where $k_s \leq l + 2 - 2k_0$, $k = 0, 1, \ldots, n$, $0 \leq k_0 \leq n$ and $s = 3 - s_{(13.28)}(j)$, the component $V_{(j)}^n(x) = V_{(j)(13.26)}^{n0}(x)$, $x \in \overline{D}$ satisfies the estimate

$$\left| \frac{\partial^k}{\partial x_1^{k_1} \partial x_2^{k_2}} V_{(j)}^{ni}(x) \right| \leq M \varepsilon^{2(i-1) + 2k_0 - k_{(j)}} \exp\left(-m\varepsilon^{-1} r(x, \Gamma_j)\right), \quad x \in \overline{D},$$

$$k \leq K, \quad j = 1, 2, 3, 4,$$

where $i = 1, 2$, $k_{(j)} = k_{(j)(13.30)}$ and $m = m_{(13.30)}$; for the component $v_{V_{(j)}}^n(x) = v_{V_{(j)(13.26)}}^{n0}(x)$, $x \in \overline{D}$, the estimate (13.30) remains. The components $\varphi_{kV_{(j)(13.31)}}^1(x)$, $x \in \Gamma$ are restrictions of $\varphi_{kV_{(j)(13.28)}}^{10}(x)$, $x \in \Gamma_{(j)}$ to the set Γ_j, i.e., $\varphi_{kV_{(j)}}^1(x) = \varphi_{kV_{(j)}}^{10}(x)$, $x \in \Gamma_j$. In the case of condition (13.31) the component $V_{k_0(j)}^{20}(x)$, $x \in \Gamma_{(j)}$, does not appear in the decomposition of the singular component $V_{(j)}^0(x)$. $\blacksquare$

13.4.3. We now estimate the corner boundary layer. Write the function $V_{(j,j+1)}^0(x)$, $x \in \overline{D}_{(j,j+1)}$ as the sum

$$V_{(j,j+1)}^0(x) = \sum_{k=0}^{n} \varepsilon^{2k} V_{k(j,j+1)}^0(x) + v_{V_{(j,j+1)}}^{n0}(x) \equiv$$

$$\equiv V_{(j,j+1)}^{n0}(x) + v_{V_{(j,j+1)}}^{n0}(x), \quad x \in \overline{D}_{(j,j+1)}.$$

The components $V_{k(j,j+1)}^{10}(x)$ and $V_{k(j,j+1)}^{20}(x)$, $x \in \overline{D}_{(j)}$ of $V_{k(j,j+1)}^0(x)$ are solutions of the following problems on the quarter-plane $\overline{D}_{(j,j+1)}$:

$$L_{(13.32)}^{10} V_{0(j,j+1)}^{10}(x) = 0, \quad x \in D_{(j,j+1)}, \tag{13.32a}$$

$$V_{0(j,j+1)}^{10}(x) = \varphi_{0V_{(j,j+1)}}^{10}(x), \quad x \in \Gamma_{(j,j+1)};$$

$$L_{(13.32)}^{20} V_{1(j,j+1)}^{20}(x) = c^{21,0}(x) V_{0(j,j+1)}^{10}(x), \quad x \in D_{(j,j+1)};$$

$$L_{(13.32)}^{10} V_{1(j,j+1)}^{10}(x) = c^{12,0}(x) V_{1(j,j+1)}^{20}(x), \quad x \in D_{(j,j+1)},$$

$$V_{1(j,j+1)}^{10}(x) = \varphi_{1V_{(j,j+1)}}^{10}(x), \quad x \in \Gamma_{(j,j+1)};$$

$$L_{(13.32)}^{20} V_{k(j,j+1)}^{20}(x) = c^{22,0}(x) V_{k-1,(j,j+1)}^{20}(x) + c^{21,0}(x) V_{k-1,(j,j+1)}^{10}(x),$$

$$x \in D_{(j,j+1)};$$

$$L_{(13.32)}^{10} V_{k(j,j+1)}^{10}(x) = c^{12,0}(x) V_{k(j,j+1)}^{20}(x), \quad x \in D_{(j,j+1)},$$

$$V_{k(j,j+1)}^{10}(x) = \varphi_{kV_{(j,j+1)}}^{10}(x), \quad x \in \Gamma_{(j,j+1)}, \quad k \geq 2.$$

Here

$$L^{10}_{(13.32)} \equiv \varepsilon^2 \left(\frac{\partial^2}{\partial x_1^2} + \frac{\partial^2}{\partial x_2^2} \right) - c^{11,0}(x), \quad L^{20}_{(13.32)} \equiv \varepsilon^2 \left(\frac{\partial^2}{\partial x_1^2} + \frac{\partial^2}{\partial x_2^2} \right). \quad (13.32b)$$

The components $V^{20}_{k(j,j+1)}(x)$, $k \geq 1$ are not specified on the boundary $\Gamma_{(j,j+1)}$. The components $V^{10}_{k(j,\,j+1)}(x)$, $k \geq 0$ and $V^{20}_{k(j,\,j+1)}(x)$, for $k \geq 1$, decrease at the rate $\mathcal{O} \left(\exp \left(-m_* \varepsilon^{-1} r(x, \Gamma_j \cap \Gamma_{j+1}) \right) \right)$ as $r(x, \Gamma_j \cap \Gamma_{j+1})$ grows. Set $V^{20}_{0(j,\,j+1)}(x) = 0$, $x \in \overline{D}_{(j,\,j+1)}$. The functions $\varphi^{10}_{k\,V_{(j,\,j+1)}}(x)$ are sufficiently smooth and satisfy the condition

$$\varphi^{10}_{k\,V_{(j,\,j+1)}}(x) = \varphi^1(x) - \left[U^1_0(x) + V^1_{0(j)}(x) + V^1_{0(j+1)}(x) \right], \quad k = 0, \quad (13.32c)$$

$$\varphi^{10}_{k\,V_{(j,\,j+1)}}(x) = - \left[U^1_k(x) + V^1_{k(j)}(x) + V^1_{k(j+1)}(x) \right], \quad k \geq 1, \quad x \in \Gamma_j \cup \Gamma_{j+1}.$$

In the case of conditions (13.13), (13.29), for the component $V^0_{(j,\,j+1)}(x)$, $x \in \overline{D}_{(j,\,j+1)}$ we obtain the estimate

$$\left| \frac{\partial^k}{\partial x_1^{k_1} \partial x_2^{k_2}} V^0_{(j,\,j+1)}(x) \right| \leq M \varepsilon^{K-k}, \quad x \in \overline{D}_{(j,\,j+1)}, \quad k \leq K. \quad (13.33)$$

13.4.4. Next, we estimate the function $u^{(1)}_{(13.16)}(x)$, $x \in \overline{D}$ in the representation (13.16). Consider the function $\mathbf{w}^0(x)$, i.e., the truncated expansion series in powers of ε^2

$$\mathbf{w}^0(x) = \sum_{k=0}^{n} \varepsilon^{2k} \mathbf{w}^0_k(x) = \mathbf{U}^n_{\mathbf{w}^0}(x) + \mathbf{V}^n_{\mathbf{w}^0}(x), \quad x \in \overline{D}, \quad (13.34a)$$

where $\mathbf{U}^n_{\mathbf{w}^0}(x)$ and $\mathbf{V}^n_{\mathbf{w}^0}(x)$ are the regular and singular components of the function $\mathbf{w}^0(x)$. The singular component $\mathbf{V}^n_{\mathbf{w}^0}(x)$ is represented as the sum of the regular boundary layers

$$\mathbf{V}^n_{\mathbf{w}^0}(x) = \sum_{j=1}^{4} \mathbf{V}^n_{\mathbf{w}^0(j)}(x), \quad x \in \overline{D}. \quad (13.34b)$$

Here

$$\mathbf{U}^n_{\mathbf{w}^0}(x) = \mathbf{U}^n_{(13.19)}(x), \quad \mathbf{V}^n_{\mathbf{w}^0(j)}(x) = \mathbf{V}^n_{(j)(13.27)}(x), \quad x \in \overline{D};$$

and for the component $\mathbf{w}^0_k(x)$, we have the representation

$$\mathbf{w}^0_k(x) = \mathbf{U}_{k(13.19)}(x) + \sum_{j=1}^{4} \mathbf{V}_{k(j)(13.27)}(x), \quad x \in \overline{D}, \quad k = 0, 1, \ldots, n. \quad (13.34c)$$

Write the function $\mathbf{u}^{(1)}_{(13.16)}(x)$, $x \in \overline{D}$ in the form

$$\mathbf{u}^{(1)}(x) = \mathbf{u}(x) - \mathbf{w}^0(x) + \mathbf{v}_{\mathbf{u}^{(1)}}(x), \quad x \in \overline{D},$$

where by virtue of the relations (13.16), (13.19), (13.27), one has

$$\mathbf{v}_{\mathbf{u}^{(1)}}(x) = -\left[\mathbf{v}_{\mathbf{U}}^n(x) + \sum_{j=1}^{4}\left(\mathbf{v}_{\mathbf{V}_{(j)}}^n(x) + \mathbf{V}_{(j,\,j+1)}(x)\right)\right], \quad x \in \overline{D}.$$

For the function $\mathbf{u}^{(1)}(x)$, we have the estimates

$$\left|\mathbf{u}^{(1)}(x)\right| \leq M\,\varepsilon^2, \qquad \left|L\,\mathbf{u}^{(1)}(x)\right| \leq M\varepsilon^{2n}, \quad x \in \overline{D}.$$

Thus, the function $\mathbf{w}^0(x)$ is the first (main) term in the decomposition of the solution $\mathbf{u}(x)$ of the problem (13.2).

In a similar way, we construct the function $\mathbf{w}^1(x)$

$$\mathbf{w}^1(x) = \sum_{k=1}^{n} \varepsilon^{2k}\, \mathbf{w}_k^1(x), \quad x \in \overline{D},$$

i.e., the first term in the decomposition of the function $\mathbf{u}^{(1)}(x)$. In an analogous way, the functions $\mathbf{w}^p(x)$, for $p = 2, \ldots, n$, are constructed that correspond to $\mathbf{u}^{(p)}(x)$ (the "remainder" terms in the representation (13.16) with the refined regular and singular components of the solution).

Write the function $\mathbf{w}^p(x)$ as the decomposition (similar to (13.34)):

$$\mathbf{w}^p(x) = \mathbf{U}_{\mathbf{w}^p}^n(x) + \mathbf{V}_{\mathbf{w}^p}^n(x), \quad \mathbf{V}_{\mathbf{w}^p}^n(x) = \sum_{j=1}^{4} \mathbf{V}_{\mathbf{w}^p(j)}^n(x), \quad x \in \overline{D},$$

$$p = 0, 1, \ldots, n.$$

13.4.5. Next, we construct a decomposition of the function $\mathbf{u}(x)$. Note that on the domain boundary the following condition holds:

$$\mathbf{U}_{\mathbf{w}}(x) + \mathbf{V}_{\mathbf{w}}(x) + \mathbf{v}_*(x) = \boldsymbol{\varphi}(x) + \boldsymbol{\varphi}^*(x), \quad x \in \Gamma, \tag{13.35a}$$

$$\mathbf{U}_{\mathbf{w}}(x) = \sum_{p=0}^{n} \mathbf{U}_{\mathbf{w}^p}^n(x), \quad \mathbf{V}_{\mathbf{w}}(x) = \sum_{j=1}^{4} \mathbf{V}_{\mathbf{w}(j)}(x), \tag{13.35b}$$

$$\mathbf{V}_{\mathbf{w}(j)}(x) = \sum_{p=0}^{n} \mathbf{V}_{\mathbf{w}^p(j)}^n(x), \quad x \in \overline{D}, \quad j = 1, 2, 3, 4;$$

$\mathbf{v}_*(x)$, $x \in \overline{D}$, is the remainder term that includes both the remainder terms obtained in the construction of the regular components $\mathbf{U}_{\mathbf{w}^p}^n(x)$ and the singular components $\mathbf{V}_{\mathbf{w}^p(j)}^n(x)$, and the corner boundary layers. The function

$\varphi^*(x)$, $x \in \Gamma$, appears because the regular and the corner layers on the set D are constructed based on the layers considered on the unbounded sets $\overline{D}_{(j)}$ and $\overline{D}_{(j,j+1)}$. These boundary layers decrease exponentially when moving away from the boundaries of the sets $\overline{D}_{(j)}$ and $\overline{D}_{(j,j+1)}$; therefore the function $\varphi^*(x)$ is less than any power of the parameter ε, i.e., $|\varphi^*(x)| = o(\varepsilon^\nu)$, where ν is an arbitrary constant. For the solution $\mathbf{u}^*(x)$ of the problem

$$L\mathbf{u}^*(x) = \mathbf{0}, \quad x \in D, \qquad \mathbf{u}^*(x) = -\varphi^*(x), \quad x \in \Gamma,$$

we have the estimate

$$\left| \frac{\partial^k}{\partial x_1^{k_1} \partial x_2^{k_2}} \mathbf{u}^*(x) \right| \le M, \quad x \in \overline{D}, \quad k \le K.$$

The solution of the problem (13.2), (13.5) can be written in the form

$$\mathbf{u}(x) = \mathbf{U}^d(x) + \mathbf{V}^d(x), \quad \mathbf{V}^d(x) = \sum_{j=1}^{4} \mathbf{V}_{(j)}^d(x), \quad x \in \overline{D}, \; j = 1, 2, 3, 4. \quad (13.35c)$$

Here

$$\mathbf{U}^d(x) = \mathbf{U_w}(x) + \mathbf{v}_*(x) + \mathbf{u}^*(x), \quad \mathbf{V}^d(x) = \mathbf{V_w}(x), \quad \mathbf{V}_{(j)}^d(x) = \mathbf{V}_{\mathbf{w}(j)}(x).$$

Taking into account (13.22), (13.30), (13.33) and the similar estimates for the functions $\mathbf{U}_{\mathbf{w}^p}^n(x)$ and $\mathbf{V}_{\mathbf{w}^p(j)}^n(x)$ with $p \ge 1$, we obtain, in the case of (13.13) and (13.29), the following estimates for the components in (13.35):

$$\left| \frac{\partial^k}{\partial x_1^{k_1} \partial x_2^{k_2}} \mathbf{U}^d(x) \right| \le M, \tag{13.36}$$

$$\left| \frac{\partial^k}{\partial x_1^{k_1} \partial x_2^{k_2}} \mathbf{V}_{(j)}^{di}(x) \right| \le M\varepsilon^{2(i-1)-k_{(j)}} \exp\left(-m\varepsilon^{-1} r(x, \Gamma_j)\right), \quad x \in \overline{D}, \; k \le K,$$

where $i = 1, 2, \; j = 1, 2, 3, 4, \; k_{(j)} = k_{(j)\,(13.30)}$, and $m = m_{(13.30)}$.

Theorem 13.4.1 *Let the data of the boundary value problem* (13.2), (13.5) *satisfy the conditions* (13.13), (13.29) *for $K \ge 2$. Then the components $\mathbf{U}^d(x)$ and $\mathbf{V}^d(x)$ in the representation* (13.35) *satisfy the estimates* (13.36).

13.5   *A priori* estimates for the problem (13.2) under the conditions (13.4), (13.6)

In this section, we show the estimates of the solutions of the boundary value problem (13.2), (13.1) in the case of the conditions (13.4) and (13.6); for the derivation of these estimates see [187].

13.5.1. In the case of the condition (13.4), we write the solution of the problem (13.2) in the form similar to (13.35c):

$$\mathbf{u}(x) = \mathbf{U}^d(x) + \mathbf{V}^d(x), \quad \mathbf{V}^d(x) = \sum_{j=1}^{4} \mathbf{V}_{(j)}^d(x), \quad x \in \overline{D}. \qquad (13.37)$$

Using a technique similar to that given in Section 13.4, under the condition (13.13), where

$$l \geq K - 2, \quad l_1 \geq 2K - 2, \qquad (13.38)$$

we obtain the estimates for the components in the representation (13.37):

$$\left| \frac{\partial^k}{\partial x_1^{k_1} \partial x_2^{k_2}} \mathbf{U}^d(x) \right| \leq M \left[1 + \varepsilon^{K-k-2} \right], \qquad (13.39)$$

$$\left| \frac{\partial^k}{\partial x_1^{k_1} \partial x_2^{k_2}} \mathbf{V}_{(j)}^d(x) \right| \leq M \varepsilon^{-k_{(j)}} \exp\left(-m \varepsilon^{-1} r(x, \Gamma_j) \right),$$

$$x \in \overline{D}, \quad k \leq K, \quad j = 1, 2, 3, 4, \quad k_{(j)} = k_{(j)(13.30)}, \quad m = m_{(13.30)}.$$

Theorem 13.5.1 *Let the data of the boundary value problem* (13.2), (13.4) *satisfy the conditions* (13.13), (13.38) *for* $K \geq 2$. *Then the components* $\mathbf{U}^d(x)$ *and* $\mathbf{V}^d(x)$ *in the representation* (13.37) *satisfy the estimates* (13.39).

13.5.2. In the case of the condition (13.6), we write the solution of the problem (13.2) as the sum of the functions

$$\mathbf{u}(x) = \mathbf{U}^d(x) + \overline{\mathbf{V}}^d(x) + \tilde{\mathbf{V}}^d(x), \qquad (13.40)$$

$$\overline{\mathbf{V}}^d(x) = \sum_{j=1}^{4} \overline{\mathbf{V}}_{(j)}^d(x), \quad \tilde{\mathbf{V}}^d(x) = \sum_{j=1}^{4} \tilde{\mathbf{V}}_{(j)}^d(x), \quad x \in \overline{D},$$

where $\mathbf{U}^d(x)$ is the regular part of the solution; $\overline{\mathbf{V}}^d(x)$ and $\tilde{\mathbf{V}}^d(x)$ are slow and fast boundary layers.

More complicated constructions similar to those considered in Section 13.4, under the condition (13.13), where

$$l \geq K - 2, \quad l_1 > 3K - 4, \qquad (13.41)$$

allow us to obtain the following estimates for the components in the representation (13.40)

$$\left| \frac{\partial^k}{\partial x_1^{k_1} \partial x_2^{k_2}} \, \mathbf{U}^d(x) \right| \le M \left[1 + \mu^{K-k-2} \right],$$

$$\left| \frac{\partial^k}{\partial x_1^{k_1} \partial x_2^{k_2}} \, \overline{\mathbf{V}}_{(j)}^d(x) \right| \le M \, \mu^{-k_{(j)}} \exp\left(-m\mu^{-1} r(x, \Gamma_j) \right), \qquad (13.42)$$

$$\left| \frac{\partial^k}{\partial x_1^{k_1} \partial x_2^{k_2}} \, \widetilde{V}_{(j)}^{di}(x) \right| \le M \left(\varepsilon \, \mu^{-1} \right)^{2(i-1)} \varepsilon^{-k_{(j)}} \exp\left(-m\varepsilon^{-1} r(x, \Gamma_j) \right),$$

$$x \in \overline{D}, \quad k \le K, \quad i = 1, 2,$$

where $V_{(j)}^{d1}(x)$ and $V_{(j)}^{d2}(x)$ are the first and second components of the vector function $\mathbf{V}_{(j)}^d(x)$, for $j = 1, 2, 3, 4$, $k_{(j)} = k_{(j)\,(13.30)}$, and $m = m_{(13.30)}$.

Note that the component $V_{(j)}^{d1}(x)$ has more strong singularity compared with $V_{(j)}^{d2}(x)$. The estimate of the function $V_{(j)}^{d1}(x)$ is similar to that of $\mathbf{V}_{(j)}^d(x)$ in the representation (13.35c). Note that the estimates (13.36) and (13.39) follow from (13.42).

Theorem 13.5.2 *Let the data of the boundary value problem (13.2), (13.6) satisfy the conditions (13.13), (13.41) for $K \ge 2$. Then the components $\mathbf{U}^d(x)$, $\overline{\mathbf{V}}^d(x)$, and $\widetilde{\mathbf{V}}^d(x)$ in (13.40) satisfy the estimates (13.42).*

13.6 The classical finite difference scheme

13.6.1. When constructing a finite difference scheme for the problem (13.2), we shall use classical finite difference approximations on rectangular grids (see, e.g., [108]). On the set $\overline{D}$ we introduce the grid

$$\overline{D}_h = \overline{D}_{h(13.43)} = \overline{\omega}_1 \times \overline{\omega}_2. \qquad (13.43)$$

Here $\overline{\omega}_s$ is a (in general, nonuniform) mesh on the interval $[0, d_s]$. Set $h_s^i = x_s^{i+1} - x_s^i$, where $x_s^i, x_s^{i+1} \in \overline{\omega}_s$, $h_s = \max_i h_s^i$, $h = \max_s h_s$, for $s = 1, 2$. Assume that $h \le MN^{-1}$, where $N = \min_s N_s$, for $s = 1, 2$, and $N_s + 1$ is the number of nodes in the mesh $\overline{\omega}_s$.

To solve the problem on the grid $\overline{D}_h$, we consider the difference scheme

$$\Lambda \, \mathbf{z}(x) = \mathbf{f}(x), \quad x \in D_h, \qquad \mathbf{z}(x) = \boldsymbol{\varphi}(x), \quad x \in \Gamma_h. \qquad (13.44a)$$

Here $D_h = D \cap \overline{D}_h$, $\Gamma_h = \Gamma \cap \overline{D}_h$,

$$\Lambda = \Lambda_{(13.44)}(\varepsilon_1, \varepsilon_2) \equiv \begin{pmatrix} \Lambda_0^1 & 0 \\ 0 & \Lambda_0^2 \end{pmatrix} - C(x) \equiv \left\{ A(\varepsilon_1, \varepsilon_2) \sum_{s=1,2} \delta_{\overline{x_s} \hat{x_s}} - C(x) \right\},$$

$\mathbf{z}(x) = (z^1(x), z^2(x))^T$, $x \in \overline{D}_h$. In scalar representation, the difference scheme takes the form

$$\Lambda^i \mathbf{z}(x) = f^i(x), \quad x \in D_h, \quad z^i(x) = \varphi^i(x), \quad x \in \Gamma_h, \quad i = 1, 2. \quad (13.44b)$$

Here the operator Λ^i is defined by the relation

$$\Lambda^i \mathbf{z}(x) = \Lambda_0^i z^i(x) - \sum_{j=1,2} c^{ij}(x) z^j(x),$$

$$\Lambda_0^i z^i(x) = \Lambda_0^i(\varepsilon_i) z^i(x) \equiv \varepsilon_i^2 \sum_{s=1,2} \delta_{\overline{x_s} \hat{x_s}} z^i(x), \quad i = 1, 2,$$

where $\delta_{\overline{x_s} \hat{x_s}} v(x) = v_{\overline{x_s} \hat{x_s}}(x)$ are the second-order difference derivatives on nonuniform grids (see [108]).

13.6.2. To study convergence of the difference scheme (13.44), (13.43), we use the maximum principle (see [108]); assume that the solution of the boundary value problem (13.2) satisfies the estimates of Theorem 13.3.1.

Note that the operators

$$\Lambda_{(13.45)}^i \equiv \Lambda_0^i - c^{ii}(x), \quad x \in D_h, \quad i = 1, 2 \quad (13.45)$$

are monotone (see [108]). Taking into account the estimate

$$|z^i(x)| \leq m \max_{\overline{D}_h} |z^{3-i}(x)| + M \left[\max_{\overline{D}_h} |f^i(x)| + \max_{\Gamma_h} |\varphi^i(x)| \right],$$

$$x \in \overline{D}_h, \quad i = 1, 2,$$

where $m < 1$, and by virtue of (13.3), we obtain the estimate

$$|\mathbf{z}(x)| \leq M \left[\max_{\overline{D}_h} |\mathbf{f}(x)| + \max_{\Gamma_h} |\boldsymbol{\varphi}(x)| \right], \quad x \in \overline{D}_h.$$

Taking into account the *a priori* estimates for the problem (13.2), for the solutions of the difference scheme (13.44), (13.43) we establish the following estimates, respectively, under the conditions (13.4), (13.5), and (13.6):

$$|\mathbf{u}_{(13.2,13.4)}(x) - \mathbf{z}(x)|, \ |\mathbf{u}_{(13.2,13.5)}(x) - \mathbf{z}(x)| \leq M \varepsilon^{-1} N^{-1}, \ x \in \overline{D}_h, \ (13.46)$$

$$|\mathbf{u}_{(13.2, 13.6)}(x) - \mathbf{z}(x)| \leq M \left[\varepsilon^{-1} + \mu^{-1} \right] N^{-1}, \quad x \in \overline{D}_h. \quad (13.47)$$

On the uniform grid

$$\overline{n}_h \quad (13.48)$$

we have the estimates

$$|\mathbf{u}_{(13.2,13.4)}(x) - \mathbf{z}(x)|, \ |\mathbf{u}_{(13.2,13.5)}(x) - \mathbf{z}(x)| \leq M\varepsilon^{-2}N^{-2}, \ x \in \overline{D}_h; \ (13.49)$$

$$|\mathbf{u}_{(13.2,\,13.6)}(x) - \mathbf{z}(x)| \leq M\left[\varepsilon^{-2} + \mu^{-2}\right]N^{-2}, \quad x \in \overline{D}_h. \qquad (13.50)$$

Theorem 13.6.1 *For the solutions of the boundary value problem (13.2) subject to the conditions (13.4), (13.5), and (13.6), let the estimates of Theorems 13.5.1, 13.4.1, and 13.5.2 be, respectively, satisfied for $K = 4$. Then the difference scheme (13.44), (13.43) converges for fixed values of the vector parameter ε. The solutions of the scheme (13.44), (13.43) (of the scheme (13.44), (13.48)) satisfy the estimates (13.46) and (13.47) (estimates (13.49) and (13.50)) in the case of the conditions (13.4), (13.5), and (13.6), respectively.*

13.7   The special finite difference scheme

The estimates of Theorems 13.5.1, 13.4.1, and 13.5.2 imply that the derivatives of the solution in a neighborhood of the boundary Γ grow unboundedly as the parameters ε_1 and ε_2 (or one of them) tend to zero. In the case of the conditions (13.4) and (13.5), the boundary layers are rather simple—they depend only on the parameter ε. In the case of the condition (13.6), the boundary layers are the double ones that are defined by the parameters ε and μ, respectively. When solving the problem (13.2) subject to the conditions (13.4)–(13.6), we use piecewise-uniform grids that condense in a neighborhood of the boundary.

13.7.1. Let us construct special difference schemes for the problem (13.2) subject to the conditions (13.4) and (13.5). On the set $\overline{D}$ we introduce the grid

$$\overline{D}_h = \overline{D}_h^S = \overline{\omega}_1^S \times \overline{\omega}_2^S, \qquad (13.51)$$

where $\overline{\omega}_s^S = \overline{\omega}_s^S(\sigma_s)$ is a piecewise-uniform mesh on the interval $[0, d_s]$. The mesh sizes in $\overline{\omega}_s^S$ are $h_s^{(1)} = 4\sigma_s N_s^{-1}$ on the sets $[0, \sigma_s]$ and $[d_s - \sigma_s, d_s]$, and $h_s^{(2)} = 2(d_s - 2\sigma_s)N_s^{-1}$ on $[\sigma_s, d_s - \sigma_s]$. The value σ_s is specified by

$$\sigma_s = \sigma_s(\varepsilon, N_s) = \min\left[4^{-1}d_s, \ M\varepsilon \ln N_s\right], \quad s = 1, 2, \quad M = 2m_{(13.30)}^{-1}.$$

To solve the problem (13.2), we use the difference scheme

$$\Lambda_{(13.52)}\mathbf{z}(x) = \mathbf{f}(x), \quad x \in D_h, \qquad \mathbf{z}(x) = \boldsymbol{\varphi}(x), \quad x \in \Gamma_h, \qquad (13.52)$$

where $\Lambda_{(13.52)} \equiv \Lambda_{(13.44)}(\varepsilon_1 = \varepsilon_2 = \varepsilon)$ and $\Lambda_{(13.52)} \equiv \Lambda_{(13.44)}(\varepsilon_1 = \varepsilon, \varepsilon_2 = 1)$ in the case of the conditions (13.4) and (13.5), respectively.

Taking into account the estimates of Theorems 13.5.1 and 13.4.1 in the case of the conditions (13.4) and (13.5), we establish the ε-uniform convergence of the difference scheme (13.52), (13.51)

$$|\mathbf{u}(x) - \mathbf{z}(x)| \leq M N^{-2} \ln^2 N, \quad x \in \overline{D}_h. \tag{13.53}$$

Theorem 13.7.1 *Let for the components in the representations (13.37) and (13.35) of the solution to the boundary value problem (13.2) under the conditions (13.4) and (13.5) the estimates of Theorems 13.5.1 and 13.4.1 be satisfied for $K = 4$. Then the solution of the difference scheme (13.52), (13.51) converges to the solution of the boundary value problem ε-uniformly. The discrete solutions satisfy the estimate (13.53).*

13.7.2. For the boundary value problem (13.2) subject to the condition (13.6), we use a special grid for which the refinement rule is determined by the parameters ε and μ. On the set $\overline{D}$ we introduce the grid

$$\overline{D}_h = \overline{D}_h^S = \overline{\omega}_1^S \times \overline{\omega}_2^S. \tag{13.54}$$

Here $\overline{\omega}_s^S = \overline{\omega}_s^S(\sigma_s^1, \sigma_s^2)$ is a piecewise-uniform mesh on the interval $[0, d_s]$. We divide the interval $[0, d_s]$ into five parts $[0, \sigma_s^1]$, $[\sigma_s^1, \sigma_s^2]$, $[\sigma_s^2, d_s - \sigma_s^2]$, $[d_s - \sigma_s^2, d_s - \sigma_s^1]$, and $[d_s - \sigma_s^1, d_s]$. On each of these parts the mesh size is constant. The mesh sizes are $h_s^{(1)} = 8\sigma_s^1 N_s^{-1}$ on $[0, \sigma_s^1]$ and $[d_s - \sigma_s^1, d_s]$, $h_s^{(2)} = 8(\sigma_s^2 - \sigma_s^1)N_s^{-1}$ on $[\sigma_s^1, \sigma_s^2]$ and $[d_s - \sigma_s^2, d_s - \sigma_s^1]$, and $h_s^{(3)} = 4(d_s - 2\sigma_s^2)N_s^{-1}$ on $[\sigma_s^2, d_s - \sigma_s^2]$. The values σ_s^1 and σ_s^2 are specified by

$$\sigma_s^1 = \sigma_s^1(\varepsilon, N_s) = \min\left[8^{-1}d_s, M^1\varepsilon \ln N_s\right],$$

$$\sigma_s^2 = \sigma_s^1 + \hat{\sigma}_s^2, \quad \hat{\sigma}_s^2 = \hat{\sigma}_s^2(\mu, N_s) = \min\left[8^{-1}d_s, M^2\mu \ln N_s\right], \quad s = 1, 2,$$

where $M^1 = M^2 = 2m_{(13.30)}^{-1}$.

To solve the problem (13.2) under the condition (13.6), we use the difference scheme

$$\Lambda_{(13.55)}\mathbf{z}(x) = \mathbf{f}(x), \quad x \in D_h, \quad \mathbf{z}(x) = \boldsymbol{\varphi}(x), \quad x \in \Gamma_h, \tag{13.55}$$

where $\Lambda_{(13.52)} \equiv \Lambda_{(13.44)}(\varepsilon_1 = \varepsilon, \varepsilon_2 = \mu)$ and $\overline{D}_h = \overline{D}_{h(13.54)}$.

Taking into account the estimates of Theorem 13.5.2, we find the ε-uniform estimate for the difference scheme (13.55), (13.54):

$$|\mathbf{u}(x) - \mathbf{z}(x)| \leq M N^{-2} \ln^2 N, \quad x \in \overline{D}_h. \tag{13.56}$$

Theorem 13.7.2 *For the components in the representation (13.40) of the solution to the boundary value problem (13.2) under the condition (13.6), let the estimates of Theorem 13.5.2 be satisfied for $K = 4$. Then the solution of the difference scheme (13.55), (13.54) converges to the solution of the boundary value problem (ε, μ)-uniformly. For the discrete solutions the estimate (13.56) holds.*

13.7.3. To solve the difference scheme (13.44), (13.43), we use an iteration method in which the grid components $z^1(x)$ and $z^2(x)$, for $x \in \overline{D}_h$, are computed independently at each iteration step. The boundary value problem (13.2), (13.1) is approximated by the difference scheme

$$\Lambda \mathbf{z}^{[k]}(x) = \mathbf{F}(\mathbf{z}^{[k-1]}(x), \mathbf{f}(x)), \quad x \in D_h, \tag{13.57}$$

$$\mathbf{z}^{[k]}(x) = \boldsymbol{\varphi}(x), \quad x \in \Gamma_h,$$

$$\mathbf{z}^{[0]}(x) = \boldsymbol{\varphi}^0(x), \quad x \in \overline{D}_h, \quad k = 1, 2, 3, \ldots.$$

Here

$$\Lambda = \Lambda_{(13.57)}(\varepsilon_1, \varepsilon_2) \equiv A(\varepsilon_1, \varepsilon_2) \sum_{s=1,2} \delta_{\overline{x_s} \widehat{x_s}} - C_1(x),$$

$$\mathbf{F}(\mathbf{z}(x), \mathbf{f}(x)) \equiv C_2(x)\, \mathbf{z}(x) + \mathbf{f}(x),$$

$$C_1(x) = \begin{pmatrix} c^{11}(x) & 0 \\ 0 & c^{22}(x) \end{pmatrix}, \quad C_2(x) = \begin{pmatrix} 0 & c^{12}(x) \\ c^{21}(x) & 0 \end{pmatrix};$$

and $\varphi^0(x)$, $x \in \overline{D}_h$ is a bounded function. The components $z^{i[k]}(x)$, $x \in \overline{D}_h$, for $i = 1, 2$, are found from the split system of difference equations

$$\Lambda^1 z^{1[k]}(x) \equiv \left\{ \varepsilon_1^2 \sum_{s=1,2} \delta_{\overline{x_s}\widehat{x_s}} - c^{11}(x) \right\} z^{1[k]}(x) = c^{12}(x)\, z^{2[k-1]}(x) - f^1(x),$$

$$\Lambda^2 z^{2[k]}(x) \equiv \left\{ \varepsilon_2^2 \sum_{s=1,2} \delta_{\overline{x_s}\widehat{x_s}} - c^{22}(x) \right\} z^{2[k]}(x) = c^{21}(x)\, z^{1[k-1]}(x) - f^2(x),$$

$$x \in D_h.$$

As $k \to \infty$, the solution of the difference scheme (13.57), (13.43) converges to the solution of the difference scheme (13.44), (13.43) (ε, μ)-uniformly with the estimate

$$|\mathbf{z}_{(13.44, 13.43)}(x) - \mathbf{z}^{[k]}_{(13.57, 13.43)}(x)| \leq M q^k, \quad x \in \overline{D}_h.$$

Here $q \leq m$ with $m = m_{(13.3)}$, and the constant M is independent of k.

As $N, k \to \infty$, the solution of the difference scheme (13.57) on the grids (13.51) and (13.54) converges to the solution of the boundary value problem (13.2), (13.1) ε-uniformly, i.e.,

$$|\mathbf{u}(x) - \mathbf{z}^{[k]}(x)| \leq M[N^{-2} \ln^2 N + q^k], \quad x \in \overline{D}_h, \quad k = 1, 2, 3, \ldots. \tag{13.58}$$

The number of iterations k_0 that are required to solve the boundary value problem (13.2), (13.1) with the estimate

$$|\mathbf{u}(x) - \mathbf{z}^{[k_0]}(x)| \leq M N^{-2} \ln^2 N, \quad x \in \overline{D}_h, \tag{13.59a}$$

satisfies the ε-uniform estimate

$$k_0 \leq 2 (\ln q^{-1})^{-1} \ln N \leq M \ln N. \tag{13.59b}$$

Theorem 13.7.3 *Let the hypotheses of Theorems 13.7.1 and 13.7.2 be fulfilled. Then the solution of the difference scheme* (13.57) *on the grid* (13.51) *in the case of the conditions* (13.4) *and* (13.5), *and on the grids* (13.54) *in the case of condition* (13.6) *converges to the solution of the boundary value problem ε-uniformly as N, $k \to \infty$. The discrete solutions satisfy the estimates* (13.58) *and* (13.59).

13.8   Generalizations

13.8.1. In the case of the boundary value problem for the system of quasilinear equations

$$L\,\mathbf{u}(x) = \mathbf{g}(x, \mathbf{u}(x)), \quad x \in D, \quad \mathbf{u}(x) = \boldsymbol{\varphi}(x), \quad x \in \Gamma, \qquad (13.60a)$$

where $\overline{D} = \overline{D}_{(13.1)}$ and $L_{(13.60)} = L_{(13.2)}(\varepsilon_1, \varepsilon_2)$, we assume that the function $\mathbf{g}(x, \mathbf{u})$, $x \in \overline{D}$, $\mathbf{u} = (u^1, u^2)^T$, $(u^1, u^2) \in R^2$, is sufficiently smooth, bounded, and satisfies the conditions

$$c^{ii}(x) \geq c_0, \quad mc^{ii}(x) \geq |c^{ij}(x)| + \sum_{s=1,2} \max_{\mathbf{u} \in R^2} \left| \frac{\partial}{\partial u_s} g^i(x, \mathbf{u}) \right|, \qquad (13.60b)$$

$$x \in \overline{D}, \quad i, j = 1, 2, \quad i \neq j, \quad c_0 > 0, \quad m = m_{(13.60)} < 1.$$

To solve the problem (13.60), (13.1), we use the difference scheme

$$\Lambda\,\mathbf{z}(x) = \mathbf{g}(x, \mathbf{z}(x)), \quad x \in D_h, \quad \mathbf{z}(x) = \boldsymbol{\varphi}(x), \quad x \in \Gamma_h, \qquad (13.61)$$

where $\overline{D}_h = \overline{D}_{h(13.43)}$ and $\Lambda_{(13.61)} \equiv \Lambda_{(13.44)}(\varepsilon_1, \varepsilon_2)$.

For the difference scheme (13.61) on the grids (13.43), (13.51), and (13.54), the statements on convergence of grid solutions similar to those in Theorems 13.6.1, 13.7.1–13.7.3, are true.

13.8.2. The technique proposed in this Chapter for the construction and investigation of ε-uniformly convergent difference schemes for the problem (13.2) allows us to construct ε-uniformly converging schemes for systems of equations with variable coefficients multiplied by the higher-order derivatives and for systems of p equations with $p > 2$.

Chapter 14

Survey

Here we sketch the state in the development of special difference schemes for, in my thinking, "advanced" problems with boundary layers. First, we discuss some applications of special finite difference schemes to solve problems with boundary layers. Next, we outline the use of special difference schemes to approximate boundary value problems for parabolic equations with piecewise-smooth and discontinuous initial-boundary conditions. Approximations of derivatives are also discussed. We touch on an approach in the construction of difference schemes based on *a posteriori* adaptive meshes. For an elliptic problem in an unbounded domain, we consider an approach to construct difference schemes on meshes with a finite number of mesh nodes whose solutions converge ε-uniformly on prescribed bounded subdomains. For an elliptic problem in a rectangle for a convection-diffusion equation with the perturbation vector-parameter $\overline{\varepsilon}$, we give compatibility conditions that guarantee smoothness of the solution and its components which are required for the construction of $\overline{\varepsilon}$-uniformly convergent schemes.

14.1 Application of special numerical methods to mathematical modeling problems

Mathematical modelling of applied problems leads us to nonlinear problems with rather complicated boundary layers.

As was already shown in [124, 130, 148, 180], for *linear elliptic and parabolic equations in the presence of layers that are not regular*, there are no ε-uniformly convergent schemes based on the fitted operator method. In the case of *differential equations that are not linear*, there are no schemes of the above type even for one-dimensional problems [137, 139]. Similar difficulties in the construction of special numerical methods for *equations with partial derivatives* are discussed in [138, 41, 86, 87, 33]. For *semilinear ordinary differential equations*, such difficulties in the construction of fitted operator schemes are considered in [84, 32, 33]; a number of results for schemes on piecewise-uniform meshes were given recently in [31] and in the bibliography therein.

It should be noted that solutions of problems concerned with flow of a viscous fluid past bodies, as a rule, have *singularities such as a parabolic boundary layer*; see, e.g., [113, 93]. A number of results in the development of finite difference schemes based on the condensing mesh technique for problems of this type are given in [150, 156, 163].

In the book [33], for the classical *nonlinear Prandtl problem* of flow past a semi-infinite flat plate, a finite difference scheme on piecewise-uniform meshes, which is solved by a nonlinear solver, was constructed. Here, a *technique for experimental study* parameters of the ε-uniform convergence of the constructed schemes has been also elaborated, where $\varepsilon = Re^{-1}$, Re is the Reynolds number. This technique for the experimental study of finite difference schemes was advanced in [34, 35]. In [85, 20, 21, 22], for some *representative* (model) *flow problems* from [113], special *difference methods similar to that designed in* [33] were developed. The results obtained in numerical modelling of nonlinear flow problems have demonstrated the efficiency of the approach, designed in [33], to the construction and study of special schemes.

A new direction in the development of robust numerical methods is opened by [142, 149, 160] in which ε-uniformly convergent difference schemes are constructed for a boundary value problem for *the viscous Burgers equation*, that is, a quasilinear parabolic equation in the case when the initial condition has a discontinuity of the first kind. When the parameter ε equals zero, this parabolic equation degenerates into a quasilinear hyperbolic equation. The problem under study describes the decay of the initial discontinuity in the case of convective substance transfer with slow diffusion (low viscosity) when the convection rate is determined by the concentration of substance (see, for example, [101, 70, 111] and the references therein). In [142, 149, 160], for the construction of schemes both the method of condensing grids and the fitted operator method were used, as well as in the case of *parabolic equations with a discontinuous initial condition* (see [128, 42]).

Other interesting applications of robust numerical methods in mathematical modelling of some processes were considered in [65, 64] (heat transfer in some technologies), [98] (hydrogen diffusion process), [21, 8, 54, 9] (flow problems), [71] (problems in financial mathematics). In [7], a robust numerical method for solving a boundary value problem for a singularly perturbed delay parabolic equation was constructed; such methods can be interesting, e.g., for mathematical biology [92].

Here, we also outline some *aspects of numerical methods* related to *computation of the solution* of special finite difference schemes using computers. Numerical methods for solving various problems for singularly perturbed elliptic equations and *appropriate solvers* are considered in [33]. In the case of boundary value problems for singularly perturbed *parabolic equations with several spatial variables*, the use of *the method of fractional steps* (see the books [216, 108] for regular problems) allows one to reduce the numerical solution of multidimensional problems to the solution of one-dimensional problems. This reduction provides the efficiency of a numerical method, see, e.g., [141, 153, 23] and the bibliography therein.

The acceleration of computation for singularly perturbed boundary value problems can be achieved by *parallelizing the computations* using *a domain decomposition method*, see, e.g., [201, 202, 146].

Note that *the technique based on domain decomposition* is effectively used in order *to construct* monotone ε-uniformly convergent *difference schemes* for sufficiently wide classes of singularly perturbed problems (see, e.g., Chapters 2–7 and also [138]). Domain decomposition finite difference schemes were developed, e.g., in [48, 90, 77, 78, 174, 53] (in [48, 53] schemes of improved accuracy order were considered, and in [174] conditioning of decomposition schemes was studied).

The *behaviour of decomposition methods* for parabolic *convection-diffusion* equations significantly *differs* from that for *reaction-diffusion* equations. In the case of parabolic reaction-diffusion equations, parallelization of ε-uniformly convergent schemes (as well as for regular problems) leads to the acceleration in solving the decomposed grid problem; moreover, the error caused by the decomposition of the discrete problem is essentially less than the error caused by the discretization of the initial boundary value problem [146]. For parabolic *convection-diffusion equations*, for small ε *the sequential schemes* turn out to be *more efficient than the schemes*, because of a significant accumulation of the decomposition error in the latter [206]. Thus, the development of efficient domain decomposition methods for singularly perturbed problems (in particular, for convection-diffusion problems) that allow one to obtain *solutions of improved convergence order* and *to accelerate the numerical solution of the problem* by parallel implementation of the computational process is a challenge to the researchers and still awaiting study.

14.2   Numerical methods for problems with piecewise-smooth and nonsmooth boundary functions

Lowering the smoothness of the boundary function leads to a decrease in the convergence rate of schemes. But this decrease of the convergence rate essentially depends on that part of the boundary where the smoothness of the boundary function is reduced. So, in the case of a one-dimensional problem for a parabolic reaction-diffusion equation with sufficiently smooth data satisfying the compatibility condition (when $u \in C^{4,2}(\overline{G})$), the known scheme on a piecewise-uniform mesh in space (we call it *the basic scheme*) converges ε-uniformly at the rate $\mathcal{O}\left(N^{-2}\ln^2 N + N_0^{-1}\right)$. In [189] it was shown that the *absence of compatibility of the data* (on the boundary of the domain for $t = 0$) and/or the *discontinuity of the first-order temporal derivative* of the boundary function (on the lateral part of the boundary) only *unessentially reduce the convergence rate* of the scheme on piecewise-uniform meshes. In this case, the basic scheme converges ε-uniformly at the rate $\mathcal{O}\left(N^{-2}\ln^3 N + N_0^{-1}\ln N_0\right)$.

If *the first-order spatial derivative* of the initial function has *a discontinuity*, the basic scheme converges ε-uniformly only at the rate $\mathcal{O}\left(N^{-1} + N_0^{-1/2}\right)$; a *significant loss in the order* of the ε-uniform convergence rate takes place. Thus, the convergence rate *crucially* depends on the type of *nonsmoothness in the initial-boundary conditions*. For this problem, *an improved scheme* can be constructed, i.e., a scheme on *the meshes condensing in* both *boundary and interior layers* that allow us to obtain discrete solutions which converge *conditionally ε-uniformly* at the rate $\mathcal{O}\left(N^{-2}\ln^2 N + N_0^{-1}\ln N_0\right)$ provided that the parameter ε satisfies the additional condition $\varepsilon = \mathcal{O}\left(N^{-1} + N_0^{-1/2}\right)$. On the other hand, *the unconditional ε-uniform* convergence rate of the improved scheme is the same as it is for the basic scheme, i.e., $\mathcal{O}\left(N^{-1} + N_0^{-1/2}\right)$.

The *technique of additive splitting of singularities* is efficient when constructing ε-uniformly convergent schemes in the case of problems with insufficiently smooth problem solutions. For a parabolic reaction-diffusion equation with *discontinuous initial-boundary conditions* special difference schemes that converge at the rate $\mathcal{O}\left(N^{-2}\ln^3 N + N_0^{-1}\ln N_0\right)$ were constructed in [191]. When constructing the schemes, *the method of additive splitting of singularities* (generated by discontinuities of the boundary function) and also *the condensing mesh method* (with piecewise-uniform meshes condensing in neighborhoods of boundary layers) were used. The singular part of the solution (the interior-layer-type functions) is *decomposed into nonsmooth singular components of the solution* generated by discontinuities not only of the boundary function but also of its low-order derivatives.

The construction of special schemes for *problems with discontinuous initial-boundary conditions* based on *the fitted operator method* was considered in [164, 42, 144] and on *the method of additive splitting of singularities* in [64, 145].

Boundary value problems for elliptic equations with nonsmooth solutions were considered, e.g., in [138, 155, 3, 4, 5, 6]; see also Remark 14.6.2 in Section 14.6 of this chapter.

14.3   On the approximation of solutions and derivatives

In applied problems, it is often required to find the first-order spatial derivatives, e.g., the heat (diffusive) flux; the first-order derivative of the component of the flow velocity along the streamlined surface determines the boundary layer separation, and the second-order derivative determines the stability of the laminar flow for large Reynolds numbers Re [113]. The problem of finding the derivatives is complicated in the case when the problem data are not sufficiently smooth.

In [184], a boundary value problem on an interval for a parabolic convection-diffusion equation is considered in the case when the first-order derivative of the initial function has a discontinuity at the point x_0. When $\varepsilon \to 0$, a boundary layer with the typical width ε appears in a neighborhood of that part of the boundary through which the convective flow leaves the domain; we call this part the *output part of the boundary*, or briefly, the *output boundary*. Also, a *transient* (moving in time) *layer* with the typical width $\varepsilon^{1/2}$ appears in a neighborhood of the characteristic of the reduced equation outgoing from the point $(x_0, 0)$. The transient (interior) layer is weak, i.e., its first-order derivative in x is bounded ε-uniformly. The *diffusion flux*, i.e., the product $\varepsilon \, (\partial/\partial x) \, u(x,t)$ (we call it the *normalized* or *scaled derivative*) is continuous on the set $\overline{G}^*$, where $\overline{G}^* = \overline{G} \setminus \{(x_0, 0)\}$. The derivative $(\partial/\partial x) \, u(x,t)$ is bounded ε-uniformly on $\overline{G}^*$ but only outside the $(\varepsilon \ln \varepsilon^{-1})$-neighborhood of the output boundary, while the diffusion flux is bounded ε-uniformly on the whole set $\overline{G}^*$.

Using the method of special grids that condense in a neighborhood of the boundary layer and the method of additive splitting of singularities (more precisely, of a *transient-layer type singularity*) special finite difference schemes have been constructed in [184]. This scheme allows to approximate the solution and its first scaled derivative ε-uniformly on $\overline{G}^*$, and also the derivative $(\partial/\partial x) \, u(x,t)$ itself outside an m-neighborhood of the output boundary. The special schemes also make it possible to construct approximations to *the first-order temporal derivative* $(\partial/\partial t) \, u(x,t)$ and to *the scaled second-order space derivative* $\varepsilon^2 \, (\partial^2/\partial x^2) \, u(x,t)$ that converge ε-uniformly on $\overline{G}^*$. Note that these derivatives grow at a rate of $(\varepsilon^{-1/2} \mid x-x_0 \mid + t^{1/2})^{-1}$ as $(x,t) \to (x_0, 0)$.

In [144, 145, 164] parabolic problems with *discontinuous initial conditions* were considered. To find approximations to the diffusion fluxes, the method of additive splitting of the singularity which is generated by the discontinuity of the initial function was used.

The technique from [184] was used in [71] to construct an ε-uniformly convergent difference scheme based on the additive splitting method, for the Black–Scholes equation with a piecewise-smooth initial function. Such a problem arises in financial mathematics (see [215]). In addition to the solution of the problem, here it is necessary to find its *first-order derivative*. This problem is reduced, by a transformation of variables, to the Cauchy problem for a singularly perturbed parabolic equation in the variables x, t with the perturbation parameter ε, where $\varepsilon \in (0, 1]$; the first-order x-derivative of the initial function has a discontinuity. In [71], using the method of additive splitting of a singularity of the interior-layer type, a special difference scheme has been constructed that approximates ε-uniformly *the solution* of the problem and its *first-order derivative in x* with the convergence orders close to 1 and 0.5, respectively. The efficiency of the constructed scheme has been illustrated by numerical experiments.

In the case of singularly perturbed problems, the derivative of their solutions (in the direction orthogonal to the boundary) grows without bound in a neighborhood of the boundary for $r \ll \varepsilon \ln \varepsilon^{-1}$, and the scaled derivatives tend to zero for $r \gg \varepsilon$ where r is the distance to the boundary. Thus, *the derivatives of the solution* to the problem and *the scaled derivatives* both are *not adequate* (at least for $\varepsilon \ll r \ll \varepsilon \ln \varepsilon^{-1}$) *for quantitative description of the derivatives of solutions* to singularly perturbed boundary value problems; see, e.g., [173, 171].

In [177] for the case of a one-dimensional (in space) problem for a parabolic convection-diffusion equation, the approximation errors of solutions and derivatives are examined in the *ρ-metric*, which is *adequate for quantitative description of solutions and their derivatives* for problems with boundary layers. In this metric, *the errors of the solution and its temporal derivative* $(\partial/\partial t)$ are defined by *the absolute errors*, and *the error in the derivative* $(\partial/\partial x)u(x,t)$ is determined by *the relative error* (with respect to a majorant function for this derivative) *in the boundary layer* and by *the absolute error outside it*. In [177] it was shown that, in the case of classical finite difference approximations of the boundary value problem, *there are no meshes on which the scheme converges ε-uniformly in the ρ-metric*. Conditions imposed on the parameters of piecewise-uniform meshes are obtained under which *the schemes converge in the ρ-metric almost ε-uniformly* , i.e., at the rate $\mathcal{O}\left(\varepsilon^{-\nu} N^{-1} + N_0^{-1}\right)$, where $\nu > 0$ (specifying the meshes to be constructed) *can be chosen arbitrarily small*.

14.4   On difference schemes on adaptive meshes

For *regular boundary value problems*, methods were well developed to improve the accuracy of discrete solutions, in which the sequential refinement mesh is used on subdomains where the computed solution is not sufficiently accurate, i.e., *methods on a posteriori condensing meshes* (see, e.g., [17, 75]). The subdomains in which local mesh refinement is required are determined using *indicators*, i.e., functionals of the solutions (for example, the solution gradient) of intermediate discrete problems. The use of numerical methods on *a posteriori* adapted meshes allows us to improve the global accuracy of the approximate solution (see, e.g., [36, 81] and the bibliography therein).

The direct application of *an adaptive mesh technique* that was *developed for regular problems* to singularly perturbed problems *does not make it possible to significantly reduce the dependence of an error of the discrete solution on the parameter ε* as compared to classical methods [176, 203, 183].

Difference schemes on *a priori* adapted locally uniform meshes were considered in [196, 193], and also in Chapter 11.

Schemes on *a posteriori* adapted locally uniform meshes were studied in [157] for an elliptic equation, in [166, 165, 203, 185, 190] for parabolic equations (in [203] for a problem with a moving interior layer), and in [154] for an ordinary differential equation. In [157, 166, 165, 203, 190], *a posteriori* adapted meshes were constructed on the basis of the *solution gradients* (in [203] we passed to the coordinate system in which the layer becomes stationary), in [154] using *majorant functions* for discrete boundary layers, and in [185] using *solutions on embedded meshes*. A boundary value problem for a semilinear parabolic equation was considered in [190].

Note that, in schemes on adaptive *a priori and a posteriori locally uniform meshes*, when solving the boundary value subproblems, *one can use efficient numerical algorithms* developed to solve boundary value problems on uniform meshes (see, e.g., [110]).

Unlike schemes on piecewise-uniform meshes that converge ε-uniformly, *schemes on a posteriori adapted locally uniform meshes* converge *almost ε-uniformly*, namely, under the condition $N_1^{-1} \ll \varepsilon^{\nu}$, where N_1+1 is the number of nodes in the local uniform mesh with respect to x (in the direction orthogonal to the output boundary). The value $\nu = \nu(K)$, for large K, can be made sufficiently small, where K is the number of iterative loops of mesh refinement (for the mesh in x) in the adaptive mesh. Nevertheless, such schemes *do not converge ε-uniformly*, they do not converge under the condition $\varepsilon \ll N^{-1/\nu_0}$, where $\nu_0 < \nu(K)$ is sufficiently small. In this case we will say that the scheme does not converge for *too small values of the parameter ε*.

We discuss some *error estimates* of the discrete solutions *for the schemes on a posteriori adapted meshes*.

In the case of a problem on an interval for a parabolic convection-diffusion equation, the scheme on an *a posteriori* adaptive mesh based on *solutions on embedded meshes* from [185] possesses a better convergence rate in comparison to schemes based on *solution gradients* from [166, 190].

For solutions of *the finite difference scheme on an a posteriori adaptive embedded mesh*, in [185] such an estimate was derived:

$$|u(x,t) - z(x,t)| \leq M \left[N_1^{-1} \ln^2 N_1 + N_0^{-1} \ln N_0 + \varepsilon^{-1} N_1^{-K} \ln^{K-1} N_1 \right],$$

$$(x,t) \in \overline{G}_h, \tag{14.1}$$

where $N_1 + 1$ and $N_0 + 1$ are the numbers of mesh points with respect to x and t, $M = M(K)$. Here we assumed that the solution of a parabolic problem and its regular and singular components are sufficiently smooth.

Estimate (14.1) for the error in the solution of the finite difference scheme, derived on basis of *a priori estimates of solutions to the boundary value problem*, belongs to *a priori estimates of the solution errors*.

Such an estimate has a scientific value for theoretical research, but is not suitable for numerical studying.

Here, we consider an example of the *recipe to construct a posteriori error estimates* of the discrete solution.

Let $z(x,t)$, $(x,t) \in \overline{G}_h$, and $z^1(x,t)$, $(x,t) \in \overline{G}_h^1$, be discrete solutions on *the basis of embedded meshes*, and let $\overline{z}(x,t)$, $(x,t) \in \overline{G}$, and $\overline{z}^1(x,t)$, $(x,t) \in \overline{G}$, be their interpolants. These interpolants allow us to find a *posteriori* estimate for the order of the solution error at each point of the set $\overline{G}$ in the boundary layer region.

We introduce the *a posteriori* defined auxiliary function based on discrete solutions

$$\delta^0(x,t) = \max_{x \le x_1} \left| \overline{z}(x_1,t) - \overline{z}^1(x_1,t) \right|, \quad (x,t) \in \overline{G}. \tag{14.2}$$

The solution of the problem on *a posteriori* embedded meshes satisfies the estimate

$$\left| u(x,t) - \overline{z}(x,t) \right| \le M \left[\delta^0(x,t) + N_1^{-1} + N_0^{-1} \right], \quad (x,t) \in \overline{G}. \tag{14.3}$$

The values N_1^{-1} and N_0^{-1} characterize *the main term of an a priori error estimate for the regular component of the solution*.

The solution error bound, defined by (14.3), belongs to *a priori / a posteriori error estimates*.

Since the values N_1^{-1}, N_0^{-1}, in general, are small with respect to maximum on $\overline{G}$ the value $\delta^0(x,t)$, the solution error bound, defined by (14.3), *belongs, in fact, to a posteriori error estimates*.

In the case when $\max \left[\ln^{-2} N_1, \ln^{-1} N_0 \right]$ is a value of order $\mathcal{O}\left(\max_{\overline{G}} |V(x,t)| \right)$, where $V(x,t)$ is the boundary layer function, we have the estimates

$$\left| u(x,t) - \overline{z}(x,t) \right| \le M \left[\delta^1(x) + N_1^{-1} + N_0^{-1} \right], \quad (x,t) \in \overline{G}, \tag{14.4a}$$

$$\left| u(x,t) - \overline{z}(x,t) \right| \le M \delta^2, \quad (x,t) \in \overline{G}, \tag{14.4b}$$

where $\delta^1(x) = \max_t \delta^0_{(14.2)}(x,t)$, $\delta^2 = \max_{\overline{G}} \delta^0_{(14.2)}(x,t)$. Thus, under the condition above, the value δ^2, up to constant-multiplier, is an error estimate to the discrete solution on embedded meshes. The solution error estimate defined by (14.4b) belongs to *a posteriori error estimates*.

Thus, the estimate (14.4) allows us to *control the computational process on a posteriori adaptive meshes*.

The approach involved in the construction of *almost ε-uniformly convergent* finite difference schemes on *a priori* or *a posteriori* adapted *locally uniform meshes* can be used when constructing efficient numerical methods for sufficiently wide classes of boundary value problems with dominant convection.

The *almost ε-uniformly convergent* difference schemes constructed on (*a priori* or *a posteriori* adapted) *locally uniform meshes* have solutions whose *error bounds* are *weakly sensitive* to the value of the parameter ε. In short, we call such schemes *weakly sensitive schemes*. These schemes can be considered as *alternatives* to *special schemes that converge ε-uniformly*, and to *classical* *schemes* that converge for $h \ll \varepsilon$, where h is the maximal mesh step-size in the direction orthogonal to the boundary.

In weakly sensitive schemes, the *solution* of the discrete problem on a coarse (uniform) mesh is *corrected only locally*, that is, on relatively small subdomains (their boundaries pass through nodes of the coarse mesh). The discrete *problem is solved on uniform meshes*. This provides the efficiency of computations. Controlled (by the number of iterations K in the refinement process), the *weak dependence of the solution error on the parameter ε* allows us to obtain discrete solutions that *converge on the whole grid domain* when *the values of the parameter ε are not too small*.

14.5 On the design of constructive difference schemes for an elliptic convection-diffusion equation in an unbounded domain

In the case of boundary value problems for elliptic equations in unbounded domains, such a difficulty appears in the construction of numerical methods admissible for computations. The domain of dependence of the solution considered on a bounded subdomain is unbounded. For such problems one can easily write down a finite difference scheme on rectangular meshes. These meshes have an infinite number of nodes and are inadmissible for actual computations on a computer. By this reason, the difference schemes constructed in such a way belong to *nonconstructive* (i.e., formal) *finite difference schemes*. The computation of solutions on a computer is possible only for schemes on meshes with a finite number of nodes; such schemes belong to *constructive finite difference schemes*. Thus, here a question on the *existence of constructive finite difference schemes* for problems in unbounded domains arises. If constructive finite difference schemes exist, it is required to construct such type schemes, and in particular, schemes that converge ε-uniformly.

Here, for a boundary value problem for an elliptic convection-diffusion equation in an unbounded domain, i.e., on the quarter plane, we discuss the existence of constructive finite difference schemes that converge ε-uniformly. The exposition follows accordingly to the results in [186].

14.5.1 Problem formulation in an unbounded domain. The task of computing the solution in a bounded domain

14.5.1.1. In the quarter plane $\overline{D}$, where

$$\overline{D} = D \cup \Gamma, \quad D = \{x : x_s \in (0, \infty), \quad s = 1, 2\}, \tag{14.5}$$

we consider the Dirichlet problem for the singularly perturbed elliptic convection-diffusion equation

$$L_{(14.6)}\, u(x) = f(x), \quad x \in D, \quad u(x) = \varphi(x), \quad x \in \Gamma. \tag{14.6}$$

Here

$$L \equiv \varepsilon \sum_{s=1,2} a_s(x) \frac{\partial^2}{\partial x_s^2} + \sum_{s=1,2} b_s(x) \frac{\partial}{\partial x_s} - c(x),$$

the functions $a_s(x)$, $b_s(x)$, $c(x)$, and $f(x)$ are assumed to be sufficiently smooth on $\overline{D}$, $s = 1, 2$, the function $\varphi(x)$ is sufficiently smooth on the sides Γ_j, $j = 1, 2$ and is continuous on Γ; $\Gamma = \Gamma_1 \cup \Gamma_2$; $\Gamma_s = \overline{\Gamma}_s$, the side Γ_s is orthogonal to the x_s-axis, $s = 1, 2$. We assume that the following conditions are satisfied:

$$a_0 \le a_s(x) \le a^0, \quad b_0 \le b_s(x) \le b^0, \quad c_0 \le c(x) \le c^0, \quad a_0, b_0, c_0 > 0;$$

$$|f(x)| \le M, \quad x \in \overline{D}; \quad |\varphi(x)| \le M, \quad x \in \Gamma. \tag{14.7}$$

By a solution of the boundary value problem, we mean its classical solution, i.e., a function $u \in C^2(D) \cap C(\overline{D})$ that is bounded on $\overline{D}$ and satisfies the differential equation on D and the boundary condition on Γ (see the problem formulation in unbounded domains, e.g., in the textbook [205]).

For simplicity, we suppose that the compatibility conditions ensuring the required smoothness of the solution for each fixed value of the parameter ε are fulfilled on the set $\Gamma^c = \Gamma_1 \cap \Gamma_2$ of "corner points"; here $\Gamma^c = \{(0, 0)\}$ (see, e.g., [170, 180, 186] and Section 14.6 in this chapter).

When the parameter ε tends to zero, boundary layers appear in a neighborhood of the boundary Γ.

14.5.1.2. In the case of problems in unbounded domains it is appropriate to use the following approach to the development of constructive numerical methods. We are interested in finding a solution of the problem (14.6), (14.5) on some *prescribed bounded domain* $\overline{D}^0$ in $\overline{D}$. Let the domain $\overline{D}^0$ be a rectangle defined by its lower-left and upper-right vertices $d^1 = (d_1^1, d_2^1)$ and $d^2 = (d_1^2, d_2^2)$, where d^1 is an arbitrary point of $\overline{D}$

$$\overline{D}^0 = \overline{D}^0(d^1, d^2), \quad \overline{D}^0 = D^0 \cup \Gamma^0. \tag{14.8}$$

Thus, we have $\overline{D}^0 = [d_1^1, d_1^2] \times [d_2^1, d_2^2]$, $d^2 = d^1 + d^0$, $d^0 = (d_1^0, d_2^0)$. We call d^1 and d^2 the *characteristic-parameters of the set* $\overline{D}^0$.

We need to construct a numerical method that allows us to approximate the solution of problem (14.6), (14.5) on the set $\overline{D}^0$. The accuracy of the discrete solution on $\overline{D}^0$ (just as the values d_s^0, $s = 1, 2$) can depend on the parameter ε and on the values of N_1, N_2 that define the numbers of mesh points used (in x_1 and x_2).

When constructing an *numerical method convergent ε-uniformly on* $\overline{D}^0$, it is required that the size of the set $\overline{D}^0$ and the accuracy of the discrete solution (on $\overline{D}^0$) be independent of the parameter ε and be defined only by the values of N_1, N_2. It is desirable that the values d_s^0 be allowed to increase as N_1, N_2 increase.

Our aim is to *justify* that the boundary value problem (14.6), (14.5) *admits to construct constructive difference schemes that converge ε-uniformly* on the set $\overline{D}^0_{(14.8)}$.

14.5.2 Domain of essential dependence for solutions of the boundary value problem

We study the behavior of disturbances of solutions to the problem (14.6), (14.5) considered on some set $\overline{D}^0$ in $\overline{D}$, which are generated by disturbances of the solutions outside the set $\overline{D}^0$. We consider how the finite variation of the solution of the problem (14.6), (14.5) on $\overline{D}$, however far from its subset $\overline{D}^0$, influences the solution to the problem on this set $\overline{D}^0$.

14.5.2.1. In the case of the problem (14.6), (14.5) we are interested to find its solution on the domain $\overline{D}^0_{(14.8)}$ in $\overline{D}$.

Let the set $\overline{D}^0_{(14.8)}$ belong to the rectangle $\overline{D}^{[0]} \subset \overline{D}$ defined by the vertices $\widehat{d}^1$ and $\widehat{d}^2$; $\widehat{d}^i \in \overline{D}$, $\widehat{d}^i = (\widehat{d}_1^i, \widehat{d}_2^i)$, $i = 1, 2$:

$$\overline{D}^{[0]} = \overline{D}^{[0]}\left(\widehat{d}^1, \widehat{d}^2\right) \equiv \overline{D}^0_{(14.8)}\left(\widehat{d}^1, \widehat{d}^2\right), \quad \overline{D}^{[0]} = D^{[0]} \cup \Gamma^{[0]}. \qquad (14.9a)$$

Here

$$\widehat{d}^i = \widehat{d}^i(d^i, \eta^i), \quad d^i = d^i_{(14.8)}, \quad \eta^i = (\eta_1^i, \eta_2^i), \quad \eta_j^i \geq 0, \quad i, j = 1, 2;$$

$$\widehat{d}_s^1 = \max[d_s^1 - \eta_s^1, 0], \quad s = 1, 2, \quad \widehat{d}^2 = d^2 + \eta^2;$$

the values $(\eta_1^1, \eta_1^2) \equiv \overline{\eta}_{(1)}$ and $(\eta_2^1, \eta_2^2) \equiv \overline{\eta}_{(2)}$ determine the neighborhood of the set $\overline{D}^0$ in the x_1- and x_2-directions, respectively. Thus, the set $\overline{D}^{[0]}$ contains the set $\overline{D}^0$ with its $\{\overline{\eta}_{(1)}, \overline{\eta}_{(2)}\}$-neighborhood; for $\overline{\eta}_{(i)} = (\eta_i^1, \eta_i^2)$ we also use the notation $\overline{\eta}_{(i)} = (\eta_{(i)1}, \eta_{(i)2})$, $i = 1, 2$ so that:

$$\overline{D}^{[0]} = \overline{D}^{[0]}\left(\overline{D}^0; \overline{\eta}_{(1)}, \overline{\eta}_{(2)}\right). \qquad (14.9b)$$

We call $\overline{D}^{[0]}_{(14.9)}$ the *test domain*, and $\overline{\eta}_{(1)}$ and $\overline{\eta}_{(2)}$ the *characteristic-parameters* of the test domain

$$\overline{D}^{[0]}_{(14.9)} = \overline{D}^{[0]}_{(14.9)}\left(\overline{D}^0; \overline{\eta}_1, \overline{\eta}_2\right), \quad \overline{D}^0 \subseteq \overline{D}^{[0]}\left(\overline{D}^0; \overline{\eta}_1, \overline{\eta}_2\right) \subset \overline{D}.$$

Let $u^{[0]}(x)$, $x \in \overline{D}^{[0]}$, be the solution of the following problem:

$$L u^{[0]}(x) = f(x), \quad x \in D^{[0]}, \qquad (14.10a)$$

$$u^{[0]}(x) = \varphi(x), \quad x \in \Gamma^{[0]} \cap \Gamma, \qquad (14.10b)$$

$$u^{[0]}(x) = 0, \qquad x \in \Gamma^{[0]} \setminus \Gamma. \qquad (14.10c)$$

We now estimate $u(x) - u^{[0]}(x)$ for $x \in \overline{D}^0$.

Owing to condition (14.7), the solution of the boundary value problem is bounded on $\overline{D}$ ε-uniformly.

The estimate $|u(x) - u^{[0]}(x)|$ on the set $\overline{D}^0$ depends essentially on the mutual disposition of the sets $\Gamma^{[0]}$ and Γ. Using the majorant functions that are equal to 1 on the set $\Gamma^{[0]} \setminus \Gamma$ and nonnegative on $\Gamma^{[0]} \cap \Gamma$, we justify that the following estimate is valid:

$$|u(x) - u^{[0]}(x)| \leq M \left\{ \beta_0(\eta_{(1)1}, \eta_{(2)1}) + \max_{i=1,2} \left[\exp(-m^i \, \eta_{(i)2}) \right] \right\}, \quad (14.11)$$

$$x \in \overline{D}^0.$$

Here

$$\beta_0(\eta_{(1)1}, \eta_{(2)1}) =$$

$$= \begin{cases} \max\limits_{i=1,2} \left[\exp(-m_i \, \varepsilon^{-1} \, \eta_{(i)1}) \right] & \textit{for } \Gamma^{[0]} \cap \Gamma_1, \ \Gamma^{[0]} \cap \Gamma_2 = \emptyset, \\ \exp(-m_1 \, \varepsilon^{-1} \, \eta_{(1)1}) & \textit{for } \Gamma^{[0]} \cap \Gamma_1 = \emptyset, \ \Gamma^{[0]} \cap \Gamma_2 \neq \emptyset, \\ \exp(-m_2 \, \varepsilon^{-1} \, \eta_{(2)1}) & \textit{for } \Gamma^{[0]} \cap \Gamma_1 \neq \emptyset, \ \Gamma^{[0]} \cap \Gamma_2 = \emptyset, \\ 0, & \textit{for } \Gamma^{[0]} \cap \Gamma_1, \ \Gamma^{[0]} \cap \Gamma_2 \neq \emptyset; \end{cases}$$

m^i and m_i are arbitrary constants in the intervals $(0, m^{i0})$ and $(0, m_i^0)$, respectively, with

$$m^{i0} = \min \left\{ 2^{-1/2} \min_{\overline{D}}^{1/2} [a_i^{-1}(x) \, c(x)], 2^{-1} \min_{\overline{D}} [b_i^{-1}(x) \, c(x)] \right\},$$

$$m_i^0 = \min_{\overline{D}} [a_i^{-1}(x) b_i(x)], \quad i = 1, 2.$$

The estimate (14.11) (up to constant factors m^i multiplying $\eta_{(i)2}$, $i = 1, 2$) is unimprovable with respect to the values of $\overline{\eta}_{(1)}$, $\overline{\eta}_{(2)}$, and ε.

Let $\eta_{(i)s} \to \infty$, for $i, s = 1, 2$, so that $\overline{D}^{[0]}(\overline{D}^0; \overline{\eta}_{(1)}, \overline{\eta}_{(2)}) \subset \overline{D}$. Then from the estimate (14.11) it follows that the solution of the boundary value problem (14.10), (14.9) converges on $\overline{D}^0$ to the solution of the boundary value problem (14.6), (14.5) ε-uniformly.

If the function $u^{[0]}(x)$, $x \in \overline{D}^{[0]}$, satisfies the condition

$$u^{[0]}(x) = u_{(14.6;\, 14.5)}(x), \quad x \in \Gamma^{[0]} \setminus \Gamma,$$

instead of the condition (14.10c), then $u^{[0]}(x) = u_{(14.6;14.5)}(x)$, $x \in \overline{D}^{[0]}$; here $u_{(14.6;14.5)}(x)$, $x \in \overline{D}$, is the solution of the problem (14.6), (14.5). Thus, the solution of the problem (14.10), (14.9) is the solution of a perturbed problem generated by the perturbed data of the problem (14.6), (14.5), namely, by a finite change in its solution (the function $u_{(14.6;14.5)}(x)$) outside the set

$$\widehat{D}^{[0]} = D^{[0]} \bigcup \left\{ \overline{D}^{[0]} \cap \Gamma \right\}. \quad (14.12)$$

It follows from the estimate (14.11) that the perturbations of the solution on the set $\overline{D}^0$ caused by the perturbation of the data above are small when the distance R^0 between the sets $\overline{D}^0$ and $\overline{D} \setminus \widehat{D}^{[0]}_{(14.12)}$ is sufficiently large. The perturbation of the solution $\left(\text{the function } u^{[0]}(x) - u(x)\right)$ on the set $\overline{D}^0$ decreases exponentially as R^0 increases. It turns out that the domain of "essential" dependence of the solution (i.e., when the perturbations of the solution are "essentially" different from zero) on the set $\overline{D}^0$ is bounded, although the domain of dependence for the solution of the problem (14.6), (14.5) on the set $\overline{D}^0$ is the whole set $\overline{D}$.

14.5.2.2. For the solution of the problem (14.6), (14.5) considered on set $\overline{D}^0$, we present an estimate for the domain of "essential" dependence.

First, we give some *definitions*.

Domain of essential dependence. Let $\overline{D}^\nabla$ be a subset of $\overline{D}$ that includes the set $\overline{D}^0$. We denote by $u^\nabla(x)$, $x \in \overline{D}^\nabla$, the solution of the perturbed problem

$$Lu^\nabla(x) = f(x), \quad x \in D^\nabla,$$

$$u^\nabla(x) = \varphi(x), \quad x \in \Gamma^\nabla \cap \Gamma, \quad u^\nabla(x) = 0, \quad x \in \Gamma^\nabla \setminus \Gamma.$$

We assume that the solution of the perturbed problem equals zero on a part of the boundary $\Gamma^\nabla \setminus \Gamma$ when the data of the problem (14.6), (14.5) are disturbed. Let for given set $\overline{D}^0$ and a sufficiently small number $\beta > 0$ there exist a set $\overline{D}^\nabla$ such that the function $u^\nabla(x)$, $x \in \overline{D}^\nabla$, considered on the set $\overline{D}^0$, satisfies the estimate

$$|u(x) - u^\nabla(x)| \leq M\beta, \quad x \in \overline{D}^0.$$

We denote this set $\overline{D}^\nabla$ by $\overline{D}^\wedge$. In this case, we say that the set $\overline{D}^\wedge$ is *a domain of essential dependence of the solution* of the problem (14.6), (14.5) on *the set* $\overline{D}^0$ with the *perturbation threshold* on set $\Gamma^\wedge \setminus \Gamma$ equal to β (or briefly, $\overline{D}^\wedge$ is a *domain of dependence of the set* $\overline{D}^0$ with *threshold* β). Thus,

$$\overline{D}^\wedge = \overline{D}^\wedge(\overline{D}^0, \beta).$$

Characteristic-parameters of the domain of essential dependence. It is possible to introduce a domain of dependence on the basis of the test domain $\overline{D}^{[0]}_{(14.9)} = \overline{D}^{[0]}\left(\overline{D}^0; \overline{\eta}_{(1)}, \overline{\eta}_{(2)}\right)$ for suitable values of characteristic-parameters $\overline{\eta}_{(i)}$, $i = 1, 2$, which define a neighborhood of the set $\overline{D}^0$. We denote by $\eta^*_{(i)}$, for $i = 1, 2$, the characteristic-parameters $\overline{\eta}_{(i)}$ (see (14.9)) such that the test domain $\overline{D}^{[0]}_{(14.9)}$ is a domain of essential dependence $\overline{D}^\wedge(\overline{D}^0, \beta)$. We call $\eta^*_{(1)}$ and $\eta^*_{(2)}$ the *characteristic-parameters of the domain of essential dependence* $\overline{D}^\wedge(\overline{D}^0, \beta)$. For the domain of essential dependence constructed above, we designate:

$$\overline{D}^{[0]\wedge}_{(14.13)} = \overline{D}^{[0]\wedge}_{(14.13)}\left(\overline{D}^0, \beta; \eta^*_{(1)}, \eta^*_{(2)}\right) \equiv \qquad (14.13)$$

$$\equiv \overline{D}^{[0]}_{(14.9)}\left(\overline{D}^0; \overline{\eta}_{(i)} = \eta^*_{(i)}, \ i = 1, 2\right).$$

By virtue of estimate (14.11), we can find estimates for the characteristic-parameters $\eta^*_{(i)}$, $i = 1, 2$, under which the estimate

$$|u(x) - u^{[0]}(x)| \le M\,\beta, \quad x \in \overline{D}^0$$

is valid. For the parameters $\eta^*_{(i)}$, $i = 1, 2$, we have the following estimates:

$$\eta^*_{(i)1} \le \min\left[M_i\,\varepsilon_1\,\ln\beta^{-1}, d_i^1\right] \le M_i\,\varepsilon_1\,\ln\beta^{-1}, \qquad (14.14)$$

$$\eta^*_{(i)2} \le M^i\,\ln\beta^{-1}, \quad i = 1, 2.$$

Here $M_i = (m_{i(14.11)})^{-1}$, $M^i = (m^i_{(14.11)})^{-1}$, $i = 1, 2$.

The components $\eta^*_{(i)s}$, $i, s = 1, 2$ of the characteristic-parameters $\eta^*_{(1)}$ and $\eta^*_{(2)}$ depend weakly on the threshold β and grow unboundedly as $\beta \to 0$. From the unimprovability of the estimate (14.11), it follows that the estimates (14.14) for the component parameters $\eta^*_{(i)s}$ that define the "sizes" of the set $\overline{D}^{[0]\wedge}_{(14.13)} \setminus D^0$, i.e., a neighborhood of the set $\overline{D}^0$ from the domain of dependence, are unimprovable with respect to ε and $\ln\beta^{-1}$.

Symmetrical domain of essential dependence. The estimate of the components $\eta^*_{(i)s}$, for $s = 1, 2$, depends essentially on the parameter ε. It is convenient to consider such domains of dependence for which the "sizes" of the set $\overline{D}^{[0]\wedge}_{(14.13)} \setminus D^0$ are controlled only by a single parameter that is independent of ε. We denote by η^* such *a scalar parameter* η^0, independent of ε, for which the set $\overline{D}^{[0]}_{(14.9)}$ under the condition $\eta_{(i)s(14.9b)} = \eta^0$, for $i, s = 1, 2$, is the domain of dependence $\overline{D}^\wedge = \overline{D}^\wedge(\overline{D}^0, \beta)$. We call the domain of dependence $\overline{D}^\wedge = \overline{D}^\wedge(\eta^*)$ the *rectangular domain of essential dependence symmetrical with respect to the set $\overline{D}^0$*, or in short, the *symmetrical domain of essential dependence*.

We call η^* *the characteristic-parameter* of the symmetrical domain of dependence $\overline{D}^\wedge(\overline{D}^0, \beta)$.

The characteristic-parameter η^* that defines the symmetrical domain of dependence

$$\overline{D}^\wedge = \overline{D}^\wedge(\eta^*) \equiv \overline{D}^{[0]}_{(14.9)}(\eta_{(i)s} = \eta^*, \ i, s = 1, 2),$$

satisfies the estimate

$$\eta^* \le M\,\ln\beta^{-1}, \qquad (14.15)$$

where $M = \max\limits_{i,j} [M_1^i, M_2^j]$, with $M_1^i = M_{1(14.14)}^i$ and $M_2^j = M_{2(14.14)}^j$. The estimate (14.15) is unimprovable with respect to $\ln \beta^{-1}$.

Thus, the estimate of the characteristic-parameter η^* of the symmetrical domain of dependence $\overline{D}^\wedge(\overline{D}^0, \beta)$ is independent of the parameter ε and the characteristic-parameters d^1 and d^2 of the set $\overline{D}^0$; and it weakly depends on β and grows without bound as $\beta \to 0$.

In a similar way one can justify that, in the case of *formal* (nonconstructive) *schemes*, the properties of their discrete *domains of essential dependence* are similar to the properties of the domains of essential dependence for the boundary value problem.

The weak dependence of the domains of essential dependence for the differential and the formal discrete problems on the threshold β and their ε-uniform boundedness for fixed values of β allow us to compose *constructive difference schemes* that converge ε-uniformly on the *prescribed rectangular domain* $\overline{D}^0$, see [186].

The constructive difference schemes from [186] converge ε-uniformly on rectangle $\overline{D}^0$ under the condition $d_i^0 \ll N_i$, $d_i^0 = d_{i(14.8)}^0$, for $i = 1, 2$. For the problem (14.6), (14.5), the *rate* of ε-uniform *convergence of the constructive* difference *schemes* on the rectangle $\overline{D}^0$ in the case of $d_i^0 = \mathcal{O}(1)$, for $i = 1, 2$, *up to a logarithmic factor, is the same* as for the singularly perturbed convection-diffusion problem (14.6) on the *bounded domain* $\overline{D}$.

14.5.3   Generalizations

14.5.3.1. When the condition $c(x) \geq c_0 > 0$, for $x \in \overline{D}$, is violated, a domain of essential dependence of the solution to the boundary value problem (and the discrete problem) is, in general, not bounded (if the ε-entropy $H_\varepsilon(\mathcal{U})$ of the set $\mathcal{U}$, i.e., the set of solutions to the boundary value problem (14.6), (14.5) whose data satisfy the condition (14.7), is infinite; for this, see [16, 63]). For such problems, the technique considered here to develop constructive difference schemes is directly inapplicable. But if the right hand-side of the equation and the boundary function under the condition $c(x) \equiv 0$, $x \in \overline{D}$, are such that the solution of the boundary value problem tends to 0, e.g., uniformly in x_2 as $x_1 \to \infty$ (in this case, the ε-entropy $H_\varepsilon(\mathcal{U})$ of the set $\mathcal{U}$, in general, can be infinite), then the developed technique allows us to construct ε-uniformly convergent constructive difference schemes.

14.5.3.2. The given technique allows us to justify the property of ε-uniform boundedness for domains of essential dependence of solutions to differential and formal discrete problems for other types of singularly perturbed problems in unbounded domains. If this property is valid and the sizes of such domains depend weakly on the threshold β, it is possible to construct ε-uniformly convergent constructive numerical methods.

14.6   Compatibility conditions for a boundary value problem on a rectangle for an elliptic convection-diffusion equation with a perturbation vector parameter

To construct and study numerical methods for regular boundary value problems on nonsmooth domains, it is quite often required that the solution is sufficiently smooth. For the Laplace equation on a rectangle, simple criteria for data compatibility that provide the smoothness of the solution of the Dirichlet problem are known for a set of angles $\pi/k, k = 2, 3, \ldots$. Such criteria for other angles have integral character, see, e.g., [38, 66] and the discussions in [80].

When constructing ε-uniformly convergent schemes on a rectangle, besides the smoothness of the solution for boundary value problems, the smoothness of its regular and singular components is required (the singular component is the sum of regular and corner boundary layers). Here such a question arises: does a solution of the problem exist with its regular component and singular component including both regular and corner boundary layers that are smooth? If this is valid, then such questions appear: what are precise compatibility conditions, and how such conditions could be expressed in simple form?

To find such compatibility conditions for representative classes of singularly perturbed boundary value problems is an important task for development of numerical methods that converge ε-uniformly in the maximum norm.

In the present section we consider a boundary value problem on a rectangle for an elliptic convection-diffusion equation with a vector parameter $\overline{\varepsilon}$, where $\overline{\varepsilon} = (\varepsilon_1, \varepsilon_2)$, ε_1 is a parameter multiplying the terms with highest derivatives, and ε_2 is a parameter multiplying one of the convective terms. The character and intensity of the boundary layers appearing here depends on the ratio between the parameters $\varepsilon_1, \varepsilon_2$. Such boundary layers may be regular, parabolic, hyperbolic, or corner elliptic.

For this problem, we give compatibility conditions that guarantee smoothness of the solution and its components that are required to construct and justify $\overline{\varepsilon}$-uniformly convergent schemes.

Difference schemes that converge $\overline{\varepsilon}$-uniformly for the problem under study when considered in the first quarter-plane are given in [170, 178]. The solution of the above problem but on a rectangle has a more various composition of boundary layers of different types as compared to the problem considered in [170, 178]. Such a boundary value problem on a rectangle and on a vertical half-strip was considered in [180]. Boundary value problems with a vector parameter were considered only in rather a few publications; see, e.g., [148, 97, 25] and the bibliography therein.

A number of results concerning the analysis of compatibility conditions for singularly perturbed problems with one perturbation parameter can be seen, e.g., in [30, 24, 39] and in the Appendix in [87]. Sufficient conditions for compatibility of input data for regular elliptic equations in two-dimensional domains with corner points are thoroughly discussed in [80] (see also the bibliography therein).

14.6.1 Problem formulation

In the rectangle $\overline{D}$, where

$$\overline{D} = D \cup \Gamma, \quad D = \{x : x_s \in (0, d_s), \quad s = 1, 2\}, \qquad (14.16)$$

consider the Dirichlet problem for the singularly perturbed equation

$$L u(x) = f(x), \quad x \in D, \quad u(x) = \varphi(x), \quad x \in \Gamma. \qquad (14.17)$$

Here

$$L \equiv \varepsilon_1 \sum_{s=1,2} a_s(x) \frac{\partial^2}{\partial x_s^2} + b_1(x) \frac{\partial}{\partial x_1} + \varepsilon_2 \, b_2(x) \frac{\partial}{\partial x_2} - c(x),$$

the functions $a_s(x)$, $b_s(x)$, $c(x)$, and $f(x)$ are assumed to be sufficiently smooth on $\overline{D}$, $s = 1, 2$, and the function $\varphi(x)$ is sufficiently smooth on the sides of D and continuous on Γ. We assume also that the conditions

$$a_0 \leq a_s(x) \leq a^0, \quad b_0 \leq b_s(x) \leq b^0, \quad c_0 \leq c(x) \leq c^0, \quad a_0, b_0, c_0 > 0;$$

$$|f(x)| \leq M, \quad x \in \overline{D}; \quad |\varphi(x)| \leq M, \quad x \in \Gamma$$

are satisfied. The parameters ε_1 and ε_2 (the components of $\overline{\varepsilon}$) take arbitrary values in the intervals $(0, 1]$ and $[-1, 1]$, respectively. The range of values of the vector parameter $\overline{\varepsilon}$ is denoted by $E_{\overline{\varepsilon}}$:

$$E_{\overline{\varepsilon}} = \{\overline{\varepsilon} = (\varepsilon_1, \varepsilon_2) : \varepsilon_1 \in (0, 1], \varepsilon_2 \in [-1, 1]\}.$$

Denote the sides of the rectangle D by Γ_j, so that $\Gamma = \cup \Gamma_j$, $j = 1, 2, 3, 4$; Γ_s and Γ_{s+2} are orthogonal to the axis x_s, $s = 1, 2$; the sides Γ_1 and Γ_2 pass through the point $(0, 0)$; moreover, Γ_j are closed sets, and Γ^c is the set of corner points.

By a solution of the boundary value problem, we mean its classical solution, i.e., a function $u \in C^2(D) \cap C(\overline{D})$ that is bounded on $\overline{D}$ and satisfies the differential equation in D and the boundary condition on Γ. The smoothness of the solution depends on the smoothness of the input data and also on the additional conditions imposed on them.

When ε_1 tends to zero, boundary layers appear in a neighborhood of a part of the boundary Γ. The *type of the layers* in the neighborhood of Γ_j, for $j = 1, 2, 4$, and Γ^c, and their *properties* are *controlled by the vector-parameter* $\overline{\varepsilon}$, for $\overline{\varepsilon} \in E_{\overline{\varepsilon}}$.

For $\varepsilon_2 = 0$, the sides Γ_j, for $j = 2, 4$, are *characteristic parts* of the boundary Γ. In a neighborhood of these sides, *parabolic boundary layers* appear as

$\varepsilon_1 \to 0$. Such a *characteristic effect* of these sides is still observed under the condition $|\varepsilon_2| = \mathcal{O}\left(\varepsilon_1^{1/2}\right)$, $\varepsilon_1 = o(1)$, $\bar{\varepsilon} \in E_{\bar{\varepsilon}}$. Under the condition $\varepsilon_2 < 0$, $\varepsilon_1^{1/2} \ll |\varepsilon_2| \ll 1$, $\varepsilon_1 = o(1)$, $\bar{\varepsilon} \in E_{\bar{\varepsilon}}$, *parabolic layers* are *transformed* into *hyperbolic* ones.

Along with the problem (14.17) in the rectangle (14.16), we consider a *model problem* for the same equation (14.17) in the vertical half-strip $\overline{D}$, i.e.,

$$\overline{D} = D \cup \Gamma, \tag{14.18}$$

$$D = \{x : x_1 \in (0, d_1), \ x_2 \in (0, \infty)\}, \quad \Gamma = \cup \Gamma_j, \quad j = 1, 2, 3.$$

14.6.2   Compatibility conditions

We now give compatibility conditions imposed on the data of the problem (14.17), (14.16) that guarantee sufficient smoothness of the solution and its components which are required to construct and justify $\bar{\varepsilon}$-uniformly convergent schemes.

In [212], for the Dirichlet problem on the rectangle

$$\triangle w(x) = F(x), \quad x \in D, \quad w(x) = \Phi(x), \quad x \in \Gamma,$$

necessary and sufficient conditions for the inclusion $w \in C^{l+2+\alpha}(\overline{D})$ for $l \geq 0$ and $0 < \alpha < 1$ are obtained.

These results in the case of the problem (14.17), (14.16) allow us to write down compatibility conditions providing the required smoothness of the solution on $\overline{D}$.

In the case

$$F \in C^{l+\alpha}(\overline{D}); \quad \Phi(x) = 0, \quad x \in \Gamma,$$

this condition has the form

$$\sum_{i=0}^{q} (-1)^i \frac{\partial^{2q}}{\partial x_1^{2(q-i)} \partial x_2^{2i}} F(x) = 0, \quad x \in \Gamma^c, \tag{14.19}$$

where $0 \leq q \leq [l/2]$, $[l/2]$ is the integer part of $l/2$.

Applying this result to the problem (14.17), (14.16), we obtain compatibility conditions.

Assume that the following conditions are fulfilled:

$$a_s(x) \equiv 1, \quad b_s(x) = b_s = const, \quad c(x) \equiv 0, \quad x \in \overline{D}, \quad s = 1, 2,$$

$$f \in C^{l+\alpha}(\overline{D}) \ \ for \ \ l \geq 0, \ \ 0 < \alpha < 1.$$

Furthermore, let

$$\frac{\partial^{i+j}}{\partial x_1^i x_2^j} f(x) = 0, \quad x \in \Gamma^c, \ \ i + j \leq l, \quad \varphi(x) = 0, \quad x \in \Gamma.$$

Then, the problem (14.17), (14.16) has a solution $u(x)$ such that

$$u \in C^{l+2+\alpha}(\overline{D}).$$

In order to verify this fact, we define the function

$$v_1(x) = u(x) \exp(\beta_1 x_1 + \beta_2 x_2), \quad x \in \overline{D}.$$

Then, for

$$\beta_1 = 2^{-1} \varepsilon_1^{-1} b_1, \quad \beta_2 = 2^{-1} \varepsilon_1^{-1} \varepsilon_2 b_2,$$

the function $v_1(x)$ is a solution of the problem

$$\triangle v_1(x) = F_1(x), \quad x \in D, \quad v_1(x) = \Phi_1(x), \quad x \in \Gamma,$$

where

$$F_1(x) = 4^{-1} \varepsilon_1^{-2} (b_1^2 + \varepsilon_2^2 b_2^2) v_1(x) + \varepsilon_1^{-1} f(x) \times$$
$$\times \exp(2^{-1} \varepsilon_1^{-1} b_1 x_1 + 2^{-1} \varepsilon_1^{-1} \varepsilon_2 b_2 x_2), \quad x \in \overline{D},$$
$$\Phi_1(x) = \varphi(x) \exp(2^{-1} \varepsilon_1^{-1} b_1 x_1 + 2^{-1} \varepsilon_1^{-1} \varepsilon_2 b_2 x_2), \quad x \in \Gamma.$$

The function $F_1(x)$ satisfies the condition

$$F_1(x) = 0, \quad x \in \Gamma^c$$

(note that $v_1 \in C^\alpha(\overline{D})$). Then, it follows from [212] that $v_1 \in C^{2+\alpha}(\overline{D})$. Therefore, we have $F_1 \in C^{2+\alpha}(\overline{D})$, and

$$\frac{\partial^2}{\partial x_1^2} F_1(x) - \frac{\partial^2}{\partial x_2^2} F_1(x) = 0, \quad x \in \Gamma^c.$$

Now, it follows from [212] that $v_1 \in C^{4+\alpha}(\overline{D})$. Proceeding in a similar fashion, we find that $v_1 \in C^{l+2+\alpha}(\overline{D})$ and, therefore, $u \in C^{l+2+\alpha}(\overline{D})$.

In the problem (14.17), (14.16), we can pass from the condition $\varphi(x) \neq 0$, $\varphi \in C^{l+2+\alpha}(\Gamma)$ to the homogeneous boundary condition by making the standard change of variables.

According to [69], under the conditions

$$a_s, b_s, c, f \in C^{l+\alpha}(\overline{D}), \quad \varphi \in C^{l+2+\alpha}(\Gamma_j), \quad \varphi \in C(\Gamma), \quad (14.20a)$$
$$j = 1, 2, 3, 4, \quad s = 1, 2, \quad l \geq 0, \quad \alpha \in (0, 1)$$

the solution of the problem (14.17), (14.16) satisfies the inclusion

$$u \in C^{l+2+\alpha}(\overline{D} \setminus D^{\delta/2}),$$

where D^δ is an δ-neighborhood of the set Γ^c (δ is independent of ε). Taking this inclusion into account and assuming that

$$a_s(x) = a(x^*), \quad x \in \overline{D}, \quad x^* \in \Gamma^c, \quad r(x, x^*) \le \delta, \quad s = 1, 2, \quad (14.20b)$$

$$r(x, x^*) \text{ is the distance between the points } x, x^*;$$

$$\frac{\partial^k}{\partial x_1^{k_1} \partial x_2^{k_2}} b_s(x) = 0, \quad 1 \le k \le l-1 \text{ for } l \ge 2,$$

$$\frac{\partial^k}{\partial x_1^{k_1} \partial x_2^{k_2}} c(x) = 0, \quad 1 \le k \le l-2 \text{ for } l \ge 3,$$

for $l < 2$ and $l < 3$ ($l \ge 0$) *restrictions on the derivatives*

of the functions $b_s(x)$ *and* $c(x)$, $x \in \Gamma^c$ *are not imposed*,

it can be shown that the condition (14.19) is necessary and sufficient for the inclusion

$$u \in C^{l+2+\alpha}(\overline{D}).$$

In (14.19), $F(x)$ is a function satisfying conditions corresponding to the data of the boundary value problem (14.17), (14.16).

If conditions (14.20a), (14.20b) are fulfilled, the following condition is sufficient for the inclusion $u \in C^{l+2+\alpha}(\overline{D})$:

$$\frac{\partial^k}{\partial x_1^{k_1} \partial x_2^{k_2}} f(x) = 0, \quad k \le l,$$

$$\frac{\partial^{k_1}}{\partial x_1^{k_1}} \varphi(x) = \frac{\partial^{k_2}}{\partial x_2^{k_2}} \varphi(x) = 0, \quad k_1, k_2 \le l+2, \quad x \in \Gamma^c. \quad (14.20c)$$

This sufficient compatibility condition is simpler than the necessary and sufficient compatibility condition for the data of the boundary value problem (14.17), (14.16) under assumptions (14.20a), (14.20b). In the case of problem (14.17), (14.18), sufficient conditions are similar to those for problem (14.17), (14.16).

To simplify the presentation, taking into account the constructions described above, we will assume that the following condition is fulfilled:

The data of problem (14.17), (14.16) (problem (14.17), (14.18))
satisfy condition (14.20) (where $j = 1, 2, 3$ *in the case of problem*
(14.17), (14.18)), which ensures the smoothness of the solution to
the boundary value problem. When constructing a priori bounds
for the regular and singular components in representations of the
solutions of the boundary value problem, we will assume that the

following condition is fulfilled in addition to (14.20):

$$a_s, \, b_s, \, c, \, f \in C^{l_1+\alpha}(\overline{D}), \quad \varphi \in C^{l_1+\alpha}(\Gamma_j), \quad \varphi \in C(\Gamma); \qquad (14.21)$$

$$\frac{\partial^k}{\partial x_1^{k_1} \partial x_2^{k_2}} f(x) = 0, \quad x \in \Gamma^{c+}, \quad k \le l_1;$$

$$\frac{\partial^{k_1}}{\partial x_1^{k_1}} \varphi(x) = \frac{\partial^{k_2}}{\partial x_2^{k_2}} \varphi(x) = 0, \quad x \in \Gamma^{c+}, \quad k_1, k_2 \le l_1, \quad l_1 \ge l;$$

$$\Gamma^{c+} = \left\{ \begin{matrix} \Gamma^c_{34} & \text{for } \varepsilon_2 > 0 \\ \Gamma^c_{23} & \text{for } \varepsilon_2 < 0 \\ \Gamma^c_{23} \cup \Gamma^c_{34} & \text{for } \varepsilon_2 = 0 \end{matrix} \right\} \text{ for problem (14.17), (14.16),}$$

$$\Gamma^{c+} = \left\{ \begin{matrix} \emptyset & \text{for } \varepsilon_2 > 0 \\ \Gamma^c_{23} & \text{for } \varepsilon_2 \le 0 \end{matrix} \right\} \text{ for problem (14.17), (14.18).}$$

This ensures the smoothness of the regular and singular components of the solution (in the case of problem (14.17), (14.18), $j = 1, 2, 3$ in (14.21)).

In the case of condition (14.20), (14.21), where

$$l \ge K - 2, \quad l_1 \ge 3K - 4, \qquad (14.22)$$

for $K = 3$ we have estimates for the regular and singular components of solutions to problems (14.17), (14.16) and (14.17), (14.18), which are required to construct and to justify $\bar{\varepsilon}$-uniformly convergent schemes, see [180].

Remark 14.6.1 Using the technique of [138, 87], it can be shown that in case of problem (14.17), (14.16) *in the presence of parabolic boundary layers,* i.e., *under the condition*

$$|\varepsilon_2| = \mathcal{O}\left(\varepsilon_1^{1/2}\right), \quad \varepsilon_1 = o(1), \quad \bar{\varepsilon} \in E_{\bar{\varepsilon}},$$

there are *no schemes based on the fitted operator method that converge $\bar{\varepsilon}$-uniformly.* In a similar way, one can justify that *in the presence of hyperbolic boundary layers,* i.e., *under the condition*

$$\varepsilon_2 < 0, \quad \varepsilon_1^{1/2} \ll |\varepsilon_2| \ll 1, \quad \varepsilon_1 = o(1), \quad \bar{\varepsilon} \in E_{\bar{\varepsilon}},$$

also there exist *no schemes of the fitted operator method that converge $\bar{\varepsilon}$-uniformly.*

Remark 14.6.2 If the problem data are *not sufficiently smooth*, the condition (14.22) is violated. The difference schemes on *simplest meshes*, that are piecewise-uniform (see [180]), *continue* to be $\bar{\varepsilon}$-*uniformly convergent.* For example, in the case $a_s,\, b_s,\, c,\, f \in C^\alpha(\overline{D})$, $\varphi \in C^{2+\alpha}C(\Gamma_j)$, $\varphi \in C(\Gamma)$, $\alpha \in (0,1)$, $s = 1,2,\ j = 1,2,3,4,$ *the technique used in* [138, 164] allows one to establish *the* $\bar{\varepsilon}$-*uniform convergence* of the schemes at the rate $\mathcal{O}\left(N^{-\nu}\right)$, where *the order of convergence* $\nu = \nu(\alpha)$, in general, is *small.*

References

[1] Alekseevskii M.V. (1980). On a difference scheme for a singularly perturbed parabolic equation. *Erevan. Gos. Univ. Uchen. Zap. Estestv. Nauki*, (1), 3–13 (in Russian).

[2] Allen D.N., Southwell R.V. (1955). Relaxation methods applied to determine the motion, in two dimensions, of viscous fluid past a fixed cylinder. *Quart. J. Mech. Appl. Math.*, **8**, (2), 129–145.

[3] Andreev V.B. (2005). Grid approximation of nonsmooth solutions of singularly perturbed equations. *Proceedings of 10th International Conference "Mathematical Modelling and Analysis" and 2nd International Conference "Computational Methods in Applied Mathematics"*, 207–212, Technica, Vilnius.

[4] Andreev V.B. (2006). On the accuracy of grid approximations of nonsmooth solutions of a singularly perturbed reaction-diffusion equation in the square. *Differential Equations*, **42**, (7), 954–966.

[5] Andreev V.B. (2008). Uniform grid approximation of nonsmooth solutions to the mixed boundary value problem for a singularly perturbed reaction-diffusion equation in a rectangle. *Comp. Math. Math. Phys.*, **48**, (1), 85–108.

[6] Andreev V.B., Kopteva N.V. (2008). Pointwise approximation of corner singularities for a singularly perturbed reaction-diffusion equation in an *L*-shaped domain. *Math. Comp.*, **77** (to appear).

[7] Ansari A.R., Bakr S.A., Shishkin G.I. (2007). A parameter-robust finite difference method for singularly perturbed delay parabolic partial differential equations. *J. Comput. Appl. Math.*, **205**, (1), 552–566.

[8] Ansari A.R., Hegarty A.F., Shishkin G.I. (2003). Parameter-uniform numerical methods for a laminar jet problem. *Internat. J. Numer. Methods Fluids*, **43**, (8), 937–952.

[9] Ansari A.R., Hossain B., Koren B., Shishkin G.I. (2006). Robust numerical methods for boundary-layer equations for a model problem of flow over a symmetric curved surface. *Math. Model. Anal.*, **11**, (4), 365–378.

[10] Babenko K.I. (1986). *Foundations of Numerical Analysis*. Nauka, Moscow (in Russian).

[11] Bagaev B.M. (1981). *A Variation-Difference Method for Solving Elliptic Equations with a Small Parameter Multiplying the Higher Derivatives.* Ph. D. Thesis, Krasnoyarsk (in Russian).

[12] Bagaev B.M., Shaidurov V.V. (1977). A variation-difference solution of an equation with a small parameter. *Meth. Numer. Appl. Math.*, Novosibirsk, 89–99 (in Russian).

[13] Bagaev B.M., Shaidurov V.V. (1998). *Grid Methods for Solving Problems with a Boundary Layer.* Part 1. Nauka, Sibirskoe Predpriyatie RAN, Novosibirsk (in Russian).

[14] Bagaev B.M., Karepova E.D., Shaidurov V.V. (2001). *Grid Methods for Solving Problems with a Boundary Layer.* Part 2. Nauka, Novosibirsk (in Russian).

[15] Bakhvalov N.S. (1969). On the optimization of methods for solving boundary value problems in the presence of a boundary layer. *USSR Comput. Maths. Math. Phys.*, **9**, (4), 139–166.

[16] Bakhvalov N.S. (1973). *Numerical Methods.* Nauka, Moscow (in Russian); English Translation: Bakhvalov N. (1976). *Numerical Methods.* Mir Publishers, Moscow.

[17] Birkhoff G., Lynch R.E. (1984). *Numerical Solution of Elliptic Problems.* SIAM, Philadelphia, PA.

[18] Böhmer K., Stetter H.J., eds. (1984). *Defect Correction Methods. Theory and Applications (Oberwolfach, 1983).* Computing Supplementum, **5**. Springer-Verlag, Vienna.

[19] Budak B.M., Samarskii A.A., Tikhonov A.N. (1980). *Problems of Mathematical Physics.* Nauka, Moscow (in Russian); Spanish Translation: Budak B.M., Samarski A.A., Tijonov A.N. (1984) Problemas de la fisica matematica. Tomo 1, 2. Mir, Moscow.

[20] Butler J.S., Miller J.J.H., Shishkin G.I. (2003) A Reynolds-uniform numerical method for Prandtl's boundary layer problem for flow past a wedge. *Internat. J. Numer. Methods Fluids*, **43**, (8), 903–914.

[21] Butler J.S., Miller J.J.H., Shishkin G.I. (2004). A Reynolds-uniform numerical method for Prandtl's boundary layer problem for flow past a plate with mass transfer. *Internat. J. Comput. Eng. Sci.*, **5**, (2), 387–402.

[22] Butler J.S., Miller J.J.H., Shishkin G.I. (2004), A Reynolds uniform numerical method for Prandtl's boundary layer problem for flow past a three dimensional yawed wedge. *Computational Mechanics WCCM VI in conjunction with APCOM'04, Beijing, China, September 5-10, 2004*, Eds. Z.H. Yao, M.W. Yuan, W.X. Zhong, 243–247, Tsinghua University Press & Springer Verlag, Beijing

[23] Clavero C., Jorge J.C., Lisbona F., Shishkin G.I. (2000). An alternating direction scheme on a nonuniform mesh for reaction-diffusion parabolic problems. *IMA J. Numer. Anal.*, **20**, (2), 263–280.

[24] Clavero C., Gracia J.L., Lisbona F., Shishkin G.I. (2002). A robust method of improved order for convection-diffusion problems in a domain with characteristic boundaries. *ZAMM Z. Angew. Math. Mech.*, **82**, (9), 631–647.

[25] Cordero N., Cronin K., Shishkin G., Shishkina L., Stynes M. (2008). Finite difference scheme for a singularly perturbed parabolic equation in the presence of initial and boundary layers. *Math. Model. Anal.*, **13**, (2).

[26] Doolan E.P., Miller J.J.H., Shilders W.H.A. (1980). *Uniform Numerical Methods for Problem with Initial Boundary Layers*. Boole Press, Dublin.

[27] Dunne R.K., O'Riordan E., Shishkin G.I. (2001). Singularly perturbed parabolic problems on non-rectangular domains. *Numerical Analysis and its Applications (Rousse, 2000)*, Lecture Notes in Comput. Sci., **1988**, 265–272, Springer, Berlin.

[28] Dunne R.K., O'Riordan E., Shishkin G.I. (2008). Fitted mesh numerical methods for singularly perturbed elliptic problems with mixed derivatives. *IMA J. Numer. Anal.*, accepted for publication.

[29] Emel'yanov K.V. (1970). On the difference scheme for a differential equation with a small parameter multiplying the highest derivatives. *Chisl. Metody Mekh. Sploshn. Sredy*, Novosibirsk, **1**, (5), 20–30 (in Russian).

[30] Emel'yanov K.V. (1973). A difference scheme for a three-dimensional elliptic equation with a small parameter multiplying the highest derivatives. *Boundary Value Problems for Equations of Mathematical Physics*. Ural. Nauchn. Centr Akad. Nauk SSSR, Sverdlovsk, 30–42 (in Russian).

[31] Farrell P.A., O'Riordan E., Shishkin G.I. (2008) A class of singularly perturbed quasilinear differential equations with interior layers. *Math. Comp.*, accepted for publication.

[32] Farrell P.A., Miller J.J.H., O'Riordan E., Shishkin G.I. (1998). On the non-existence of ε-uniform finite difference methods on uniform meshes for semilinear two-point boundary value problems. *Math. Comp.*, **67**, (222), 603–617.

[33] Farrell P.A., Hegarty A.F., Miller J.J.H., O'Riordan E., Shishkin G.I. (2000). *Robust Computational Techniques for Boundary Layers*. Chapman & Hall / CRC, Boca Raton, FL.

[34] Farrell P.A., Hegarty A.F., Miller J.J.H., O'Riordan E., Shishkin G.I. (2002). An experimental technique for computing parameter-uniform error estimates for numerical solutions of singular perturbation problems, with an application to Prandtl's problem at high Reynolds number. *Appl. Numer. Math.*, **40**, (1-2), 143–149.

[35] Farrell P.A., Hegarty A.F., Miller J.J.H., O'Riordan E., Shishkin G.I. (2003). Computing realistic Reynolds-uniform error bounds for discrete derivatives of flow velosities in the boundary layer for Prandtl's problem. *Internat. J. Numer. Methods Fluids*, **43**, (8), 895–902.

[36] Flaherty J.E., Paslow P.J., Shephard M.S., Vasilakis J.D., eds. (1989). *Adaptive Methods for Partial Differential Equations.* SIAM, Philadelphia, PA.

[37] Friedman A. (1964). *Partial Differential Equations of Parabolic Type,* Prentice-Hall, Inc., Englewood Cliffs, NJ.

[38] Fufaev V.V. (1960). On the Dirichlet problem for regions with corners. *Soviet Math. Dokl.*, (1), 199–201.

[39] Han H., Kellogg R.B. (1990). Differentiability properties of solutions of the equation $-\varepsilon^2 \triangle u + ru = f(x,y)$ in a square. *SIAM J. Math. Anal.*, **21**, (2), 394–408.

[40] Hegarty A.F., Miller J.J.H., O'Riordan E., Shishkin G.I. (1993). Key to the computation of the transfer of a substance by convection-diffusion in a laminar fluid. *Application of Computational Methods for Boundary and Interior Layers*, J.J.H. Miller, ed., 94–107, Boole Press, Dublin.

[41] Hemker P.W., Shishkin G.I. (1994). On a class of singularly perturbed boundary value problems for which an adaptive mesh technique is necessary. *Proceedings of the Second International Colloquium on Numerical Analysis (Plovdiv, 1993)*, D.Bainov and V.Covachev, eds., 83–92, VSP, Utrecht.

[42] Hemker P.W., Shishkin G.I. (1994). Discrete approximation of singularly perturbed parabolic PDEs with a discontinuous initial condition. *Comput. Fluid Dynamics J.*, **2**, (4), 375–392.

[43] Hemker P.W., Farrell P.A., Shishkin G.I. (1996). Discrete approximations for singularly perturbed boundary value problems with parabolic layers. I. *J. Comput. Math.*, **14**, (1), 71–97.

[44] Hemker P.W., Farrell P.A., Shishkin G.I. (1996). Discrete approximations for singularly perturbed boundary value problems with parabolic layers. II. *J. Comput. Math.*, 14, (?) 183–194.

[45] Hemker P.W., Farrell P.A., Shishkin G.I. (1996). Discrete approximations for singularly perturbed boundary value problems with parabolic layers. III. *J. Comput. Math.*, **14**, (3), 273–290.

[46] Hemker P.W., Shishkin G.I., Shishkina L.P. (1997). The use of defect correction for the solution of parabolic singular perturbation problems. *ZAMM Z. Angew. Math. Mech.*, **77**, (1), 59–74.

[47] Hemker P.W., Shishkin G.I., Shishkina L.P. (2000). $\mathcal{E}$-uniform schemes with high-order time-accuracy for parabolic singular perturbation problems. *IMA J. Numer. Anal.*, **20**, (1), 99–121.

[48] Hemker P.W., Shishkin G.I., Shishkina L.P. (2000). Distributing the numerical solution of parabolic singularly perturbed problems with defect correction over independent processes. *Siberian J. Numer. Math.*, **3**, (3), 229–258.

[49] Hemker P.W., Shishkin G.I., Shishkina L.P. (2001). High-order time-accurate parallel schemes for parabolic singularly perturbed problems with convection. *Computing*, **66**, (2), 139–161.

[50] Hemker P.W., Shishkin G.I., Shishkina L.P. (2002). High-order time-accurate schemes for parabolic singular perturbation problems with convection. *Russian J. Numer. Anal. Math. Modelling*, **17**, (1), 1–24.

[51] Hemker P.W., Shishkin G.I., Shishkina L.P. (2002). High-order time-accurate schemes for parabolic singular perturbation convection-diffusion problems with Robin boundary conditions. *Comput. Methods Appl. Math.*, **2**, (1), 3–25.

[52] Hemker P.W., Shishkin G.I., Shishkina L.P. (2003). Novel defect-correction high-order, in space and time, accurate schemes for parabolic singularly perturbed convection-diffusion problems. *A posteriori* adaptive mesh technique. *Comput. Methods Appl. Math.*, **3**, (3), 387–404.

[53] Hemker P.W., Shishkin G.I., Shishkina L.P. (2004). High-order accuracy decomposition of Richardson's method for a singularly perturbed elliptic reaction-diffusion equation. *Comput. Math. Math. Phys.*, **44**, (2), 309–316.

[54] Hossain B., Ansari A.R., Shishkin G.I. (2003). On numerical methods for a boundary layer on a body of revolution. *Comput. Methods Appl. Math.*, **3**, (3), 405-416.

[55] Hundsdorfer W., Verwer J. (2003). *Numerical Solution of Time-Dependent Advection-Diffusion-Reaction Equations*, Springer-Verlag, Berlin.

[56] Il'in A.M. (1969). Differencing scheme for a differential equation with a small parameter affecting the highest derivative. *Math. Notes*, **6**, (2), 596–602.

[57] Il'in A.M. (1992). *Matching of Asymptotic Expansions of Solutions of Boundary Value Problems*. Translations of Mathematical Monographs, 102. American Mathematical Society, Providence, RI.

[58] Il'in A.M., Kalashnikov A.S., Oleinik O.A. (1962). Second-order linear equations of parabolic type. *Uspehi Mat. Nauk*, **17**, (3), 3–146 (in Russian).

[59] Il'in A.M., Lelikova E.F. (1975). The method of matching asymptotic expansions for the equation $\varepsilon \Delta u - a(x, y)u_y = f(x, y)$ in a rectangle. *Mat. Sb.* (Collection of Mathematical Problems), **96**, (4), 568–583 (in Russian).

[60] Kellogg R.B., Stynes M. (1999). n-widths and singularly perturbed boundary value problems. *SIAM J. Numer. Anal.*, **36**, (5), 1604–1620.

[61] Kellogg R.B., Stynes M. (2001). n-widths and singularly perturbed boundary value problems. II. *SIAM J. Numer. Anal.*, **39**, (2), 690–707.

[62] Kevorkian J., Cole J.D. (1981). *Perturbation Methods in Applied Mathematics*. Springer-Verlag, New York-Berlin.

[63] Kolmogorov A.N., Tikhomirov V.M. (1959). ε-entropy and ε-capacity of sets in functional spaces. *Uspehi Mat. Nauk*, **14**, (2), 3–86 (in Russian).

[64] Kolmogorov V.L., Shishkin G.I. (1997). Numerical methods for singularly perturbed boundary value problems modeling diffusion processes. *Singular Perturbation Problems in Chemical Physics*, J.J.H. Miller, ed., Advances in Chemical Physics Series, vol. XCVII, 181–362, J.Wiley & Sons, New York.

[65] Kolmogorov V.L., Shishkin G.I., Shishkina L.P. (1997). Numerical analysis in singularly perturbed boundary value problems modeling heat transfer processes. *Numerical Analysis and its Applications (Rousse, 1996)*, Lecture Notes in Comput. Sci., **1196**, 250–257, Springer, Berlin.

[66] Kondratiev V.A. (1967). Boundary value problems for elliptic equations in domains with conical or angular points. *Trudy Moskov. Mat. Obshch. (Transactions of the Moscow Math. Society)*, **16**, 209–292 (in Russian).

[67] Ladyzhenskaya O.A., Solonnikov V.A., Ural'tseva N.N. (1967). *Linear and Quasilinear Equations of Parabolic Type*, Translations of Math-

ematical Monographs, **23**, American Mathematical Society, Providence, RI.

[68] Ladyzhenskaya O.A., Ural'ceva N.N. (1968). *Linear and Quasilinear Elliptic Equations*, Academic Press, New York-London.

[69] Ladyzhenskaya O.A., Ural'ceva N.N. (1973). *Linear and Quasilinear Equations of Elliptic Type*. Nauka, Moscow (in Russian).

[70] LeVeque R.J. (1990). *Numerical Methods for Conservation Laws*. Bikhäuser Verlag, Basel.

[71] Li S., Shishkin G., Shishkina L. (2007). Approximation of the solution and its derivative for the singularly perturbed Black-Scholes equation with nonsmooth initial data. *Comp. Math. Math. Phys.*, **47**, (3), 442-462.

[72] Linß T. (2001). The necessity of Shishkin decompositions. *Appl. Math. Lett.*, **14**, (7), 891–896.

[73] Linß T., Stynes M. (2001). Asymptotic analysis and Shishkin-type decomposition for an elliptic convection-diffusion problem. *J. Math. Anal. Appl.*, **261**, (2), 604–632.

[74] Liseikin V.D. (1983). Numerical solution of a two-dimensional elliptic equation with a small parameter multiplying highest derivatives. *Chisl. Metody Mekh. Sploshn. Sredy*, Novosibirsk, **14**, (4), 110–115 (in Russian).

[75] Liseikin V.D. (1999). *Grid Generation Methods*. Springer-Verlag, Berlin.

[76] Liseikin V.D. (2001). *Layer Resolving Grids and Transformations for Singular Perturbation Problems*. VSP BV, Utrecht.

[77] MacMullen H., Miller J.J.H., O'Riordan E., Shishkin G.I. (2001). A second-order parameter-uniform overlapping Schwarz method for reaction-diffusion problems with boundary layers. *J. Comput. Appl. Math.*, **130**, (1-2), 231–244.

[78] MacMullen H., O'Riordan E., Shishkin G.I. (2002). The convergence of classical Schwarz methods applied to convection-diffusion problems with regular boundary layers. *Appl. Numer. Math.*, **43**, (3), 297–313.

[79] Marchuk G. I. (1982). *Methods of Numerical Mathematics*, 2nd edition. Springer-Verlag, New York-Berlin.

[80] Marchuk G. I., Shaidurov V. V. (1983). *Difference Methods and Their Interpolations*, Springer-Verlag, New York.

[81] McCormick S.F. (1989). *Multilevel Adaptive Methods for Partial Differential Equations*. SIAM, Philadelphia, PA.

[82] Melenk J.M. (2000). On n-widths for elliptic problems. *J. Math. Anal. Appl.*, **247**, (1), 272–289.

[83] Miller J.J.H., O'Riordan E. (1997). The necessity of fitted operators and Shishkin meshes for resolving thin layer phenomena. *CWI Quarterly*, **10**, (3-4), 207–213.

[84] Miller J.J.H., Shishkin G. I. (1991). On the construction of uniformly convergent finite difference scheme for singularly perturbed problems for quasilinear elliptic equation. *Computational Methods for Boundary and Interior Layers in Several Dimensions*, MCMXCI, 103–118, Boole Press, Dublin.

[85] Miller J.J.H., Musgrave A.P., Shishkin G.I. (2003). A Reynolds-uniform numerical method for the Prandtl solution and its derivatives for stagnation line flow. *Internat. J. Numer. Methods Fluids*, **43**, (8), 881–894.

[86] Miller J.J.H., O'Riordan E., Shishkin G.I. (1995). On the use of fitted operator methods for singularly perturbed partial differential equations. *Advanced Mathematics: Computations and Applications (Novosibirsk, 1995)*, 518–531, NCC Publ., Novosibirsk.

[87] Miller J.J.H., O'Riordan E., Shishkin G.I. (1996). *Fitted Numerical Methods for Singular Perturbation Problems. Error Estimates in the Maximum Norm for Linear Problems in One and Two Dimensions.* World Scientific, Singapore.

[88] Miller J.J.H., Mullarkey E., O'Riordan E., Shishkin G.I. (1991). A simple recipe for uniformly convergent finite-difference schemes for singularly perturbed problems. *C. R. Acad. Sci. Paris Sér. I Math.*, **312**, (8), 643–648.

[89] Miller J.J.H., O'Riordan E., Petrenko E.A., Shishkin G.I. (1993). Special finite difference methods for calculating heat fields in solid bodies with rapidly changing surface temperature. *Application of Computational Methods for Boundary and Interior Layers*, J.J.H. Miller, ed., 108–123, Boole Press, Dublin.

[90] Miller J.J.H., O'Riordan E., Shishkin G.I., Wang S. (2000). A parameter-uniform Schwarz method for a singularly perturbed reaction-diffusion problem with an interior layer. *Applied Numerical Mathematics*, **35**, (4), 323–337.

[91] Morton K.W. (1996). *Numerical Solution of Convection-Diffusion Problems.* Chapman & Hall, London.

[92] Murray J.D. (2002). *Mathematical Biology. I. An Introduction*, 3rd ed., Springer-Verlag, New York.

[93] Oleinik O.A., Samokhin V.N. (1999). *Mathematical Models in Boundary Layer Theory*. Chapman & Hall / CRC, Boca Raton, FL.

[94] O'Malley R.E., Jr. (1991). *Singular Perturbation Methods for Ordinary Differential Equations*. Springer-Verlag, New York.

[95] O'Riordan E., Shishkin G.I. (2008). Parameter uniform numerical methods for singularly perturbed elliptic problems with parabolic boundary layers. *Appl. Numer. Math.*, **58** (to appear).

[96] O'Riordan E., Shishkin G.I. (2007). A technique to prove parameter-uniform convergence for a singularly perturbed convection-diffusion equation. *J. Comput. Appl. Math.*, **206**, (1), 136–145.

[97] O'Riordan E., Pickett M.L., Shishkin G.I. (2003). Singularly perturbed problems modeling reaction-convection-diffusion equations. *Comput. Methods Appl. Math.*, **3**, (3), 424–442.

[98] Pershin I.V., Titov V.A., Shishkin G.I., Khripunov A.P., Yakovlev V.V. (1991). Mathematical modelling of hydrogen diffusion process in welding seams in presence of inclusions. *Mat. Model.*, **3**, (3), 27–35 (in Russian).

[99] Protter M.H., Weinberger H.F. (1967). *Maximum Principles in Differential Equations*. Prentice-Hall, Inc., Englewood Cliffs, NJ.

[100] Richtmyer R.D., Morton K.W. (1967). *Difference Methods for Initial-Value Problems*. 2nd edition. Interscience Publishers John Wiley & Sons, Inc., New York-London-Sydney.

[101] Rozhdestvenskii B.L., Yanenko N.N. (1983). *Systems of Quasilinear Equations and their Applications to Gas Dynamics*. Translations of Mathematical Monographs, **55**. American Mathematical Society, Providence, RI.

[102] Roos H.-G. (1985). Necessary convergence conditions for upwind schemes in the two-dimensional case. *Internat. J. Numer. Methods Engrg.*, **21**, (8), 1459–1469.

[103] Roos H.-G. (1994). Ten ways to generate the Il'in and related schemes. *J. Comput. Appl. Math.*, **53**, (1), 43–59.

[104] Roos H.-G. (1996). A note on the conditioning of upwind schemes on Shishkin meshes. *IMA J. Numer. Anal.*, **16**, (4), 529–538.

[105] Roos H.-G. (2002). Optimal convergence of basic schemes for elliptic boundary value problems with strong parabolic layers. *J. Math. Anal. Appl.*, **267**, (1), 194–208.

[106] Roos H.-G., Stynes M., Tobiska L. (1996). *Numerical Methods for Singularly Perturbed Differential Equations. Convection-Diffusion and Flow Problems*. Springer-Verlag, Berlin.

[107] Samarskii A.A. (1971). *Introduction to the Theory of Difference Schemes*. Nauka, Moscow (in Russian).

[108] Samarskii A.A. (2001). *The Theory of Difference Schemes*. Marcel Dekker, Inc., New York.

[109] Samarskii A.A., Andreev V.B. (1976). *Difference Methods for Elliptic Equations*. Nauka, Moscow (in Russian); French Translation: Samarski A., Andréev V. (1978). *Méthodes aux Différences pour Équations Elliptiques*. Mir, Moscow.

[110] Samarskii A.A., Nikolaev E.S. (1978). *Methods for the Solution of Grid Equations*. Nauka, Moscow; English Translation: Samarskii A.A., Nikolaev E.S. (1989). *Numerical Methods for Grid Equations*. Vol. I, II. Birkhäuser Verlag, Basel.

[111] Samarskii A.A., Popov Yu. P. (1992). *Difference Methods for the Solution of Problems of Gas Dinamics*. 3rd edition. Nauka, Moscow (in Russian).

[112] Samarskii A.A., Mazhukin V.I., Matus P.P., Shishkin G.I. (2001). Monotone difference schemes for equations with mixed derivatives. *Mat. Model.*, **13**, (2), 17–26 (in Russian).

[113] Schlichting H. (1979). *Boundary Layer Theory*, 7th edition. McGraw-Hill, New York.

[114] Shishkin G.I. (1978). A difference scheme for the solution of an elliptic equation with a small parameter in a region with a curvilinear boundary. *Zh. Vychisl. Mat. i Mat. Fiz.*, **18**, (6), 1466–1475 (in Russian).

[115] Shishkin G.I. (1979). Numerical solution of elliptic equations with a small parameter multiplying higher derivatives. I. Parameter-uniform convergence of the approximate solution. *Chisl. Metody Mekh. Sploshn. Sredy*, Novosibirsk, **10**, (4), 107–124 (in Russian).

[116] Shishkin G.I. (1979). Numerical solution of elliptic equations with a small parameter multiplying higher derivatives. II. Approximation of derivatives. *Chisl. Metody Mekh. Sploshn. Sredy*, Novosibirsk, **10**, (5), 127–143 (in Russian).

[117] Shishkin G. I. (1981). Numerical solution of differential equations with a small parameter multiplying the highest derivatives. *Chisl. Metody Mekh. Sploshn. Sredy*, Novosibirsk, **12**, (4), 135–147 (in Russian).

[118] Shishkin G.I. (1983). Difference scheme on a nonuniform grid for differential equations with a small parameter multiplying the highest derivative. *USSR Comput. Maths. Math. Phys.*, **23**, (3), 59–66.

[119] Shishkin G.I. (1984). A difference scheme for a fourth-order differential equation with a small parameter multiplying the highest derivative. *Soviet Math. Dokl.*, **29**, (2), 402–405.

[120] Shishkin G.I. (1984). Increasing the accuracy of solutions of difference schemes for parabolic equations with a small parameter multiplying the highest derivative. *Zh. Vychisl. Mat. i Mat. Fiz.* **24**, (6), 864–875 (in Russian).

[121] Shishkin G.I. (1984). A difference scheme for the solution of a system of differential equations with a small parameter. *Differential Equations with a Small Parameter*, 119–130, Akad. Nauk SSSR, Ural. Nauchn. Tsentr, Sverdlovsk (in Russian).

[122] Shishkin G.I. (1985). A difference scheme for a fourth-order elliptic equation with a small parameter multiplying the highest derivatives. *Differential Equations*, **21**, (12), 1465–1470.

[123] Shishkin G.I. (1986). A difference scheme for an elliptic equation with a small parameter multiplying the highest derivatives. *Dokl. Akad. Nauk SSSR*, **286**, (1), 57–61 (in Russian).

[124] Shishkin G.I. (1986). Solution of a boundary value problem for an elliptic equation with a small parameter multiplying the highest derivatives. *Zh. Vychisl. Mat. i Mat. Fiz.*, **26**, (7), 1019–1031 (in Russian).

[125] Shishkin G.I. (1987). Approximation of solutions of singularly perturbed boundary value problems with a corner boundary layer. *Zh. Vychisl. Mat. i Mat. Fiz.*, **27**, (9), 1360–1374 (in Russian).

[126] Shishkin G.I. (1988). Approximation of the solutions of singularly perturbed boundary value problems with a corner boundary layer. *Soviet Math. Dokl.* **36**, (2), 240–244.

[127] Shishkin G.I. (1988). A difference scheme for a singularly perturbed equation of parabolic type with a discontinuous initial condition. *Soviet Math. Dokl.* **37**, (3), 792–796.

[128] Shishkin G.I. (1988). A difference scheme for a singularly perturbed equation of parabolic type with a discontinuous boundary condition. *USSR Comput. Math. Math. Phys.* **28**, (6), 32–41.

[129] Shishkin G.I. (1988). Grid approximation of singularly perturbed parabolic equations with internal layers. *Soviet J. Numer. Anal. Math. Modelling*, **3**, (5), 393–407.

[130] Shishkin G. I. (1989). Approximation of solutions of singularly perturbed boundary value problems with a parabolic boundary layer. *USSR Comput. Maths. Math. Phys.*, **29**, (4), 1–10.

[131] Shishkin G.I. (1990). *Grid Approximation of Singularly Perturbed Elliptic and Parabolic Equations.* Second Doctoral thesis, Keldysh Institute, Moscow (in Russian).

[132] Shishkin G. I. (1990). Grid approximation of singularly perturbed boundary value problems with convective terms. *Soviet J. Numer. Anal. Math. Modelling,* **5**, (2), 173–187.

[133] Shishkin G. I. (1990). Grid approximation of singularly perturbed elliptic equations in domains with characteristic faces. *Soviet J. Numer. Anal. Math. Modelling,* **5**, (4-5), 327–343.

[134] Shishkin G.I. (1990). Grid approximation of singularly perturbed elliptic equations in case of limit zero-order equations degenerating at the boundary, *Soviet J. Numer. Anal. Math. Modelling,* **5**, (6), 523–548.

[135] Shishkin G.I. (1991). Grid approximation of a singularly perturbed boundary value problem for a quasilinear elliptic equation in the case of complete degeneration. *Comput. Maths. Math. Phys.,* **31**, (12), 33–46.

[136] Shishkin G.I. (1991). Grid approximation of singularly perturbed boundary value problem for the quasi-linear elliptic equation degenerating into the first-order equation. *Soviet J. Numer. Anal. Math. Modelling,* **6**, (1), 61–81.

[137] Shishkin G.I. (1991). Grid approximation of singularly parturbed boundary value problem for quasi-linear parabolic equations in case of complete degeneracy in spatial variables. *Soviet J. Numer. Anal. Math. Modelling,* **6**, (3), 243–261.

[138] Shishkin G.I. (1992). *Discrete Approximations of Singularly Perturbed Elliptic and Parabolic Equations.* Russian Academy of Sciences, Ural Section, Ekaterinburg (in Russian).

[139] Shishkin G.I. (1992). Difference approximation of a singularly perturbed boundary value problem for quasilinear elliptic equations that degenerate into a first-order equation. *Comput. Maths. Math. Phys.* **32**, (4), 467–480.

[140] Shishkin G.I. (1992). A difference scheme for a singularly perturbed parabolic equation that is degenerate on the boundary. *Comput. Maths. Math. Phys.,* **32**, (5), 621–636.

[141] Shishkin G.I. (1993). Method of splitting for singularly perturbed parabolic equations. *East-West J. Numer. Math.,* **1**, (2), 147–163.

[142] Shishkin G.I. (1995). A difference scheme for the problem of the decay of a discontinuity in the case of the viscous Burgers equation. *Dokl. Akad. Nauk,* **342**, (0), 010 017 (in Russian)

[143] Shishkin G.I. (1995). Grid approximation of singularly perturbed boundary value problems for systems of elliptic and parabolic equations. *Comput. Maths. Math. Phys.*, **35**, (4), 429–446.

[144] Shishkin G.I. (1996). Approximation of solutions and of diffusion flows in the case of singularly perturbed boundary value problems with discontinuous initial conditions. *Comp. Maths. Math. Phys.*, **36**, (9), 1233–1250.

[145] Shishkin G.I. (1997). Singularly perturbed boundary value problems with concentrated sources and discontinuous initial conditions. *Comp. Maths. Math. Phys.*, **37**, (4), 417–434.

[146] Shishkin G.I. (1997). Acceleration of the process of the numerical solution to singularly perturbed boundary value problems for parabolic equations on the basis of parallel computations. *Russian J. Numer. Anal. Math. Modelling*, **12**, (3), 271–291.

[147] Shishkin G.I. (1997). On finite difference fitted schemes for singularly perturbed boundary value problems with a parabolic boundary layer. *J. Math. Anal. Appl.*, **208**, (1), 181–204.

[148] Shishkin G.I. (1998). Grid approximation of parabolic equations with small parameter multiplying the space and time derivatives. Reaction–diffusion equations. *Math. Balkanica (N.S.)*, **12**, (1-2), 179–214.

[149] Shishkin G.I. (1998) A grid approximation for the Riemann problem in the case of the Burgers equation. *Comp. Math. Math. Phys.*, **38**, (8), 1361–1363.

[150] Shishkin G.I. (1998). Approximation of singularly perturbed elliptic equations with convective terms in the case of a flow impinging on an impermeable wall. *Comput. Maths. Math. Phys.*, **38**, (11), 1768–1782.

[151] Shishkin G.I. (1998). Finite-difference approximations for singularly perturbed elliptic equations. *Comput. Maths. Math. Phys.*, **38**, (12), 1909–1921.

[152] Shishkin G.I. (1998). Grid approximation of singularly perturbed systems of elliptic and parabolic equations with convective terms. *Differential Equations*, **34**, (12), 1693–1704.

[153] Shishkin G.I. (1999). Locally one-dimensional difference schemes for singularly perturbed parabolic equations with convective terms. *Russian J. Numer. Anal. Math. Modelling*, **14**, (6), 495–511.

[154] Shishkin G.I. (1999). Grid approximation of singularly perturbed boundary value problems on locally refined grids. Reaction-diffusion equations. *Mat. Model.*, **11**, (12), 87–104 (in Russian).

References

[155] Shishkin G.I. (1999). Grid approximation of singularly perturbed boundary value problems in a nonconvex domain with a piecewise smooth boundary. *Mat. Model.* **11**, (11), 75–90 (in Russian).

[156] Shishkin G. I. (2000). Grid approximation of the transport equation for the Prandtl problem of flow past a plate at high Reynolds numbers. *Doklady Mathematics*, **62**, (2), 166–168.

[157] Shishkin G.I. (2000). Grid approximation of singularly perturbed boundary value problems on locally condensing grids. Convection-diffusion equations. *Comput. Maths. Math. Phys.*, **40**, (5), 680–691.

[158] Shishkin G.I. (2000). Approximation of systems of elliptic convection-diffusion equations with parabolic boundary layers. *Comput. Maths. Math. Phys.*, **40**, (11), 1582–1595.

[159] Shishkin G.I. (2000). On numerical methods on adaptive meshes for a singularly perturbed reaction-diffusion equation with a moving concentrated source. *Finite Difference Schemes*, 205–214, Lith. Acad. Sci., Inst. Math. Inform., Vilnius.

[160] Shishkin G.I. (2001) Discrete approximations of the Riemann problem for the viscous Burgers equation. *Math. Proc. R. Ir. Acad.* **101A**, (1), 27–48.

[161] Shishkin G.I. (2001). Grid approximation of the Blasius equation and its derivatives. *Comput. Maths. Math. Phys.* **41**, (5), 649–664.

[162] Shishkin G. I. (2001). Mesh approximation of singularly perturbed equations with convective terms for perturbation of data. *Comput. Maths. Math. Phys.*, **41**, (5), 649–664.

[163] Shishkin G. I. (2001). A grid approximation to the transport equation in the problem on a flow past a flat plate at large Reynolds numbers. *Differential Equations*, **37**, (3), 444–453.

[164] Shishkin G.I. (2001). Approximation of singularly perturbed parabolic reaction-diffusion equations with nonsmooth data. *Comput. Methods Appl. Math.*, **1**, (3), 298–315.

[165] Shishkin G.I. (2001). Approximation of singularly perturbed reaction-diffusion equations on adaptive grids. *Mat. Model.*, **13**, (3), 103–118.

[166] Shishkin G.I. (2001). A posteriori adapted (to the solution gradient) grids in the approximation of singularly perturbed convection-diffusion equations. *Vychisl. Tekhnol.*, **6**, (1), 72–87 (in Russian).

[167] Shishkin G.I. (2002). Adaptive mesh method for a singularly perturbed parabolic equation with a moving concentrated source. *Doklady Mathematics*, **66**, (2), 175–170.

[168] Shishkin G.I. (2002). Grid approximations with an improved rate of convergence for singularly perturbed elliptic equations in domains with characteristic boundaries. *Siberian J. Numer. Math.*, **5**, (1), 71–92 (in Russian).

[169] Shishkin G.I. (2002). Piecewise-uniform grids, optimal with respect to the order of convergence, for singularly perturbed convection-diffusion equations. *Russian Math. (Iz. VUZ)*, **46**, (3), 56–68.

[170] Shishkin G.I. (2003). Grid approximation of elliptic convection-diffusion equations in an unbounded domain with boundary layers of different types. *Doklady Mathematics*, **68**, (2), 234–238.

[171] Shishkin G.I. (2003). Approximation of solutions and derivatives for a singularly perturbed elliptic convection-diffusion equation. *Comput. Maths. Math. Phys.*, **43**, (5), 641–657.

[172] Shishkin G.I. (2003). Grid approximation of a singularly perturbed parabolic equation on a composed domain with a moving interface containing a concentrated source. *Comput. Maths. Math. Phys.*, **43**, (12), 1738–1755.

[173] Shishkin G.I. (2003). Approximation of solutions and derivatives for singularly perturbed elliptic convection-diffusion equations. *Mathematical Proceedings of the Royal Irish Academy*, **103A**, Issue 2, 169–201.

[174] Shishkin G.I. (2003). On conditioning of a Schwarz method for singularly perturbed convection-diffusion equations in the case of disturbances in the data of the boundary-value problem. *Comput. Methods Appl. Math.*, **3**, (3), 459–487.

[175] Shishkin G.I. (2003). Grid approximation of improved convergence order for a singularly perturbed elliptic convection-diffusion equation. *Proc. Steklov Inst. Math.*, Asymptotic Expansions. Approximation Theory. Topology, Suppl. 1, S184–S202.

[176] Shishkin G.I. (2004). Limitations of adaptive mesh refinement techniques for singularly perturbed problems with a moving interior layer. *J. Comput. Appl. Math.*, **166**, (1), 267–280.

[177] Shishkin G.I. (2004). Discrete approximations of solutions and derivatives for a singularly perturbed parabolic convection-diffusion equation. *J. Comput. Appl. Math.*, **166**, (1), 247–266.

[178] Shishkin G.I. (2004). Grid Approximation of Singularly Perturbed Elliptic Convection-Diffusion Equations in Unbounded Domains. *TCD-MATH Report Series, the School of Mathematics*, Trinity College Dublin. Preprint TCDMATH 04–01.

386   *References*

[179] Shishkin G.I. (2005). Grid approximations of parabolic convection-diffusion equations with piecewise-smooth initial conditions. *Doklady Mathematics*, **72**, (3), 850–853.

[180] Shishkin G. I. (2005). Grid approximation of a singularly perturbed elliptic equation with convective terms in the presence of various boundary layers. *Comp. Maths. and Math. Phys.*, **45**, (1), 104–119.

[181] Shishkin G.I. (2005). Grid approximation of the domain and solution decomposition method with improved convergence rate for singularly perturbed elliptic equations in domains with characteristic boundaries. *Comp. Maths. and Math. Phys.*, **45**, (7), 1155–1171.

[182] Shishkin G.I. (2005). Robust novel high-order accurate numerical methods for singularly perturbed convection-diffusion problems. *Math. Model. Anal.*, **10**, (4), 393–412.

[183] Shishkin G.I. (2005). On an adaptive grid method for singularly perturbed elliptic reaction-diffusion equations in a domain with a curvilinear boundary. *Russian Math. (Iz. VUZ)*, **49**, (1), 69–83.

[184] Shishkin G.I. (2006). Grid approximation of singularly perturbed parabolic convection-diffusion equations subject to a piecewise smooth initial condition. *Comp. Maths. Math. Phys.*, **46**, (1), 49–72.

[185] Shishkin G.I. (2006). The use of solutions on embedded grids for the approximation of a singularly perturbed parabolic convection–diffusion equation on adapted grids. *Comp. Maths. Math. Phys.*, **46**, (9), 1539–1559.

[186] Shishkin G.I. (2006). Grid approximation of a singularly perturbed elliptic convection-diffusion equation in an unbounded domain. *Russian J. Numer. Anal. Math. Modelling*, **21**, (1), 67–94.

[187] Shishkin G.I. (2007). Approximation of systems of singularly perturbed elliptic reaction-diffusion equations with two parameters. *Comp. Math. Math. Phys.*, **47**, (5), 797–828.

[188] Shishkin G.I. (2007). Necessary conditions for ε-uniform convergence of finite difference schemes for parabolic equations with moving boundary layers. *Comp. Maths. Math. Phys.*, **47**, (10), 1636–1655.

[189] Shishkin G.I. (2007). Grid approximation of singularly perturbed parabolic reaction-diffusion equations with piecewise smooth initial-boundary conditions. *Math. Model. Anal.*, **12**, (2), 235–254.

[190] Shishkin G.I. (2007). A posteriori adaptive mesh technique with a priori error estimates for singularly perturbed semilinear parabolic convection-diffusion equations, *Internat. J. Computing Science and Mathematics*, **1**, (2/3/4), 374–395.

[191] Shishkin G.I. (2007). Grid approximation of singularly perturbed parabolic equations with piecewise continuous initial-boundary conditions. *Proc. Steklov Inst. Math.*, Suppl. 2, S213–S230.

[192] Shishkin G.I. (2008). Conditioning of finite difference schemes for a singularly perturbed convection-diffusion parabolic equation. *Comp. Maths. Math. Phys.*, **48**, (5).

[193] Shishkin G.I. (2008). Grid approximation of a parabolic convection-diffusion equation on a priori adapted grids: ε-uniformly convergent schemes. *Comp. Maths. Math. Phys.*, **48**, (6).

[194] Shishkin G.I. (2008). Grid approximation of singularly perturbed parabolic equations with moving boundary layers. *Math. Model. Anal.*, **13**, (3). http://www.vgtu.lt/rc/mma

[195] Shishkin G.I. (2008). Optimal difference schemes on piecewise-uniform meshes for a singularly perturbed parabolic convection-diffusion equation. *Math. Model. Anal.*, **13**, (1), 99–112. http://www.vgtu.lt/rc/mma

[196] Shishkin G.I. (2008). A finite difference scheme on a priori adapted meshes for a singularly perturbed parabolic convection-diffusion equation. *Numer. Math. Theory Methods Appl.*, **1**, (2), 214–234.

[197] Shishkin G.I., Shishkina L.P. (2005). A higher-order Richardson method for a quasilinear singularly perturbed elliptic reaction-diffusion equation. *Differential Equations*, **41**, (7), 1030–1039.

[198] Shishkin G.I., Shishkina L.P. (2008). Approximation of a system of singularly perturbed parabolic equations in a rectangle. *Comp. Maths. Math. Phys.*, **48** (4), 627–640.

[199] Shishkin G.I., Shishkina L.P. (2008). Robust numerical method for a system of singularly perturbed parabolic reaction-diffusion equations on a rectangle. *Math. Model. Anal.*, **13**, (2). http://www.vgtu.lt/rc/mma/

[200] Shishkin G.I., Titov V.A. (1976). A difference scheme for a differential equation with two small parameters at the derivatives. *Chisl. Metody Mekh. Sploshn. Sredy*, Novosibirsk, **7**, (2), 145–155 (in Russian).

[201] Shishkin G.I., Tselishcheva I.V. (1996). Parallel methods of solving singularly perturbed boundary value problems for elliptic equations. *Mat. Model.*, **8**, (3), 111–127 (in Russian).

[202] Shishkin G.I., Vabishchevich P.N. (1996). Parallel domain decomposition methods with the overlapping of subdomains for parabolic problems. *Math. Models and Methods in Applied Sciences*, **6**, (8), 1169–1185.

[203] Shishkin G.I., Shishkina L.P., Hemker P.W. (2004). A class of singularly perturbed convection-diffusion problems with a moving interior layer. An *a posteriori* adaptive mesh technique. *Comput. Methods Appl. Math.*, **4**, (1), 105–127.

[204] Stynes M. (2005). Steady-state convection-diffusion problems. *Acta Numerica*, **14**, 445–508, Cambridge University Press.

[205] Tikhonov A. N., Samarskii A. A. (1990). *Equations of Mathematical Physics*. Dover Publications, Inc., New York.

[206] Tselishcheva I.V., Shishkin G.I. (2008). Sequential and parallel domain decomposition methods for singularly perturbed parabolic convection-diffusion equation. *Trudy Instituta Matematiki i Mekhaniki UrO RAN*, **14**, (1), 202–220 (in Russian).

[207] Vabishchevich P.N., Shishkin G.I. (1995). Difference schemes on locally condensing grids. *Differential Equations*, **31**, (7), 1121–1126.

[208] Vasil'eva A.B., Butuzov V.F. (1973). *Asymptotic Expansions of the Solutions of Singularly Perturbed Equations*. Nauka, Moscow (in Russian).

[209] Vasil'eva A.B., Butuzov V.F. (1990). *Asymptotic Methods in the Theory of Singular Perturbations*. Vyssh. Shkola, Moscow (in Russian).

[210] Vishik M.I., Lyusternik L.A. (1960). The solution of some perturbation problems for matrices and selfadjoint or non-selfadjoint differential equations I. *Russian Math. Surveys*, **15**, (3), 1–73.

[211] Vishik M. I., Lyusternik L. A. (1961). Regular degeneration and boundary layer for linear differential equations with a small parameter. *Amer. Math. Soc. Translations*, **20**, Ser. 2, 239–364.

[212] Volkov E. A. (1965). Differentiability properties of solutions of boundary value problems for the Laplace and Poisson equations on a rectangle. *Proc. Steklov Inst. Math.*, **77**, 101–126.

[213] Volkov E. A. (1969). On the differential properties of the solutions of the Laplace and Poisson equations on a parellelepiped, and effective error estimates for the method of meshes. *Trudy Mat. Inst. Steklov*, **105**, 46–65 (in Russian).

[214] Wesseling P. (2001). *Principles of Computational Fluid Dynamics*. Springer-Verlag, Berlin.

[215] Wilmott P., Howison S., Dewynne J. (1995). *The Mathematics of Financial Derivatives*. Cambridge University Press, Cambridge.

[216] Yanenko N.N. (1971). *The Method of Fractional Steps. The Solution of Problems of Mathematical Physics in Several Variables*. Springer-Verlag, Berlin Heidelberg.

Index

Milton Keynes UK
Ingram Content Group UK Ltd.
UKHW021825071024
449327UK00021B/1429